QUANTUM FIELD THEORY

F. Mandl and G. Shaw

Department of Theoretical Physics,
The Schuster Laboratory,
The University,
Manchester

A Wiley–Interscience Publication

John Wiley & Sons

CHICHESTER · NEW YORK · BRISBANE · TORONTO · SINGAPORE

Reprinted July 1988
Reprinted January 1990

Library of Congress Cataloging in Publication Data:

Mandl, F. (Franz), date.
Quantum field theory.
'A Wiley–Interscience publication.'
Includes index.
1. Quantum field theory.
I. Shaw, Graham, date. II. Title.
QC174.45.M32 1984 530.1′43 84-5229

ISBN 0 471 10509 0 (cloth)
ISBN 0 471 90650 6 (paper)

British Library Cataloguing in Publication Data:

Mandl, F.
Quantum field theory.
1. Quantum field theory
I. Title II. Shaw, G.
530.1′43 QC174.45

ISBN 0 471 10509 0 (cloth)
ISBN 0 471 90650 6 (paper)

Filmset by Eta Services (Typesetters) Ltd, Beccles, Suffolk
Printed and bound in Great Britain by
Biddles Ltd, Guildford and King's Lynn

Contents

Preface

Our aim in writing this book has been to produce a short introduction to quantum field theory, suitable for beginning research students in theoretical and experimental physics. The main objectives are: (i) to explain the basic physics and formalism of quantum field theory, (ii) to make the reader fully proficient in perturbation theory calculations using Feynman diagrams, and (iii) to introduce the reader to gauge theories which are playing such a central role in elementary particle physics.

The theory has been applied to two areas. The beginning parts of the book deal with quantum electrodynamics (QED) where quantum field theory had its early triumphs. The last four chapters, on weak interactions, introduce non-Abelian gauge groups, spontaneous symmetry breaking and the Higgs mechanism, culminating in the Weinberg–Salam standard electro-weak theory. For reasons of space, we have limited ourselves to purely leptonic processes, but this theory is equally successful when extended to include hadrons. The recent observations of the $W^{\pm}$ and Z° bosons, with the predicted masses, lend further support to this theory, and there is every hope that it is the fundamental theory of electro-weak interactions.

The introductory nature of this book and the desire to keep it reasonably short have influenced both the level of treatment and the selection of material. We have formulated quantum field theory in terms of non-commuting operators, as this approach should be familiar to the reader from non-relativistic quantum mechanics and it brings out most clearly the physical meaning of the formalism in terms of particle creation and annihilation operators. We have only developed the formalism to the level we require in the applications. These concentrate primarily on calculations in lowest order of perturbation theory. The techniques for obtaining cross-sections, decay rates, and spin and polarization sums have been developed in detail and applied to a variety of processes, many of them of interest in current research on electro-weak interactions. After studying this material, the reader should be able to tackle confidently any process in lowest order.

Our treatment of renormalization and radiative corrections is much less complete. We have explained the general concepts of regularization and renormalization. For QED we have shown in some detail how to calculate the lowest-order radiative corrections, using dimensional regularization as well as the older cut-off techniques. The infra-red divergence and its connection with radiative corrections have similarly been discussed in lowest order only. The scope of this book precludes a serious study of higher-order corrections in QED and of the renormalization of the electro-weak theory. For the latter, the Feynman path integral formulation of quantum field theory seems almost essential. Regretfully, we were not able to provide a short and simple treatment of this topic.

This book arose out of lectures which both of us have given over many years. We have greatly benefited from discussions with students and colleagues, some of whom have read parts of the manuscript. We would like to thank all of them for their help, and particularly Sandy Donnachie who encouraged us to embark on this collaboration.

January 1984

FRANZ MANDL
GRAHAM SHAW

In this book we have always taken

$$e > 0$$

so that the charge of the electron is $(-e)$.

CHAPTER 1

Photons and the electromagnetic field

1.1 PARTICLES AND FIELDS

The concept of photons as the quanta of the electromagnetic field dates back to the beginning of this century. In order to explain the spectrum of black-body radiation Planck, in 1900, postulated that the process of emission and absorption of radiation by atoms occurs discontinuously in quanta. Einstein by 1905 had arrived at a more drastic interpretation. From a statistical analysis of the Planck radiation law and from the energetics of the photoelectric effect he concluded that it was not merely the atomic mechanism of emission and absorption of radiation which is quantized, but that electromagnetic radiation itself consists of photons. The Compton effect confirmed this interpretation.

The foundations of a systematic quantum theory of fields were laid by Dirac in 1927 in his famous paper on 'The Quantum Theory of the Emission and Absorption of Radiation'. From the quantization of the electromagnetic field one is naturally led to the quantization of any classical field, the quanta of the field being particles with well-defined properties. The interactions between these particles is brought about by other fields whose quanta are other particles. For example, we can think of the interaction between electrically charged particles, such as electrons and positrons, as being brought about by the electromagnetic field or as due to an exchange of photons. The electrons and positrons themselves can be thought of as the quanta of an electron–positron field. An important reason for quantizing such particle fields is to allow for the possibility that the number of particles changes as, for example, in the creation or annihilation of electron–positron pairs.

These and other processes of course only occur through the interactions of fields. The solution of the equations of the quantized interacting fields is extremely difficult. If the interaction is sufficiently weak, one can employ perturbation theory. This has been outstandingly successful in quantum electrodynamics, where complete agreement exists between theory and experiment to an incredibly high degree of accuracy. More recently, perturbation theory has also very successfully been applied to weak interactions.

The most important modern perturbation-theoretic technique employs Feynman diagrams which are also extremely useful in many areas other than relativistic quantum field theory. We shall later develop the Feynman diagram technique and apply it to electromagnetic and weak interactions. For this a Lorentz–covariant formulation will be essential.

In this introductory chapter we employ a simpler non-covariant approach which suffices for many applications and brings out many of the ideas of field quantization. We shall consider the important case of electrodynamics for which a complete classical theory—Maxwell's—exists. As quantum electrodynamics will be re-derived later, we shall in this chapter at times rely on plausibility arguments rather than fully justify all steps.

1.2 THE ELECTROMAGNETIC FIELD IN THE ABSENCE OF CHARGES

1.2.1 The classical field

Classical electromagnetic theory is summed up in Maxwell's equations. In the presence of a charge density $\rho(\mathbf{x}, t)$ and a current density $\mathbf{j}(\mathbf{x}, t)$, the electric and magnetic fields $\mathbf{E}$ and $\mathbf{B}$ satisfy the equations

$$\nabla \cdot \mathbf{E} = \rho \tag{1.1a}$$

$$\nabla \wedge \mathbf{B} = \frac{1}{c}\mathbf{j} + \frac{1}{c}\frac{\partial \mathbf{E}}{\partial t} \tag{1.1b}$$

$$\nabla \cdot \mathbf{B} = 0 \tag{1.1c}$$

$$\nabla \wedge \mathbf{E} = -\frac{1}{c}\frac{\partial \mathbf{B}}{\partial t} \tag{1.1d}$$

where, as throughout this book, rationalized Gaussian (c.g.s.) units are being used.‡

‡ They are also called rationalized Lorentz–Heaviside units. In these units the fine structure constant is given by $\alpha = e^2/(4\pi\hbar c) \approx 1/137$, whereas in unrationalized Gaussian units $\alpha = e^2_{\text{unrat}}/\hbar c$, i.e. $e = e_{\text{unrat}}\sqrt{(4\pi)}$. Correspondingly for the fields $\mathbf{E} = \mathbf{E}_{\text{unrat}}/\sqrt{(4\pi)}$, etc.

From the second pair of Maxwell's equations [(1.1c) and (1.1d)] follows the existence of scalar and vector potentials $\phi(\mathbf{x}, t)$ and $\mathbf{A}(\mathbf{x}, t)$, defined by

$$\mathbf{B} = \nabla \wedge \mathbf{A}, \qquad \mathbf{E} = -\nabla\phi - \frac{1}{c}\frac{\partial \mathbf{A}}{\partial t}. \tag{1.2}$$

Eqs. (1.2) do not determine the potentials uniquely, since for an arbitrary function $f(\mathbf{x}, t)$ the transformation

$$\phi \to \phi' = \phi + \frac{1}{c}\frac{\partial f}{\partial t}, \qquad \mathbf{A} \to \mathbf{A}' = \mathbf{A} - \nabla f \tag{1.3}$$

leaves the fields **E** and **B** unaltered. The transformation (1.3) is known as a gauge transformation of the second kind. Since all observable quantities can be expressed in terms of **E** and **B**, it is a fundamental requirement of any theory formulated in terms of potentials that it is gauge-invariant, i.e. that the predictions for observable quantities are invariant under such gauge transformations.

Expressed in terms of the potentials, the second pair of Maxwell's equations [(1.1c) and (1.1d)] are satisfied automatically, while the first pair [(1.1a) and (1.1b)] become

$$-\nabla^2\phi - \frac{1}{c}\frac{\partial}{\partial t}(\nabla\cdot\mathbf{A}) = \Box\phi - \frac{1}{c}\frac{\partial}{\partial t}\left(\frac{1}{c}\frac{\partial\phi}{\partial t} + \nabla\cdot\mathbf{A}\right) = \rho \tag{1.4a}$$

$$\Box\mathbf{A} + \nabla\left(\frac{1}{c}\frac{\partial\phi}{\partial t} + \nabla\cdot\mathbf{A}\right) = \frac{1}{c}\mathbf{j} \tag{1.4b}$$

where

$$\Box \equiv \frac{1}{c^2}\frac{\partial^2}{\partial t^2} - \nabla^2. \tag{1.5}$$

We now go on to consider the case of the free field, i.e. the absence of charges and currents: $\rho = 0$, $\mathbf{j} = 0$. We can then choose a gauge for the potentials such that

$$\nabla\cdot\mathbf{A} = 0. \tag{1.6}$$

The condition (1.6) defines the Coulomb or radiation gauge. A vector field with vanishing divergence, i.e. satisfying Eq. (1.6), is called a transverse field, since for a wave

$$\mathbf{A}(\mathbf{x}, t) = \mathbf{A}_0\, e^{i(\mathbf{k}\cdot\mathbf{x} - \omega t)}$$

Eq. (1.6) gives

$$\mathbf{k}\cdot\mathbf{A} = 0, \tag{1.7}$$

i.e. **A** is perpendicular to the direction of propagation **k** of the wave. In the

Coulomb gauge, the vector potential is a transverse vector. In this chapter we shall be employing the Coulomb gauge.

In the absence of charges, Eq. (1.4a) now becomes $\nabla^2\phi = 0$ with the solution, which vanishes at infinity, $\phi \equiv 0$. Hence Eq. (1.4b) reduces to the wave equation

$$\Box \mathbf{A} = 0. \tag{1.8}$$

The corresponding electric and magnetic fields are, from Eqs. (1.2), given by

$$\mathbf{B} = \nabla \wedge \mathbf{A}, \qquad \mathbf{E} = -\frac{1}{c}\frac{\partial \mathbf{A}}{\partial t}, \tag{1.9}$$

and, like **A**, are transverse fields. The solutions of Eq. (1.8) are the transverse electromagnetic waves in free space. These waves are often called the radiation field. Its energy is given by

$$H_{\text{rad}} = \tfrac{1}{2}\int (\mathbf{E}^2 + \mathbf{B}^2)\,\mathrm{d}^3\mathbf{x}. \tag{1.10}$$

In order to quantize the theory we shall want to introduce canonically conjugate coordinates (like x and p_x in non-relativistic quantum mechanics) for each degree of freedom and subject these to commutation relations. At a given instant of time t, the vector potential **A** must be specified at every point **x** in space. Looked at from this view point, the electromagnetic field possesses a continuous infinity of degrees of freedom. The problem can be simplified by considering the radiation inside a large cubic enclosure, of side L and volume $V = L^3$, and imposing periodic boundary conditions on the vector potential **A** at the surfaces of the cube. The vector potential can then be represented as a Fourier series, i.e. it is specified by the denumerable set of Fourier expansion coefficients, and we have obtained a description of the field in terms of an infinite but denumerable number of degrees of freedom. The Fourier analysis corresponds to finding the normal modes of the radiation field, each mode being described independently of the others by a harmonic oscillator equation. (All this is analogous to the Fourier analysis of a vibrating string.) This will enable us to quantize the radiation field by taking over the quantization of the harmonic oscillator from non-relativistic quantum mechanics.

With the periodic boundary conditions

$$\mathbf{A}(0, y, z, t) = \mathbf{A}(L, y, z, t), \quad \text{etc.}, \tag{1.11}$$

the functions

$$\frac{1}{\sqrt{V}}\,\boldsymbol{\varepsilon}_r(\mathbf{k})\,\mathrm{e}^{\mathrm{i}\mathbf{k}\cdot\mathbf{x}}, \qquad r = 1, 2, \tag{1.12}$$

form a complete set of transverse orthonormal vector fields. Here the wave vectors $\mathbf{k}$ must be of the form

$$\mathbf{k} = \frac{2\pi}{L}(n_1, n_2, n_3), \qquad n_1, n_2, n_3 = 0, \pm 1, \ldots, \tag{1.13}$$

so that the fields (1.12) satisfy the periodicity conditions (1.11). $\boldsymbol{\varepsilon}_1(\mathbf{k})$ and $\boldsymbol{\varepsilon}_2(\mathbf{k})$ are two mutually perpendicular real unit vectors which are also orthogonal to $\mathbf{k}$:

$$\boldsymbol{\varepsilon}_r(\mathbf{k}) \cdot \boldsymbol{\varepsilon}_s(\mathbf{k}) = \delta_{rs}, \qquad \boldsymbol{\varepsilon}_r(\mathbf{k}) \cdot \mathbf{k} = 0, \qquad r, s = 1, 2. \tag{1.14}$$

The last of these conditions ensures that the fields (1.12) are transverse, satisfying the Coulomb gauge condition (1.6) and (1.7).‡

We can now expand the vector potential $\mathbf{A}(\mathbf{x}, t)$ as a Fourier series

$$\mathbf{A}(\mathbf{x}, t) = \sum_{\mathbf{k}} \sum_{r} \left(\frac{\hbar c^2}{2V\omega_{\mathbf{k}}}\right)^{1/2} \boldsymbol{\varepsilon}_r(\mathbf{k})[a_r(\mathbf{k}, t)\, \mathrm{e}^{\mathrm{i}\mathbf{k}\cdot\mathbf{x}} + a_r^*(\mathbf{k}, t)\, \mathrm{e}^{-\mathrm{i}\mathbf{k}\cdot\mathbf{x}}], \tag{1.15}$$

where $\omega_{\mathbf{k}} = c|\mathbf{k}|$. The summations with respect to r and $\mathbf{k}$ are over both polarization states $r = 1, 2$ (for each $\mathbf{k}$) and over all allowed momenta $\mathbf{k}$. The factor to the left of $\boldsymbol{\varepsilon}_r(\mathbf{k})$ has been introduced for later convenience only. The form of the series (1.15) ensures that the vector potential is real: $\mathbf{A} = \mathbf{A}^*$. Eq. (1.15) is an expansion of $\mathbf{A}(\mathbf{x}, t)$ at each instant of time t. The time dependence of the Fourier expansion coefficients follows since $\mathbf{A}$ must satisfy the wave equation (1.8). Substituting Eq. (1.15) in (1.8) and projecting out individual amplitudes, one obtains

$$\frac{\partial^2}{\partial t^2} a_r(\mathbf{k}, t) = -\omega_{\mathbf{k}}^2 a_r(\mathbf{k}, t). \tag{1.16}$$

These are the harmonic oscillator equations of the normal modes of the radiation field. It will prove convenient to take their solutions in the form

$$a_r(\mathbf{k}, t) = a_r(\mathbf{k}) \exp(-\mathrm{i}\omega_{\mathbf{k}} t), \tag{1.17}$$

where the $a_r(\mathbf{k})$ are initial amplitudes at time $t = 0$.

Eq. (1.15) for the vector potential, with Eq. (1.17) and its complex conjugate substituted for the amplitudes a_r and a_r^*, represents our final result for the classical theory. We can express the energy of the radiation field, Eq. (1.10), in terms of the amplitudes by substituting Eqs. (1.9) and (1.15) in (1.10) and carrying out the integration over the volume V of the enclosure. In this way one obtains

$$H_{\mathrm{rad}} = \sum_{\mathbf{k}} \sum_{r} \hbar\omega_{\mathbf{k}} a_r^*(\mathbf{k}) a_r(\mathbf{k}). \tag{1.18}$$

‡ With this choice of $\boldsymbol{\varepsilon}_r(\mathbf{k})$, Eqs. (1.12) represent linearly polarized fields. By taking appropriate complex linear combinations of $\boldsymbol{\varepsilon}_1$ and $\boldsymbol{\varepsilon}_2$ one obtains circular or, in general, elliptic polarization.

Note that this is independent of time, as expected in the absence of charges and currents; we could equally have written the time-dependent amplitudes (1.17) instead, since the time dependence of a_r and of a_r^* cancels.

As already stated, we shall quantize the radiation field by quantizing the individual harmonic oscillator modes. As the interpretation of the quantized field theory in terms of photons is intimately connected with the quantum treatment of the harmonic oscillator, we shall summarize the latter.

1.2.2 Harmonic oscillator

The harmonic oscillator Hamiltonian is, in an obvious notation,

$$H_{\text{osc}} = \frac{p^2}{2m} + \tfrac{1}{2}m\omega^2 q^2,$$

with q and p satisfying the commutation relation $[q, p] = i\hbar$. We introduce the operators

$$\left.\begin{matrix} a \\ a^\dagger \end{matrix}\right\} = \frac{1}{(2\hbar m\omega)^{1/2}} (m\omega q \pm ip).$$

These satisfy the commutation relation

$$[a, a^\dagger] = 1, \tag{1.19}$$

and the Hamiltonian expressed in terms of a and $a^\dagger$ becomes:

$$H_{\text{osc}} = \tfrac{1}{2}\hbar\omega(a^\dagger a + aa^\dagger) = \hbar\omega(a^\dagger a + \tfrac{1}{2}). \tag{1.20}$$

This is essentially the operator

$$N \equiv a^\dagger a, \tag{1.21}$$

which is positive definite, i.e. for any state $|\Psi\rangle$

$$\langle\Psi|N|\Psi\rangle = \langle\Psi|a^\dagger a|\Psi\rangle = \langle a\Psi|a\Psi\rangle \geqslant 0.$$

Hence, N possesses a lowest non-negative eigenvalue

$$\alpha_0 \geqslant 0.$$

It follows from the eigenvalue equation

$$N|\alpha\rangle = \alpha|\alpha\rangle$$

and Eq. (1.19) that

$$Na|\alpha\rangle = (\alpha - 1)a|\alpha\rangle, \qquad Na^\dagger|\alpha\rangle = (\alpha + 1)a^\dagger|\alpha\rangle, \tag{1.22}$$

i.e. $a|\alpha\rangle$ and $a^\dagger|\alpha\rangle$ are eigenfunctions of N belonging to the eigenvalues $(\alpha - 1)$ and $(\alpha + 1)$, respectively. Since α_0 is the lowest eigenvalue we must have

$$a|\alpha_0\rangle = 0, \tag{1.23}$$

and since

$$a^\dagger a|\alpha_0\rangle = \alpha_0|\alpha_0\rangle$$

Eq. (1.23) implies $\alpha_0 = 0$. It follows from Eqs. (1.19) and (1.22) that the eigenvalues of N are the integers $n = 0, 1, 2, \ldots$, and that if $\langle n|n\rangle = 1$, then the states $|n \pm 1\rangle$, defined by

$$a|n\rangle = n^{1/2}|n - 1\rangle, \qquad a^\dagger|n\rangle = (n + 1)^{1/2}|n + 1\rangle, \qquad (1.24)$$

are also normed to unity. If $\langle 0|0\rangle = 1$, the normed eigenfunctions of N are

$$|n\rangle = \frac{(a^\dagger)^n}{\sqrt{n!}}|0\rangle, \qquad n = 0, 1, 2, \ldots. \qquad (1.25)$$

These are also the eigenfunctions of the harmonic oscillator Hamiltonian (1.20) with the energy eigenvalues

$$E_n = \hbar\omega(n + \tfrac{1}{2}), \qquad n = 0, 1, 2, \ldots. \qquad (1.26)$$

The operators a and $a^\dagger$ are called lowering and raising operators because of the properties (1.24). We shall see that in the quantized field theory $|n\rangle$ represents a state with n quanta. The operator a (changing $|n\rangle$ into $|n - 1\rangle$) will annihilate a quantum; similarly, $a^\dagger$ will create a quantum.

So far we have considered one instant of time, say $t = 0$. We now discuss the equations of motion in the Heisenberg picture.[‡] In this picture, the operators are functions of time. In particular

$$i\hbar \frac{da(t)}{dt} = [a(t), H_{\text{osc}}] \qquad (1.27)$$

(margin handwriting: $aa^+ - a^+a = 1$ *;* $a^+a - aa^+ = -1$ *)*

with the initial condition $a(0) = a$, the lowering operator considered so far. Since H_{osc} is time-independent, and $a(t)$ and $a^\dagger(t)$ satisfy the same commutation relation (1.19) as a and $a^\dagger$, the Heisenberg equation of motion (1.27) reduces to

$$\frac{da(t)}{dt} = -i\omega a(t)$$

(margin handwriting:
$[a(t), H] = \hbar\omega(aa^+a + \tfrac{1}{2}a) - \hbar\omega(a^+aa + \tfrac{1}{2}a)$
$\hbar\omega(aa^+a + \tfrac{1}{2}a - (aa^+ -1)a - \tfrac{1}{2}a)$
$= a(t)\hbar\omega = i\hbar \frac{da(t)}{dt}$ *)*

with the solution

$$a(t) = a\,e^{-i\omega t}. \qquad (1.28)$$

(margin handwriting: $= -i\omega a(t)$ *)*

1.2.3   The quantized radiation field

The harmonic oscillator results we have derived can at once be applied to the radiation field. Its Hamiltonian, Eq. (1.18), is a superposition of independent

---

[‡] See the appendix to this chapter (Section 1.5) for a concise development of the Schrödinger, Heisenberg and interaction pictures.

harmonic oscillator Hamiltonians (1.20), one for each mode of the radiation field. [The order of the factors in (1.18) is not significant and can be changed, since the a_r and a_r^* are classical amplitudes.] We therefore introduce commutation relations analogous to Eq. (1.19)

$$\left.\begin{aligned}[][a_r(\mathbf{k}), a_s^\dagger(\mathbf{k}')] &= \delta_{rs}\,\delta_{\mathbf{kk}'}\\ [a_r(\mathbf{k}), a_s(\mathbf{k}')] &= [a_r^\dagger(\mathbf{k}), a_s^\dagger(\mathbf{k}')] = 0\end{aligned}\right\} \tag{1.29}$$

and write the Hamiltonian (1.18) as

$$H_{\text{rad}} = \sum_{\mathbf{k}} \sum_{r} \hbar\omega_{\mathbf{k}}(a_r^\dagger(\mathbf{k})a_r(\mathbf{k}) + \tfrac{1}{2}). \tag{1.30}$$

The operators

$$N_r(\mathbf{k}) = a_r^\dagger(\mathbf{k})a_r(\mathbf{k})$$

then have eigenvalues $n_r(\mathbf{k}) = 0, 1, 2, \ldots,$ and eigenfunctions of the form (1.25)

$$|n_r(\mathbf{k})\rangle = \frac{[a_r^\dagger(\mathbf{k})]^{n_r(\mathbf{k})}}{\sqrt{n_r(k)!}}|0\rangle. \tag{1.31}$$

The eigenfunctions of the radiation Hamiltonian (1.30) are products of such states, i.e.

$$|\ldots n_r(\mathbf{k})\ldots\rangle = \prod_{\mathbf{k}_i} \prod_{r_i} |n_{r_i}(\mathbf{k}_i)\rangle, \tag{1.32}$$

with energy

$$\sum_{\mathbf{k}} \sum_{r} \hbar\omega_{\mathbf{k}}(n_r(\mathbf{k}) + \tfrac{1}{2}). \tag{1.33}$$

The interpretation of these equations is a straightforward generalization from one harmonic oscillator to a superposition of independent oscillators, one for each radiation mode $(\mathbf{k}, r)$. $a_r(\mathbf{k})$ operating on the state (1.32) will reduce the *occupation number* $n_r(\mathbf{k})$ of the mode $(\mathbf{k}, r)$ by unity, leaving all other occupation numbers unaltered, i.e. from Eq. (1.24):

$$a_r(\mathbf{k})|\ldots n_r(\mathbf{k})\ldots\rangle = [n_r(\mathbf{k})]^{1/2}|\ldots, n_r(\mathbf{k}) - 1, \ldots\rangle. \tag{1.34}$$

Correspondingly the energy (1.33) is reduced by $\hbar\omega_{\mathbf{k}} = \hbar c|\mathbf{k}|$. We interpret $a_r(\mathbf{k})$ as an annihilation (or destruction or absorption) operator which annihilates one photon in the mode $(\mathbf{k}, r)$, i.e. with momentum $\hbar\mathbf{k}$, energy $\hbar\omega_{\mathbf{k}}$ and linear polarization vector $\boldsymbol{\varepsilon}_r(\mathbf{k})$. Similarly $a_r^\dagger(\mathbf{k})$ is interpreted as a creation operator of such a photon. The assertion that $a_r(\mathbf{k})$ and $a_r^\dagger(\mathbf{k})$ are absorption and creation operators of photons with momentum $\hbar\mathbf{k}$ can be justified by calculating the momentum of the radiation field. We shall see later that the momentum operator of the field is given by

$$\mathbf{P} = \sum_{\mathbf{k}} \sum_{r} \hbar \mathbf{k} (N_r(\mathbf{k}) + \tfrac{1}{2}), \tag{1.35}$$

which leads to the above interpretation. We shall not consider the more intricate problem of the angular momentum of the photons but only mention that circular polarization states obtained by forming linear combinations

$$-\frac{1}{\sqrt{2}}[\boldsymbol{\varepsilon}_1(\mathbf{k}) + i\boldsymbol{\varepsilon}_2(\mathbf{k})], \qquad \frac{1}{\sqrt{2}}[\boldsymbol{\varepsilon}_1(\mathbf{k}) - i\boldsymbol{\varepsilon}_2(\mathbf{k})], \tag{1.36}$$

are more appropriate for this. Remembering that $(\boldsymbol{\varepsilon}_1(\mathbf{k}), \boldsymbol{\varepsilon}_2(\mathbf{k}), \mathbf{k})$ form a right-handed Cartesian coordinate system we see that these two combinations correspond to angular momentum $\pm\hbar$ in the direction $\mathbf{k}$ (analogous to the properties of the spherical harmonics $Y_1^{\pm 1}$), i.e. they represent right- and left-circular polarization: the photon behaves like a particle of spin 1. The third spin component is of course missing because of the transverse nature of the photon field.

The state of lowest energy of the radiation field is the vacuum state $|0\rangle$ in which all occupation numbers $n_r(\mathbf{k})$ are zero. According to Eqs. (1.30) or (1.33), this state has the energy $\frac{1}{2}\sum_{\mathbf{k}}\sum_r \hbar\omega_{\mathbf{k}}$. This is an infinite *constant* which is of no physical significance: we can eliminate it altogether by shifting the zero of the energy scale to coincide with the vacuum state $|0\rangle$. This corresponds to replacing Eq. (1.30) by

$$H_{\text{rad}} = \sum_{\mathbf{k}} \sum_{r} \hbar\omega_{\mathbf{k}} a_r^{\dagger}(\mathbf{k}) a_r(\mathbf{k}). \tag{1.37}$$

[The 'extra' term in Eq. (1.35) for the momentum will similarly be dropped. It actually vanishes in any case due to symmetry in the $\mathbf{k}$ summation.]

The representation (1.32) in which states are specified by the occupation numbers $n_r(\mathbf{k})$ is called the *number representation*. It is of great practical importance in calculating transitions (possibly via intermediate states) between initial and final states containing definite numbers of photons with well-defined properties. These ideas are of course not restricted to photons but apply generally to the particles of quantized fields. We shall have to modify the formalism in one respect. We have seen that the photon occupation numbers $n_r(\mathbf{k})$ can assume all values 0, 1, 2, ... Thus, photons satisfy Bose–Einstein statistics. They are *bosons*. So a modification will be required to describe particles obeying Fermi–Dirac statistics (*fermions*), such as electrons or muons, for which the occupation numbers are restricted to the values 0 and 1.

We have quantized the electromagnetic field by replacing the classical amplitudes a_r and a_r^* in the vector potential (1.15) by operators, so that the vector potential and the electric and magnetic fields become operators. In particular, the vector potential (1.15) becomes, in the Heisenberg picture [cf. Eqs. (1.28) and (1.17)], the time-dependent operator

$$\mathbf{A}(\mathbf{x}, t) = \mathbf{A}^{+}(\mathbf{x}, t) + \mathbf{A}^{-}(\mathbf{x}, t), \tag{1.38a}$$

with

$$\mathbf{A}^{+}(\mathbf{x}, t) = \sum_{\mathbf{k}} \sum_{r} \left(\frac{\hbar c^2}{2V\omega_{\mathbf{k}}} \right)^{1/2} \boldsymbol{\varepsilon}_r(\mathbf{k}) a_r(\mathbf{k})\, \mathrm{e}^{\mathrm{i}(\mathbf{k}\cdot\mathbf{x} - \omega_{\mathbf{k}} t)}, \tag{1.38b}$$

$$\mathbf{A}^{-}(\mathbf{x}, t) = \sum_{\mathbf{k}} \sum_{r} \left(\frac{\hbar c^2}{2V\omega_{\mathbf{k}}} \right)^{1/2} \boldsymbol{\varepsilon}_r(\mathbf{k}) a_r^{\dagger}(\mathbf{k})\, \mathrm{e}^{-\mathrm{i}(\mathbf{k}\cdot\mathbf{x} - \omega_{\mathbf{k}} t)}. \tag{1.38c}$$

The operator $\mathbf{A}^+$ contains only absorption operators, $\mathbf{A}^-$ only creation operators. $\mathbf{A}^+$ and $\mathbf{A}^-$ are called the positive and negative frequency parts of $\mathbf{A}$.[‡] The operators for $\mathbf{E}(\mathbf{x}, t)$ and $\mathbf{B}(\mathbf{x}, t)$ follow from Eqs. (1.9). There is an important difference between a quantized field theory and non-relativistic quantum mechanics. In the former it is the amplitudes (and hence the fields) which are operators and the position and time coordinates $(\mathbf{x}, t)$ are ordinary numbers, whereas in the latter the position coordinates (but not the time) are operators.

Finally, we note that a state with a definite number ν of photons (i.e. an eigenstate of the total photon number operator $N = \sum_{\mathbf{k}} \sum_r N_r(\mathbf{k})$) cannot be a classical field, not even for $\nu \to \infty$. This is a consequence of the fact that $\mathbf{E}$, like $\mathbf{A}$, is linear in the creation and absorption operators. Hence the expectation value of $\mathbf{E}$ in such a state vanishes. It is possible to form so-called coherent states $|c\rangle$ for which $\langle c|\mathbf{E}|c\rangle$ represents a transverse wave and for which the relative fluctuation $\Delta\mathbf{E}/\langle c|\mathbf{E}|c\rangle$ tends to zero as the number of photons in the state, $\langle c|N|c\rangle$, tends to infinity, i.e. in this limit the state $|c\rangle$ goes over into a classical state of a well-defined field.[§]

1.3 THE ELECTRIC DIPOLE INTERACTION

In the last section we quantized the radiation field. Since the occupation number operators $a_r^{\dagger}(\mathbf{k})a_r(\mathbf{k})$ commute with the radiation Hamiltonian (1.37), the occupation numbers $n_r(\mathbf{k})$ are constants of the motion for the free field. For anything 'to happen' requires interactions with charges and currents so that photons can be absorbed, emitted or scattered.

The complete description of the interaction of a system of charges (for example an atom or a nucleus) with an electromagnetic field is very complicated. In this section we shall consider the simpler and in practice

[‡] This is like in non-relativistic quantum mechanics where a time-dependence $\mathrm{e}^{-\mathrm{i}\omega t}$ with $\omega = E/\hbar > 0$ corresponds to a positive energy, i.e. a positive frequency.

[§] For a discussion of coherent states see R. Loudon, *The Quantum Theory of Light*, Clarendon Press, Oxford, 1973, pp. 148–153. See also Problem 1.1.

important special case of the interaction occurring via the electric dipole moment of the system of charges. The more complete (but still non-covariant) treatment of Section 1.4 will justify some of the points asserted in this section.

We shall consider a system of N charges $e_1, e_2, \ldots, e_N$ which can be described non-relativistically, i.e. the position of e_i, $i = 1, \ldots, N$, at time t is classically given by $\mathbf{r}_i = \mathbf{r}_i(t)$. We consider transitions between definite initial and final states of the system (e.g. between two states of an atom). The transitions are brought about by the electric dipole interaction if two approximations are valid.

Firstly it is permissible to neglect the interactions with the magnetic field.

Secondly, one may neglect the spatial variation of the electric radiation field, causing the transitions, across the system of charges (e.g. across the atom). Under these conditions the electric field

$$\mathbf{E}_{\mathrm{T}}(\mathbf{r}, t) = -\frac{1}{c}\frac{\partial \mathbf{A}(\mathbf{r}, t)}{\partial t}, \tag{1.39}$$

resulting from the transverse vector potential (1.38) of the radiation field (we are again using the Coulomb gauge $\nabla \cdot \mathbf{A} = 0$), can be calculated at *one* point somewhere inside the system of charges, instead of at the position of each charge.‡ Taking this point as the origin of coordinates $\mathbf{r} = 0$, we obtain for the interaction causing transitions the electric dipole interaction H_{I} given by

$$H_{\mathrm{I}} = -\mathbf{D} \cdot \mathbf{E}_{\mathrm{T}}(0, t) \tag{1.40}$$

where the electric dipole moment is defined by

$$\mathbf{D} = \sum_i e_i \mathbf{r}_i. \tag{1.41}$$

Transitions brought about by the interaction (1.40) in first-order perturbation theory are called *electric dipole transitions*. Since $\mathbf{E}_{\mathrm{T}}$ like $\mathbf{A}$ [Eq. (1.38)] is linear in the photon absorption and creation operators, so is H_{I}. It follows that in electric dipole transitions one photon is emitted or absorbed. In the next section it will be shown that the electric dipole approximation is valid provided the wavelength $\lambda = 2\pi/k$ of the radiation emitted or absorbed in the transition is very large compared to the linear dimensions R of the system of charges: $\lambda \gg R$. For example, for optical transitions in atoms, R is of the order of 1 Å and λ lies in the range 4000–7500 Å. Similarly, for gamma-rays emitted by nuclei, R is of the order of a few fermis (1 f $= 10^{-15}$ m) and since $\lambda/2\pi =$

‡ In Eq. (1.39) we have written $\mathbf{E}_{\mathrm{T}}$, since we now also have the Coulomb interaction between the charges which makes a contribution $-\nabla\phi$ to the electric field. [See Eqs. (1.2) and (1.4a) and Section 1.4.]

[197/(E in MeV)] f for a gamma-ray of E MeV, the electric dipole approximation is valid up to quite high gamma-ray energies.

If there are selection rules forbidding a transition in the electric dipole approximation, it might still occur via the magnetic interactions or via parts of the electric interactions which are neglected in the dipole approximation. It may happen that a transition is strictly forbidden, i.e. cannot occur in first-order perturbation theory even when the exact interaction is used as perturbation instead of H_{I} [Eq. (1.40)]. In such cases the transition can still occur in higher orders of perturbation theory or, possibly, by some quite different mechanism.‡

Let us now consider in some detail the emission and absorption of radiation in electric dipole transitions in atoms. The atom will make a transition from an initial state $|A\rangle$ to a final state $|B\rangle$ and the occupation number of one photon state will change from $n_r(\mathbf{k})$ to $n_r(\mathbf{k}) \pm 1$. The initial and final states of the system will be

$$\left.\begin{aligned} |A, n_r(\mathbf{k})\rangle &= |A\rangle|n_r(\mathbf{k})\rangle \\ |B, n_r(\mathbf{k}) \pm 1\rangle &= |B\rangle|n_r(\mathbf{k}) \pm 1\rangle \end{aligned}\right\}, \tag{1.42}$$

where the occupation numbers of the photon states which are not changed in the transition are not shown. The dipole operator (1.41) now becomes:

$$\mathbf{D} = -e\sum_i \mathbf{r}_i \equiv -e\mathbf{x}, \tag{1.43}$$

where the summation is over the atomic electrons and we have introduced the abbreviation $\mathbf{x}$. The transverse electric field $\mathbf{E}_{\mathrm{T}}(0, t)$ which occurs in the interaction (1.40) is from Eqs. (1.38)

$$\begin{aligned} \mathbf{E}_{\mathrm{T}}(0, t) &= -\frac{1}{c}\frac{\partial \mathbf{A}(0, t)}{\partial t} \\ &= \mathrm{i}\sum_{\mathbf{k}}\sum_{r}\left(\frac{\hbar\omega_{\mathbf{k}}}{2V}\right)^{1/2} \boldsymbol{\varepsilon}_r(\mathbf{k})[a_r(\mathbf{k})\,\mathrm{e}^{-\mathrm{i}\omega_{\mathbf{k}}t} - a_r^{\dagger}(\mathbf{k})\,\mathrm{e}^{\mathrm{i}\omega_{\mathbf{k}}t}]. \end{aligned}$$

Let us consider radiative emission. The transition matrix element of the interaction (1.40) between the states (1.42) then is given by

$$\begin{aligned} &\langle B, n_r(\mathbf{k}) + 1|H_{\mathrm{I}}|A, n_r(\mathbf{k})\rangle \\ &\quad = \mathrm{i}\left(\frac{\hbar\omega_{\mathbf{k}}}{2V}\right)^{1/2} \langle n_r(\mathbf{k}) + 1|a_r^{\dagger}(\mathbf{k})|n_r(\mathbf{k})\rangle\langle B|\boldsymbol{\varepsilon}_r(\mathbf{k})\cdot\mathbf{D}|A\rangle\,\mathrm{e}^{\mathrm{i}\omega_{\mathbf{k}}t} \\ &\quad = \mathrm{i}\left(\frac{\hbar\omega_{\mathbf{k}}}{2V}\right)^{1/2} [n_r(\mathbf{k}) + 1]^{1/2}\langle B|\boldsymbol{\varepsilon}_r(\mathbf{k})\cdot\mathbf{D}|A\rangle\,\mathrm{e}^{\mathrm{i}\omega_{\mathbf{k}}t} \end{aligned} \tag{1.44}$$

where the last line follows from Eq. (1.24).

‡ For selection rules for radiative transitions in atoms, see H. A. Bethe and R. W. Jackiw, *Intermediate Quantum Mechanics*, 2nd edn, Benjamin, New York, 1968, Chapter 11.

The transition probability per unit time between initial and final states (1.42) is given by time-dependent perturbation theory as

$$w = \frac{2\pi}{\hbar} |\langle B, n_r(\mathbf{k}) + 1|H_I|A, n_r(\mathbf{k})\rangle|^2 \, \delta(E_A - E_B - \hbar\omega_{\mathbf{k}}) \tag{1.45}$$

where E_A and E_B are the energies of the initial and final atomic states $|A\rangle$ and $|B\rangle$.‡ The delta function ensures conservation of energy in the transition, i.e. the emitted photon's energy $\hbar\omega_{\mathbf{k}}$ must satisfy the Bohr frequency condition

$$\omega_{\mathbf{k}} = \omega \equiv (E_A - E_B)/\hbar. \tag{1.46}$$

The delta function is eliminated in the usual way from Eq. (1.45) by integrating over a narrow group of final photon states. The number of photon states in the interval $(\mathbf{k}, \mathbf{k} + \mathrm{d}\mathbf{k})$, all in the same polarization state $(\boldsymbol{\varepsilon}_1(\mathbf{k})$ or $\boldsymbol{\varepsilon}_2(\mathbf{k}))$, is

$$\frac{V \, \mathrm{d}^3\mathbf{k}}{(2\pi)^3} = \frac{Vk^2 \, \mathrm{d}k \, \mathrm{d}\Omega}{(2\pi)^3}.^{\S} \tag{1.47}$$

From Eqs. (1.44)–(1.47) we obtain the probability per unit time for an atomic transition $|A\rangle \rightarrow |B\rangle$ with emission of a photon of wave vector in the range $(\mathbf{k}, \mathbf{k} + \mathrm{d}\mathbf{k})$ and with polarization vector $\boldsymbol{\varepsilon}_r(\mathbf{k})$:

$$w_r \, \mathrm{d}\Omega = \int \frac{Vk^2 \, \mathrm{d}k \, \mathrm{d}\Omega}{(2\pi)^3} \frac{2\pi}{\hbar} \delta(E_A - E_B - \hbar\omega_{\mathbf{k}}) \times \left(\frac{\hbar\omega_{\mathbf{k}}}{2V}\right) [n_r(\mathbf{k}) + 1] |\langle B|\boldsymbol{\varepsilon}_r(\mathbf{k}) \cdot \mathbf{D}|A\rangle|^2. \tag{1.49}$$

If we perform the integration with respect to k $(=\omega_{\mathbf{k}}/c)$ and substitute (1.43) for $\mathbf{D}$, the last expression reduces to

$$w_r \, \mathrm{d}\Omega = \frac{e^2\omega^3 \, \mathrm{d}\Omega}{8\pi^2\hbar c^3} [n_r(\mathbf{k}) + 1] |\boldsymbol{\varepsilon}_r(\mathbf{k}) \cdot \mathbf{x}_{BA}|^2 \tag{1.50}$$

where $\mathbf{x}_{BA}$ stands for the matrix element

$$\mathbf{x}_{BA} \equiv \langle B|\mathbf{x}|A\rangle = \langle B| \sum_i \mathbf{r}_i |A\rangle. \tag{1.51}$$

‡ Time-dependent perturbation theory is, for example, developed in A. S. Davydov, *Quantum Mechanics*, 2nd edn, Pergamon, Oxford, 1976, see Section 93 [Eq. (93.7)]; E. Merzbacher, *Quantum Mechanics*, 2nd edn, Wiley, New York, 1970, see Section 18.8; L. I. Schiff, *Quantum Mechanics*, 3rd edn, McGraw-Hill, New York, 1968, see Section 35.

§ Since we are using a finite normalization volume V, we should be summing over a group of allowed wave vectors $\mathbf{k}$ [see Eq. (1.13)]. For large V (strictly $V \rightarrow \infty$)

$$\frac{1}{V} \sum_{\mathbf{k}} \rightarrow \frac{1}{(2\pi)^3} \int \mathrm{d}^3\mathbf{k}. \tag{1.48}$$

The normalization volume V must of course drop out of all physically significant quantities such as transition rates etc.

The most interesting feature of Eq. (1.50) is the occurrence of the factor $[n_r(\mathbf{k}) + 1]$. $n_r(\mathbf{k})$ is the occupation number of photons in the $(\mathbf{k}, r)$ mode present initially, and thus the part of (1.50) proportional to $n_r(\mathbf{k})$ represents induced (or stimulated) emission, i.e. radiation which results from the radiation incident on the atom; classically, we can think of it as resulting from the forced oscillations of the electrons, and this term can be produced from a semiclassical theory of radiation.‡ However, even with no radiation present initially ($n_r(\mathbf{k}) = 0$), the transition probability (1.50) is different from zero. This corresponds to the spontaneous emission of radiation from an atom, and this cannot be derived from a semiclassical theory of radiation.

Eqs. (1.50) and (1.51) represent the basic result about emission of radiation in electric dipole transitions, and we only briefly indicate some consequences.

To sum over the two polarization states for a given $\mathbf{k}$, we note that $\boldsymbol{\varepsilon}_1(\mathbf{k})$, $\boldsymbol{\varepsilon}_2(\mathbf{k})$ and $\hat{\mathbf{k}} = \mathbf{k}/|\mathbf{k}|$ form an orthonormal coordinate system. Hence,

$$\begin{aligned}\sum_{r=1}^{2} |\boldsymbol{\varepsilon}_r(\mathbf{k})\cdot\mathbf{x}_{BA}|^2 &= \mathbf{x}_{BA}\cdot\mathbf{x}_{BA}^* - (\hat{\mathbf{k}}\cdot\mathbf{x}_{BA})(\hat{\mathbf{k}}\cdot\mathbf{x}_{BA}^*)\\ &= (\mathbf{x}_{BA}\cdot\mathbf{x}_{BA}^*)(1 - \cos^2\theta)\\ &= |\mathbf{x}_{BA}|^2 \sin^2\theta,\end{aligned}$$

where the last line but one defines the angle θ which the complex vector $\mathbf{x}_{BA}$ makes with $\hat{\mathbf{k}}$. Hence from Eq. (1.50)

$$\sum_{r=1}^{2} w_r\, d\Omega = \frac{e^2\omega^3}{8\pi^2\hbar c^3}\, d\Omega[n_r(\mathbf{k}) + 1]|\mathbf{x}_{BA}|^2 \sin^2\theta. \tag{1.52}$$

For spontaneous emission, the total transition probability per unit time is obtained from the last equation, with $n_r(\mathbf{k}) = 0$, by integrating over all directions. Since

$$\int \sin^2\theta\, d\Omega = \frac{8\pi}{3},$$

we obtain

$$w_{\text{total}}(A \to B) = \frac{e^2\omega^3}{3\pi\hbar c^3}\,|\mathbf{x}_{BA}|^2. \tag{1.53}$$

The life time τ of an excited atomic state $|A\rangle$ is defined as the reciprocal of the total transition probability per unit time to *all* possible final states $|B_1\rangle$, $|B_2\rangle, \ldots$, i.e.

$$\frac{1}{\tau} = \sum_n w_{\text{total}}(A \to B_n). \tag{1.54}$$

‡ See, for example, L. I. Schiff, *Quantum Mechanics*, 3rd edn, McGraw-Hill, New York, 1968, Chapter 11, or Bethe and Jackiw, referred to earlier in this section, Chapter 10.

In particular, if the state $|A\rangle$ can decay to states with non-zero total angular momentum, Eq. (1.54) must contain a summation over the corresponding magnetic quantum numbers.

The selection rules for electric dipole transitions follow from the matrix element (1.51). For example, since $\mathbf{x}$ is a vector, the states $|A\rangle$ and $|B\rangle$ must have opposite parity and the total angular momentum quantum number J of the atom and its z-component M must satisfy the selection rules

$$\Delta J = 0, \pm 1, \qquad \text{not } J = 0 \to J = 0, \qquad \Delta M = 0, \pm 1.$$

The second selection rule (not $J = 0 \to J = 0$) applies strictly to one-photon processes, not only in the electric dipole approximation. It is a consequence of the fact that there are no one-photon states with zero angular momentum. To form such a state from the spin 1 of the photon and a unit of orbital angular momentum requires all three components of the spin angular momentum, but because of the transversality of the radiation field only two of the spin components are available [compare Eq. (1.36)].

Finally, we note that very similar results hold for the absorption of radiation in electric dipole transitions. The matrix element

$$\langle B, n_r(\mathbf{k}) - 1|H_I|A, n_r(\mathbf{k})\rangle$$

corresponding to Eq. (1.44) now involves the factor $[n_r(\mathbf{k})]^{1/2}$ instead of $[n_r(\mathbf{k}) + 1]^{1/2}$. Our final result for emission, Eq. (1.50), also holds for absorption, with $[n_r(\mathbf{k}) + 1]$ replaced by $[n_r(\mathbf{k})]$, $d\Omega$ being the solid angle defining the incident radiation, and the matrix element $\mathbf{x}_{BA}$, Eq. (1.51), representing a transition from an atomic state $|A\rangle$ with energy E_A to a state $|B\rangle$ with energy $E_B > E_A$. Correspondingly the frequency ω is defined by $\hbar\omega = E_B - E_A$ instead of Eq. (1.46).

1.4 THE ELECTROMAGNETIC FIELD IN THE PRESENCE OF CHARGES

After the special case of the electric dipole interaction, we now want to consider the general interaction of moving charges and an electromagnetic field. As this problem will later be treated in a relativistically covariant way, we shall not give a rigorous complete derivation but rather stress the physical interpretation. As in the last section, the motion of the charges will again be described non-relativistically. In Section 1.4.1 we shall deal with the Hamiltonian formulation of the classical theory. This will enable us very easily to go over to the quantized theory in Section 1.4.2. In Sections 1.4.3 and 1.4.4 we shall illustrate the application of the theory for radiative transitions and Thomson scattering.

1.4.1 Classical electrodynamics

We would expect the Hamiltonian of a system of moving charges, such as an atom, in an electromagnetic field to consist of three parts: a part referring to matter (i.e. the charges), a part referring to the electromagnetic field, and a part describing the interaction between matter and field.

For a system of point masses m_i, $i = 1, \ldots, N$, with charges e_i and position coordinates $\mathbf{r}_i$, the Hamiltonian is

$$H_{\mathrm{m}} = \sum_i \frac{\mathbf{p}_i^2}{2m_i} + H_C \tag{1.55a}$$

where H_{C} is the Coulomb interaction

$$H_{\mathrm{C}} \equiv \tfrac{1}{2} \sum_{\substack{i,\, j \\ (i \neq j)}} \frac{e_i e_j}{4\pi |\mathbf{r}_i - \mathbf{r}_j|} \tag{1.55b}$$

and $\mathbf{p}_i = m_i\, \mathrm{d}\mathbf{r}_i/\mathrm{d}t$ is the kinetic momentum of the ith particle. This is the usual Hamiltonian of atomic physics, for example.

The electromagnetic field in interaction with charges is described by Maxwell's equations [Eqs. (1.1)]. We continue to use the Coulomb gauge, $\nabla \cdot \mathbf{A} = 0$, so that the electric field (1.2) decomposes into transverse and longitudinal fields

$$\mathbf{E} = \mathbf{E}_{\mathrm{T}} + \mathbf{E}_{\mathrm{L}},$$

where

$$\mathbf{E}_{\mathrm{T}} = -\frac{1}{c}\frac{\partial \mathbf{A}}{\partial t}, \qquad \mathbf{E}_{\mathrm{L}} = -\nabla\phi.$$

(A longitudinal field is defined by the condition $\nabla \wedge \mathbf{E}_{\mathrm{L}} = 0$.) The magnetic field is given by $\mathbf{B} = \nabla \wedge \mathbf{A}$.

The total energy of the electromagnetic field

$$\tfrac{1}{2} \int (\mathbf{E}^2 + \mathbf{B}^2)\, \mathrm{d}^3\mathbf{x}$$

can be written

$$\tfrac{1}{2} \int (\mathbf{E}_{\mathrm{T}}^2 + \mathbf{B}^2)\, \mathrm{d}^3\mathbf{x} + \tfrac{1}{2} \int \mathbf{E}_{\mathrm{L}}^2\, \mathrm{d}^3\mathbf{x}.$$

The last integral can be transformed, using Poisson's equation $\nabla^2\phi = -\rho$, into

$$\tfrac{1}{2} \int \mathbf{E}_{\mathrm{L}}^2\, \mathrm{d}^3\mathbf{x} = \tfrac{1}{2} \int \frac{\rho(\mathbf{x}, t)\rho(\mathbf{x}', t)}{4\pi|\mathbf{x} - \mathbf{x}'|}\, \mathrm{d}^3\mathbf{x}\, \mathrm{d}^3\mathbf{x}'. \tag{1.56}$$

Thus the energy associated with the longitudinal field is the energy of the *instantaneous* electrostatic interaction between the charges. With

$$\rho(\mathbf{x}, t) = \sum_i e_i\, \delta(\mathbf{x} - \mathbf{r}_i(t))$$

Eq. (1.56) reduces to

$$\tfrac{1}{2}\int \mathbf{E}_\mathrm{L}^2\, \mathrm{d}^3\mathbf{x} = \tfrac{1}{2}\sum_{i,j} \frac{e_i e_j}{4\pi|\mathbf{r}_i - \mathbf{r}_j|}$$

$$= \tfrac{1}{2}\sum_{\substack{i,j\\ i\neq j}} \frac{e_i e_j}{4\pi|\mathbf{r}_i - \mathbf{r}_j|} \equiv H_\mathrm{C}, \tag{1.57}$$

where in the last line we have dropped the infinite self-energy which occurs for point charges. The term H_C has already been included in the Hamiltonian H_m, Eqs. (1.55), so we must take as additional energy of the electromagnetic field that of the transverse radiation field

$$H_\mathrm{rad} = \tfrac{1}{2}\int (\mathbf{E}_\mathrm{T}^2 + \mathbf{B}^2)\, \mathrm{d}^3\mathbf{x}. \tag{1.58}$$

Eqs. (1.55) allow for the instantaneous Coulomb interaction of charges. To allow for the interaction of moving charges with an electromagnetic field, one must replace the matter-Hamiltonian (1.55a) by

$$H'_\mathrm{m} = \sum_i \frac{1}{2m_i}\left(\mathbf{p}_i - \frac{e_i}{c}\mathbf{A}_i\right)^2 + H_\mathrm{C} \tag{1.59}$$

where $\mathbf{A}_i = \mathbf{A}(\mathbf{r}_i, t)$ denotes the vector potential at the position $\mathbf{r}_i$ of the charge e_i at time t. In Eq. (1.59) $\mathbf{p}_i$ is the momentum coordinate canonically conjugate to the position coordinate $\mathbf{r}_i$, in the sense of Lagrangian mechanics, and it is related to the velocity $\mathbf{v}_i = \mathrm{d}\mathbf{r}_i/\mathrm{d}t$ of the ith particle by

$$\mathbf{p}_i = m_i\mathbf{v}_i + \frac{e_i}{c}\mathbf{A}_i.$$

It is only for $\mathbf{A} = 0$ that this conjugate momentum reduces to the kinetic momentum $m_i\mathbf{v}_i$. The justification for the form (1.59) for H'_m is that it gives the correct equations of motion for the charges (see Problem 1.2):

$$m_i\frac{\mathrm{d}\mathbf{v}_i}{\mathrm{d}t} = e_i\left[\mathbf{E}_i + \frac{\mathbf{v}_i}{c}\wedge\mathbf{B}_i\right], \tag{1.60}$$

where $\mathbf{E}_i$ and $\mathbf{B}_i$ are the electric and magnetic fields at the instantaneous position of the ith charge.‡

‡ For the Lagrangian and Hamiltonian formulations of mechanics which are here used see, for example, H. Goldstein, *Classical Mechanics*, 2nd edn, Addison-Wesley, Reading, Mass., 1980, in particular pp. 21–23 and 346.

We can regroup the terms in Eq. (1.59) as

$$H'_{\mathrm{m}} = H_{\mathrm{m}} + H_{\mathrm{I}} \tag{1.61}$$

where H_{I}, the interaction Hamiltonian of matter and field, is given by

$$\begin{aligned} H_{\mathrm{I}} &= \sum_i \left\{ -\frac{e_i}{2m_i c} (\mathbf{p}_i \cdot \mathbf{A}_i + \mathbf{A}_i \cdot \mathbf{p}_i) + \frac{e_i^2}{2m_i c^2} \mathbf{A}_i^2 \right\} \\ &= \sum_i \left\{ -\frac{e_i}{m_i c} \mathbf{A}_i \cdot \mathbf{p}_i + \frac{e_i^2}{2m_i c^2} \mathbf{A}_i^2 \right\}. \end{aligned} \tag{1.62}$$

In the quantum theory $\mathbf{p}_i$, the momentum canonically conjugate to $\mathbf{r}_i$, will become the operator $-i\hbar\nabla_i$. Nevertheless, the replacement of $\mathbf{p}_i \cdot \mathbf{A}_i$ by $\mathbf{A}_i \cdot \mathbf{p}_i$ in the second line of Eq. (1.62) is justified by our gauge condition $\nabla_i \cdot \mathbf{A}_i = 0$. Eq. (1.62) represents the general interaction of moving charges in an electromagnetic field (apart from H_{C}). It does not include the interaction of the magnetic moments, such as that due to the spin of the electron, with magnetic fields.

Combining the above results (1.55), (1.58), (1.59) and (1.62), we obtain for the complete Hamiltonian

$$H = H'_{\mathrm{m}} + H_{\mathrm{rad}} = H_{\mathrm{m}} + H_{\mathrm{rad}} + H_{\mathrm{I}}. \tag{1.63}$$

Just as this Hamiltonian leads to the correct equations of motion (1.60) for charges, so it also leads to the correct field equations (1.4), with $\nabla \cdot \mathbf{A} = 0$, for the potentials.‡

1.4.2 Quantum electrodynamics

The quantization of the system described by the Hamiltonian (1.63) is carried out by subjecting the particles' coordinates $\mathbf{r}_i$ and canonically conjugate momenta $\mathbf{p}_i$ to the usual commutation relations (e.g. in the coordinate representation $\mathbf{p}_i \to -i\hbar\nabla_i$), and quantizing the radiation field as in Section 1.2.3. The longitudinal electric field $\mathbf{E}_{\mathrm{L}}$ does not provide any additional degrees of freedom, being completely determined via the first Maxwell equation $\nabla \cdot \mathbf{E}_{\mathrm{L}} = \rho$ by the charges.

The interaction H_{I} in Eq. (1.63) is usually treated as a perturbation which causes transitions between the states of the non-interacting Hamiltonian

$$H_0 = H_{\mathrm{m}} + H_{\mathrm{rad}}. \tag{1.64}$$

The eigenstates of H_0 are again of the form

$$|A, \ldots n_r(\mathbf{k}) \ldots \rangle = |A\rangle | \ldots n_r(\mathbf{k}) \ldots \rangle,$$

with $|A\rangle$ and $| \ldots n_r(\mathbf{k}) \ldots \rangle$ eigenstates of H_{m} and H_{rad}.

‡ See W. Heitler, *The Quantum Theory of Radiation*, 3rd edn, Clarendon Press, Oxford, 1954, pp. 48–50.

Compared with the electric dipole interaction (1.40), the interaction (1.62) differs in that it contains a term quadratic in the vector potential. This results in two-photon processes in first-order perturbation theory (i.e. emission or absorption of two photons or scattering). In addition, the first term in (1.62) contains magnetic interactions and higher-order effects due to the spatial variation of $\mathbf{A}(\mathbf{x}, t)$ which are absent from the electric dipole interaction (1.40). These aspects are illustrated in the applications to radiative transitions and Thomson scattering which follow.

1.4.3 Radiative transitions in atoms

We consider transitions between two states of an atom with emission or absorption of one photon. This problem was treated in Section 1.3 in the electric dipole approximation but now we shall use the interaction (1.62).

We shall consider the emission process between the initial and final states (1.42). Using the expansion (1.38) of the vector potential, we obtain the matrix element for this transition [which results from the term linear in $\mathbf{A}$ in Eq. (1.62)]

$$\langle B, n_r(\mathbf{k}) + 1|H_{\mathrm{I}}|A, n_r(\mathbf{k})\rangle$$

$$= \frac{e}{m}\left(\frac{\hbar}{2V\omega_{\mathbf{k}}}\right)^{1/2} [n_r(\mathbf{k}) + 1]^{1/2}\langle B|\boldsymbol{\varepsilon}_r(\mathbf{k})\cdot\sum_i \mathrm{e}^{-\mathrm{i}\mathbf{k}\cdot\mathbf{r}_i}\,\mathbf{p}_i|A\rangle\,\mathrm{e}^{\mathrm{i}\omega_{\mathbf{k}}t}. \tag{1.65}$$

Using this matrix element, one calculates the transition probability per unit time as in Section 1.3. Instead of Eqs. (1.50) and (1.51), one obtains:

$$w_r\,\mathrm{d}\Omega = \frac{e^2\omega\,\mathrm{d}\Omega}{8\pi^2 m^2\hbar c^3}[n_r(\mathbf{k}) + 1]\left|\boldsymbol{\varepsilon}_r(\mathbf{k})\cdot\langle B|\sum_i \mathrm{e}^{-\mathrm{i}\mathbf{k}\cdot\mathbf{r}_i}\,\mathbf{p}_i|A\rangle\right|^2. \tag{1.66}$$

These results go over into the electric dipole approximation if in the matrix elements in Eqs. (1.65) and (1.66) we can approximate the exponential functions by unity:

$$\mathrm{e}^{-\mathrm{i}\mathbf{k}\cdot\mathbf{r}_i} \approx 1. \tag{1.67}$$

This is justified provided the wavelength $\lambda = 2\pi/k$ of the radiation emitted in the transition is very large compared to the linear dimensions R of the system of charges (in our case, of the atom): $\lambda \gg R$. The atomic wavefunctions $|A\rangle$ and $|B\rangle$ restrict the effective values of $\mathbf{r}_i$ to $r_i \lesssim R$, so that $\mathbf{k}\cdot\mathbf{r}_i \lesssim kR \ll 1$. We saw in Section 1.3 that this inequality is generously satisfied for optical atomic transitions. From the equation of motion $\mathrm{i}\hbar\dot{\mathbf{r}}_i = [\mathbf{r}_i, H]$ and Eq. (1.46)

$$\langle B|\mathbf{p}_i|A\rangle = m\langle B|\dot{\mathbf{r}}_i|A\rangle = -\mathrm{i}m\omega\langle B|\mathbf{r}_i|A\rangle.$$

Hence, in the approximation (1.67), Eqs. (1.65) and (1.66) reduce to the electric dipole form, Eqs. (1.44) and (1.50).

If selection rules forbid the transition $|A\rangle$ to $|B\rangle$ via the electric dipole interaction, it may in general still occur via higher terms in the expansion of the exponentials

$$e^{-i\mathbf{k}\cdot\mathbf{r}_i} = 1 - i\mathbf{k}\cdot\mathbf{r}_i + \cdots.$$

With the second term, the expression within the modulus sign in Eq. (1.66) becomes

$$\boldsymbol{\varepsilon}_r(\mathbf{k})\cdot\langle B|\sum_i(-i\mathbf{k}\cdot\mathbf{r}_i)\mathbf{p}_i|A\rangle = -i\sum_{\alpha=1}^{3}\sum_{\beta=1}^{3}\varepsilon_{r\alpha}(\mathbf{k})k_\beta\langle B|\sum_i r_{i\beta}p_{i\alpha}|A\rangle,$$

where α, β $(=1, 2, 3)$ label the Cartesian components of the vectors $\boldsymbol{\varepsilon}_r$, $\mathbf{k}$, $\mathbf{r}_i$ and $\mathbf{p}_i$. The matrix element can be written as the sum of an antisymmetric and a symmetric second-rank tensor

$$\langle B|\sum_i r_{i\beta}p_{i\alpha}|A\rangle = \tfrac{1}{2}\left\{\langle B|\sum_i(r_{i\beta}p_{i\alpha} - r_{i\alpha}p_{i\beta})|A\rangle + \langle B|\sum_i(r_{i\beta}p_{i\alpha} + r_{i\alpha}p_{i\beta})|A\rangle\right\}.$$

The first term contains the antisymmetric angular momentum operator and corresponds to the magnetic dipole interaction. (In practice this must be augmented by the spin part.) The symmetric term corresponds to the electric quadrupole interaction. The parity and angular momentum selection rules for the transitions brought about by these matrix elements are easily determined from their forms. We obtain in this way an expansion into electric and magnetic multipoles, i.e. photons of definite parity and angular momentum. As usual, a better procedure for such an expansion, except in the simplest cases, is to use spherical rather than Cartesian coordinates.[‡]

The result (1.66) can again be adapted to the case of absorption of radiation by replacing the factor $[n_r(\mathbf{k}) + 1]$ by $n_r(\mathbf{k})$ and the appropriate reinterpretation of the matrix element, etc.

1.4.4 Thomson scattering

As a second illustration we consider Thomson scattering, i.e. the scattering of photons of energy $\hbar\omega$ by atomic electrons, with $\hbar\omega$ large compared to the binding energies of the electrons, so that they can be considered as free electrons, but $\hbar\omega$ very small compared to the electron rest energy mc^2. In this case the energy $\hbar\omega'$ of the scattered photon is not changed: $\hbar\omega' = \hbar\omega$, since for small recoil momenta the recoil energy may be neglected.

[‡] See A. S. Davydov, *Quantum Mechanics*, 2nd edn, Pergamon, Oxford, 1976, Sections 81 and 95.

The scattering from an initial state with one photon of momentum $\hbar\mathbf{k}$ and polarization $\boldsymbol{\varepsilon}_\alpha(\mathbf{k})$ (with $\alpha = 1$ or 2) to a final state with one photon of momentum $\hbar\mathbf{k}'$ and polarization $\boldsymbol{\varepsilon}_\beta(\mathbf{k}')$ (with $\beta = 1$ or 2) can occur in first-order perturbation theory via the term in $\mathbf{A}^2$ in the interaction (1.62). It can also occur in second-order perturbation theory via the term linear in $\mathbf{A}$ in Eq. (1.62), but one can show that under our conditions the contribution of the second-order process is negligible.[‡] The operator $\mathbf{A}^2(0, t)$ can, from Eq. (1.38), be written

$$\mathbf{A}^2(0, t) = \sum_{\mathbf{k}_1\mathbf{k}_2}\sum_{r,s} \frac{\hbar c^2}{2V(\omega_1\omega_2)^{1/2}} (\boldsymbol{\varepsilon}_r(\mathbf{k}_1)\cdot\boldsymbol{\varepsilon}_s(\mathbf{k}_2))$$
$$\times [a_r(\mathbf{k}_1)\,\mathrm{e}^{-\mathrm{i}\omega_1 t} + a_r^\dagger(\mathbf{k}_1)\,\mathrm{e}^{+\mathrm{i}\omega_1 t}][a_s(\mathbf{k}_2)\,\mathrm{e}^{-\mathrm{i}\omega_2 t} + a_s^\dagger(\mathbf{k}_2)\,\mathrm{e}^{+\mathrm{i}\omega_2 t}], \tag{1.68}$$

where $\omega_r \equiv c|\mathbf{k}_r|$, $r = 1, 2$. This operator can bring about the transition from the initial state $|\mathbf{k}, \alpha\rangle$ to the final state $|\mathbf{k}', \beta\rangle$ (we use a somewhat simplified but unambiguous notation) in two ways: either of the factors in square parentheses can act to absorb the initial photon, and the other factor then creates the final photon. One then obtains the matrix element for this transition from Eq. (1.62)

$$\langle\mathbf{k}', \beta|\,\frac{e^2}{2mc^2}\,\mathbf{A}^2(0, t)|\mathbf{k}, \alpha\rangle = \frac{e^2\hbar}{2mV(\omega\omega')^{1/2}}\,\boldsymbol{\varepsilon}_\alpha(\mathbf{k})\cdot\boldsymbol{\varepsilon}_\beta(\mathbf{k}')\,\mathrm{e}^{\mathrm{i}(\omega'-\omega)t}$$

where $\omega = c|\mathbf{k}|$ and $\omega' = c|\mathbf{k}'|$. The transition probability per unit time for a photon, initially in the state $|\mathbf{k}, \alpha\rangle$, to be scattered into an element of solid angle $\mathrm{d}\Omega$ in the direction $\mathbf{k}'$, and with polarization $\boldsymbol{\varepsilon}_\beta(\mathbf{k}')$, is given by

$$\begin{aligned} w_{\alpha\to\beta}(\mathbf{k}')\,\mathrm{d}\Omega &= \frac{2\pi}{\hbar}\int \frac{Vk'^2\,\mathrm{d}k'\,\mathrm{d}\Omega}{(2\pi)^3}\,\delta(\hbar\omega' - \hbar\omega) \\ &\quad\times \left(\frac{e^2\hbar}{2mV\omega}\right)^2 [\boldsymbol{\varepsilon}_\alpha(\mathbf{k})\cdot\boldsymbol{\varepsilon}_\beta(\mathbf{k}')]^2 \\ &= \frac{c}{V}\left(\frac{e^2}{4\pi mc^2}\right)^2 [\boldsymbol{\varepsilon}_\alpha(\mathbf{k})\cdot\boldsymbol{\varepsilon}_\beta(\mathbf{k}')]^2\,\mathrm{d}\Omega \end{aligned}$$

where $|\mathbf{k}'| = |\mathbf{k}|$. Dividing this transition probability per unit time by the incident photon flux (c/V), one obtains the corresponding differential cross-section

$$\sigma_{\alpha\to\beta}(\mathbf{k}')\,\mathrm{d}\Omega = r_0^2[\boldsymbol{\varepsilon}_\alpha(\mathbf{k})\cdot\boldsymbol{\varepsilon}_\beta(\mathbf{k}')]^2\,\mathrm{d}\Omega, \tag{1.69}$$

‡ See J. J. Sakurai, *Advanced Quantum Mechanics*, Addison-Wesley, Reading, Mass., 1967, p. 51.

where the classical electron radius has been introduced by

$$r_0 = \frac{e^2}{4\pi mc^2} = 2.818 \text{ f}. \tag{1.70}$$

For an unpolarized incident photon beam, the unpolarized differential cross-section (i.e. the final polarization state is not observed) is obtained from Eq. (1.69) by summing over final and averaging over initial polarization states. We introduce the abbreviations $\boldsymbol{\varepsilon}_\alpha \equiv \boldsymbol{\varepsilon}_\alpha(\mathbf{k})$ and $\boldsymbol{\varepsilon}'_\beta \equiv \boldsymbol{\varepsilon}_\beta(\mathbf{k}')$. Since $\boldsymbol{\varepsilon}_1$, $\boldsymbol{\varepsilon}_2$ and $\hat{\mathbf{k}} = \mathbf{k}/|\mathbf{k}|$ form an orthonormal coordinate system,

$$\sum_{\alpha=1}^{2} (\boldsymbol{\varepsilon}_\alpha \cdot \boldsymbol{\varepsilon}'_\beta)^2 = 1 - (\hat{\mathbf{k}} \cdot \boldsymbol{\varepsilon}'_\beta)^2.$$

Similarly

$$\sum_{\beta=1}^{2} (\hat{\mathbf{k}} \cdot \boldsymbol{\varepsilon}'_\beta)^2 = 1 - (\hat{\mathbf{k}} \cdot \hat{\mathbf{k}}')^2 = \sin^2\theta$$

where θ is the angle between the directions $\mathbf{k}$ and $\mathbf{k}'$ of the incident and scattered photons, i.e. the angle of scattering. From the last two equations

$$\tfrac{1}{2} \sum_{\alpha=1}^{2} \sum_{\beta=1}^{2} (\boldsymbol{\varepsilon}_\alpha \cdot \boldsymbol{\varepsilon}'_\beta)^2 = \tfrac{1}{2}(2 - \sin^2\theta) = \tfrac{1}{2}(1 + \cos^2\theta) \tag{1.71}$$

and hence the unpolarized differential cross-section for scattering through an angle θ is from Eq. (1.69) given as

$$\sigma(\theta)\, \mathrm{d}\Omega = \tfrac{1}{2} r_0^2 (1 + \cos^2\theta)\, \mathrm{d}\Omega. \tag{1.69a}$$

Integrating over angles, we obtain the total cross-section for Thomson scattering

$$\sigma_{\text{total}} = \frac{8\pi}{3} r_0^2 = 6.65 \times 10^{-25} \text{ cm}^2. \tag{1.72}$$

1.5 APPENDIX: THE SCHRÖDINGER, HEISENBERG AND INTERACTION PICTURES

These three pictures (abbreviated S.P., H.P. and I.P.) are three different ways of describing the time development of a system. We shall derive the H.P. and the I.P. from the S.P. Quantities in these three pictures will be distinguished by the labels S, H and I.

In the S.P. the time dependence is carried by the states according to the Schrödinger equation

$$\mathrm{i}\hbar \frac{\mathrm{d}}{\mathrm{d}t} |A, t\rangle_{\mathrm{S}} = H |A, t\rangle_{\mathrm{S}}. \tag{1.73}$$

This can formally be solved in terms of the state of the system at an arbitrary initial time t_0

$$|A, t\rangle_S = U|A, t_0\rangle_S \tag{1.74}$$

where U is the unitary operator:

$$U \equiv U(t, t_0) = e^{-iH(t-t_0)/\hbar}. \tag{1.75}$$

By means of U we can carry out a unitary transformation of states and operators (O) from the S.P. to the H.P. in which we define

$$|A, t\rangle_H = U^\dagger |A, t\rangle_S = |A, t_0\rangle_S \tag{1.76}$$

and

$$O^H(t) = U^\dagger O^S U. \tag{1.77}$$

At $t = t_0$, states and operators in the two pictures are the same. We see from Eq. (1.76) that in the H.P. state vectors are constant in time; the time-dependence is carried by the Heisenberg operators. From Eq. (1.77)

$$\mathrm{H}^H = \mathrm{H}^S \equiv \mathrm{H}. \tag{1.78}$$

Since the transformation from the S.P. to the H.P. is unitary, it ensures the invariance of matrix elements and commutation relations:

$${}_S\langle B, t|O^S|A, t\rangle_S = {}_H\langle B, t|O^H(t)|A, t\rangle_H, \tag{1.79}$$

and if O and P are two operators for which $[O^S, P^S] = \text{const.}$, then $[O^H(t), P^H(t)]$ equals the same constant.

Differentation of Eq. (1.77) gives the Heisenberg equation of motion

$$i\hbar \frac{d}{dt} O^H(t) = [O^H(t), H]. \tag{1.80}$$

For an operator which is time-dependent in the S.P. (corresponding to a quantity which classically has an explicit time dependence), Eq. (1.80) is augmented to

$$i\hbar \frac{d}{dt} O^H(t) = i\hbar \frac{\partial}{\partial t} O^H(t) + [O^H(t), H]. \tag{1.81}$$

We shall not be considering such operators.

The I.P. arises if the Hamiltonian is split into two parts

$$H = H_0 + H_I. \tag{1.82}$$

In quantum field theory H_I will describe the interaction between two fields, themselves described by H_0. [Note that the suffix I on H_I stands for 'interaction'. It does not label a picture. Eq. (1.82) holds in any picture.] The

I.P. is related to the S.P. by the unitary transformation

$$U_0 \equiv U_0(t, t_0) = e^{-iH_0(t-t_0)/\hbar} \tag{1.83}$$

i.e.

$$|A, t\rangle_I = U_0^\dagger |A, t\rangle_S \tag{1.84}$$

and

$$O^I(t) = U_0^\dagger O^S U_0 . \tag{1.85}$$

Thus the relation between I.P. and S.P. is similar to that between H.P. and S.P., but with the unitary transformation U_0 involving the non-interacting Hamiltonian H_0, instead of U involving the total Hamiltonian H. From Eq. (1.85):

$$H_0^I = H_0^S \equiv H_0 . \tag{1.86}$$

Differentiating Eq. (1.85) gives the differential equation of motion of operators in the I.P.:

$$i\hbar \frac{d}{dt} O^I(t) = [O^I(t), H_0] . \tag{1.87}$$

Substituting Eq. (1.84) into the Schrödinger equation (1.73), one obtains the equation of motion of state vectors in the I.P.

$$i\hbar \frac{d}{dt} |A, t\rangle_I = H_I^I(t) |A, t\rangle_I \tag{1.88}$$

where

$$H_I^I(t) = e^{iH_0(t-t_0)/\hbar} H_I^S e^{-iH_0(t-t_0)/\hbar} . \tag{1.89}$$

PROBLEMS

1.1 The radiation field inside a cubic enclosure, which contains no charges, is specified by the state

$$|c\rangle = \exp(-\tfrac{1}{2}|c|^2) \sum_{n=0}^{\infty} \frac{c^n}{\sqrt{n!}} |n\rangle$$

where $c = |c| e^{i\delta}$ is any complex number and $|n\rangle$ is the state (1.31) in which there are n photons with wave vector $\mathbf{k}$ and polarization vector $\boldsymbol{\varepsilon}_r(\mathbf{k})$ present, and no others. Derive the following properties of the state $|c\rangle$.

(i) $|c\rangle$ is normalized: $\langle c|c\rangle = 1$.

(ii) $|c\rangle$ is an eigenstate of the destruction operator $a_r(\mathbf{k})$ with the complex eigenvalue c:

$$a_r(\mathbf{k})|c\rangle = c|c\rangle .$$

(iii) The mean number $\bar{N}$ of photons in the enclosure in the state $|c\rangle$ is given by

$$\bar{N} = \langle c|N|c\rangle = |c|^2 \tag{A}$$

where N is the total photon number operator.

(iv) The root-mean-square fluctuation ΔN in the number of photons in the enclosure in the state $|c\rangle$ is given by

$$(\Delta N)^2 = \langle c|N^2|c\rangle - \bar{N}^2 = |c|^2. \tag{B}$$

(v) The expectation value of the electric field **E** in the state $|c\rangle$ is given by

$$\langle c|\mathbf{E}|c\rangle = -\boldsymbol{\varepsilon}_r(\mathbf{k})2\left(\frac{\hbar\omega_{\mathbf{k}}}{2V}\right)^{1/2}|c|\sin(\mathbf{k}\cdot\mathbf{x} - \omega_{\mathbf{k}}t + \delta) \tag{C}$$

where V is the volume of the enclosure.

(vi) The root-mean-square fluctuation ΔE of the electric field in the state $|c\rangle$ is given by

$$(\Delta E)^2 = \langle c|\mathbf{E}^2|c\rangle - \langle c|\mathbf{E}|c\rangle^2 = \frac{\hbar\omega_{\mathbf{k}}}{2V}. \tag{D}$$

We noted in Section 1.2.3 that the expectation value of **E** in a state with a definite number of photons is zero, so that such a state cannot represent a classical field, even for very large photon numbers. In contrast, it follows from Eqs. (A)–(D) that the relative fluctuation in photon numbers

$$\frac{\Delta N}{\bar{N}} = N^{-1/2}$$

tends to zero as $\bar{N} \to \infty$, and that the fluctuation ΔE becomes negligible for large field strengths, i.e. $|c\rangle$ goes over into a classical state in which the field is well defined as $\bar{N} \to \infty$. The state $|c\rangle$ is called a coherent state and represents the closest quantum-mechanical approach to a classical electromagnetic field. (For a full discussion, see the book by Loudon, quoted at the end of Section 1.2.)

1.2 The Lagrangian of a particle of mass m and charge q, moving in an electromagnetic field, is given by

$$L(\mathbf{x}, \dot{\mathbf{x}}) = \tfrac{1}{2}m\dot{\mathbf{x}}^2 + \frac{q}{c}\mathbf{A}\cdot\dot{\mathbf{x}} - q\phi$$

where $\mathbf{A} = \mathbf{A}(\mathbf{x}, t)$ and $\phi = \phi(\mathbf{x}, t)$ are the vector and scalar potentials of the electromagnetic field at the position **x** of the particle at time t.

(i) Show that the momentum conjugate to **x** is given by

$$\mathbf{p} = m\dot{\mathbf{x}} + \frac{q}{c}\mathbf{A} \tag{A}$$

(i.e. the conjugate momentum **p** is not the kinetic momentum $m\dot{\mathbf{x}}$, in general) and that Lagrange's equations reduce to the equations of motion of the particle [compare Eq. (1.60)]

$$m\frac{\mathrm{d}}{\mathrm{d}t}\dot{\mathbf{x}} = q\left[\mathbf{E} + \frac{1}{c}\dot{\mathbf{x}} \wedge \mathbf{B}\right], \tag{B}$$

where **E** and **B** are the electric and magnetic fields at the instantaneous position of the charge.

(ii) Derive the corresponding Hamiltonian [compare Eq. (1.59)]

$$H = \frac{1}{2m}\left(\mathbf{p} - \frac{q}{c}\mathbf{A}\right)^2 + q\phi,$$

and show that the resulting Hamilton equations again lead to Eqs. (A) and (B).

1.3 For Thomson scattering of an unpolarized beam of photons, obtain the differential cross-section for scattering through an angle θ, with the scattered radiation being linearly polarized in a given direction. By considering two mutually perpendicular such directions, use your result to re-derive Eq. (1.69a) for the unpolarized differential cross-section.

Show that for $\theta = 90°$, the scattered beam is 100 per cent linearly polarized in the direction of the normal to the plane of scattering.

CHAPTER 2

Lagrangian field theory

In the last chapter we quantized the electromagnetic field by Fourier analysing the classical field into normal modes and imposing harmonic oscillator commutation relations on the normal coordinates. We shall now take the fields at each point in space as the dynamical variables and quantize these directly. This approach generalizes the classical mechanics of a system of particles, and its quantization, to a continuous system, i.e. to fields.‡ One introduces a Lagrangian (actually, as we shall see, it is a Lagrangian density) from which the field equations follow by means of Hamilton's principle. One introduces momenta conjugate to the fields and imposes canonical commutation relations directly on the fields and the conjugate momenta. This formalism provides a systematic quantization procedure for any classical field theory derivable from a Lagrangian. Since this approach is equivalent to that of the last chapter, one can only obtain bosons in this way and a different formalism will be needed for fermions.

Another difference from Chapter 1 is that the theory will now be developed in a manifestly relativistically covariant form, and in Section 2.1 we shall define our relativistic notation. The classical Lagrangian field theory will be developed in Section 2.2, to be quantized in Section 2.3. An important feature of a Lagrangian field theory is that all its symmetry properties and the consequent conservation laws are contained in the Lagrangian density. We shall consider some of these aspects in Section 2.4.

‡ The relevant Lagrangian and Hamiltonian mechanics is, for example, developed in H. Goldstein, *Classical Mechanics*, 2nd edn, Addison-Wesley, Reading, Mass., 1980, Chapters 2 and 8, or in L. D. Landau and E. M. Lifshitz, *Mechanics*, Pergamon, Oxford, 1960, Sections 1–7 and Section 40.

2.1 RELATIVISTIC NOTATION

We shall write x^μ ($\mu = 0, 1, 2, 3$) for the space–time four-vector with the time component $x^0 = ct$ and the space coordinates x^j ($j = 1, 2, 3$), i.e. $x^\mu = (ct, \mathbf{x})$. The components of four-vectors will be labelled by Greek indices, the components of spatial three-vectors by Latin indices.

By means of the metric tensor $g_{\mu\nu}$, with components

$$\left.\begin{aligned} g_{00} = -g_{11} = -g_{22} = -g_{33} = +1 \\ g_{\mu\nu} = 0 \quad \text{if } \mu \neq \nu \end{aligned}\right\}, \tag{2.1}$$

we define the covariant vector x_μ from the contravariant x^μ:

$$x_\mu = \sum_{\nu=0}^{3} g_{\mu\nu} x^\nu \equiv g_{\mu\nu} x^\nu. \tag{2.2}$$

In the last expression we have used the summation convention: repeated Greek indices, one contravariant and one covariant, are summed. From Eqs. (2.1) and (2.2) we have $x_\mu = (ct, -\mathbf{x})$.

We also define the contravariant metric tensor $g^{\lambda\mu}$ by

$$g^{\lambda\mu} g_{\mu\nu} = g^\lambda_\nu = \delta^\lambda_\nu \tag{2.3}$$

where δ^λ_ν is the usual Kronecker delta: $\delta^\lambda_\nu = 1$ if $\lambda = \nu$, and $\delta^\lambda_\nu = 0$ if $\lambda \neq \nu$. From Eqs. (2.1) and (2.3) $g^{\mu\nu} = g_{\mu\nu}$.

A Lorentz transformation

$$x^\mu \to x'^\mu = \Lambda^\mu{}_\nu x^\nu \tag{2.4}$$

leaves

$$x^\mu x_\mu = (x^0)^2 - \mathbf{x}^2 \tag{2.5}$$

invariant, i.e. $x'^\mu x'_\mu = x^\mu x_\mu$ is a scalar quantity. Hence

$$\Lambda^{\lambda\mu} \Lambda_{\lambda\nu} = \delta^\mu_\nu. \tag{2.6}$$

(In addition the matrix $\Lambda^{\lambda\mu}$ must be real to ensure the reality of the space–time coordinates.)

A four-component object s^μ (s_μ) transforming like x^μ (x_μ) under Lorentz transformations, and hence with $s^\mu s_\mu$ invariant, is a contravariant (covariant) four-vector. An example is the energy–momentum vector $p^\mu = (E/c, \mathbf{p})$. When no confusion can result, we shall often omit the tensor indices, e.g. we may write x for x^μ or x_μ.

The scalar product of two four-vectors a and b can be written in various ways:

$$ab = a^\mu b_\mu = a_\mu b^\mu = g_{\mu\nu} a^\mu b^\nu = \cdots = a^0 b^0 - \mathbf{a}\cdot\mathbf{b}. \tag{2.7}$$

Like $x^2 = x^\mu x_\mu$, so the scalar product ab is an invariant under Lorentz transformations.

The four-dimensional generalization of the gradient operator $\boldsymbol{\nabla}$ transforms like a four-vector. If $\phi(x)$ is a scalar function, so is

$$\delta\phi = \frac{\partial\phi}{\partial x^\mu}\,\delta x^\mu,$$

and hence

$$\frac{\partial\phi}{\partial x^\mu} \equiv \partial_\mu\phi \equiv \phi_{,\mu} \tag{2.8a}$$

is a covariant four-vector. Similarly

$$\frac{\partial\phi}{\partial x_\mu} \equiv \partial^\mu\phi \equiv \phi^{,\mu} \tag{2.8b}$$

is a contravariant four-vector. Note that indices following a comma denote differentiation. Finally, we note that the operator $\Box$ is a scalar:

$$\partial^\mu\partial_\mu = \frac{1}{c^2}\frac{\partial^2}{\partial t^2} - \boldsymbol{\nabla}^2 \equiv \Box. \tag{2.9}$$

2.2 CLASSICAL LAGRANGIAN FIELD THEORY

We consider a system which requires several fields $\phi_r(x)$, $r = 1, \ldots, N$, to specify it. The index r may label components of the same field [for example, the components of the vector potential $\mathbf{A}(x)$] or it may refer to different independent fields. We restrict ourselves to theories which can be derived by means of a variational principle from an action integral involving a Lagrangian density

$$\mathscr{L} = \mathscr{L}(\phi_r, \phi_{r,\alpha}) \tag{2.10}$$

where the derivative $\phi_{r,\alpha}$ is defined by Eq. (2.8a). The Lagrangian density (2.10), depending on the fields and their first derivatives only, is not the most general case possible, but it covers all theories discussed in this book and greatly simplifies the formalism.

We define the action integral $S(\Omega)$ for an arbitrary region Ω of the four-dimensional space–time continuum by

$$S(\Omega) = \int_\Omega \mathrm{d}^4x\,\mathscr{L}(\phi_r, \phi_{r,\alpha}), \tag{2.11}$$

where d^4x stands for the four-dimensional element $\mathrm{d}x^0\,\mathrm{d}^3\mathbf{x}$.

We now postulate that the equations of motion, i.e. the field equations, are obtained from the following variational principle which is closely analogous to Hamilton's principle in mechanics. For an arbitrary region Ω, we consider variations of the fields,

$$\phi_r(x) \to \phi_r(x) + \delta\phi_r(x), \tag{2.12}$$

which vanish on the surface $\Gamma(\Omega)$ bounding the region Ω

$$\delta\phi_r(x) = 0 \quad \text{on } \Gamma(\Omega). \tag{2.13}$$

The fields ϕ_r may be real or complex. In the case of a complex field $\phi(x)$, the fields $\phi(x)$ and $\phi^*(x)$ are treated as two independent fields. Alternatively, a complex field $\phi(x)$ can be decomposed into a pair of real fields which are then treated as independent fields. We now demand that for an arbitrary region Ω and the variation (2.12–2.13) the action (2.11) has a stationary value, i.e.

$$\delta S(\Omega) = 0. \tag{2.14}$$

Calculating $\delta S(\Omega)$ from Eq. (2.11), we obtain[‡]

$$\begin{aligned}\delta S(\Omega) &= \int_\Omega \mathrm{d}^4x \left\{\frac{\partial \mathscr{L}}{\partial \phi_r}\,\delta\phi_r + \frac{\partial \mathscr{L}}{\partial \phi_{r,\alpha}}\,\delta\phi_{r,\alpha}\right\} \\ &= \int_\Omega \mathrm{d}^4x \left\{\frac{\partial \mathscr{L}}{\partial \phi_r} - \frac{\partial}{\partial x^\alpha}\left(\frac{\partial \mathscr{L}}{\partial \phi_{r,\alpha}}\right)\right\}\delta\phi_r + \int_\Omega \mathrm{d}^4x\,\frac{\partial}{\partial x^\alpha}\left(\frac{\partial \mathscr{L}}{\partial \phi_{r,\alpha}}\,\partial\phi_r\right),\end{aligned} \tag{2.15}$$

where the last line is obtained by partial integration, since

$$\delta\phi_{r,\alpha} = \frac{\partial}{\partial x^\alpha}\,\delta\phi_r.$$

The last term in Eq. (2.15) can be converted to a surface integral over the surface $\Gamma(\Omega)$ using Gauss's divergence theorem in four dimensions. Since $\delta\phi_r = 0$ on Γ, this surface integral vanishes. If $\delta S(\Omega)$ is to vanish for arbitrary regions Ω and arbitrary variations $\delta\phi_r$, Eq. (2.15) leads to the Euler–Lagrange equations

$$\frac{\partial \mathscr{L}}{\partial \phi_r} - \frac{\partial}{\partial x^\alpha}\left(\frac{\partial \mathscr{L}}{\partial \phi_{r,\alpha}}\right) = 0, \qquad r = 1, \ldots, N. \tag{2.16}$$

These are the equations of motion of the fields.

In order to quantize this classical theory by the canonical formalism of non-relativistic quantum mechanics we must introduce conjugate variables. We are dealing with a system with a continuously infinite number of degrees of freedom, corresponding to the values of the fields ϕ_r, considered as functions of time, at each point of space $\mathbf{x}$. We shall again approximate the

[‡] In Eq. (2.15) and thereafter summations over repeated indices r and α, occurring in products, is implied.

system by one having a countable number of degrees of freedom and ultimately go to the continuum limit.

Consider the system at a fixed instant of time t and decompose the three-dimensional space, i.e. the flat space-like surface t = const., into small cells of equal volume $\delta\mathbf{x}_i$, labelled by the index $i = 1, 2, \ldots$ We approximate the values of the fields within each cell by their values at, say, the centre of the cell $\mathbf{x} = \mathbf{x}_i$. The system is now described by the discrete set of generalized coordinates:

$$q_{ri}(t) \equiv \phi_r(i, t) \equiv \phi_r(\mathbf{x}_i, t), \qquad r = 1, \ldots N, \qquad i = 1, 2, \ldots \tag{2.17}$$

which are the values of the fields at the discrete lattice sites $\mathbf{x}_i$. If we also replace the spatial derivatives of the fields by their difference coefficients between neighbouring sites, we can write the Lagrangian of the discrete system as

$$L(t) = \sum_i \delta\mathbf{x}_i \mathscr{L}_i(\phi_r(i, t), \dot{\phi}_r(i, t), \phi_r(i', t)) \tag{2.18}$$

where the dot denotes differentiation with respect to time. The Lagrangian density in the ith cell, $\mathscr{L}_i$, depends on the fields at the neighbouring lattice sites i' on account of the approximation of the spatial derivatives. We define momenta conjugate to q_{ri} in the usual way as

$$p_{ri}(t) = \frac{\partial L}{\partial \dot{q}_{ri}} \equiv \frac{\partial L}{\partial \dot{\phi}_r(i, t)} \equiv \pi_r(i, t)\, \delta\mathbf{x}_i \tag{2.19}$$

where

$$\pi_r(i, t) \equiv \frac{\partial \mathscr{L}_i}{\partial \dot{\phi}_r(i, t)}. \tag{2.20}$$

The Hamiltonian of the discrete system is then given by

$$\begin{aligned} H &= \sum_i p_{ri}\dot{q}_{ri} - L \\ &= \sum_i \delta\mathbf{x}_i\{\pi_r(i, t)\dot{\phi}_r(i, t) - \mathscr{L}_i\}. \end{aligned} \tag{2.21}$$

With a view to going to the limit $\delta\mathbf{x}_i \to 0$, i.e. letting the cell size and the lattice spacing shrink to zero, we define the fields conjugate to $\phi_r(x)$ as

$$\pi_r(x) = \frac{\partial \mathscr{L}}{\partial \dot{\phi}_r}. \tag{2.22}$$

In the limit as $\delta\mathbf{x}_i \to 0$, $\pi_r(i, t)$ tends to $\pi_r(\mathbf{x}_i, t)$, and the discrete Lagrangian and Hamiltonian functions (2.18) and (2.21) become

$$L(t) = \int \mathrm{d}^3\mathbf{x}\, \mathscr{L}(\phi_r, \phi_{r,\alpha}) \tag{2.23}$$

and

$$H = \int \mathrm{d}^3\mathbf{x}\mathscr{H}(x), \tag{2.24}$$

where the Hamiltonian density $\mathscr{H}(x)$ is defined by

$$\mathscr{H}(x) = \pi_r(x)\dot{\phi}_r(x) - \mathscr{L}(\phi_r, \phi_{r,\alpha}), \tag{2.25}$$

and the integrations in Eqs. (2.23) and (2.24) are over all space, at time t. With our Lagrangian density which does not depend explicitly on the time, the Hamiltonian H is of course constant in time. The conservation of energy will be proved in Section 2.4, where the expressions (2.24) and (2.25) for the Hamiltonian will also be re-derived.

As an example, consider the Lagrangian density

$$\mathscr{L} = \tfrac{1}{2}(\phi_{,\alpha}\phi_{,}^{\alpha} - \mu^2\phi^2) \tag{2.26}$$

for a single real field $\phi(x)$, with μ a constant which has the dimensions $(\text{length})^{-1}$. In the next chapter we shall see that the quanta of this field are spinless neutral bosons with Compton wavelength μ^{-1}, i.e. particles of mass $(\hbar\mu/c)$. The equation of motion (2.16) for this field is the Klein–Gordon equation

$$(\Box + \mu^2)\phi(x) = 0, \tag{2.27}$$

the conjugate field (2.22) is

$$\pi(x) = \frac{1}{c^2}\dot{\phi}(x) \tag{2.28}$$

and the Hamiltonian density (2.25) is

$$\mathscr{H}(x) = \tfrac{1}{2}[c^2\pi^2(x) + (\nabla\phi)^2 + \mu^2\phi^2]. \tag{2.29}$$

2.3 QUANTIZED LAGRANGIAN FIELD THEORY

It is now easy to go from the classical to the quantum field theory by interpreting the conjugate coordinates and momenta of the discrete lattice approximation, Eqs. (2.17) and (2.19), as Heisenberg operators and subjecting these to the usual canonical commutation relations:

$$\left.\begin{aligned} [\phi_r(j,t), \pi_s(j',t)] &= \mathrm{i}\hbar\,\frac{\delta_{rs}\delta_{jj'}}{\delta\mathbf{x}_j} \\ [\phi_r(j,t), \phi_s(j',t)] = [\pi_r(j,t), \pi_s(j',t)] &= 0 \end{aligned}\right\}. \tag{2.30}$$

If we let the lattice spacing go to zero, Eqs. (2.30) go over into the commutation relations for the fields:

$$\left.\begin{aligned}[\phi_r(\mathbf{x}, t), \pi_s(\mathbf{x}', t)] &= i\hbar\, \delta_{rs}\, \delta(\mathbf{x} - \mathbf{x}') \\ [\phi_r(\mathbf{x}, t), \phi_s(\mathbf{x}', t)] &= [\pi_r(\mathbf{x}, t), \pi_s(\mathbf{x}', t)] = 0\end{aligned}\right\}, \tag{2.31}$$

since in the limit, as $\delta\mathbf{x}_j \to 0$, $\delta_{jj'}/\delta\mathbf{x}_j$ becomes the three-dimensional Dirac delta function $\delta(\mathbf{x} - \mathbf{x}')$, the points $\mathbf{x}$ and $\mathbf{x}'$ lying in the jth and j'th cell, respectively. Note that the canonical commutation relations (2.31) involve the fields at the same time; they are equal-time commutation relations. In the next chapter we shall obtain the commutators of the fields at different times.

For the Klein–Gordon field (2.26), Eqs. (2.31) reduce to the commutation relations:

$$\left.\begin{aligned}[\phi(\mathbf{x}, t), \dot{\phi}(\mathbf{x}', t)] &= i\hbar c^2 \delta(\mathbf{x} - \mathbf{x}') \\ [\phi(\mathbf{x}, t), \phi(\mathbf{x}', t)] &= [\dot{\phi}(\mathbf{x}, t), \dot{\phi}(\mathbf{x}', t)] = 0\end{aligned}\right\}. \tag{2.32}$$

In the next chapter we shall study the Klein–Gordon field in detail.

2.4 SYMMETRIES AND CONSERVATION LAWS

It follows from the Heisenberg equation of motion of an operator $O(t)$

$$i\hbar \frac{\mathrm{d}O(t)}{\mathrm{d}t} = [O(t), H]$$

(we are not considering operators with explicit time-dependence) that O is a constant of the motion provided

$$[O, H] = 0.$$

Constants of the motion generally stem from invariance properties of systems under groups of transformations, e.g. translational and rotational invariance lead to conservation of linear and angular momentum, respectively. Such transformations lead to equivalent descriptions of the system; for example, referred to two frames of reference related by a Lorentz transformation. Quantum-mechanically, two such descriptions must be related by a unitary transformation U under which states and operators transform according to

$$|\Psi\rangle \to |\Psi'\rangle = U|\Psi\rangle, \qquad O \to O' = UOU^\dagger. \tag{2.33}$$

The unitarity of the transformation ensures two things. Firstly, operator equations are covariant, i.e. have the same form whether expressed in terms of the original or the transformed operators. In particular, this will be true of the commutation relations of the fields and of the equations of motion, e.g.

Maxwell's equations will be covariant with respect to Lorentz transformations. Secondly, under a unitary transformation, amplitudes and hence observable predictions are invariant.

If one deals with continuous transformations, the unitary operator U can be written

$$U = e^{i\alpha T} \tag{2.34}$$

where $T = T^{\dagger}$ and α is a real continuously variable parameter. For $\alpha = 0$, U goes over into the unit operator. For an infinitesimal transformation

$$U \approx 1 + i\delta\alpha T$$

and Eq. (2.33) becomes

$$O' = O + \delta O = (1 + i\delta\alpha T)O(1 - i\delta\alpha T)$$

i.e.

$$\delta O = i\delta\alpha[T, O]. \tag{2.35}$$

If the theory is invariant under this transformation, the Hamiltonian H will be invariant, $\delta H = 0$, and taking $O = H$ in Eq. (2.35) we obtain $[T, H] = 0$, i.e. T is a constant of the motion.

For a field theory derived from a Lagrangian density $\mathscr{L}$, one can construct conserved quantities from the invariance of $\mathscr{L}$ under symmetry transformations. We shall show that for such a theory, the invariance of $\mathscr{L}$ leads to equations of the form

$$\frac{\partial f^{\alpha}}{\partial x^{\alpha}} = 0 \tag{2.36}$$

where the f^{α} are functions of the field operators and their derivatives. If we define

$$F^{\alpha}(t) = \int d^3\mathbf{x} f^{\alpha}(\mathbf{x}, t), \tag{2.37}$$

where integration is over all space, then the continuity equation (2.36) gives

$$\frac{1}{c}\frac{dF^0(t)}{dt} = -\int d^3\mathbf{x} \sum_{j=1}^{3} \frac{\partial}{\partial x^j} f^j(\mathbf{x}, t) = 0 \tag{2.36a}$$

where the last step follows by transforming the integral into a surface integral by means of Gauss's divergence theorem and assuming (as always) that the fields, and hence the f^j, tend to zero sufficiently fast at infinity.[‡] Hence

$$F^0 = \int d^3\mathbf{x} f^0(\mathbf{x}, t) \tag{2.38}$$

[‡] If one employs a finite normalization volume for the system, as we did in the last chapter, the surface integral vanishes on account of the periodic boundary conditions.

is a conserved quantity. With $T = F^0$, the corresponding unitary operator is then given by Eq. (2.34).

The interpretation of the four-vector f^α follows from Eqs. (2.36)–(2.38). f^0/c and f^j are the three-dimensional volume and current densities of the conserved quantity F^0/c. Eq. (2.36a), applied to a finite three-dimensional volume V bounded by a surface S, then states that the rate of decrease of F^0/c within V equals the current of F^0/c flowing out through S. Correspondingly, the four-vector $f^\alpha(x)$, satisfying the conservation equation (2.36), is called a *conserved current*. (Strictly speaking, one should call it a four-current density.) The result, that the invariance of the Lagrangian density $\mathscr{L}$ under a continuous one-parameter set of transformations implies a conserved quantity, is known as Noether's theorem.

We apply these ideas to the transformation

$$\phi_r(x) \to \phi_r'(x) = \phi_r(x) + \delta\phi_r(x) \tag{2.39}$$

of the fields. The change induced in $\mathscr{L}$ is given by

$$\delta\mathscr{L} = \frac{\partial\mathscr{L}}{\partial\phi_r}\,\delta\phi_r + \frac{\partial\mathscr{L}}{\partial\phi_{r,\alpha}}\,\delta\phi_{r,\alpha} = \frac{\partial}{\partial x^\alpha}\left(\frac{\partial\mathscr{L}}{\partial\phi_{r,\alpha}}\,\delta\phi_r\right),$$

where the last step follows since $\phi_r(x)$ satisfies the field equations (2.16), and summations over repeated indices r and α are implied as previously. If $\mathscr{L}$ is invariant under the transformation (2.39) so that $\delta\mathscr{L} = 0$, the last equation reduces to the continuity equation (2.36) with

$$f^\alpha = \frac{\partial\mathscr{L}}{\partial\phi_{r,\alpha}}\,\delta\phi_r$$

and the constant of the motion, from Eqs. (2.38) and (2.22), is

$$F^0 = c\int \mathrm{d}^3\mathbf{x}\,\pi_r(x)\,\delta\phi_r(x). \tag{2.40}$$

An important particular case of the above arises for complex fields ϕ_r, i.e. non-Hermitian operators in the quantized theory. ϕ_r and $\phi_r^\dagger$ are then treated as independent fields, as discussed earlier. Suppose $\mathscr{L}$ is invariant under the transformation

$$\left.\begin{aligned} \phi_r &\to \phi_r' = \mathrm{e}^{\mathrm{i}\varepsilon}\phi_r \approx (1+\mathrm{i}\varepsilon)\phi_r \\ \phi_r^\dagger &\to \phi_r^{\dagger\prime} = \mathrm{e}^{-\mathrm{i}\varepsilon}\phi_r^\dagger \approx (1-\mathrm{i}\varepsilon)\phi_r^\dagger \end{aligned}\right\} \tag{2.41}$$

where ε is a real parameter, and the right-hand expressions result for very small ε. From Eqs. (2.41)

$$\delta\phi_r = \mathrm{i}\varepsilon\phi_r, \qquad \delta\phi_r^\dagger = -\mathrm{i}\varepsilon\phi_r^\dagger,$$

and Eq. (2.40) becomes

$$F^0 = \mathrm{i}\varepsilon c \int \mathrm{d}^3\mathbf{x}[\pi_r(x)\phi_r(x) - \pi_r^\dagger(x)\phi_r^\dagger(x)].$$

Since F^0 multiplied by any constant is also conserved, we shall, instead of F^0, consider

$$Q = -\frac{\mathrm{i}q}{\hbar} \int \mathrm{d}^3\mathbf{x}[\pi_r(x)\phi_r(x) - \pi_r^\dagger(x)\phi_r^\dagger(x)], \tag{2.42}$$

where q is a constant to be determined later. The reason for this change is that $\pm q$ will turn out to be the electric charges of the particles represented by the fields.

We evaluate the commutator $[Q, \phi_r(x)]$. Since ϕ_r and $\phi_r^\dagger$ are independent fields, ϕ_r commutes with all fields except π_r [see Eq. (2.31)]. Hence taking $(x')^0 = x^0 = ct$,

$$[Q, \phi_r(x)] = -\frac{\mathrm{i}q}{\hbar} \int \mathrm{d}^3\mathbf{x}'[\pi_s(x'), \phi_r(x)]\phi_s(x'),$$

and using the commutation relations (2.31) one obtains

$$[Q, \phi_r(x)] = -q\phi_r(x). \tag{2.43}$$

From this relation one easily verifies that if $|Q'\rangle$ is an eigenstate of Q with the eigenvalue Q', then $\phi_r(x)|Q'\rangle$ is also an eigenstate of Q belonging to the eigenvalue $(Q' - q)$, and correspondingly $\phi_r^\dagger(x)|Q'\rangle$ belongs to $(Q' + q)$. In the next chapter we shall see that, consistent with these results, ϕ_r and $\phi_r^\dagger$ are linear in creation and absorption operators, with ϕ_r absorbing particles of charge $(+q)$ or creating particles of charge $(-q)$, while $\phi_r^\dagger$ absorbs particles of charge $(-q)$ or creates particles of charge $(+q)$. Hence, we interpret the operator Q, Eq. (2.42), as the charge operator. We have therefore shown that charge is conserved ($\mathrm{d}Q/\mathrm{d}t = 0$, $[Q, H] = 0$), provided the Lagrangian density $\mathscr{L}$ is invariant with respect to the transformation (2.41), which is known as a global phase transformation [corresponding to the fact that the phase ε in Eq. (2.41) is independent of x] or as a gauge transformation of the first kind. We see from Eq. (2.42) that we require complex, i.e. non-Hermitian, fields to represent particles with charge. Real, i.e. Hermitian, fields represent uncharged particles. Interpreting Eq. (2.42) as an operator involves the usual ambiguity as to the order of factors. We shall have to choose these so that for the vacuum state $|0\rangle$, in which no particles are present, $Q|0\rangle = 0$. We shall return to this point in the next chapter.

The unitary transformation corresponding to the phase transformation (2.41) can, from Eq. (2.34), be written

$$U = \mathrm{e}^{\mathrm{i}\alpha Q}. \tag{2.44}$$

Hence for infinitesimally small α, we obtain from Eq. (2.33)

$$\begin{aligned}\phi'_r &= e^{i\alpha Q}\phi_r e^{-i\alpha Q} \\ &= \phi_r + i\alpha[Q, \phi_r] = (1 - i\alpha q)\phi_r\end{aligned} \tag{2.45}$$

where the last line follows from Eq. (2.43). Comparing Eqs. (2.45) and (2.41), we see that they are consistent if we take $\varepsilon = -\alpha q$.

Although we have talked of electric charge, with which one is familiar, this analysis applies equally to other types of charge, such as hypercharge.

Conservation of energy and momentum and of angular momentum follows from the invariance of the Lagrangian density $\mathscr{L}$ under translations and rotations. Since these transformations form a continuous group we need only consider infinitesimal transformations. Any finite transformation can be built up through repeated infinitesimal transformations. In four dimensions these are given by

$$x_\alpha \to x'_\alpha \equiv x_\alpha + \delta x_\alpha = x_\alpha + \varepsilon_{\alpha\beta}x^\beta + \delta_\alpha, \tag{2.46}$$

where δ_α is an infinitesimal displacement and $\varepsilon_{\alpha\beta}$ is an infinitesimal antisymmetric tensor, $\varepsilon_{\alpha\beta} = -\varepsilon_{\beta\alpha}$, to ensure the invariance of $x_\alpha x^\alpha$ under homogeneous Lorentz transformations ($\delta_\alpha = 0$).

The transformation (2.46) will induce a transformation in the fields which we assume to be

$$\phi_r(x) \to \phi'_r(x') = \phi_r(x) + \tfrac{1}{2}\varepsilon_{\alpha\beta}S^{\alpha\beta}_{rs}\phi_s(x). \tag{2.47}$$

In this section, summation over repeated indices r, s, labelling fields, as well as over Lorentz indices α, β, is implied. Here x and x' label the *same* point in space–time referred to the two frames of reference, and ϕ_r and ϕ'_r are the field components referred to these two coordinate systems. The coefficients $S^{\alpha\beta}_{rs}$ in Eq. (2.47) are antisymmetric in α and β, like $\varepsilon_{\alpha\beta}$, and are determined by the transformation properties of the fields. For example, for the vector potential $A_\alpha(x)$, Eq. (2.47) reduces to the transformation law of a vector.

Invariance under the transformations (2.46) and (2.47) means that the Lagrangian density expressed in terms of the new coordinates and fields has the same functional form as when expressed in the original coordinates and fields:

$$\mathscr{L}(\phi_r(x), \phi_{r,\alpha}(x)) = \mathscr{L}(\phi'_r(x'), \phi'_{r,\alpha}(x')). \tag{2.48}$$

(Here $\phi'_{r,\alpha}(x') \equiv \partial\phi'_r(x')/\partial x'^\alpha$.) From Eq. (2.48) the covariance of the field equations, etc. follows; i.e. they will have the same form expressed in terms of either the original or the transformed coordinates and fields.

The conservation laws follow by expressing the right-hand side of Eq. (2.48) in terms of the original coordinates and fields by means of Eqs. (2.46) and (2.47). We shall first state and discuss these results, postponing their derivation to the end of this section.

For a translation (i.e. $\varepsilon_{\alpha\beta} = 0$) one obtains the four continuity equations

$$\frac{\partial \mathscr{T}^{\alpha\beta}}{\partial x^\alpha} = 0, \tag{2.49}$$

where

$$\mathscr{T}^{\alpha\beta} \equiv \frac{\partial \mathscr{L}}{\partial \phi_{r,\alpha}} \frac{\partial \phi_r}{\partial x_\beta} - \mathscr{L} g^{\alpha\beta}, \tag{2.50}$$

and the four conserved quantities are

$$cP^\alpha \equiv \int \mathrm{d}^3\mathbf{x}\, \mathscr{T}^{0\alpha} = \int \mathrm{d}^3\mathbf{x} \left\{ c\pi_r(x) \frac{\partial \phi_r(x)}{\partial x_\alpha} - \mathscr{L} g^{0\alpha} \right\}. \tag{2.51}$$

P^α is the energy–momentum four-vector, with

$$\left.\begin{aligned} cP^0 &= \int \mathrm{d}^3\mathbf{x} \{\pi_r(x)\dot{\phi}_r(x) - \mathscr{L}(\phi_r, \phi_{r,\alpha})\} \\ &= \int \mathrm{d}^3\mathbf{x}\, \mathscr{H} = H \end{aligned}\right\} \tag{2.51a}$$

being the Hamiltonian, Eqs. (2.24) and (2.25), and

$$P^j = \int \mathrm{d}^3\mathbf{x}\, \pi_r(x) \frac{\partial \phi_r(x)}{\partial x_j} \tag{2.51b}$$

being the momentum components of the fields. This interpretation will be confirmed when we come to express these operators in the number representation. Correspondingly $\mathscr{T}^{\alpha\beta}$ is called the energy–momentum tensor.

For a rotation (i.e. $\delta_\alpha = 0$) Eqs. (2.46)–(2.48) give the continuity equations

$$\frac{\partial \mathscr{M}^{\alpha\beta\gamma}}{\partial x^\alpha} = 0, \tag{2.52}$$

where

$$\mathscr{M}^{\alpha\beta\gamma} \equiv \frac{\partial \mathscr{L}}{\partial \phi_{r,\alpha}} S_{rs}^{\beta\gamma} \phi_s(x) + [x^\beta \mathscr{T}^{\alpha\gamma} - x^\gamma \mathscr{T}^{\alpha\beta}], \tag{2.53}$$

and the six conserved quantities (note that $\mathscr{M}^{\alpha\beta\gamma} = -\mathscr{M}^{\alpha\gamma\beta}$) are

$$\begin{aligned} cM^{\alpha\beta} &= \int \mathrm{d}^3\mathbf{x}\, \mathscr{M}^{0\alpha\beta} \\ &= \int \mathrm{d}^3\mathbf{x} \{[x^\alpha \mathscr{T}^{0\beta} - x^\beta \mathscr{T}^{0\alpha}] + c\pi_r(x) S_{rs}^{\alpha\beta} \phi_s(x)\}. \end{aligned} \tag{2.54}$$

For two space-like indices $(i, j = 1, 2, 3)$, M^{ij} is the angular momentum operator of the field (M^{12} being the z-component, etc.). Remembering that $\mathscr{T}^{0i}/c$ is the momentum density of the field [see Eqs. (2.51)], we interpret the term in square brackets in Eq. (2.54) as the orbital angular momentum and the last term as the intrinsic spin angular momentum.

We return to the derivation of the continuity equations (2.49) and (2.52).‡ The variation of a function $\phi_\alpha(x)$ with the argument unchanged was defined in Eq. (2.39) as

$$\delta\phi_r(x) \equiv \phi'_r(x) - \phi_r(x). \tag{2.55a}$$

In addition, we now define the variation

$$\delta_{\mathrm{T}}\phi(x) \equiv \phi'_r(x') - \phi_r(x) \tag{2.55b}$$

which results from changes of both the form and the argument of the function. We can then write

$$\begin{aligned}\delta_{\mathrm{T}}\phi_r(x) &= [\phi'_r(x') - \phi_r(x')] + [\phi_r(x') - \phi_r(x)]\\ &= \delta\phi_r(x') + \frac{\partial\phi_r}{\partial x_\beta}\,\delta x_\beta,\end{aligned} \tag{2.56}$$

where δx_β is given by Eq. (2.46). To first order in small quantities this can be written

$$\delta_{\mathrm{T}}\phi_r(x) = \delta\phi_r(x) + \frac{\partial\phi_r}{\partial x_\beta}\,\delta x_\beta. \tag{2.57}$$

We can similarly write Eq. (2.48) as

$$\begin{aligned}0 &= \mathscr{L}(\phi'_r(x'), \phi'_{r,\alpha}(x')) - \mathscr{L}(\phi_r(x), \phi_{r,\alpha}(x))\\ &= \delta\mathscr{L} + \frac{\partial\mathscr{L}}{\partial x^\alpha}\,\delta x^\alpha.\end{aligned} \tag{2.58}$$

For $\delta\mathscr{L}$ we obtain, since $\phi_r(x)$ satisfies the field equations (2.16),

$$\begin{aligned}\delta\mathscr{L} &= \frac{\partial\mathscr{L}}{\partial\phi_r}\,\delta\phi_r + \frac{\partial\mathscr{L}}{\partial\phi_{r,\alpha}}\,\delta\phi_{r,\alpha}\\ &= \frac{\partial}{\partial x^\alpha}\left\{\frac{\partial\mathscr{L}}{\partial\phi_{r,\alpha}}\,\delta\phi_r\right\} = \frac{\partial}{\partial x^\alpha}\left\{\frac{\partial\mathscr{L}}{\partial\phi_{r,\alpha}}\left[\delta_{\mathrm{T}}\phi_r - \frac{\partial\phi_r}{\partial x_\beta}\,\delta x_\beta\right]\right\}.\end{aligned} \tag{2.59}$$

We combine Eqs. (2.58) and (2.59) to obtain the continuity equations

$$\frac{\partial f^\alpha}{\partial x^\alpha} = 0, \tag{2.60}$$

‡ The reader may omit the rest of this section as the details of this derivation will not be required later on.

where

$$f^\alpha \equiv \frac{\partial \mathscr{L}}{\partial \phi_{r,\alpha}} \delta_{\mathrm{T}}\phi_r - \mathscr{T}^{\alpha\beta}\, \delta x_\beta \tag{2.61}$$

with $\mathscr{T}^{\alpha\beta}$ given by Eq. (2.50).

We first consider translations, i.e. $\varepsilon_{\alpha\beta} = 0$, so that from Eqs. (2.46) and (2.47) $\delta x_\beta = \delta_\beta$ and $\delta_{\mathrm{T}}\phi_r = 0$. Eq. (2.61) reduces to $f^\alpha = -\mathscr{T}^{\alpha\beta}\,\delta x_\beta$, and, since the four displacements δ_β are independent of each other, Eq. (2.60) reduces to the four continuity equations (2.49) for energy and momentum conservation.

Finally, we consider rotations, i.e. $\delta_\alpha = 0$. From Eqs. (2.46) and (2.47) for δx_β and $\delta_{\mathrm{T}}\phi_r$, and the antisymmetry of $\varepsilon_{\alpha\beta}$, Eq. (2.61) becomes

$$f^\alpha = \tfrac{1}{2}\varepsilon_{\beta\gamma}\mathscr{M}^{\alpha\beta\gamma} \tag{2.62}$$

where $\mathscr{M}^{\alpha\beta\gamma}$ is the tensor defined in Eq. (2.53). Since the rotations $\varepsilon_{\beta\gamma}$ are independent of each other, Eq. (2.60) reduces to the continuity equations (2.52).

PROBLEMS

2.1 Show that replacing the Lagrangian density $\mathscr{L} = \mathscr{L}(\phi_r, \phi_{r,\alpha})$ by

$$\mathscr{L}' = \mathscr{L} + \partial_\alpha \Lambda^\alpha(x),$$

where $\Lambda^\alpha(x)$, $\alpha = 0, \ldots, 3$, are arbitrary functions of the fields $\phi_r(x)$, does not alter the equations of motion.

2.2 The real Klein–Gordon field is described by the Hamiltonian density (2.29). Use the commutation relations (2.31) to show that

$$[H, \phi(x)] = -i\hbar c^2 \pi(x), \qquad [H, \pi(x)] = i\hbar(\mu^2 - \nabla^2)\phi(x),$$

where H is the Hamiltonian of the field.

From this result and the Heisenberg equations of motion for the operators $\phi(x)$ and $\pi(x)$, show that

$$\dot{\phi}(x) = c^2\pi(x), \qquad (\Box + \mu^2)\phi(x) = 0.$$

2.3 Show that the Lagrangian density

$$\mathscr{L} = -\tfrac{1}{2}[\partial_\alpha\phi_\beta(x)][\partial^\alpha\phi^\beta(x)] + \tfrac{1}{2}[\partial_\alpha\phi^\alpha(x)][\partial_\beta\phi^\beta(x)] + \frac{\mu^2}{2}\phi_\alpha(x)\phi^\alpha(x)$$

for the real vector field $\phi^\alpha(x)$ leads to the field equations

$$[g_{\alpha\beta}(\Box + \mu^2) - \partial_\alpha\partial_\beta]\phi^\beta(x) = 0,$$

and that the field $\phi^\alpha(x)$ satisfies the Lorentz condition

$$\partial_\alpha\phi^\alpha(x) = 0.$$

2.4 Use the commutation relations (2.31) to show that the momentum operator of the fields

$$P^j = \int d^3\mathbf{x}\,\pi_r(x)\frac{\partial\phi_r(x)}{\partial x_j} \tag{2.51b}$$

satisfies the equations

$$[P^j, \phi_r(x)] = -i\hbar \frac{\partial \phi_r(x)}{\partial x_j}, \qquad [P^j, \pi_r(x)] = -i\hbar \frac{\partial \pi_r(x)}{\partial x_j}.$$

Hence show that any operator $F(x) = F(\phi_r(x), \pi_r(x))$, which can be expanded in a power series in the field operators $\phi_r(x)$ and $\pi_r(x)$, satisfies

$$[P^j, F(x)] = -i\hbar \frac{\partial F(x)}{\partial x_j}.$$

Note that we can combine these equations with the Heisenberg equation of motion for the operator $F(x)$

$$[H, F(x)] = -i\hbar c \frac{\partial F(x)}{\partial x_0}$$

to obtain the covariant equations of motion

$$[P^\alpha, F(x)] = -i\hbar \frac{\partial F(x)}{\partial x_\alpha},$$

where $P^0 = H/c$.

2.5 Under a translation of coordinates

$$x_\alpha \to x'_\alpha = x_\alpha + \delta_\alpha \qquad (\delta_\alpha = \text{a constant four-vector})$$

a scalar field $\phi(x_\alpha)$ remains invariant:

$$\phi'(x'_\alpha) = \phi(x_\alpha), \quad \text{i.e. } \phi'(x_\alpha) = \phi(x_\alpha - \delta_\alpha).$$

Show that the corresponding unitary transformation

$$\phi(x) \to \phi'(x) = U\phi(x)U^\dagger$$

is given by $U = \exp[-i\delta_\alpha P^\alpha/\hbar]$, where P^α is the energy–momentum four-vector of the field, Eqs. (2.51). (You may find the results of the previous problem helpful.)

CHAPTER 3

The Klein–Gordon field

In Chapter 1 we quantized the electromagnetic field by Fourier analysing it and imposing harmonic oscillator commutation relations on the Fourier expansion coefficients. This approach naturally led to photons. In the last chapter a different procedure, the canonical quantization formalism, led directly to quantized field operators. We shall now Fourier analyse these field operators and we shall see that the Fourier coefficients, which are now also operators, satisfy the same commutation relations as the absorption and creation operators of the number representation. In this way the interpretation in terms of field quanta is regained.

In this chapter we shall consider relativistic material particles of spin 0. Photons which are much more complicated on account of their transverse polarization will be treated in Chapter 5.

3.1 THE REAL KLEIN–GORDON FIELD

For particles of rest mass m, energy and momentum are related by

$$E^2 = m^2c^4 + c^2\mathbf{p}^2. \tag{3.1}$$

If the particles can be described by a single scalar wavefunction $\phi(x)$, the prescription of non-relativistic quantum mechanics

$$\mathbf{p} \to -i\hbar\nabla, \qquad E \to i\hbar\, \partial/\partial t \tag{3.2}$$

leads to the Klein–Gordon equation (2.27):

$$(\Box + \mu^2)\phi(x) = 0 \tag{3.3}$$

($\mu \equiv mc/\hbar$). The interpretation of Eq. (3.3) as a single-particle equation leads to difficulties. These are related to defining a positive-definite particle density and to the two signs of the energy E which result from Eq. (3.1). We shall not discuss these difficulties but only mention that they are typical of relativistic single-particle equations. We shall see that such difficulties do not occur in the many-particle theories which result when the fields, such as the Klein–Gordon field $\phi(x)$, are quantized.‡

We know from Eq. (2.54) for the angular momentum of the field that a single scalar field possesses orbital but no spin angular momentum, i.e. it represents particles of spin 0. Hence, the Klein–Gordon equation affords the appropriate description of π-mesons (pions) and K-mesons, both of which have spin 0.

We shall now consider a real scalar field $\phi(x)$, satisfying the Klein–Gordon equation (3.3). Such a field corresponds to electrically neutral particles. Charged particles, described by a complex field, will be dealt with in the next section.

We know from Section 2.2 that the Klein–Gordon equation (3.3) can be derived from the Lagrangian density

$$\mathscr{L} = \tfrac{1}{2}(\phi_{,\alpha}\phi_{,}^{\ \alpha} - \mu^2\phi^2), \tag{3.5}$$

and that the field conjugate to ϕ is

$$\pi(x) = \frac{\partial \mathscr{L}}{\partial \dot{\phi}} = \frac{1}{c^2}\dot{\phi}(x). \tag{3.6}$$

On quantization the real field ϕ becomes a Hermitian operator, $\phi^\dagger = \phi$, satisfying the equal-time commutation relations (2.32):

$$\left.\begin{aligned} [\phi(\mathbf{x}, t), \dot{\phi}(\mathbf{x}', t)] &= i\hbar c^2\delta(\mathbf{x} - \mathbf{x}') \\ [\phi(\mathbf{x}, t), \phi(\mathbf{x}', t)] = [\dot{\phi}(\mathbf{x}, t), &\dot{\phi}(\mathbf{x}', t)] = 0 \end{aligned}\right\}. \tag{3.7}$$

To establish contact with particles, we expand $\phi(x)$ in a complete set of solutions of the Klein–Gordon equation:

$$\phi(x) = \phi^+(x) + \phi^-(x) \tag{3.8a}$$

where

$$\phi^+(x) = \sum_{\mathbf{k}} \left(\frac{\hbar c^2}{2V\omega_{\mathbf{k}}}\right)^{1/2} a(\mathbf{k})\, \mathrm{e}^{-ikx} \tag{3.8b}$$

and

$$\phi^-(x) = \sum_{\mathbf{k}} \left(\frac{\hbar c^2}{2V\omega_{\mathbf{k}}}\right)^{1/2} a^\dagger(\mathbf{k})\, \mathrm{e}^{ikx}. \tag{3.8c}$$

‡ This is often referred to as second quantization, in contrast to the derivation of the single-particle wave equations by means of the substitution (3.2).

These equations are analogous to Eqs. (1.38) for the photon field. The summations are again over the wave vectors **k** allowed by the periodic boundary conditions, but k^0 and $\omega_{\mathbf{k}}$ are now given by

$$k^0 = \frac{1}{c}\,\omega_{\mathbf{k}} = +(\mu^2 + \mathbf{k}^2)^{1/2}, \tag{3.9a}$$

i.e. k is the wave four-vector of a particle of mass $m = \mu\hbar/c$, momentum $\hbar\mathbf{k}$ and energy

$$E = \hbar\omega_{\mathbf{k}} = +[m^2c^4 + c^2(\hbar\mathbf{k})^2]^{1/2}. \tag{3.9b}$$

The fact that each operator $a(\mathbf{k})$ occurs paired with its adjoint $a^\dagger(\mathbf{k})$ in Eqs. (3.8) ensures that ϕ is Hermitian.

From Eqs. (3.8) and the commutation relations (3.7), one easily obtains the commutation relations for the operators $a(\mathbf{k})$ and $a^\dagger(\mathbf{k})$. We shall leave the details for the reader (see Problem 3.1) and only quote the important result:

$$\left.\begin{aligned} [a(\mathbf{k}), a^\dagger(\mathbf{k}')] &= \delta_{\mathbf{k}\mathbf{k}'} \\ [a(\mathbf{k}), a(\mathbf{k}')] = [a^\dagger(\mathbf{k}), a^\dagger(\mathbf{k}')] &= 0 \end{aligned}\right\}. \tag{3.10}$$

These are precisely the harmonic oscillator commutation relations [Eqs. (1.19) and (1.29)] and all the results can at once be taken over from Section 1.2. In particular, the operators

$$N(\mathbf{k}) = a^\dagger(\mathbf{k})a(\mathbf{k}) \tag{3.11}$$

have as their eigenvalues the occupation numbers

$$n(\mathbf{k}) = 0, 1, 2, \ldots, \tag{3.12}$$

and, correspondingly, $a(\mathbf{k})$ and $a^\dagger(\mathbf{k})$ are the annihilation and creation operators of particles with momentum $\hbar\mathbf{k}$ and energy $\hbar\omega_{\mathbf{k}}$, given by Eq. (3.9).

The Hamiltonian and momentum operators of the Klein–Gordon field are, from Eqs. (2.51), (3.5) and (3.6), given by

$$H = \int \mathrm{d}^3\mathbf{x}\tfrac{1}{2}\left[\frac{1}{c^2}\dot{\phi}^2 + (\nabla\phi)^2 + \mu^2\phi^2\right] \tag{3.13}$$

and

$$\mathbf{P} = -\int \mathrm{d}^3\mathbf{x}\,\frac{1}{c^2}\,\dot{\phi}\nabla\phi. \tag{3.14}$$

Substituting the expansion (3.8) in the last two equations one obtains

$$H = \sum_{\mathbf{k}} \hbar\omega_{\mathbf{k}}(a^\dagger(\mathbf{k})a(\mathbf{k}) + \tfrac{1}{2}), \tag{3.15}$$

$$\mathbf{P} = \sum_{\mathbf{k}} \hbar\mathbf{k}(a^\dagger(\mathbf{k})a(\mathbf{k}) + \tfrac{1}{2}), \tag{3.16}$$

confirming our interpretation of $[a^\dagger(\mathbf{k})a(\mathbf{k})]$ as the number operator for particles with wave vector **k**. From the last two equations one also sees directly that the momentum **P** is a constant of the motion for the free Klein–Gordon field. (Nevertheless we prefer the discussion of Section 2.4 because it reveals the fundamental and general connection between symmetries and conservation laws.)

From Eq. (3.15) we see that the state of lowest energy, the ground state, of the Klein–Gordon field is the vacuum state $|0\rangle$ in which no particles are present (all $n(\mathbf{k}) = 0$). We can also characterize this state by

$$a(\mathbf{k})|0\rangle = 0, \quad \text{all } \mathbf{k}, \tag{3.17a}$$

or, expressed in terms of the field operators (3.8), by

$$\phi^+(x)|0\rangle = 0, \quad \text{all } x. \tag{3.17b}$$

The vacuum has the infinite energy $\frac{1}{2}\sum_{\mathbf{k}} \hbar\omega_{\mathbf{k}}$. As discussed for the radiation field, only energy differences are observable. Hence, this infinite constant is harmless and easily removed by measuring all energies relative to the vacuum state.

One can avoid the explicit occurrence of such infinite constants by normal ordering of operators. In a *normal product*, all absorption operators stand to the right of all creation operators in each product of operators. Denoting the normal product by N(...) we have, for example,

$$\mathrm{N}(a(\mathbf{k}_1)a(\mathbf{k}_2)a^\dagger(\mathbf{k}_3)) = a^\dagger(\mathbf{k}_3)a(\mathbf{k}_1)a(\mathbf{k}_2), \tag{3.18}$$

and

$$\begin{aligned}\mathrm{N}[\phi(x)\phi(y)] &= \mathrm{N}[(\phi^+(x) + \phi^-(x))(\phi^+(y) + \phi^-(y))] \\ &= \mathrm{N}[\phi^+(x)\phi^+(y)] + \mathrm{N}[\phi^+(x)\phi^-(y)] \\ &\quad + \mathrm{N}[\phi^-(x)\phi^+(y)] + \mathrm{N}[\phi^-(x)\phi^-(y)] \\ &= \phi^+(x)\phi^+(y) + \phi^-(y)\phi^+(x) \\ &\quad + \phi^-(x)\phi^+(y) + \phi^-(x)\phi^-(y),\end{aligned} \tag{3.19}$$

where the order of the factors has been interchanged in the second term, i.e. all positive frequency parts ϕ^+ (which contain only absorption operators) stand to the right of all negative frequency parts ϕ^- (which contain only creation operators).‡ Normal ordering does not fix the order of absorption or creation operators, each amongst themselves, but since each of these commute amongst themselves such different ways of writing a normal

‡ This definition of the normal product will be modified when fermions are introduced. Another notation commonly used for the normal product $\mathrm{N}(AB\ldots L)$ is $:AB\ldots L:$.

product are equal; for example, expression (3.18) also equals $a^\dagger(\mathbf{k}_3)a(\mathbf{k}_2)a(\mathbf{k}_1)$. Hence in arranging a product of operators in normal order, one simply treats them as though all commutators vanish.

Clearly, the vacuum expectation value of any normal product vanishes. We redefine the Lagrangian density $\mathscr{L}$ and all observables, such as the energy–momentum or angular momentum of the field, or their densities, as normal products. We are free to do this, as it merely corresponds to a particular order of factors before quantization. With observables defined as normal products, their vacuum expectation values vanish. In particular, Eqs. (3.15) and (3.16) become

$$P^\alpha = (H/c, \mathbf{P}) = \sum_{\mathbf{k}} \hbar k^\alpha a^\dagger(\mathbf{k})a(\mathbf{k}). \tag{3.20}$$

From the vacuum state $|0\rangle$ one constructs particle states in the same way as was done for photons in Section 1.2. For example, one-particle states are linear superpositions of

$$a^\dagger(\mathbf{k})|0\rangle, \qquad \text{all } \mathbf{k}; \tag{3.21a}$$

two-particle states are linear superpositions of

$$a^\dagger(\mathbf{k})a^\dagger(\mathbf{k}')|0\rangle, \qquad \text{all } \mathbf{k} \text{ and } \mathbf{k}' \neq \mathbf{k}, \tag{3.21b}$$

and

$$\frac{1}{\sqrt{2}}[a^\dagger(\mathbf{k})]^2|0\rangle, \qquad \text{all } \mathbf{k}, \tag{3.21c}$$

and so on. With the vacuum state normalized, i.e. $\langle 0|0\rangle = 1$, the states (3.21) are also normalized. That is the purpose of the factor $1/\sqrt{2}$ in Eq. (3.21c). Similar factors occur for more than two particles.

The particles of the Klein–Gordon field are bosons; the occupation numbers can take on any value $n(\mathbf{k}) = 0, 1, 2, \ldots$. Eq. (3.21b) illustrates another aspect of boson states: they are symmetric under interchange of particle labels. Since all creation operators commute with each other, we have

$$a^\dagger(\mathbf{k})a^\dagger(\mathbf{k}')|0\rangle = a^\dagger(\mathbf{k}')a^\dagger(\mathbf{k})|0\rangle. \tag{3.22}$$

3.2 THE COMPLEX KLEIN–GORDON FIELD

We shall now extend the treatment of the last section to the complex Klein–Gordon field. The new feature this introduces is, as we know from Section 2.4, that we can associate a conserved charge with the field. For the real field this was not possible. We shall concentrate on this aspect of a conserved charge. In other respects the real and complex fields are very similar, and we shall only quote the main results leaving verification to the reader.

For the complex Klein–Gordon field, the Lagrangian density (3.5) is replaced by

$$\mathscr{L} = \mathrm{N}(\phi^{\dagger}_{,\alpha}\phi^{,\alpha} - \mu^2\phi^{\dagger}\phi), \tag{3.23}$$

where we have at once written the quantized operator as a normal product. The field ϕ and its adjoint $\phi^{\dagger}$, treated as independent fields, lead to the Klein–Gordon equations

$$(\Box + \mu^2)\phi(x) = 0, \qquad (\Box + \mu^2)\phi^{\dagger}(x) = 0. \tag{3.24}$$

The fields conjugate to ϕ and $\phi^{\dagger}$ are

$$\pi(x) = \frac{1}{c^2}\dot{\phi}^{\dagger}(x), \qquad \pi^{\dagger}(x) = \frac{1}{c^2}\dot{\phi}(x), \tag{3.25}$$

and the equal-time commutation relations (2.31) become

$$\left.\begin{aligned} [\phi(\mathbf{x}, t), \dot{\phi}^{\dagger}(\mathbf{x}', t)] &= i\hbar c^2\delta(\mathbf{x} - \mathbf{x}') \\ [\phi(\mathbf{x}, t), \phi(\mathbf{x}', t)] = [\phi(\mathbf{x}, t), \phi^{\dagger}(\mathbf{x}', t)] &= [\dot{\phi}(\mathbf{x}, t), \dot{\phi}(\mathbf{x}', t)] \\ = [\dot{\phi}(\mathbf{x}, t), \dot{\phi}^{\dagger}(\mathbf{x}', t)] &= [\phi(\mathbf{x}, t), \dot{\phi}(\mathbf{x}', t)] = 0 \end{aligned}\right\}. \tag{3.26}$$

Analogously to Eqs. (3.8), we write the Fourier expansions of the fields as

$$\phi(x) = \phi^{+}(x) + \phi^{-}(x) = \sum_{\mathbf{k}} \left(\frac{\hbar c^2}{2V\omega_{\mathbf{k}}}\right)^{1/2} [a(\mathbf{k})\,\mathrm{e}^{-ikx} + b^{\dagger}(\mathbf{k})\,\mathrm{e}^{ikx}] \tag{3.27a}$$

and

$$\phi^{\dagger}(x) = \phi^{\dagger +}(x) + \phi^{\dagger -}(x) = \sum_{\mathbf{k}} \left(\frac{\hbar c^2}{2V\omega_{\mathbf{k}}}\right)^{1/2} [b(\mathbf{k})\,\mathrm{e}^{-ikx} + a^{\dagger}(\mathbf{k})\,\mathrm{e}^{ikx}]. \tag{3.27b}$$

($\phi^{\dagger +}$ and $\phi^{\dagger -}$ are the positive and negative frequency parts of $\phi^{\dagger}$.)

From Eqs. (3.26) and (3.27a), one obtains the commutation relations

$$[a(\mathbf{k}), a^{\dagger}(\mathbf{k}')] = [b(\mathbf{k}), b^{\dagger}(\mathbf{k}')] = \delta_{\mathbf{k}\mathbf{k}'}, \tag{3.28a}$$

and the commutator of any other pair of operators vanishes, i.e.

$$[a(\mathbf{k}), a(\mathbf{k}')] = [b(\mathbf{k}), b(\mathbf{k}')] = [a(\mathbf{k}), b(\mathbf{k}')] = [a^{\dagger}(\mathbf{k}), b(\mathbf{k}')] = 0. \tag{3.28b}$$

From the commutation relations (3.28) it follows that we can interpret $a(\mathbf{k})$ and $a^{\dagger}(\mathbf{k})$, and $b(\mathbf{k})$ and $b^{\dagger}(\mathbf{k})$, as absorption and creation operators of two types of particles—we shall call them a-particles and b-particles—and

$$\mathrm{N}_a(\mathbf{k}) = a^{\dagger}(\mathbf{k})a(\mathbf{k}), \qquad \mathrm{N}_b(\mathbf{k}) = b^{\dagger}(\mathbf{k})b(\mathbf{k}), \tag{3.29}$$

as the corresponding number operators with eigenvalues 0, 1, 2, ... Hence a number representation can be set up as before, with states containing a- and b-particles generated by means of the creation operators $a^{\dagger}$ and $b^{\dagger}$ from the vacuum state $|0\rangle$ which is now defined by

$$a(\mathbf{k})|0\rangle = b(\mathbf{k})|0\rangle = 0, \quad \text{all } \mathbf{k}, \tag{3.30a}$$

or equivalently by

$$\phi^+(x)|0\rangle = \phi^{\dagger +}(x)|0\rangle = 0, \quad \text{all } x. \tag{3.30b}$$

Expressed in terms of the absorption and creation operators, the energy–momentum operator (2.51) of the complex Klein–Gordon field assumes the form we expect

$$P^\alpha = (H/c, \mathbf{P}) = \sum_{\mathbf{k}} \hbar k^\alpha(\mathrm{N}_a(\mathbf{k}) + \mathrm{N}_b(\mathbf{k})). \tag{3.31}$$

We now turn to the charge. From the invariance of the Lagrangian density (3.23) under the phase transformation (2.41) follows the conservation of charge Q, Eq. (2.42), which now takes the form

$$Q = \frac{-iq}{\hbar c^2} \int \mathrm{d}^3\mathbf{x} \mathrm{N}[\dot{\phi}^\dagger(x)\phi(x) - \dot{\phi}(x)\phi^\dagger(x)]. \tag{3.32}$$

The corresponding charge–current density is given by

$$s^\alpha(x) = (c\rho(x), \mathbf{j}(x)) = \frac{-iq}{\hbar} \mathrm{N}\left[\frac{\partial\phi^\dagger}{\partial x_\alpha}\phi - \frac{\partial\phi}{\partial x_\alpha}\phi^\dagger\right], \tag{3.33}$$

which obviously satisfies the continuity equation

$$\frac{\partial s^\alpha(x)}{\partial x^\alpha} = 0. \tag{3.34}$$

Expressed in terms of creation and absorption operators, Eq. (3.32) becomes

$$Q = q \sum_{\mathbf{k}} [\mathrm{N}_a(\mathbf{k}) - \mathrm{N}_b(\mathbf{k})] \tag{3.35}$$

which clearly commutes with the Hamiltonian H, Eq. (3.31).

It follows from Eq. (3.35) that one must associate charges $+q$ and $-q$ with a- and b-particles, respectively. Apart from the sign of the charge, a- and b-particles have identical properties. Furthermore, the theory is completely symmetric between them, as one sees from Eqs. (3.27)–(3.35). Interchanging a and b merely changes the sign of Q. This result is not restricted to spin 0 bosons but holds generally. The occurrence of antiparticles in association with all particles of non-zero charge is a fundamental feature of relativistic quantum field theory which is fully vindicated by experiment.

An example of a particle–antiparticle pair is the pair of charged pi-mesons. Taking $q = e(>0)$, one can identify the π^+- and π^--mesons with the a- and b-particles of the complex Klein–Gordon field. On the other hand, for a real field the charge operator Q, Eqs. (3.32) or (3.35), is identically zero, and such a field corresponds to a neutral meson, such as the π^0.

The above considerations are not restricted to electric charge. The invariance of the Lagrangian density $\mathscr{L}$ under phase transformations would allow conservation of other additive quantities which by analogy one would call some kind of charge other than electric. The above argument would lead to the occurrence of pairs of particles and antiparticles differing from each other in the sign of this new kind of charge. Because of this, even electrically neutral particles may possess antiparticles. This situation does occur in nature. The electrically neutral pseudo-scalar K^0-meson possesses an antiparticle, the $\bar{K}^0$-meson, which is also electrically neutral. K^0 and $\bar{K}^0$ possess opposite hypercharge, $Y = \pm 1$, and are represented by a complex Klein–Gordon field ϕ. Hypercharge is very nearly conserved (unlike electric charge which is always exactly conserved) which is why it is a useful concept. To be specific, hypercharge is conserved in the strong interactions which are responsible for nuclear forces and associated production of strange particles, but it is not conserved in the weak interactions (about 10^{12} times weaker than the strong interactions) responsible for the decay of strange particles.

Instead of treating the complex Klein–Gordon field directly in terms of ϕ and $\phi^\dagger$ as independent fields, as we have done, one can define two real Klein–Gordon fields ϕ_1 and ϕ_2 by

$$\phi = \frac{1}{\sqrt{2}}(\phi_1 + \mathrm{i}\phi_2), \qquad \phi^\dagger = \frac{1}{\sqrt{2}}(\phi_1 - \mathrm{i}\phi_2), \tag{3.36}$$

and use these as independent fields. We shall not give the development in terms of the real fields as the two approaches are closely related and very similar. Since the fields ϕ_1 and ϕ_2 are real, the creation and annihilation operators associated with them cannot describe charged particles, and it is only linear combinations of them, corresponding to the complex fields (3.36), which describe charged particles. Consequently, when dealing with conserved charges, it is in general more natural to work directly with the complex fields.

3.3 COVARIANT COMMUTATION RELATIONS

While the equations of motion obtained using the Lagrangian formalism are manifestly covariant, this is not so obvious for the field commutation relations derived by the canonical formalism, since these single out equal times. Taking the real Klein–Gordon field as a typical example, we shall illustrate the covariance of the commutation relations by calculating the commutator $[\phi(x), \phi(y)]$ for two arbitrary space–time points x and y. Since this commutator is a scalar, it must equal an invariant function.

Writing $\phi = \phi^+ + \phi^-$, we note that

$$[\phi^+(x), \phi^+(y)] = [\phi^-(x), \phi^-(y)] = 0, \tag{3.37}$$

since $\phi^+(\phi^-)$ contains only absorption (creation) operators. Hence

$$[\phi(x), \phi(y)] = [\phi^+(x), \phi^-(y)] + [\phi^-(x), \phi^+(y)], \tag{3.38}$$

and we need only evaluate the first commutator on the right-hand side of this equation. From Eqs. (3.8) one obtains

$$\begin{aligned}[\phi^+(x), \phi^-(y)] &= \frac{\hbar c^2}{2V} \sum_{\mathbf{k}\mathbf{k}'} \frac{1}{(\omega_{\mathbf{k}}\omega_{\mathbf{k}'})^{1/2}} [a(\mathbf{k}), a^\dagger(\mathbf{k}')] \,\mathrm{e}^{-ikx+ik'y} \\ &= \frac{\hbar c^2}{2(2\pi)^3} \int \frac{\mathrm{d}^3\mathbf{k}}{\omega_{\mathbf{k}}} \mathrm{e}^{-ik(x-y)}\end{aligned} \tag{3.39}$$

where we have taken the limit $V \to \infty$ [see Eq. (1.48)], and in the last integral $k_0 = \omega_{\mathbf{k}}/c$. We introduce the definition

$$\Delta^+(x) \equiv \frac{-ic}{2(2\pi)^3} \int \frac{d^3\mathbf{k}}{\omega_{\mathbf{k}}} \mathrm{e}^{-ikx}, \qquad k_0 = \omega_{\mathbf{k}}/c, \tag{3.40}$$

since this and related functions will occur repeatedly.‡ Eq. (3.39) can then be written

$$[\phi^+(x), \phi^-(y)] = i\hbar c\Delta^+(x-y), \tag{3.41}$$

and

$$[\phi^-(x), \phi^+(y)] = -i\hbar c\Delta^+(y-x) \equiv i\hbar c\Delta^-(x-y), \tag{3.42}$$

defining the function $\Delta^-(x)$. From Eqs. (3.41), (3.42) and (3.38) we obtain the commutation relation

$$[\phi(x), \phi(y)] = i\hbar c\Delta(x-y) \tag{3.43}$$

with $\Delta(x)$ defined by

$$\Delta(x) \equiv \Delta^+(x) + \Delta^-(x) = \frac{-c}{(2\pi)^3} \int \frac{\mathrm{d}^3\mathbf{k}}{\omega_{\mathbf{k}}} \sin kx. \tag{3.44}$$

We see that $\Delta(x)$ is a real odd function, as required by the commutation relation (3.43), which (like $\Delta^{\pm}$) satisfies the Klein–Gordon equation

$$(\Box_x + \mu^2)\,\Delta(x-y) = 0. \tag{3.45}$$

The Δ-function (3.44) can be written

$$\Delta(x) = \frac{-\mathrm{i}}{(2\pi)^3} \int \mathrm{d}^4k\, \delta(k^2 - \mu^2)\varepsilon(k_0)\,\mathrm{e}^{-ikx} \tag{3.46}$$

‡ There is no generally accepted definitions of these Δ-functions, with different definitions differing by constant factors, so care is required in using the literature.

where $d^4k = d^3\mathbf{k}\, dk_0$, the k_0-integration is over $-\infty < k_0 < \infty$, and $\varepsilon(k_0)$ is defined by

$$\varepsilon(k_0) = \frac{k_0}{|k_0|} = \begin{cases} +1, & \text{if } k_0 > 0 \\ -1, & \text{if } k_0 < 0. \end{cases} \tag{3.47}$$

The equivalence of the definitions (3.44) and (3.46) is easily established; for example, by writing the δ-function in Eq. (3.46) as

$$\delta(k^2 - \mu^2) = \delta[k_0^2 - (\omega_{\mathbf{k}}/c)^2] = \frac{c}{2\omega_{\mathbf{k}}}\left[\delta\left(k_0 + \frac{\omega_{\mathbf{k}}}{c}\right) + \delta\left(k_0 - \frac{\omega_{\mathbf{k}}}{c}\right)\right] \tag{3.48}$$

and performing the k_0-integration.

The invariance of $\Delta(x)$ under proper Lorentz transformations (i.e. involving neither space nor time reflections) is obvious from Eq. (3.46), since each factor in the integrand is Lorentz-invariant [$\varepsilon(k_0)$ is invariant since proper Lorentz transformations do not interchange past and future].

The Lorentz-invariance of $\Delta(x)$ enables one to give a new interpretation to the equal-time commutation relation

$$[\phi(\mathbf{x}, t), \phi(\mathbf{y}, t)] = i\hbar c\, \Delta(\mathbf{x} - \mathbf{y}, 0) = 0 \tag{3.49}$$

which we had earlier [Eq. (3.7)].‡ The invariance of $\Delta(x - y)$ implies that

$$[\phi(x), \phi(y)] = i\hbar c\, \Delta(x - y) = 0, \quad \text{for } (x - y)^2 < 0, \tag{3.50}$$

i.e. the fields at any two points x and y with space-like separation commute. Hence if the field is a physical observable, measurements of the fields at two points with space-like separation must not interfere with each other. This is known as microcausality, since for a space-like separation, however small, a signal would have to travel with a velocity greater than the speed of light in order to cause interference, contrary to the special theory of relativity. When discussing the connnection between spin and statistics, at the end of Section 4.3, we shall see that the microcausality condition (3.50) is of fundamental importance even if the field itself is not a physical observable.

A particularly useful way of representing Δ-functions is as contour integrals in the complex k_0-plane. The functions $\Delta^{\pm}(x)$ are given by

$$\Delta^{\pm}(x) = \frac{-1}{(2\pi)^4}\int_{C^{\pm}} \frac{d^4k\, e^{-ikx}}{k^2 - \mu^2} \tag{3.51}$$

with the contours C^+ and C^-, for Δ^+ and Δ^- respectively, shown in Fig. 3.1. Performing the contour integrations, one picks up the residues from one or other of the poles at $k_0 = \pm\omega_{\mathbf{k}}/c$, and Eq. (3.51) reduces to the definitions (3.40) and (3.42) of $\Delta^{\pm}(x)$. The function $\Delta(x)$, Eq. (3.44), is represented by the

‡ This also follows directly from Eq. (3.44) since for $x^0 = 0$ the integrand is an odd function of $\mathbf{k}$.

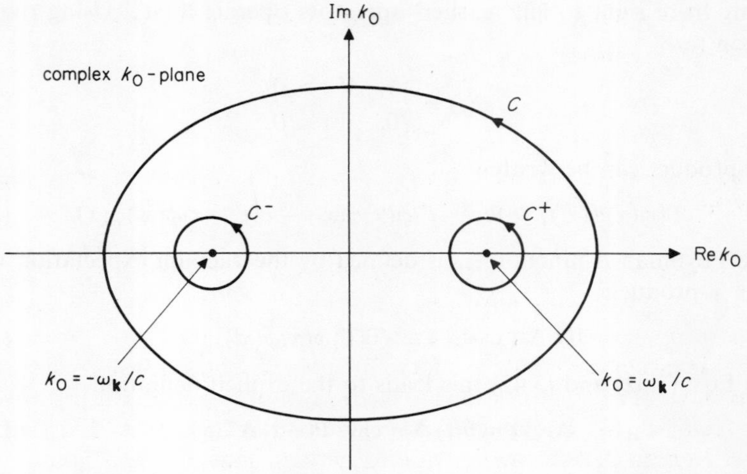

Fig. 3.1. Contours for the integral representation (3.51) of the functions $\Delta^{\pm}(x)$ and $\Delta(x)$.

same integral (3.51) with the contour C shown in Fig. 3.1. Other Δ-functions are obtained by a different choice of contour.

3.4 THE MESON PROPAGATOR

We shall now derive and discuss a Δ-function which is of great importance in quantum field theory. Its power and utility, particularly for the development of a systematic covariant perturbation theory, was first fully realized by Feynman. We shall again consider the real Klein–Gordon field.

To start with, we note that the Δ^+-function can be written as the vacuum expectation value of a product of two field operators. We have from Eq. (3.41) that

$$\begin{aligned} i\hbar c\,\Delta^+(x-x') &= \langle 0|[\phi^+(x), \phi^-(x')]|0\rangle = \langle 0|\phi^+(x)\phi^-(x')|0\rangle \\ &= \langle 0|\phi(x)\phi(x')|0\rangle. \end{aligned} \tag{3.52}$$

We define the *time-ordered* or T-product by

$$\mathrm{T}\{\phi(x)\phi(x')\} = \begin{cases} \phi(x)\phi(x'), & \text{if } t > t' \\ \phi(x')\phi(x), & \text{if } t' > t \end{cases} \tag{3.53}$$

$(t \equiv x^0/c$, etc.), i.e. the operators are written in chronological order with time running from right to left: 'earlier' operators operate 'first'.[‡] Using the step function

$$\theta(t) = \begin{cases} 1, & \text{if } t > 0 \\ 0, & \text{if } t < 0, \end{cases} \tag{3.54}$$

the T-product can be written

$$\mathrm{T}\{\phi(x)\phi(x')\} = \theta(t - t')\phi(x)\phi(x') + \theta(t' - t)\phi(x')\phi(x). \tag{3.55}$$

The Feynman Δ-function Δ_F is defined by the vacuum expectation value of this T-product:

$$i\hbar c\,\Delta_F(x - x') \equiv \langle 0|\mathrm{T}\{\phi(x)\phi(x')\}|0\rangle. \tag{3.56}$$

From Eqs. (3.52) and (3.42) this leads to the explicit definition

$$\Delta_F(x) = \theta(t)\,\Delta^+(x) - \theta(-t)\,\Delta^-(x). \tag{3.57a}$$

Thus

$$\Delta_F(x) = \pm\Delta^{\pm}(x). \quad \text{if } t \gtrless 0. \tag{3.57b}$$

We would like to be able to visualize the meaning of Δ_F, Eq. (3.56). For $t > t'$, this vacuum expectation value becomes $\langle 0|\phi(x)\phi(x')|0\rangle$. We can think of this expression as representing a meson being created at x', travelling to x, and being annihilated at x. The corresponding expression for $t' > t$, $\langle 0|\phi(x')\phi(x)|0\rangle$, admits a similar interpretation as a meson created at x, propagating to x' where it is absorbed. These two situations are illustrated schematically in Fig. 3.2. The dashed lines represent the propagation of the

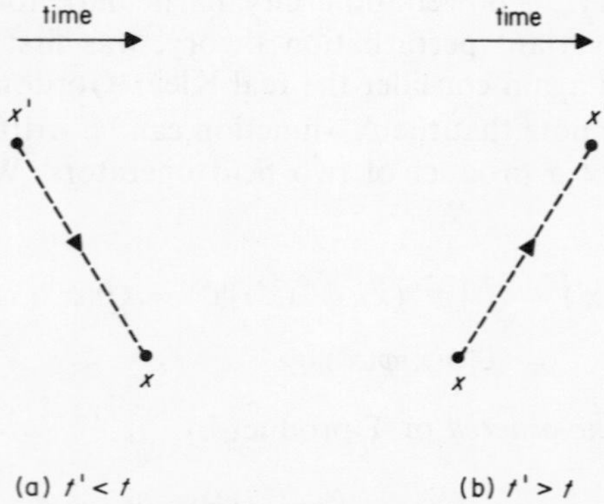

Fig. 3.2. The meson propagator (3.56).

[‡] The definition (3.53) will be modified for fermions.

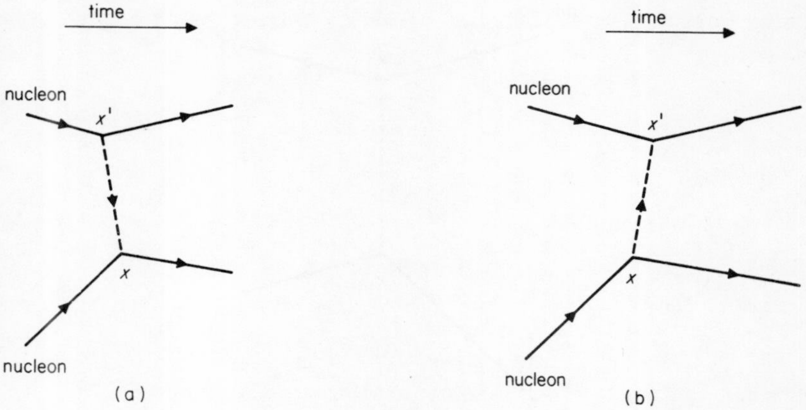

Fig. 3.3. Contribution from one-meson exchange to nucleon–nucleon scattering. (a) $t' < t$; (b) $t' > t$.

meson in the direction of the arrow, from x' to x or vice versa. Hence Δ_F, or the vacuum expectation value (3.56), is referred to as the *Feynman propagator* for the mesons of the Klein–Gordon field. We shall, briefly, call it the *meson propagator*, to distinguish it from the fermion and photon propagators to be introduced later.

To illustrate how these propagators arise, we shall consider qualitatively nucleon–nucleon scattering. In this process there will be two nucleons but no mesons present in the initial and final states (i.e. before and after the scattering). The scattering, i.e. the interaction, corresponds to the exchange of virtual mesons between the nucleons. The simplest such process is the one-meson exchange, schematically illustrated in Fig. 3.3. The continuous lines represent the nucleons, the dashed lines the mesons. As before, two situations arise according to whether $t > t'$ or $t' > t$. In the actual calculation, all values of x and x' are integrated over, corresponding to emission and absorption of the meson occurring at any two space–time points.

It is interesting to note that the division into the two types of process (a) and (b) of Fig. 3.3, depending on whether $t > t'$ or $t' > t$, is not Lorentz-invariant for $(x - x')$ a space-like separation. In this case what constitutes 'later' and what 'earlier' depends on the frame of reference. On the other hand, considering both cases *together* leads to the covariant Feynman propagator (3.56), which we represent by the single diagram in Fig. 3.4. No time-ordering is implied in this diagram and correspondingly there is no arrow on the meson line.

We have here introduced the ideas of *Feynman graphs* or diagrams. We shall deal with these fully later and shall see that they are a most useful way of

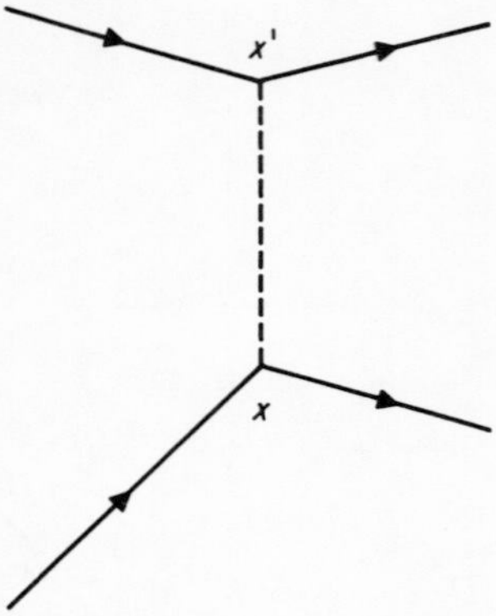

Fig. 3.4. Feynman graph for the one-meson contribution to nucleon–nucleon scattering.

picturing the mathematics. But the reader must be warned not to take this pictorial description of the mathematics as a literal description of a process in space and time. For example, our naive interpretation of the meson propagator would imply that, for $(x - x')$ a space-like separation, the meson travels between the two points with a speed greater than the velocity of light. It is however possible to substantiate the above description if, instead of considering propagation between two points x and x', one calculates the probability for emission and absorption in two appropriately chosen four-dimensional regions.‡

In the following we shall frequently require a representation of the meson propagator not in coordinate but in momentum space. This is given by the following integral representation, similar to Eq. (3.51) for $\Delta^{\pm}(x)$:

$$\Delta_F(x) = \frac{1}{(2\pi)^4} \int_{C_F} \frac{d^4k\, e^{-ikx}}{k^2 - \mu^2} \tag{3.58}$$

where the contour C_F is shown in Fig. 3.5. To verify Eq. (3.58), we evaluate the contour integral. For $x^0 > 0$, we must complete the contour C_F in the lower half k_0-plane [since $\exp(-ik_0x^0) \to 0$ for $k_0 \to -i\infty$], and comparing Eqs. (3.58) and (3.51) we obtain $\Delta_F(x) = \Delta^+(x)$, in agreement with Eq. (3.57b). For $x^0 < 0$, completion of the contour in the upper half k_0-plane similarly leads to agreement with Eq. (3.57b).

‡ See the article by G. Källén in *Encyclopedia of Physics*, vol. V, part 1, Springer, Berlin, 1958, Section 23. An English translation of this article, entitled *Quantum Electrodynamics*, has been published by Springer, New York, 1972, and by Allen & Unwin, London, 1972.

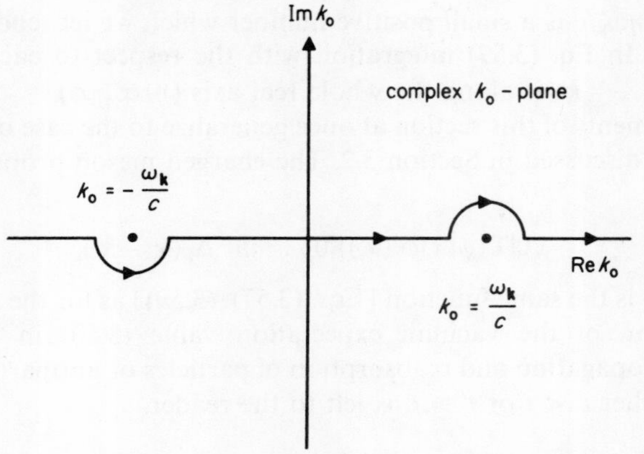

Fig. 3.5. The contour C_F for the meson propagator Δ_F, Eq. (3.58).

Instead of deforming the contour as in Fig. 3.5, we can move the poles an infinitesimal distance η off the real axis, as shown in Fig. 3.6, and perform the k_0-integration along the whole real axis, i.e. we replace Eq. (3.58) by

$$\Delta_F(x) = \frac{1}{(2\pi)^4} \int \frac{\mathrm{d}^4k\, \mathrm{e}^{-ikx}}{k_0^2 - \left(\dfrac{\omega_{\mathbf{k}}}{c} - \mathrm{i}\eta\right)^2}$$

$$= \frac{1}{(2\pi)^4} \int \frac{\mathrm{d}^4k\, \mathrm{e}^{-ikx}}{k^2 - \mu^2 + \mathrm{i}\varepsilon} \tag{3.59}$$

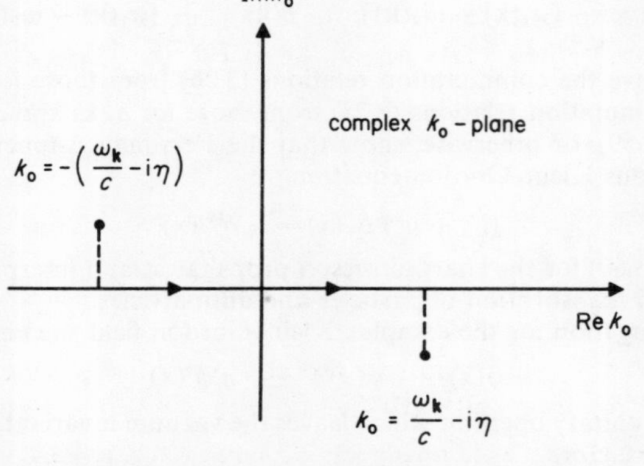

Fig. 3.6. Contour and displaced poles for the meson propagator Δ_F, Eq. (3.59).

where $\varepsilon = 2\eta\omega_{\mathbf{k}}/c$ is a small positive number which we let tend to zero after integration. In Eq. (3.59) integration with the respect to each of the four variables $k_0, \ldots, k_3$ is along the whole real axis $(-\infty, \infty)$.

The arguments of this section at once generalize to the case of the complex scalar field, discussed in Section 3.2. The charged meson propagator is now given by

$$\langle 0|\mathrm{T}\{\phi(x)\phi^{\dagger}(x')\}|0\rangle = i\hbar c\,\Delta_{\mathrm{F}}(x - x'), \qquad (3.60)$$

where $\Delta_{\mathrm{F}}(x)$ is the same function [Eqs. (3.57)–(3.59)] as for the real field. The interpretation of the vacuum expectation value (3.60) in terms of the emission, propagation and reabsorption of particles or antiparticles, depending on whether $t' < t$ or $t' > t$, is left to the reader.

PROBLEMS

3.1 From the expansion (3.8) for the real Klein–Gordon field $\phi(x)$ derive the following expression for the absorption operator $a(\mathbf{k})$:

$$a(\mathbf{k}) = \frac{1}{(2\hbar c^2 V\omega_{\mathbf{k}})^{1/2}} \int \mathrm{d}^3\mathbf{x}\, \mathrm{e}^{ikx}(\mathrm{i}\dot{\phi}(x) + \omega_{\mathbf{k}}\phi(x)).$$

Hence derive the commutation relations (3.10) for the creation and annihilation operators from the commutation relations (3.7) for the fields.

3.2 With the complex Klein–Gordon fields $\phi(x)$ and $\phi^{\dagger}(x)$ expressed in terms of two independent real Klein–Gordon fields $\phi_1(x)$ and $\phi_2(x)$ by Eqs. (3.36), and with $\phi_r(x)$ expanded in the form

$$\phi_r(x) = \sum_{\mathbf{k}} \left(\frac{\hbar c^2}{2V\omega_{\mathbf{k}}}\right)^{1/2} [a_r(\mathbf{k})\,\mathrm{e}^{-ikx} + a_r^{\dagger}(\mathbf{k})\,\mathrm{e}^{ikx}], \quad r = 1, 2,$$

show that

$$a(\mathbf{k}) = \frac{1}{\sqrt{2}}[a_1(\mathbf{k}) + \mathrm{i}a_2(\mathbf{k})], \qquad b(\mathbf{k}) = \frac{1}{\sqrt{2}}[a_1(\mathbf{k}) - \mathrm{i}a_2(\mathbf{k})].$$

Hence derive the commutation relations (3.26) from those for the real fields, and the commutation relations (3.28) from those for $a_r(\mathbf{k})$ and $a_r^{\dagger}(\mathbf{k})$, $r = 1, 2$.

3.3 From Eq. (3.59), or otherwise, show that the Feynman Δ-function satisfies the inhomogeneous Klein–Gordon equation

$$(\Box + \mu^2)\,\Delta_{\mathrm{F}}(x) = -\delta^{(4)}(x).$$

3.4 Derive Eq. (3.60) for the charged meson propagator, and interpret it in terms of emission and reabsorption of particles and antiparticles.

3.5 Charge conjugation for the complex Klein–Gordon field $\phi(x)$ is defined by

$$\phi(x) \to \mathscr{C}\phi(x)\mathscr{C}^{-1} = \eta_c\phi^{\dagger}(x) \qquad \text{(A)}$$

where $\mathscr{C}$ is a unitary operator which leaves the vacuum invariant, $\mathscr{C}|0\rangle = |0\rangle$, and η_c is a phase factor.

Show that under the transformation (A) the Lagrangian density (3.23) is invariant and the charge–current density (3.33) changes sign.

Derive

$$\mathscr{C}a(\mathbf{k})\mathscr{C}^{-1} = \eta_c b(\mathbf{k}), \qquad \mathscr{C}b(\mathbf{k})\mathscr{C}^{-1} = \eta_c^* a(\mathbf{k})$$

for the absorption operators, and hence show that

$$\mathscr{C}|a, \mathbf{k}\rangle = \eta_c^*|b, \mathbf{k}\rangle, \qquad \mathscr{C}|b, \mathbf{k}\rangle = \eta_c|a, \mathbf{k}\rangle$$

where $|a, \mathbf{k}\rangle$ denotes the state with a single a-particle of momentum $\mathbf{k}$ present, etc.

($\mathscr{C}$ is called the charge conjugation operator. It interchanges particles and antiparticles: $a \leftrightarrow b$. The phase η_c is arbitrary and is usually set equal to unity, $\eta_c = 1$.)

3.6 The parity transformation (i.e. space inversion) of the Hermitian Klein–Gordon field $\phi(x)$ is defined by

$$\phi(\mathbf{x}, t) \to \mathscr{P}\phi(\mathbf{x}, t)\mathscr{P}^{-1} = \eta_P\phi(-\mathbf{x}, t) \tag{A}$$

where the parity operator $\mathscr{P}$ is a unitary operator which leaves the vacuum invariant, $\mathscr{P}|0\rangle = |0\rangle$, and $\eta_P = \pm 1$ is called the intrinsic parity of the field. Show that the parity transformation leaves the Lagrangian density (3.5) invariant.

Show that

$$\mathscr{P}|\mathbf{k}_1, \mathbf{k}_2, \ldots, \mathbf{k}_n\rangle = \eta_P^n|-\mathbf{k}_1, -\mathbf{k}_2, \ldots, -\mathbf{k}_n\rangle$$

for an arbitrary n-particle state.

For any operators A and B

$$e^{i\alpha A}B\,e^{-i\alpha A} = \sum_{n=0}^{\infty} \frac{(i\alpha)^n}{n!} B_n$$

where

$$B_0 = B, \qquad B_n = [A, B_{n-1}], \qquad n = 1, 2, \ldots,$$

holds identically. Hence prove that

$$\mathscr{P}_1 a(\mathbf{k})\mathscr{P}_1^{-1} = ia(\mathbf{k}), \qquad \mathscr{P}_2 a(\mathbf{k})\mathscr{P}_2^{-1} = -i\eta_P a(-\mathbf{k}),$$

where the $a(\mathbf{k})$ are the annihilation operators of the field, and $\mathscr{P}_1$ and $\mathscr{P}_2$ are given by

$$\mathscr{P}_1 = \exp\left[-i\frac{\pi}{2}\sum_{\mathbf{k}} a^\dagger(\mathbf{k})a(\mathbf{k})\right], \qquad \mathscr{P}_2 = \exp\left[i\frac{\pi}{2}\eta_P\sum_{\mathbf{k}} a^\dagger(\mathbf{k})a(-\mathbf{k})\right].$$

Hence show that the operator $\mathscr{P} = \mathscr{P}_1\mathscr{P}_2$ is unitary and satisfies Eq. (A), i.e. it gives an explicit expression for the parity operator $\mathscr{P}$.

CHAPTER 4

The Dirac field

We now wish to consider systems of particles which satisfy the Pauli exclusion principle, i.e. which obey Fermi–Dirac statistics, so-called *fermions*. We saw in Chapter 2 that the canonical quantization formalism necessarily leads to bosons. On the other hand, the harmonic oscillator quantization, used in Chapter 1, allows an *ad hoc* modification which leads to Fermi–Dirac statistics. This modification was first introduced in 1928 by Jordan and Wigner and consists in replacing the commutation relations between absorption and creation operators by anticommutation relations. We shall develop this general formalism in Section 4.1.

In the remainder of this chapter, this formalism will be applied to the Dirac equation, i.e. to relativistic material particles of spin $\frac{1}{2}$. One of the distinctions between bosons and fermions is that the former always have integral spin (0, 1, ...) whereas the latter must have half-integral spin ($\frac{1}{2}, \frac{3}{2}, \ldots$). We shall see that this connection between spin and statistics is an essential feature of relativistic quantum field theory.

4.1 THE NUMBER REPRESENTATION FOR FERMIONS

In Sections 1.2.2 and 1.2.3 we derived a number representation for bosons from the quantization of a system of independent harmonic oscillators. We shall now modify this formalism so as to obtain a number representation for fermions.

The essence of our earlier treatment can be stated as follows. We had operators $a_r, a_r^\dagger, r = 1, 2, \ldots$, satisfying the commutation relations

$$[a_r, a_s^\dagger] = \delta_{rs}, \qquad [a_r, a_s] = [a_r^\dagger, a_s^\dagger] = 0, \tag{4.1}$$

and defined operators

$$N_r = a_r^\dagger a_r. \tag{4.2}$$

It then follows from the operator identity

$$[AB, C] = A[B, C] + [A, C]B \tag{4.3}$$

that

$$[N_r, a_s] = -\delta_{rs} a_s, \qquad [N_r, a_s^\dagger] = \delta_{rs} a_s^\dagger. \tag{4.4}$$

The interpretation of a_r, $a_r^\dagger$ and N_r as absorption, creation and number operators, follows from Eqs. (4.2) and (4.4). In particular, N_r has the eigenvalues $n_r = 0, 1, 2, \ldots$. The vacuum state $|0\rangle$ is defined by

$$a_r|0\rangle = 0, \quad \text{all } r, \tag{4.5}$$

and other states are built up from the vacuum state as linear superpositions of states of the form

$$(a_{r_1}^\dagger)^{n_1}(a_{r_2}^\dagger)^{n_2}\ldots|0\rangle. \tag{4.6}$$

It is a remarkable fact that there is an alternative way of deriving the relations (4.4). Define the *anticommutator* of two operators A and B by

$$[A, B]_+ \equiv AB + BA. \tag{4.7}$$

We then have a second operator identity, analogous to (4.3),

$$[AB, C] = A[B, C]_+ - [A, C]_+ B. \tag{4.8}$$

Suppose now that the operators $a_r, a_r^\dagger, r = 1, 2, \ldots$, instead of satisfying the commutation relations (4.1) satisfy the anticommutation relations

$$[a_r, a_s^\dagger]_+ = \delta_{rs}, \qquad [a_r, a_s]_+ = [a_r^\dagger, a_s^\dagger]_+ = 0; \tag{4.9}$$

in particular

$$(a_r)^2 = (a_r^\dagger)^2 = 0. \tag{4.9a}$$

One verifies from Eqs. (4.2), (4.8) and (4.9) that for the anticommuting operators [i.e. satisfying Eqs. (4.9)] the *same* commutation relations (4.4) hold which were previously derived for the commuting operators [i.e. satisfying Eqs. (4.1)]. This again leads to the interpretation of a_r, $a_r^\dagger$ and N_r as absorption, creation and number operators but, from Eqs. (4.9), we now have

$$N_r^2 = a_r^\dagger a_r a_r^\dagger a_r = a_r^\dagger(1 - a_r^\dagger a_r)a_r = N_r, \tag{4.10}$$

whence

$$N_r(N_r - 1) = 0; \tag{4.10a}$$

e., for the anticommuting creation and absorption operators, the number perator N_r has the eigenvalues $n_r = 0$ and $n_r = 1$ only, i.e. we are dealing ith Fermi–Dirac statistics.

The vacuum state $|0\rangle$ is again defined by Eq. (4.5). The state in which one article is in the state r is

$$|1_r\rangle = a_r^\dagger|0\rangle. \tag{4.11}$$

or the two-particle states we have from the anticommutation relations (4.9) hat for $r \neq s$

$$|1_r 1_s\rangle = a_r^\dagger a_s^\dagger|0\rangle = -a_s^\dagger a_r^\dagger|0\rangle = -|1_s 1_r\rangle \tag{4.12}$$

.e. the state is antisymmetric under interchange of particle labels as required or fermions. For $r = s$, we have

$$|2_r\rangle = (a_r^\dagger)^2|0\rangle = 0, \tag{4.13}$$

thus regaining the earlier result that two particles cannot be in the same single-particle state.

In conclusion we would like to note the fundamental difference, in spite of their superficial similarity, in the derivation of the boson and fermion results of this section. The boson commutation relations (4.1) are a direct consequence of the canonical commutation relations of non-relativistic quantum mechanics [compare the derivation of Eq. (1.19)]. We have no such foundation for the fermion anticommutation relations (4.9).

4.2 THE DIRAC EQUATION

We shall now consider the classical field theory of the Dirac equation, in preparation for going over to the quantized field theory in the next section.‡ The Dirac equation describes material particles of spin $\frac{1}{2}$. We shall see that in the quantum field theory antiparticles again necessarily occur, e.g. for electrons these are the positrons. Because of our later applications to quantum electrodynamics, we shall for definiteness speak of electrons and positrons in this chapter but the theory is equally applicable to other spin $\frac{1}{2}$ material particles such as nucleons.

The Dirac equation for particles of rest mass m

$$i\hbar \frac{\partial\psi(x)}{\partial t} = [c\boldsymbol{\alpha}\cdot(-i\hbar\boldsymbol{\nabla}) + \beta mc^2]\psi(x)$$

‡ The reader is assumed to be familiar with the elementary theory of the Dirac equation, as discussed in, for example, L. I. Schiff, *Quantum Mechanics*, 3rd edn, McGraw-Hill, New York, 1968, pp. 472–488. Further results of the Dirac theory, which will be needed later, are derived and summarized in Appendix A.

can be written

$$i\hbar\gamma^{\mu}\frac{\partial\psi(x)}{\partial x^{\mu}} - mc\psi(x) = 0 \tag{4.14}$$

where

$$\gamma^0 = \beta, \qquad \gamma^i = \beta\alpha_i, \qquad i = 1, 2, 3,$$

are Dirac 4 × 4 matrices which satisfy the anticommutation relations

$$[\gamma^{\mu}, \gamma^{\nu}]_+ = 2g^{\mu\nu} \tag{4.15}$$

and the Hermiticity conditions $\gamma^{0\dagger} = \gamma^0$ and $\gamma^{j\dagger} = -\gamma^j$ for $j = 1, 2, 3$, which can be combined into

$$\gamma^{\mu\dagger} = \gamma^0\gamma^{\mu}\gamma^0.^{\ddagger} \tag{4.16}$$

Correspondingly, $\psi(x)$ is a spinor wavefunction with four components $\psi_{\alpha}(x)$, $\alpha = 1, \ldots, 4$. The indices labelling spinor components and matrix elements will usually be suppressed.[§] Although it is at times convenient to use a particular matrix representation this is generally not necessary. We shall formulate the theory in a representation-free way and only assume that the γ-matrices satisfy the anticommutation and Hermiticity relations (4.15) and (4.16). This will facilitate use of the most convenient representation in a given situation.

The adjoint field $\bar{\psi}(x)$ is defined by

$$\bar{\psi}(x) = \psi^{\dagger}(x)\gamma^0 \tag{4.18}$$

and satisfies the adjoint Dirac equation

$$i\hbar\frac{\partial\bar{\psi}(x)}{\partial x^{\mu}}\gamma^{\mu} + mc\bar{\psi}(x) = 0. \tag{4.19}$$

The Dirac equations (4.14) and (4.19) can be derived from the Lagrangian density

$$\mathscr{L} = c\bar{\psi}(x)\left[i\hbar\gamma^{\mu}\frac{\partial}{\partial x^{\mu}} - mc\right]\psi(x) \tag{4.20}$$

by varying the action integral (2.11) independently with respect to the fields ψ_{α} and $\bar{\psi}_{\alpha}$. From Eq. (4.20) one obtains for the conjugate fields of ψ_{α} and $\bar{\psi}_{\alpha}$

$$\pi_{\alpha}(x) = \frac{\partial\mathscr{L}}{\partial\dot{\psi}_{\alpha}} = i\hbar\psi^{\dagger}_{\alpha}, \qquad \bar{\pi}_{\alpha}(x) = \frac{\partial\mathscr{L}}{\partial\dot{\bar{\psi}}_{\alpha}} \equiv 0. \tag{4.21}$$

‡ These conditions ensure the Hermiticity of the Dirac Hamiltonian

$$H = c\gamma^0\gamma^j(-i\hbar\,\partial/\partial x^j) + mc^2\gamma^0. \tag{4.17}$$

§ In case of doubt, the reader should write the indices out explicitly; e.g. Eq. (4.14) becomes

$$\sum_{\beta=1}^{4} i\hbar\gamma^{\mu}_{\alpha\beta}\frac{\partial\psi_{\beta}(x)}{\partial x^{\mu}} - mc\psi_{\alpha}(x) = 0, \qquad \alpha = 1, \ldots, 4. \tag{4.14a}$$

The Hamiltonian and the momentum of the Dirac field are, from Eqs. (2.51), (4.20) and (4.21), given by

$$H = \int d^3\mathbf{x}\bar{\psi}(x)\left[-i\hbar c\gamma^j \frac{\partial}{\partial x^j} + mc^2\right]\psi(x) \tag{4.22}$$

and

$$\mathbf{P} = -i\hbar \int d^3\mathbf{x}\psi^\dagger(x)\nabla\psi(x). \tag{4.23}$$

Eq. (4.22) of course also follows from the usual definition of the Hamiltonian density (2.25) applied to the present case.

The angular momentum of the Dirac field follows similarly from Eq. (2.54). The transformation of the field under an infinitesimal Lorentz transformation, i.e. Eq. (2.47), is in the case of the Dirac field given by

$$\psi_\alpha(x) \to \psi'_\alpha(x') = \psi_\alpha(x) - \frac{i}{4}\varepsilon_{\mu\nu}\sigma^{\mu\nu}_{\alpha\beta}\psi_\beta(x), \tag{4.24}$$

where summation over $\mu, \nu = 0, \ldots, 3$ and $\beta = 1, \ldots, 4$ is implied, and where $\sigma^{\mu\nu}_{\alpha\beta}$ is the (α, β) matrix element of the 4×4 matrix

$$\sigma^{\mu\nu} \equiv \frac{i}{2}[\gamma^\mu, \gamma^\nu]. \tag{4.25a}$$

Eq. (4.24) is derived in Appendix A, Eq. (A.60). Eq. (2.54) now gives for the angular momentum of the Dirac field

$$\mathbf{M} = \int d^3\mathbf{x}\psi^\dagger(x)[\mathbf{x} \wedge (-i\hbar\nabla)]\psi(x) + \int d^3\mathbf{x}\psi^\dagger(x)\left(\frac{\hbar}{2}\boldsymbol{\sigma}\right)\psi(x), \tag{4.26}$$

where the 4×4 matrices

$$\boldsymbol{\sigma} = (\sigma^{23}, \sigma^{31}, \sigma^{12}) \tag{4.25b}$$

are the generalizations for the Dirac theory of the 2×2 Pauli spin matrices. We see that the two terms in Eq. (4.26) represent the orbital and spin angular momenta of particles of spin $\frac{1}{2}$.

The Lagrangian density (4.20) is invariant under the phase transformation (2.41). Hence Eq. (2.42) leads to the conserved charge

$$Q = q\int d^3\mathbf{x}\psi^\dagger(x)\psi(x), \tag{4.27}$$

and the charge–current density

$$s^\alpha(x) = (c\rho(x), \mathbf{j}(x)) = cq\bar{\psi}(x)\gamma^\alpha\psi(x) \tag{4.28}$$

satisfies the continuity equation (i.e. conservation equation)

$$\frac{\partial s^\alpha}{\partial x^\alpha} = 0. \tag{4.29}$$

The continuity equation also follows directly from the Dirac equations (4.14) and (4.19).

In order to quantize the Dirac field in the next section, we shall expand it in a complete set of solutions of the Dirac equation and then impose appropriate anticommutation relations on the expansion coefficients. To conclude this section, we shall therefore specify a complete orthonormal set of solutions of the Dirac equation (4.14).

We shall again consider a cubic enclosure, of volume V, with periodic boundary conditions. A complete set of plane wave states can then be defined as follows. For each momentum $\mathbf{p}$, allowed by the periodic boundary conditions, and positive energy

$$cp_0 = E_{\mathbf{p}} = +(m^2c^4 + c^2\mathbf{p}^2)^{1/2}, \tag{4.30}$$

the Dirac equation (4.14) possesses four independent solutions. These will be written

$$u_r(\mathbf{p})\frac{\mathrm{e}^{-ipx/\hbar}}{\sqrt{V}}, \qquad v_r(\mathbf{p})\frac{\mathrm{e}^{ipx/\hbar}}{\sqrt{V}}, \qquad r = 1, 2 \tag{4.31}$$

i.e. $u_r(\mathbf{p})$ and $v_r(\mathbf{p})$ are constant spinors satisfying the equations

$$(\not{p} - mc)u_r(\mathbf{p}) = 0, \qquad (\not{p} + mc)v_r(\mathbf{p}) = 0, \qquad r = 1, 2. \tag{4.32}$$

Here we introduce the very convenient notation $\not{A}$ (called A slash) which is defined for any four-vector A_μ by

$$\not{A} \equiv \gamma^\mu A_\mu. \tag{4.33}$$

Because of their time-dependence, the solutions (4.31) involving u_r and v_r are referred to as positive and negative energy solutions respectively. We shall use these terms merely as a way of labelling the u- and v-solutions. We shall not follow up their interpretation in the single-particle theory, the resulting difficulties and the reinterpretation in terms of the hole theory. We shall see that the second quantization of the theory (i.e. when ψ and $\psi^\dagger$ become operators) leads directly to the interpretation in terms of particles and antiparticles without the intellectual contortions of the hole theory.‡

The two-fold degeneracies of the two positive and the two negative energy solutions for a given momentum $\mathbf{p}$ result from the possible spin orientations. For the Dirac equation only the longitudinal spin components (i.e. parallel to

‡ This remark is in no way meant to denigrate Dirac's tremendous intellectual achievement of inventing the hole theory originally.

$\pm\mathbf{p}$) are constants of the motion, and we shall choose these spin eigenstates for the solutions (4.31). With

$$\sigma_{\mathbf{p}} = \frac{\boldsymbol{\sigma}\cdot\mathbf{p}}{|\mathbf{p}|}, \tag{4.34}$$

where $\boldsymbol{\sigma}$ is defined in Eqs. (4.25a) and (4.25b), we then choose the spinors in Eqs. (4.31) so that

$$\sigma_{\mathbf{p}} u_r(\mathbf{p}) = (-1)^{r+1} u_r(\mathbf{p}), \qquad \sigma_{\mathbf{p}} v_r(\mathbf{p}) = (-1)^r v_r(\mathbf{p}), \qquad r = 1, 2. \tag{4.35}$$

The asymmetry in labelling u- and v-spinors will be convenient for labelling the spin properties of particles and antiparticles.

We normalize the spinors u_r and v_r so that

$$u_r^\dagger(\mathbf{p}) u_r(\mathbf{p}) = v_r^\dagger(\mathbf{p}) v_r(\mathbf{p}) = \frac{E_{\mathbf{p}}}{mc^2}. \tag{4.36}$$

They then satisfy the orthonormality relations

$$\left.\begin{aligned} u_r^\dagger(\mathbf{p}) u_s(\mathbf{p}) = v_r^\dagger(\mathbf{p}) v_s(\mathbf{p}) &= \frac{E_{\mathbf{p}}}{mc^2}\,\delta_{rs} \\ u_r^\dagger(\mathbf{p}) v_s(-\mathbf{p}) &= 0 \end{aligned}\right\}, \tag{4.37}$$

and the states (4.31) form a complete orthonormal set of solutions of the free-particle Dirac equation, normalized to $E_{\mathbf{p}}/mc^2$ in a volume V. These and other properties of the plane wave solutions (4.31) are discussed further in Appendix A.

4.3 SECOND QUANTIZATION

In order to quantize the Dirac field we expand it in terms of the complete set of plane wave states (4.31):

$$\begin{aligned} \psi(x) &= \psi^+(x) + \psi^-(x) \\ &= \sum_{r\mathbf{p}} \left(\frac{mc^2}{VE_{\mathbf{p}}}\right)^{1/2} [c_r(\mathbf{p}) u_r(\mathbf{p})\, \mathrm{e}^{-\mathrm{i}px/\hbar} + d_r^\dagger(\mathbf{p}) v_r(\mathbf{p})\, \mathrm{e}^{\mathrm{i}px/\hbar}] \end{aligned} \tag{4.38a}$$

and hence the conjugate field $\bar{\psi} = \psi^\dagger\gamma^0$ has the expansion

$$\begin{aligned} \bar{\psi}(x) &= \bar{\psi}^+(x) + \bar{\psi}^-(x) \\ &= \sum_{r\mathbf{p}} \left(\frac{mc^2}{VE_{\mathbf{p}}}\right)^{1/2} [d_r(\mathbf{p}) \bar{v}_r(\mathbf{p})\, \mathrm{e}^{-\mathrm{i}px/\hbar} + c_r^\dagger(\mathbf{p}) \bar{u}_r(\mathbf{p})\, \mathrm{e}^{\mathrm{i}px/\hbar}] \end{aligned} \tag{4.38b}$$

where $\bar{u}_r = u_r^\dagger\gamma^0$, etc. The summations in Eqs. (4.38) are over the allowed momenta $\mathbf{p}$ and the spin states, labelled by $r = 1, 2$.‡ The factors $(mc^2/VE_{\mathbf{p}})^{1/2}$ will be convenient for the subsequent interpretation of the expansion coefficients. We have written $c_r^\dagger$ and $d_r^\dagger$ for two of these, anticipating that they will become operators on second quantization.

Eqs. (4.38) are closely analogous to the expansions of the complex Klein–Gordon field, Eqs. (3.27). However, the Dirac equation describes spin $\frac{1}{2}$ particles, such as electrons, which obey the Pauli principle and Fermi–Dirac statistics. Following the treatment in Section 4.1, we shall therefore impose the following *anti*commutation relations on the expansion coefficients:

$$[c_r(\mathbf{p}), c_s^\dagger(\mathbf{p}')]_+ = [d_r(\mathbf{p}), d_s^\dagger(\mathbf{p}')]_+ = \delta_{rs}\,\delta_{\mathbf{pp}'}, \tag{4.39a}$$

and all other anticommutators vanish, i.e. with $c_r \equiv c_r(\mathbf{p})$, $c_s \equiv c_s(\mathbf{p}')$, etc.:

$$\left.\begin{aligned} &[c_r, c_s]_+ = [c_r^\dagger, c_s^\dagger]_+ = [d_r, d_s]_+ = [d_r^\dagger, d_s^\dagger]_+ = 0 \\ &[c_r, d_s]_+ = [c_r, d_s^\dagger]_+ = [c_r^\dagger, d_s]_+ = [c_r^\dagger, d_s^\dagger]_+ = 0 \end{aligned}\right\}. \tag{4.39b}$$

If we define the operators

$$N_r(\mathbf{p}) = c_r^\dagger(\mathbf{p})c_r(\mathbf{p}), \qquad \bar{N}_r(\mathbf{p}) = d_r^\dagger(\mathbf{p})d_r(\mathbf{p}), \tag{4.40}$$

the interpretation of c_r, $c_r^\dagger$, N_r and d_r, $d_r^\dagger$, $\bar{N}_r$ as absorption, creation and number operators of two kinds of particles, both fermions, follows from the anticommutation relations (4.39), analogously to the development in Section 4.1.

The vacuum state $|0\rangle$ is defined by

$$c_r(\mathbf{p})|0\rangle = d_r(\mathbf{p})|0\rangle = 0, \qquad \text{all } \mathbf{p}, \text{ and } r = 1, 2, \tag{4.41}$$

or, equivalently, by

$$\psi^+(x)|0\rangle = \bar{\psi}^+(x)|0\rangle = 0, \quad \text{all } x. \tag{4.42}$$

States containing particles are generated from the vacuum state by means of the creation operators. As in Section 4.1, one sees that these states have all the properties characteristic of fermions [i.e. equations analogous to Eqs. (4.12) and (4.13) hold].

To obtain the physical properties of the particles associated with the c- and d-operators, we express the constants of the motion in terms of them. (The reader should be able to have a good guess at most of these properties.) In Section 4.2, we derived expressions for the energy, momentum, angular momentum and charge of the Dirac field [see Eqs. (4.22), (4.23), (4.26) and

‡ We have chosen particular spin states u_r and v_r, but it should be clear to the reader that one may equally well use any other orthonormal spin states. The following arguments remain valid; only the interpretation of the spin properties of the states has to be modified.

(4.27)]. However, these operators do not necessarily have the value zero for the vacuum state. We found a similar situation for the Klein–Gordon field [see Eqs. (3.15) and (3.16)]. As in the latter case, we automatically measure quantities relative to the vacuum state if we redefine the expressions for the constants of the motion with the operators ordered as normal products (i.e. in any product, absorption operators occur to the right of creation operators) so that vacuum values necessarily vanish.

For fermions we must modify our earlier definition of the normal product. In arranging a product of boson operators in normal order, one treats them as though all commutators vanish [see Eqs. (3.18) and (3.19)]. For fermion operators, one treats them as though all *anti*commutators vanish, e.g. with $\psi_\alpha \equiv \psi_\alpha(x)$ and $\psi_\beta \equiv \psi_\beta(x')$, etc., one has

$$\begin{aligned} \mathrm{N}(\psi_\alpha\psi_\beta) &= \mathrm{N}[(\psi_\alpha^+ + \psi_\alpha^-)(\psi_\beta^+ + \psi_\beta^-)] \\ &= \psi_\alpha^+\psi_\beta^+ - \psi_\beta^-\psi_\alpha^+ + \psi_\alpha^-\psi_\beta^+ + \psi_\alpha^-\psi_\beta^- \end{aligned} \tag{4.43}$$

which should be compared with Eq. (3.19) for bosons. Similar results hold if in Eq. (4.43) one or both operators are replaced by their adjoint operators, $\bar{\psi}$, or for products of more than two fields.‡

With the expressions for the constants of the motion, i.e. Eqs. (4.22), (4.23) and (4.26)–(4.28), modified to be normal products, e.g.

$$H = \int \mathrm{d}^3\mathbf{x}\, \mathrm{N}\left\{\bar{\psi}(x)\left[-i\hbar c\gamma^j \frac{\partial}{\partial x^j} + mc^2\right]\psi(x)\right\}, \tag{4.22a}$$

etc., we substitute the expansions (4.38) for ψ and $\bar{\psi}$. Using the orthonormality properties of the single-particle states (4.31) and, in the calculation of H, that they are solutions of the Dirac equation, we obtain for the energy, momentum and charge operators

$$H = \sum_{r\mathbf{p}} E_\mathbf{p}[N_r(\mathbf{p}) + \bar{N}_r(\mathbf{p})] \tag{4.44}$$

$$\mathbf{P} = \sum_{r\mathbf{p}} \mathbf{p}[N_r(\mathbf{p}) + \bar{N}_r(\mathbf{p})] \tag{4.45}$$

$$Q = -e\sum_{r\mathbf{p}} [N_r(\mathbf{p}) - \bar{N}_r(\mathbf{p})]. \tag{4.46}$$

‡ The fermion operators ψ and $\bar{\psi}$ are non-commuting quantities not only through their dependence on the absorption and creation operators but also as four-component spinors. Hence care is required in changing the order of operators; e.g. if O is a 4×4 matrix (such as a product of γ-matrices) then

$$\mathrm{N}(\bar{\psi}O\psi) = \bar{\psi}_\alpha^+ O_{\alpha\beta}\psi_\beta^+ - \psi_\beta^- O_{\alpha\beta}\bar{\psi}_\alpha^+ + \bar{\psi}_\alpha^- O_{\alpha\beta}\psi_\beta^+ + \bar{\psi}_\alpha^- O_{\alpha\beta}\psi_\beta^-, \tag{4.43a}$$

i.e. suppressing the spinor indices the second term on the right-hand side is $-\psi^- O^{\mathrm{T}}\bar{\psi}^+$, where O^{T} is the transposed matrix: $O^{\mathrm{T}}_{\beta\alpha} = O_{\alpha\beta}$. In case of doubt, the reader should write the spinor indices out explicitly.

In the last equation we have taken the parameter q to be the charge of the electron: $q = -e < 0$. Hence identifying the mass m in the Dirac equation with the mass of the electron, we can interpret the particles associated with the c- and d-operators as electrons and positrons, respectively.

To identify the spin properties, we calculate the spin angular momentum in the states $c_r^\dagger(\mathbf{p})|0\rangle$ and $d_r^\dagger(\mathbf{p})|0\rangle$, containing one electron or one positron of momentum $\mathbf{p}$. From Eqs. (4.26) and (4.34) we define the longitudinal spin operator, i.e. in the direction of motion $\mathbf{p}$, by

$$S_{\mathbf{p}} = \frac{\hbar}{2} \int d^3\mathbf{x} \mathrm{N}[\psi^\dagger(x)\sigma_{\mathbf{p}}\psi(x)]. \tag{4.47}$$

It is left to the reader to verify that

$$S_{\mathbf{p}} c_r^\dagger(\mathbf{p})|0\rangle = (-1)^{r+1} \frac{\hbar}{2} c_r^\dagger(\mathbf{p})|0\rangle,$$

$$S_{\mathbf{p}} d_r^\dagger(\mathbf{p})|0\rangle = (-1)^{r+1} \frac{\hbar}{2} d_r^\dagger(\mathbf{p})|0\rangle, \qquad r = 1, 2. \tag{4.48}$$

We see from Eqs. (4.48) that in both the electron state $c_r^\dagger(\mathbf{p})|0\rangle$ and the positron state $d_r^\dagger(\mathbf{p})|0\rangle$ the spin component in the direction of motion has the value $+\hbar/2$ for $r = 1$, and the value $-\hbar/2$ for $r = 2$. We refer to these two spin states, i.e. spin parallel and antiparallel to the direction of motion, as having positive (right-handed) and negative (left-handed) *helicity* respectively. (Right- and left-handed here specifies the screw sense of the spin in the direction of motion.) We shall call $S_{\mathbf{p}}$ the helicity operator of a spin $\frac{1}{2}$ particle (whether electron or positron) with momentum $\mathbf{p}$.

It follows from Eqs. (4.44)–(4.46) and (4.48) that, as for the complex Klein–Gordon field, the theory is completely symmetric between particles (electrons) and antiparticles (positrons). These have the same properties except for the reversal of the sign of the electric charge. (As a result other electromagnetic properties such as the magnetic moments have opposite signs.)

The symmetry of the theory between particles and antiparticles is not obvious from the expansions of the field operators ψ and $\bar{\psi}$, Eqs. (4.38). This is due to the fact that we have not chosen a specific spinor representation and in most representations the positive and negative energy spinors will look very different. The expansions (4.38) only manifest the particle–antiparticle symmetry for representations of a particular kind, known as Majorana representations. Labelling the γ-matrices in a Majorana representation with the subscript M, the defining property of a Majorana representation is that

$$\gamma_{\mathrm{M}}^{\mu *} = -\gamma_{\mathrm{M}}^{\mu}, \qquad \mu = 0, \ldots, 3, \tag{4.49}$$

where the asterisk denotes complex conjugation, i.e. all four γ-matrices are pure imaginary. A particular Majorana representation is given in Appendix A, Eqs. (A.79). Here we only require the defining property (4.49).

We see from Eq. (4.49) that in a Majorana representation the operator

$$\left(i\hbar\gamma_{\mathrm{M}}^{\mu}\frac{\partial}{\partial x^{\mu}} - mc\right)$$

is real. Hence if ψ_{M} is a solution of the Dirac equation in a Majorana representation, so is its complex conjugate ψ_{M}^{*}. It follows that if we denote the positive energy solutions (4.31) by

$$u_{\mathrm{M}r}(\mathbf{p})\frac{\mathrm{e}^{-\mathrm{i}px/\hbar}}{\sqrt{V}}, \qquad r = 1, 2, \tag{4.50a}$$

in a Majorana representation, then the corresponding negative energy solutions are

$$u_{\mathrm{M}r}^{*}(\mathbf{p})\frac{\mathrm{e}^{\mathrm{i}px/\hbar}}{\sqrt{V}}, \qquad r = 1, 2. \tag{4.50b}$$

Hence the expansions (4.38) become, in aMajorana representation,

$$\left.\begin{aligned}\psi_{\mathrm{M}}(x) &= \sum_{r\mathbf{p}}\left(\frac{mc^2}{VE_{\mathbf{p}}}\right)^{1/2}[c_r(\mathbf{p})u_{\mathrm{M}r}(\mathbf{p})\,\mathrm{e}^{-\mathrm{i}px/\hbar} + d_r^{\dagger}(\mathbf{p})u_{\mathrm{M}r}^{*}(\mathbf{p})\,\mathrm{e}^{\mathrm{i}px/\hbar}]\\ \psi_{\mathrm{M}}^{\dagger}(x) &= \sum_{r\mathbf{p}}\left(\frac{mc^2}{VE_{\mathbf{p}}}\right)^{1/2}[d_r(\mathbf{p})u_{\mathrm{M}r}(\mathbf{p})\,\mathrm{e}^{-\mathrm{i}px/\hbar} + c_r^{\dagger}(\mathbf{p})u_{\mathrm{M}r}^{*}(\mathbf{p})\,\mathrm{e}^{\mathrm{i}px/\hbar}]\end{aligned}\right\}. \tag{4.51}$$

In the last equation we gave the expansion for $\psi^{\dagger}$, rather than $\bar{\psi}$, to bring out the complete symmetry between particles and antiparticles. The absorption operators $c_r(\mathbf{p})$ and $d_r(\mathbf{p})$ are multiplied by the same single-particle wavefunctions and thus are absorption operators of particles and antiparticles in the same single-particle state, i.e. with the same momentum, energy and helicity. The same is true of the creation operators. Having used a Majorana representation to manifest the particle–antiparticle symmetry of the field operators, we shall now revert to the representation-free formulation of Eqs. (4.31) and (4.38) in which this symmetry is masked.

The anticommutation relations (4.39) for the creation and absorption operators imply anticommutation relations for the Dirac field operators ψ and $\bar{\psi}$. From Eqs. (4.39) and the expansions (4.38) of the fields, one obtains

$$[\psi_{\alpha}(x), \psi_{\beta}(y)]_{+} = [\bar{\psi}_{\alpha}(x), \bar{\psi}_{\beta}(y)]_{+} = 0, \tag{4.52a}$$

$$[\psi_{\alpha}^{\pm}(x), \bar{\psi}_{\beta}^{\mp}(y)]_{+} = \mathrm{i}\left(\mathrm{i}\gamma^{\mu}\frac{\partial}{\partial x^{\mu}} + \frac{mc}{\hbar}\right)_{\alpha\beta}\Delta^{\pm}(x-y) \tag{4.52b}$$

where $\Delta^{\pm}(x)$ are the invariant Δ-functions introduced for the Klein–

Gordon equation, Eqs. (3.40) and (3.42). Eqs. (4.52a) are obvious. The derivation of Eqs. (4.52b) is left as an exercise for the reader.

Omitting suffixes, i.e. considered as a 4 × 4 matrix equation, we can write Eqs. (4.52b) as

$$[\psi^{\pm}(x), \bar{\psi}^{\mp}(y)]_{+} = \mathrm{i}S^{\pm}(x-y) \tag{4.53a}$$

where the 4 × 4 matrix functions $S^{\pm}(x)$ are defined by

$$S^{\pm}(x) = \left(\mathrm{i}\gamma^{\mu}\frac{\partial}{\partial x^{\mu}} + \frac{mc}{\hbar}\right)\Delta^{\pm}(x). \tag{4.54a}$$

From the last two equations

$$[\psi(x), \bar{\psi}(y)]_{+} = \mathrm{i}S(x-y) \tag{4.53b}$$

where, analogously to $\Delta(x) = \Delta^{+}(x) + \Delta^{-}(x)$, we defined

$$S(x) = S^{+}(x) + S^{-}(x) = \left(\mathrm{i}\gamma^{\mu}\frac{\partial}{\partial x^{\mu}} + \frac{mc}{\hbar}\right)\Delta(x). \tag{4.54b}$$

From the representations of the various Δ-functions, obtained in Chapter 3, Eqs. (4.54) provide representations of the corresponding S-functions. For example, from the integral representation (3.51) for $\Delta^{\pm}(x)$, we can write Eqs. (4.54a) as

$$S^{\pm}(x) = \frac{-\hbar}{(2\pi\hbar)^4}\int_{C^{\pm}} \mathrm{d}^4p\, \mathrm{e}^{-\mathrm{i}px/\hbar}\frac{\not{p}+mc}{p^2-m^2c^2}, \tag{4.55a}$$

where the contours $C^{\pm}$ in the complex p_0-plane are anticlockwise closed paths enclosing the poles at $p_0 = \pm(E_{\mathbf{p}}/c)$, corresponding to Fig. 3.1 for the complex $k_0(=p_0/c)$-plane. Since

$$(\not{p} \pm mc)(\not{p} \mp mc) = p^2 - m^2c^2, \tag{4.56}$$

the last equation is often abbreviated into the symbolic form

$$S^{\pm}(x) = \frac{-\hbar}{(2\pi\hbar)^4}\int_{C^{\pm}} \mathrm{d}^4p \frac{\mathrm{e}^{-\mathrm{i}px/\hbar}}{\not{p}-mc}. \tag{4.55b}$$

We conclude this section with a brief discussion on the connection between spin and statistics of particles. In this section we quantized the Dirac equation according to the anticommutation relations (4.39) in order to obtain Fermi–Dirac statistics for electrons. It is interesting to ask what the consequences would be if we quantize the Dirac equation according to Bose–Einstein statistics, i.e. by replacing all the anticommutators in Eqs. (4.39) by commutators. With this change, the energy of the field, again calculated from Eq. (4.22a), is not given by Eq. (4.44) but by

$$H = \sum_{r\mathbf{p}} E_{\mathbf{p}}[N_r(\mathbf{p}) - \bar{N}_r(\mathbf{p})]. \tag{4.56}$$

We are now dealing with Bose–Einstein statistics, and the occupation number operators $N_r(\mathbf{p})$ and $\bar{N}_r(\mathbf{p})$ can take on all values 0, 1, 2, ... Hence the Hamiltonian (4.56) does not possess a lower bound. If we demand the existence of a state of lowest energy (i.e. a stable ground state), we must quantize the Dirac equation according to Fermi–Dirac statistics.

One may similarly ask what the consequences are of quantizing the Klein–Gordon field according to Fermi–Dirac statistics. In Section 3.3 we referred to microcausality, i.e. the requirement that two observables $A(x)$ and $B(y)$ must be compatible if $(x - y)$ is a space-like interval, i.e.

$$[A(x), B(y)] = 0, \quad \text{for } (x - y)^2 < 0. \tag{4.57}$$

We have seen that the observables of the fields, such as the energy–momentum densities or the charge–current densities, are bilinear in the field operators [see, for example, Eqs. (3.13), (3.14), (3.33), (4.22), (4.23) and (4.28)]. Using the operator identities (4.3) and (4.8), one can show that for Eq. (4.57) to hold for such bilinear observables, the fields themselves must either commute or anticommute for $(x - y)$ a space-like interval. For the real Klein–Gordon field we must have either

$$[\phi(x), \phi(y)] = 0, \quad \text{for } (x - y)^2 < 0,$$

or

$$[\phi(x), \phi(y)]_+ = 0, \quad \text{for } (x - y)^2 < 0.$$

We know that the first of these relations holds if the Klein–Gordon field is quantized according to Bose–Einstein statistics [compare Eq. (3.50)]. It is easy to show that neither relation holds if we quantize according to Fermi–Dirac statistics, i.e. replace the commutators by anticommutators in the commutation relations (3.10). Hence, the requirement of microcausality forces us to quantize the Klein–Gordon field according to Bose–Einstein statistics.

These conclusions generalize to interacting particles and other spin values. Particles with integral spin must be quantized according to Bose–Einstein statistics, particles with half-integral spin according to Fermi–Dirac statistics. The 'wrong' spin-statistics connections lead to the two types of difficulties we found above. This spin-statistics theorem, to which no exception is known in nature, represents an impressive success for relativistic quantum field theory.

4.4 THE FERMION PROPAGATOR

In Section 3.4 we introduced the meson propagator. Corresponding to Eq. (3.56), we now define the Feynman fermion propagator as

$$\langle 0|\mathrm{T}\{\psi(x)\bar{\psi}(x')\}|0\rangle \tag{4.58}$$

where spinor indices have again been suppressed. For fermion fields, the time-ordered product is defined by

$$\left.\begin{aligned}\mathrm{T}\{\psi(x)\bar{\psi}(x')\} &= \theta(t-t')\psi(x)\bar{\psi}(x') - \theta(t'-t)\bar{\psi}(x')\psi(x)\\ &= \begin{cases}\psi(x)\bar{\psi}(x'), & \text{if } t > t'\\ -\bar{\psi}(x')\psi(x), & \text{if } t' > t\end{cases}\end{aligned}\right\} \tag{4.59}$$

(where $t = x^0/c$, etc.). This definition differs by a factor (-1) in the $t' > t$ term from the corresponding boson definition, Eqs. (3.53) and (3.55). This change in sign reflects the anticommutation property of fermion fields. (A similar difference occurred in the definition of the normal products of boson and fermion fields.)

In order to calculate the fermion propagator (4.58), using Eq. (4.59), we note that

$$\begin{aligned}\langle 0|\psi(x)\bar{\psi}(x')|0\rangle &= \langle 0|\psi^+(x)\bar{\psi}^-(x')|0\rangle\\ &= \langle 0|[\psi^+(x), \bar{\psi}^-(x')]_+|0\rangle = \mathrm{i}S^+(x-x'),\end{aligned} \tag{4.60a}$$

where we used Eq. (4.53a); similarly

$$\langle 0|\bar{\psi}(x')\psi(x)|0\rangle = \mathrm{i}S^-(x-x'). \tag{4.60b}$$

Combining Eqs. (4.58)–(4.60), we obtain the fermion propagator:

$$\langle 0|\mathrm{T}\{\psi(x)\bar{\psi}(x')\}|0\rangle = \mathrm{i}S_\mathrm{F}(x-x') \tag{4.61}$$

where $S_\mathrm{F}(x)$ is defined, analogously to Eqs. (3.57) and (4.54), by

$$S_\mathrm{F}(x) = \theta(t)S^+(x) - \theta(-t)S^-(x) = \left(\mathrm{i}\gamma^\mu\frac{\partial}{\partial x^\mu} + \frac{mc}{\hbar}\right)\Delta_\mathrm{F}(x). \tag{4.62}$$

Corresponding to the integral representation (3.59) for $\Delta_\mathrm{F}(x)$, $S_\mathrm{F}(x)$ can be written

$$S_\mathrm{F}(x) = \frac{\hbar}{(2\pi\hbar)^4}\int \mathrm{d}^4p\, \mathrm{e}^{-\mathrm{i}px/\hbar}\frac{\not{p} + mc}{p^2 - m^2c^2 + \mathrm{i}\varepsilon}, \tag{4.63}$$

where the integration in the complex p_0-plane is along the whole real axis: $-\infty < p_0 < \infty$. (Compare Fig. 3.6.)

As for the meson propagator, it is useful to visualize the fermion propagator in terms of Feynman diagrams. (As mentioned before, one must not take this interpretation too literally.)

For $t' < t$, the contribution to the fermion propagator (4.61) stems from the term (4.60a) and thus leads to the interpretation of (4.61) as creation of an electron at x', its propagation to x and its annihilation at x. On the other hand for $t < t'$, the contribution to (4.61) comes from (4.60b) and is pictured as the emission of a positron at x, and its propagation to x' where it is annihilated. The two cases are illustrated in Fig. 4.1. Note that in both

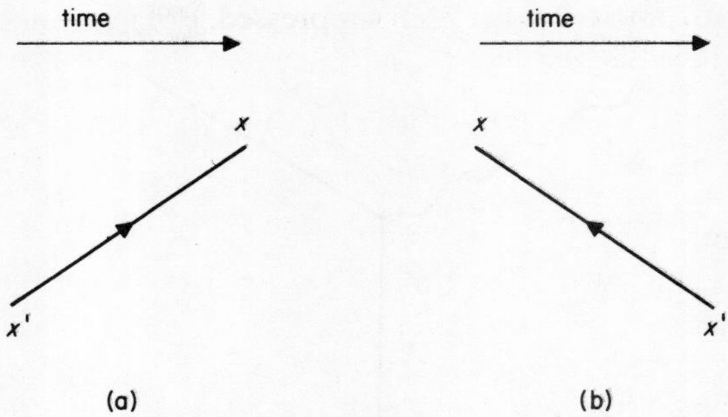

Fig. 4.1. (a) $t' < t$: electron propagated from x' to x;
(*b*) $t' > t$: positron propagated from x to x'.

diagrams the arrow on the fermion line points *from* the vertex associated with the $\bar{\psi}$-field (x') *to* the vertex associated with the ψ-field (x), i.e. the arrow runs in the same direction as time for electrons, in the opposite direction for positrons.

These ideas are illustrated in Fig. 4.2, which shows two of the leading contributions, in lowest order of perturbation theory, to Compton scattering by electrons. Fig. 4.2(a) represents an electron propagating in the direction of the arrow on the fermion line, emitting the final photon at the vertex x' and absorbing the initial photon at x. Fig. 4.2(b) represents the corresponding

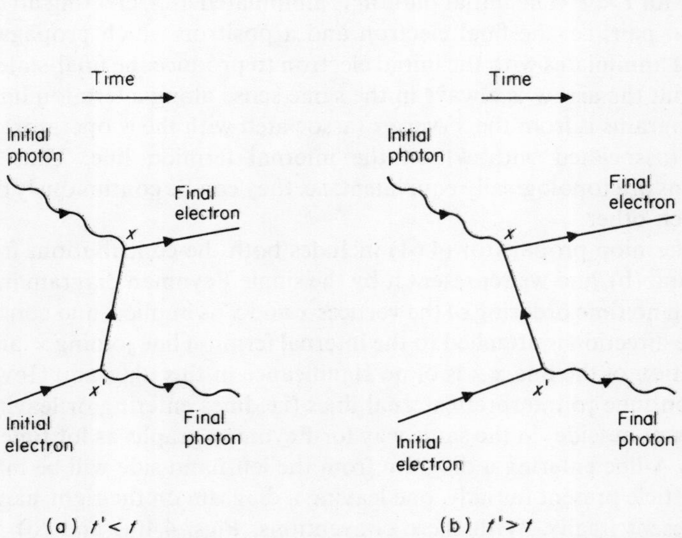

Fig. 4.2. Contributions to Compton scattering: time-ordered graphs.

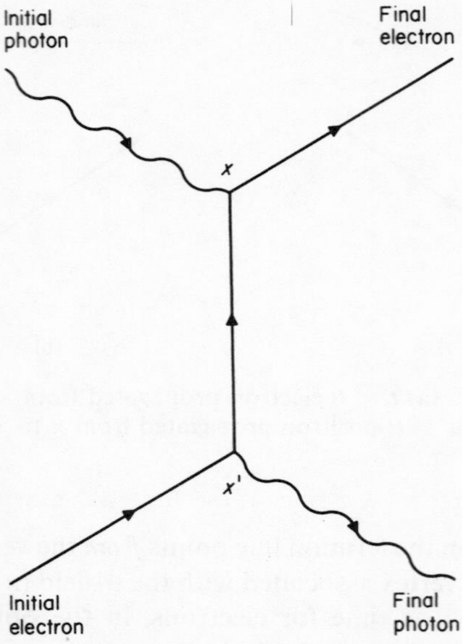

Fig. 4.3. Contribution to Compton scattering: Feynman graph corresponding to the time-ordered graphs of Fig. 4.2.

process for $t < t'$. The initial photon is annihilated at x, creating an electron–positron pair, i.e. the final electron and a positron which propagates to x' where it annihilates with the initial electron to produce the final-state photon. Note that the arrow is always in the same sense along a fermion line, and in *both* diagrams is from the x'-vertex (associated with the $\bar{\psi}$ operator) to the x-vertex (associated with ψ) on the internal fermion line. Thus the two diagrams are topologically equivalent, i.e. they can be continuously deformed into each other.

The fermion propagator (4.61) includes both the contributions from Figs. 4.2(a) and (b), and we represent it by the single Feynman diagram in Fig. 4.3, in which no time ordering of the vertices x and x' is implied, and consequently no time direction is attached to the internal fermion line joining x' and x. The orientation of the line $x'x$ is of no significance in this diagram. However, we shall continue to interpret external lines (i.e. lines entering or leaving a diagram from outside) in the same way for Feynman graphs as for time-ordered graphs. A line entering a diagram from the left-hand side will be interpreted as a particle present initially, one leaving a diagram on the right-hand side as one present finally. With these conventions, Figs. 4.4(a) and (b) represent Compton scattering by electrons and positrons, without further labelling

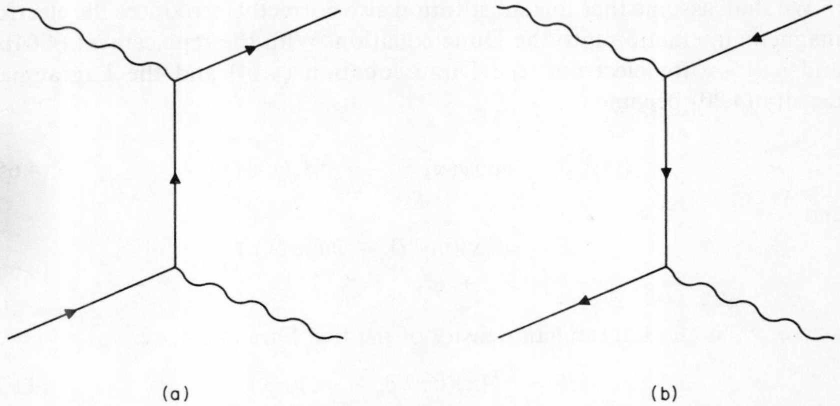

Fig. 4.4. Compton scattering: (a) by electrons; (b) by positrons.

being necessary. The arrows on the fermion lines are required to distinguish electrons (arrows on external lines from left to right) from positrons (arrows on external lines from right to left). Although no arrows are required on the photon lines, we shall at times use arrows on external photon lines to emphasize initially and finally present quanta. In the following we shall frequently use such Feynman diagrams.

4.5 THE ELECTROMAGNETIC INTERACTION AND GAUGE INVARIANCE

We shall now consider the interaction of relativistic electrons with an electromagnetic field, specified by the scalar and vector potentials $\phi(x)$ and $\mathbf{A}(x)$. For this purpose we shall take over the procedure which is successful in non-relativistic quantum mechanics. In the latter case, making the substitution

$$i\hbar \frac{\partial}{\partial t} \to i\hbar \frac{\partial}{\partial t} - q\phi(x), \qquad -i\hbar\boldsymbol{\nabla} \to -i\hbar\boldsymbol{\nabla} - \frac{q}{c}\mathbf{A}(x), \tag{4.64a}$$

in the free-particle Schrödinger equation leads to the correct wave equation for a particle of charge q in this field. [The corresponding classical result is contained in Eq. (1.59).]

The substitution (4.64a) is usually referred to as the 'minimal substitution'. In terms of the four-vector potential $A^\mu(x) = (\phi, \mathbf{A})$, the minimal substitution takes the explicitly covariant form

$$\partial_\mu \equiv \frac{\partial}{\partial x^\mu} \to D_\mu = \left[\partial_\mu + \frac{iq}{\hbar c} A_\mu(x)\right]. \tag{4.64b}$$

We shall assume that this substitution also correctly introduces the electromagnetic interaction into the Dirac equation. With the replacement (4.64b), and $q = -e$ for electrons, the Dirac equation (4.14) and the Lagrangian density (4.20) become

$$(i\hbar\gamma^\mu \partial_\mu - mc)\psi(x) = -\frac{e}{c}\gamma^\mu A_\mu(x)\psi(x), \tag{4.65}$$

and

$$\left.\begin{aligned}\mathscr{L} &= c\bar{\psi}(x)(i\hbar\gamma^\mu D_\mu - mc)\psi(x)\\ &= \mathscr{L}_0 + \mathscr{L}_\mathrm{I}\end{aligned}\right\} \tag{4.66}$$

where $\mathscr{L}_0$ is the Lagrangian density of the free Dirac field, i.e.

$$\mathscr{L}_0 = c\bar{\psi}(x)(i\hbar\gamma^\mu \partial_\mu - mc)\psi(x), \tag{4.67}$$

and $\mathscr{L}_\mathrm{I}$ is the interaction Lagrangian density

$$\mathscr{L}_\mathrm{I} = e\bar{\psi}(x)\gamma^\mu\psi(x)A_\mu(x), \tag{4.68}$$

which couples the conserved current $s^\mu(x) = c(-e)\bar{\psi}\gamma^\mu\psi$, Eq. (4.28), to the electromagnetic field.

To obtain the complete Lagrangian density for electrodynamics, we must add to Eq. (4.66) the Lagrangian density $\mathscr{L}_\mathrm{rad}$ of the radiation field, i.e. of the electromagnetic field in the absence of charges. This division is analogous to that of the Hamiltonian in Chapter 1, Eqs. (1.61)–(1.63). $\mathscr{L}_\mathrm{rad}$ depends on the potential $A_\mu(x)$ only, and we shall study it in the next chapter.

We know that it is only the electromagnetic fields **E** and **B** which have physical significance, not the potential A_μ itself, i.e. the theory must be invariant under the gauge transformations of the potentials, Eqs. (1.3). The latter can be written in the covariant form

$$A_\mu(x) \to A'_\mu(x) = A_\mu(x) + \partial_\mu f(x) \tag{4.69a}$$

where $f(x)$ is an arbitrary function. The invariance of the theory under gauge transformations follows from that of the Lagrangian density. $\mathscr{L}_\mathrm{rad}$ has this invariance property as we shall see in the next chapter. However, applying (4.69a) to Eq. (4.66), we obtain

$$\mathscr{L} \to \mathscr{L}' = \mathscr{L} + e\bar{\psi}(x)\gamma^\mu\psi(x)\,\partial_\mu f(x) \tag{4.70}$$

i.e. $\mathscr{L}$ is not gauge-invariant with respect to (4.69a). We can restore gauge invariance by demanding that coupled with the gauge transformation (4.69a) of the electromagnetic potentials, the Dirac fields transform according to

$$\left.\begin{aligned}\psi(x) &\to \psi'(x) = \psi(x)\,\mathrm{e}^{ief(x)/\hbar c}\\ \bar{\psi}(x) &\to \bar{\psi}'(x) = \bar{\psi}(x)\,\mathrm{e}^{-ief(x)/\hbar c}\end{aligned}\right\}. \tag{4.69b}$$

Under the coupled transformations (4.69a) and (4.69b), the Lagrangian densities (4.67) and (4.68) transform according to

$$\mathscr{L}_0 \to \mathscr{L}'_0 = \mathscr{L}_0 - e\bar{\psi}(x)\gamma^\mu\psi(x)\,\partial_\mu f(x) \tag{4.71a}$$

$$\mathscr{L}_I \to \mathscr{L}'_I = \mathscr{L}_I + e\bar{\psi}(x)\gamma^\mu\psi(x)\,\partial_\mu f(x). \tag{4.71b}$$

Consequently $\mathscr{L} = \mathscr{L}_0 + \mathscr{L}_I$ remains invariant under the coupled transformations.

Eqs. (4.69b) are called a *local* phase transformation since the phase factors depend on x. In the special case that $f(x) = const.$, Eqs. (4.69b) reduce to a *global* phase transformation, considered in Section 2.4 where we saw that invariance under a global phase transformation leads to a conserved charge. We have now seen that gauge invariance of the theory requires invariance when simultaneously transforming the electromagnetic potentials according to the gauge transformation (4.69a) and the Dirac fields according to the local phase transformation (4.69b), and we shall in future refer to these coupled transformations as gauge transformations. It can be shown that the application of Noether's theorem to the invariance with respect to these (coupled) gauge transformations does not lead to a new conservation law, but only reproduces the conservation of charge.

In what follows we shall assume that Eq. (4.68) gives the correct interaction of quantum electrodynamics. One could try and add other gauge-invariant and Lorentz-invariant local interaction terms, but these are generally excluded by another general restriction—renormalizability of the theory—which we shall study in Chapter 9. The ultimate justification for taking Eq. (4.68) as the correct interaction lies, of course, in the complete agreement between some of physics' most precise experiments and theoretical predictions based on this interaction.

PROBLEMS

4.1 From Eq. (4.53b), or otherwise, derive the equal-time anticommutation relation

$$[\psi(x), \bar{\psi}(y)]_+|_{x_0=y_0} = \gamma^0\,\delta(\mathbf{x} - \mathbf{y}).$$

4.2 Show that the functions $S(x)$ and $S_F(x)$ are solutions of the homogeneous Dirac equation and of an inhomogeneous Dirac equation respectively.

4.3 Show that the charge–current density operator

$$s^\mu(x) = -ec\bar{\psi}(x)\gamma^\mu\psi(x)$$

of the Dirac equation satisfies the relation

$$[s^\mu(x), s^\nu(y)] = 0, \quad \text{for } (x - y)^2 < 0.$$

This relation shows that the charge–current densities, which are observable quantities, at two different space–time points x and y, are compatible provided the interval $(x - y)$ is space-like, as required by microcausality.

4.4 Show that if in the expansion (3.8) of the real Klein–Gordon field $\phi(x)$ we impose the anticommutation relations

$$[a(\mathbf{k}), a^\dagger(\mathbf{k}')]_+ = \delta_{\mathbf{kk}'}, \qquad [a(\mathbf{k}), a(\mathbf{k}')]_+ = [a^\dagger(\mathbf{k}), a^\dagger(\mathbf{k}')]_+ = 0,$$

then for $(x - y)$ a space-like interval:

$$[\phi(x), \phi(y)] \neq 0 \quad \text{and} \quad [\phi(x), \phi(y)]_+ \neq 0.$$

[We know from the discussion at the end of Section 4.3 that either the commutator or the anticommutator of $\phi(x)$ and $\phi(y)$ must vanish for $(x - y)$ space-like, if the bilinear observables constructed from ϕ are to satisfy the microcausality condition (4.57).]

4.5 For a Dirac field, the transformations

$$\psi(x) \to \psi'(x) = \exp(i\alpha\gamma_5)\psi(x), \qquad \psi^\dagger(x) \to \psi^{\dagger\prime}(x) = \psi^\dagger(x)\exp(-i\alpha\gamma_5),$$

where α is an arbitrary real parameter, are called chiral phase transformations.

Show that the Lagrangian density (4.20) is invariant under chiral phase transformations in the zero-mass limit $m = 0$ only, and that the corresponding conserved current in this limit is the axial vector current $J^\alpha_A(x) \equiv \bar{\psi}(x)\gamma^\alpha\gamma_5\psi(x)$.

Deduce the equations of motion for the fields

$$\psi_L(x) \equiv \tfrac{1}{2}(1 - \gamma_5)\psi(x), \qquad \psi_R(x) \equiv \tfrac{1}{2}(1 + \gamma_5)\psi(x)$$

for non-vanishing mass, and show that they decouple in the limit $m = 0$. Hence show that the Lagrangian density

$$\mathscr{L}(x) = i\hbar c\bar{\psi}_L(x)\gamma^\mu\,\partial_\mu\psi_L(x)$$

describes zero-mass fermions with negative helicity only, and zero-mass antifermions with positive helicity only. (This field is called the Weyl field and can be used to describe the neutrinos in weak interactions if they have zero mass.)

CHAPTER 5

Photons: covariant theory

In our discussion of the electromagnetic field in Chapter 1, we saw that only the transverse radiation field corresponds to independent dynamical degrees of freedom, and we only quantized this transverse field. On the other hand, the instantaneous Coulomb interaction between charges is fully determined by the charge distribution and, in the formulation of Chapter 1, is treated as a classical potential. This formulation of quantum electrodynamics is closely related to the classical theory and so facilitates interpretation in familiar terms. However, the decomposition of the fields into transverse and longitudinal components is clearly frame-dependent and so hides the Lorentz-invariance of the theory.

An explicitly Lorentz-covariant formulation of the theory is essential for a complete development of quantum electrodynamics. This is required to establish the renormalizability of the theory, i.e. the possibility of carrying out calculations to all orders of perturbation theory with finite self-consistent results, and it is very helpful in practice in calculating such higher-order radiative corrections.

We shall therefore in this chapter develop a covariant theory starting, in Section 5.1, from an explicitly covariant formulation of classical electrodynamics in which all four components of the four-vector potential $A^{\mu}(x) = (\phi, \mathbf{A})$ are treated on an equal footing. This corresponds to introducing more dynamical degrees of freedom than the system possesses and these will later have to be removed by imposing suitable constraints.

The quantized theory, derived in Section 5.2 by quantizing all four components of the four-vector potential $A^{\mu}(x)$, looks on the face of it very different from the theory of Chapter 1. However, the two formulations are equivalent, as we shall illustrate when discussing the photon propagator in Section 5.3.

5.1 THE CLASSICAL FIELDS

To express Maxwell's equations in covariant form, we introduce the antisymmetric field tensor

$$F^{\mu\nu}(x) = \begin{pmatrix} 0 & E_x & E_y & E_z \\ -E_x & 0 & B_z & -B_y \\ -E_y & -B_z & 0 & B_x \\ -E_z & B_y & -B_x & 0 \end{pmatrix} \quad \begin{matrix} \nu \to 0,1,2,3 \\ \mu \downarrow 0,1,2,3 \end{matrix} \tag{5.1}$$

In terms of $F^{\mu\nu}$ and the charge–current density $s^\mu(x) = (c\rho(x), \mathbf{j}(x))$, Maxwell's equations (1.1) become

$$\partial_\nu F^{\mu\nu}(x) = \frac{1}{c} s^\mu(x) \tag{5.2}$$

$$\partial^\lambda F^{\mu\nu}(x) + \partial^\mu F^{\nu\lambda}(x) + \partial^\nu F^{\lambda\mu}(x) = 0. \tag{5.3}$$

Since $F^{\mu\nu}$ is antisymmetric, Eq. (5.2) at once gives

$$\partial_\mu s^\mu(x) = 0, \tag{5.4}$$

i.e. consistency requires conservation of the current to which the electromagnetic field is coupled.

The field $F^{\mu\nu}$ can be expressed in terms of the four-vector potential $A^\mu(x) = (\phi, \mathbf{A})$ by

$$F^{\mu\nu}(x) = \partial^\nu A^\mu(x) - \partial^\mu A^\nu(x) \tag{5.5}$$

which is identical with Eqs. (1.2). In terms of the potentials, Eqs. (5.3) are satisfied identically, and Eqs. (5.2) become

$$\Box A^\mu(x) - \partial^\mu(\partial_\nu A^\nu(x)) = \frac{1}{c} s^\mu(x). \tag{5.6}$$

These equations are Lorentz-covariant, and they are also invariant under the gauge transformation

$$A^\mu(x) \to A'^\mu(x) = A^\mu(x) + \partial^\mu f(x). \tag{5.7}$$

The field equations (5.6) can be derived from the Lagrangian density

$$\mathscr{L} = -\tfrac{1}{4}F_{\mu\nu}(x)F^{\mu\nu}(x) - \frac{1}{c} s_\mu(x)A^\mu(x) \tag{5.8}$$

by treating the four components $A^{\mu}(x)$ as the independent fields in the variational principle (2.11)–(2.14). The form of this Lagrangian density ensures the correct behaviour of the field equations (5.6) under Lorentz and gauge transformations.‡

Unfortunately, the Lagrangian density (5.8) is not suitable for carrying out the canonical quantization. Eq. (5.8) leads to the conjugate fields

$$\pi^{\mu}(x) = \frac{\partial \mathscr{L}}{\partial \dot{A}_{\mu}} = -\frac{1}{c} F^{\mu 0}(x).$$

The antisymmetry of $F^{\mu\nu}$ then implies $\pi^{0}(x) \equiv 0$, and this is plainly incompatible with the canonical commutation relations (2.31) which we wish to impose.

A Lagrangian density which is suitable for quantization, first proposed by Fermi, is

$$\mathscr{L} = -\tfrac{1}{2}(\partial_{\nu} A_{\mu}(x))(\partial^{\nu} A^{\mu}(x)) - \frac{1}{c} s_{\mu}(x) A^{\mu}(x). \tag{5.10}$$

From Eq. (5.10) one obtains the conjugate fields

$$\pi^{\mu}(x) = \frac{\partial \mathscr{L}}{\partial \dot{A}_{\mu}} = -\frac{1}{c^{2}} \dot{A}^{\mu}(x) \tag{5.11}$$

which are now all non-vanishing so that the canonical quantization formalism can be applied.

The Lagrangian density (5.10) leads to the field equations

$$\Box A^{\mu}(x) = \frac{1}{c} s^{\mu}(x). \tag{5.12}$$

Comparison with Eqs. (5.6) shows that Eqs. (5.12) are only equivalent to Maxwell's equations if the potential $A^{\mu}(x)$ satisfies the constraint

$$\partial_{\mu} A^{\mu}(x) = 0. \tag{5.13}$$

Hence to carry out the quantization but end up with Maxwell's equations, we must in the first place quantize the theory for the general Lagrangian density (5.10), ignoring the constraint (5.13), and after quantization impose Eq. (5.13) or an equivalent constraint as a subsidiary condition. We shall consider this point in detail in the next section.

‡ $\mathscr{L}$ is clearly Lorentz-invariant. Under the gauge transformation (5.7)

$$\mathscr{L} \to \mathscr{L} - \frac{1}{c} s_{\mu}(x)\, \partial^{\mu} f(x) = \mathscr{L} - \frac{1}{c} \partial^{\mu}[s_{\mu}(x) f(x)], \tag{5.9}$$

on account of current conservation. Although $\mathscr{L}$ is not invariant, it follows from Eqs. (2.11)–(2.14) that adding a four-divergence to the Lagrangian density does not alter the field equations, i.e. their gauge invariance is ensured by the form of $\mathscr{L}$. (See Problem 2.1.)

In the classical theory, starting from potentials $A^\mu(x)$ in an arbitrary gauge, we can always perform a gauge transformation (5.7) so that the transformed potentials $A'^\mu(x)$ satisfy the subsidiary condition (5.13). We achieve this by choosing the function $f(x)$ in Eq. (5.7) as a solution of

$$\partial_\mu A^\mu(x) + \Box f(x) = 0. \tag{5.14}$$

The subsidiary condition (5.13) does not specify the potentials uniquely. If the potentials $A^\mu(x)$ satisfy Eq. (5.13), so will any potentials $A'^\mu(x)$ obtained by the gauge transformation (5.7), provided the gauge function $f(x)$ satisfies

$$\Box f(x) = 0. \tag{5.15}$$

The subsidiary condition (5.13) is called the Lorentz condition. Its imposition represents a restriction on the choice of gauge. Any gauge in which Eq. (5.13) holds is called a Lorentz guage.

Using a Lorentz gauge has some important advantages. Firstly, the Lorentz condition (5.13) is a Lorentz-covariant constraint. This is in contrast to the condition for the Coulomb gauge, Eq. (1.6),

$$\boldsymbol{\nabla}\cdot\mathbf{A} = 0,$$

which decomposes fields into transverse and longitudinal components and so is manifestly frame-dependent. Secondly, the field equations (5.12) in a Lorentz gauge are much simpler than the corresponding Eqs. (5.6) in a general gauge. In particular, in the free field case ($s^\mu(x) = 0$) Eqs. (5.12) reduce to

$$\Box A^\mu(x) = 0. \tag{5.16}$$

Eq. (5.16) is the limit of the Klein–Gordon equation (3.3) for particles with mass zero. This will enable us to adapt many of our earlier results when considering the covariant quantization of the electromagnetic field.

Eq. (5.16) enables us to expand the free electromagnetic field $A^\mu(x)$ in a complete set of solutions of the wave equation, in close analogy to the expansion (3.8) for the Klein–Gordon field:

$$A^\mu(x) = A^{\mu+}(x) + A^{\mu-}(x) \tag{5.16a}$$

where

$$A^{\mu+}(x) = \sum_{r\mathbf{k}} \left(\frac{\hbar c^2}{2V\omega_{\mathbf{k}}}\right)^{1/2} \varepsilon_r^\mu(\mathbf{k}) a_r(\mathbf{k})\, \mathrm{e}^{-ikx} \tag{5.16b}$$

and

$$A^{\mu-}(x) = \sum_{r\mathbf{k}} \left(\frac{\hbar c^2}{2V\omega_{\mathbf{k}}}\right)^{1/2} \varepsilon_r^\mu(\mathbf{k}) a_r^\dagger(\mathbf{k})\, \mathrm{e}^{ikx}. \tag{5.16c}$$

The summations in these equations are over wave vectors $\mathbf{k}$ allowed by the periodic boundary conditions, and

$$k^0 = \frac{1}{c}\,\omega_{\mathbf{k}} = |\mathbf{k}|. \tag{5.17}$$

The summation over r, from $r = 0$ to $r = 3$, corresponds to the fact that for the four-vector field $A^\mu(x)$ there exist, for each $\mathbf{k}$, four linearly independent polarization states. These are described by the polarization vectors $\varepsilon_r^\mu(\mathbf{k})$, $r = 0, \ldots, 3$, which we choose to be real, and which satisfy the orthonormality and completeness relations

$$\varepsilon_r(\mathbf{k})\varepsilon_s(\mathbf{k}) = \varepsilon_{r\mu}(\mathbf{k})\varepsilon_s^\mu(\mathbf{k}) = -\zeta_r\,\delta_{rs}, \qquad r, s = 0, \ldots, 3, \tag{5.18}$$

$$\sum_r \zeta_r \varepsilon_r^\mu(\mathbf{k})\varepsilon_r^\nu(\mathbf{k}) = -g^{\mu\nu}, \tag{5.19}$$

where

$$\zeta_0 = -1, \qquad \zeta_1 = \zeta_2 = \zeta_3 = 1. \tag{5.20}$$

The classical potentials $A^\mu(x)$, $\mu = 0, \ldots, 3$, are of course real quantities. Anticipating their interpretation in the quantized theory as operators, we have denoted the expansion coefficients in Eqs. (5.16) by a_r and $a_r^\dagger$.

Eqs. (5.16) should be compared with Eqs. (1.38). The latter expand the radiation field in terms of two transverse polarization states for each value of $\mathbf{k}$ and, in addition, we had the instantaneous Coulomb interaction between charges. Eqs. (5.16) give an expansion of the *total* field $A^\mu(x)$ in terms of four polarization states for each value of $\mathbf{k}$. We shall see in Section 5.3 that the two extra polarization states provide a covariant description of the instantaneous Coulomb interaction.

For many purposes one only requires the properties (5.18) and (5.19) of the polarization vectors. However a specific choice of polarization vectors in one given frame of reference often facilitates the interpretation. We shall choose these vectors as

$$\varepsilon_0^\mu(\mathbf{k}) = n^\mu \equiv (1, 0, 0, 0), \tag{5.21a}$$

$$\varepsilon_r^\mu(\mathbf{k}) = (0, \boldsymbol{\varepsilon}_r(\mathbf{k})), \qquad r = 1, 2, 3, \tag{5.21b}$$

where $\boldsymbol{\varepsilon}_1(\mathbf{k})$ and $\boldsymbol{\varepsilon}_2(\mathbf{k})$ are mutually orthogonal unit vectors which are also orthogonal to $\mathbf{k}$, and

$$\boldsymbol{\varepsilon}_3(\mathbf{k}) = \mathbf{k}/|\mathbf{k}|, \tag{5.22a}$$

i.e.

$$\mathbf{k}\cdot\boldsymbol{\varepsilon}_r(\mathbf{k}) = 0, \quad r = 1, 2; \qquad \boldsymbol{\varepsilon}_r(\mathbf{k})\cdot\boldsymbol{\varepsilon}_s(\mathbf{k}) = \delta_{rs}, \quad r, s = 1, 2, 3. \tag{5.22b}$$

ε_1^μ and ε_2^μ are called transverse, ε_3^μ longitudinal polarizations, and ε_0^μ scalar or time-like polarization.

For later use we note that $\varepsilon_3^\mu(\mathbf{k})$ can be written in the covariant form

$$\varepsilon_3^\mu(\mathbf{k}) = \frac{k^\mu - (kn)n^\mu}{[(kn)^2 - k^2]^{1/2}}. \tag{5.22c}$$

This expression comes about since $(kn)n^{\mu}$ subtracts off the time-like component of k^{μ}, and the denominator makes ε_3^{μ} a space-like unit vector. We have not set $k^2 = 0$ in Eq. (5.22c), as it would be for a real photon, since we shall later require the more general case $k^2 \neq 0$.

Real polarization vectors correspond to linear polarization. To describe circular or elliptic polarization would require complex polarization vectors and corresponding modifications of Eqs. (5.18) and (5.19).

5.2 COVARIANT QUANTIZATION

We now apply the canonical formalism of Chapter 2 to quantize the free electromagnetic field, using the Lagrangian density (5.10) with $s_{\mu}(x) = 0$ and, in the first place, ignoring the Lorentz condition (5.13). With the fields $\pi^{\mu}(x)$ conjugate to $A_{\mu}(x)$ given by Eqs. (5.11), the equal-time commutation relations (2.31) become

$$\left.\begin{aligned} [A^{\mu}(\mathbf{x}, t), A^{\nu}(\mathbf{x}', t)] = 0, \qquad [\dot{A}^{\mu}(\mathbf{x}, t), \dot{A}^{\nu}(\mathbf{x}', t)] = 0, \\ [A^{\mu}(\mathbf{x}, t), \dot{A}^{\nu}(\mathbf{x}', t)] = -i\hbar c^2 g^{\mu\nu}\, \delta(\mathbf{x} - \mathbf{x}') \end{aligned}\right\}. \tag{5.23}$$

Apart from the factor $(-g^{\mu\nu})$, these equations are identical with the commutation relations (3.7) of four independent Klein–Gordon fields, and each component $A^{\mu}(x)$ satisfies the wave equation (5.16) which is the limit of the Klein–Gordon equation (3.3) for particles of mass zero. [Both these points can be appreciated by comparing the Lagrangian densities (5.10) and (3.5).] This similarity enables us to take over earlier mathematical results although their physical interpretation will have to be re-examined taking into account the factor $(-g^{\mu\nu})$.

In Section 3.3, we derived the covariant commutation relations (3.43) for the Klein–Gordon field. From these we can at once write down the covariant commutation relations for the $A^{\mu}(x)$:

$$[A^{\mu}(x), A^{\nu}(x')] = i\hbar c D^{\mu\nu}(x - x'), \tag{5.24}$$

where

$$D^{\mu\nu}(x) = \lim_{m \to 0} [-g^{\mu\nu}\, \Delta(x)], \tag{5.25}$$

and $\Delta(x)$ is the invariant Δ-function (3.44).

The Feynman photon propagator is similarly given by

$$\langle 0|\mathrm{T}\{A^{\mu}(x)A^{\nu}(x')\}|0\rangle = i\hbar c D_{\mathrm{F}}^{\mu\nu}(x - x'), \tag{5.26}$$

where

$$D_{\mathrm{F}}^{\mu\nu}(x) = \lim_{m \to 0} [-g^{\mu\nu}\, \Delta_{\mathrm{F}}(x)] = \frac{-g^{\mu\nu}}{(2\pi)^4} \int \frac{\mathrm{d}^4 k\, \mathrm{e}^{-ikx}}{k^2 + i\varepsilon}, \tag{5.27}$$

as is seen from Eqs. (3.56) and (3.59). The photon propagator will be discussed fully in the next section.

To gain the photon interpretation of the quantized fields, we substitute the field expansions (5.16) in the commutation relations (5.23), with the result

$$\left.\begin{aligned}[][a_r(\mathbf{k}), a_s^\dagger(\mathbf{k}')] &= \zeta_r\,\delta_{rs}\,\delta_{\mathbf{k}\mathbf{k}'}\\ [a_r(\mathbf{k}), a_s(\mathbf{k}')] = [a_r^\dagger(\mathbf{k}), a_s^\dagger(\mathbf{k}')] &= 0\end{aligned}\right\}. \tag{5.28}$$

From Eq. (5.20) $\zeta_r = 1$ for $r = 1, 2, 3$, so that for these values of r Eqs. (5.28) are the standard boson commutation relations (3.10) leading to the usual number representation for transverse photons ($r = 1, 2$) and longitudinal photons ($r = 3$). For $r = 0$ (scalar photons) $\zeta_0 = -1$, and it consequently looks as though the usual roles of absorption and creation operators must be interchanged for $a_0(\mathbf{k})$ and $a_0^\dagger(\mathbf{k})$. However, effecting only this change results in other difficulties, and the standard formalism must be modified more radically. Of the several procedures available, we shall follow that due to Gupta and to Bleuler.

In the Gupta–Bleuler theory, the operators $a_r(\mathbf{k})$, $r = 1, 2, 3$ *and* 0, are interpreted as absorption operators, $a_r^\dagger(\mathbf{k})$, $r = 1, 2, 3$ *and* 0, as creation operators for transverse, longitudinal *and* scalar photons. The vacuum state $|0\rangle$ is defined as the state in which there are no photons of any kind present, i.e.

$$a_r(\mathbf{k})|0\rangle = 0, \quad \text{all } \mathbf{k}, \qquad r = 0, \ldots, 3, \tag{5.29a}$$

or, equivalently,

$$A^{\mu+}(x)|0\rangle = 0, \quad \text{all } x, \qquad \mu = 0, \ldots, 3. \tag{5.29b}$$

The operators $a_r^\dagger(\mathbf{k})$ operating on the vacuum state $|0\rangle$ create the one-photon states

$$|1_{\mathbf{k}r}\rangle = a_r^\dagger(\mathbf{k})|0\rangle \tag{5.30}$$

in which one transverse ($r = 1, 2$), longitudinal ($r = 3$) or scalar ($r = 0$) photon of momentum $\mathbf{k}$ is present.

To justify this interpretation of the operators a_r and $a_r^\dagger$, we consider the Hamiltonian operator of the field. From Eq. (2.51a) this is given by

$$H = \int \mathrm{d}^3\mathbf{x}\,\mathrm{N}[\pi^\mu(x)\dot{A}_\mu(x) - \mathscr{L}(x)], \tag{5.31}$$

which, as usual, is to be taken as a normal product. On substituting the free Lagrangian density corresponding to Eq. (5.10), Eq. (5.11) for $\pi^\mu(x)$ and the expansions (5.16) for the fields, Eq. (5.31) becomes

$$H = \sum_{r\mathbf{k}} \hbar\omega_{\mathbf{k}}\zeta_r a_r^\dagger(\mathbf{k})a_r(\mathbf{k}). \tag{5.32}$$

Despite the minus sign ($\zeta_0 = -1$) associated with the scalar photons in Eq. (5.32), this energy is positive definite. For example, for the one-photon states (5.30) one easily obtains, using the commutation relations (5.28),

$$H|1_{\mathbf{k}r}\rangle = \sum_{\mathbf{q}s} \hbar\omega_{\mathbf{q}}\zeta_s a_s^\dagger(\mathbf{q})a_s(\mathbf{q})a_r^\dagger(\mathbf{k})|0\rangle$$
$$= \hbar\omega_{\mathbf{k}} a_r^\dagger(\mathbf{k})|0\rangle, \qquad r = 0, \ldots, 3,$$

i.e. the energy has the positive value $\hbar\omega_{\mathbf{k}}$ for transverse, longitudinal and scalar photons. Correspondingly, we must define the number operators by

$$N_r(\mathbf{k}) = \zeta_r a_r^\dagger(\mathbf{k})a_r(\mathbf{k}), \tag{5.33}$$

and these definitions, together with the commutation relations (5.28), lead to consistent number representations for all types of photons.

Although the formalism, as far as we have developed it, seems satisfactory, there are some difficulties which show up if we calculate the normalization of photon states. For example, the norm of the state (5.30) is

$$\langle 1_{\mathbf{k}r}|1_{\mathbf{k}r}\rangle = \langle 0|a_r(\mathbf{k})a_r^\dagger(\mathbf{k})|0\rangle = \zeta_r\langle 0|0\rangle = \zeta_r$$

(if we normalize $|0\rangle$ to $\langle 0|0\rangle = 1$), and for a scalar photon this norm is negative. More generally, one can show that for any state containing an odd number of scalar photons the norm is negative. At first sight this looks like a serious difficulty, since the probability interpretation of quantum mechanics depends on states having positive norms. However, no scalar or longitudinal photons have ever been observed. Both these points are related to the fact that so far we have ignored the Lorentz condition (5.13), so that our theory is not yet equivalent to Maxwell's equations. We must now try and impose the Lorentz condition.

Unfortunately, we cannot simply take the Lorentz condition (5.13) as an operator identity. Eq. (5.13) is incompatible with the commutation relations (5.24), since

$$[\partial_\mu A^\mu(x), A^\nu(x')] = i\hbar c\partial_\mu D^{\mu\nu}(x - x')$$

and this is not identically zero.

This problem was resolved by Gupta and Bleuler by replacing the Lorentz condition (5.13) by the weaker condition

$$\partial_\mu A^{\mu+}(x)|\Psi\rangle = 0, \tag{5.34}$$

involving absorption operators only. Eq. (5.34) is a restriction on the states which are allowed by the theory. From Eq. (5.34) and its adjoint

$$\langle\Psi|\partial_\mu A^{\mu-}(x) = 0$$

it follows that the Lorentz condition holds for expectation values:

$$\langle\Psi|\partial_\mu A^\mu(x)|\Psi\rangle = \langle\Psi|\partial_\mu A^{\mu+}(x) + \partial_\mu A^{\mu-}(x)|\Psi\rangle = 0. \tag{5.35}$$

This ensures that the Lorentz condition and hence Maxwell's equations hold as the classical limit of this theory.

In order to understand the meaning of the subsidiary condition (5.34), we express it in momentum space. On substituting Eqs. (5.16b) and (5.21), (5.22) for $A_\mu^+(x)$ and $\varepsilon_r^\mu(\mathbf{k})$, we obtain the conditions

$$[a_3(\mathbf{k}) - a_0(\mathbf{k})]|\Psi\rangle = 0, \quad \text{all } \mathbf{k}. \tag{5.36}$$

This is a constraint on the linear combinations of longitudinal and scalar photons, for each value of **k**, that may be present in a state. It places no restriction on the transverse photons that may be present.

The effect of the subsidiary condition (5.36) becomes apparent if we calculate the expectation value of the energy of an allowed state $|\Psi\rangle$. Since from Eq. (5.36) and its adjoint we have

$$\langle\Psi|a_3^\dagger(\mathbf{k})a_3(\mathbf{k}) - a_0^\dagger(\mathbf{k})a_0(\mathbf{k})|\Psi\rangle = \langle\Psi|a_3^\dagger(\mathbf{k})[a_3(\mathbf{k}) - a_0(\mathbf{k})]|\Psi\rangle = 0,$$

it follows from Eq. (5.32) that

$$\langle\Psi|H|\Psi\rangle = \langle\Psi|\sum_{\mathbf{k}}\sum_{r=1}^{2} \hbar\omega_{\mathbf{k}} a_r^\dagger(\mathbf{k})a_r(\mathbf{k})|\Psi\rangle, \tag{5.37}$$

i.e. only the transverse photons contribute to the expectation value of the energy as a consequence of the subsidiary condition. The same is true for all other observables.

Thus, as a result of the subsidiary condition, in free space observable quantities will involve transverse photons only. This explains our earlier assertion that longitudinal and scalar photons are not observed as free particles. Only transverse photons are so observed, corresponding to the two degrees of freedom (for each **k**) of the radiation field, which we found in the non-covariant formalism of Chapter 1 where we worked in the Coulomb gauge. In the covariant treatment, although they don't show up as free particles, the presence of longitudinal and scalar photons is not ruled out altogether. Of the resulting additional two degrees of freedom (for each **k**), one is removed by the subsidiary condition (5.36). The other can be shown to correspond to the arbitrariness in choice of Lorentz gauge. More specifically, one can show that altering the allowed admixtures of longitudinal and scalar photons is equivalent to a gauge transformation between two potentials both of which are in Lorentz gauges. (See Problems 5.2 and 5.3.)

For free fields (i.e. no charges present), it is then simplest to work in a gauge such that the vacuum is represented by the state $|0\rangle$ in which no photons of any kind are present [see Eq. (5.29a)]. But the vacuum could also be described by any state containing no transverse and only allowed admixtures of scalar and longitudinal photons. This description would merely correspond to a different choice of Lorentz gauge. The situation is entirely analogous for states containing transverse photons.

For the electromagnetic field in the presence of charges, the situation is more complicated. We can no longer ignore the longitudinal and scalar photons. When discussing the photon propagator in the next section, we shall see that longitudinal and scalar photons play an important role as virtual particles in intermediate states and provide a covariant description of the instantaneous Coulomb interaction of Chapter 1. However, in this case too, one need consider only transverse photons in initial and final states of scattering processes. This corresponds to a particular choice of gauge and the fact that one can consider particles initially and finally, when they are far apart, as free. In Section 6.2 we shall return to this idea of switching the interaction between colliding particles on and off adiabatically as they approach and as they move apart.

We have developed the Gupta–Bleuler formalism only to the limited extent to which it is needed in applications. It is possible to develop a more complete systematic formalism in which states with negative norm do not appear as a blemish in Hilbert space but occur in a self-consistent manner in a function space with an indefinite metric. For most purposes, this complete formalism is not required.‡

5.3 THE PHOTON PROPAGATOR

In Section 3.4 we interpreted the Klein–Gordon propagator (3.56) as the exchange of a virtual meson in an intermediate state. We now expect a similar interpretation for the photon propagator (5.26) but, corresponding to the four-vector nature of the field $A^\mu(x)$ and the resulting four independent polarization states, we expect the exchange of four kinds of photons, two corresponding to transverse polarization and one each to longitudinal and scalar polarization. This description differs markedly from that of Chapter 1 where only transverse radiation occurred but no longitudinal or scalar radiation. Instead we had the instantaneous Coulomb interaction between charges. We shall see that these two descriptions are indeed equivalent.

To establish this interpretation in terms of photon exchange we consider the momentum space propagator $D_F^{\mu\nu}(k)$, related to the configuration space propagator $D_F^{\mu\nu}(x)$, Eq. (5.27), by

$$D_F^{\mu\nu}(x) = \frac{1}{(2\pi)^4} \int d^4k D_F^{\mu\nu}(k)\, e^{-ikx}. \qquad (5.38)$$

‡ The interested reader is referred to S. N. Gupta, *Quantum Electrodynamics*, Gordon and Breach, New York, 1977; G. Källén, *Quantum Electrodynamics*, Springer, New York, 1972, and Allen & Unwin, London, 1972; or J. M. Jauch and F. Rohrlich, *The Theory of Photons and Electrons*, 2nd edn, Springer, New York, 1976, Section 6.3.

From Eqs. (5.27) and (5.19) we obtain

$$D_{\mathrm{F}}^{\mu\nu}(k) = \frac{-g^{\mu\nu}}{k^2 + \mathrm{i}\varepsilon} = \frac{1}{k^2 + \mathrm{i}\varepsilon} \sum_r \zeta_r \varepsilon_r^\mu(\mathbf{k})\varepsilon_r^\nu(\mathbf{k}). \tag{5.39}$$

In order to interpret this expression, we use the special frame of reference in which the polarization vectors $\varepsilon_r^\mu(\mathbf{k})$ are given by Eqs. (5.21) and (5.22). The last equation then becomes:

$$D_{\mathrm{F}}^{\mu\nu}(k) = \frac{1}{k^2 + \mathrm{i}\varepsilon} \left\{ \sum_{r=1}^{2} \varepsilon_r^\mu(\mathbf{k})\varepsilon_r^\nu(\mathbf{k}) + \frac{[k^\mu - (kn)n^\mu][k^\nu - (kn)n^\nu]}{(kn)^2 - k^2} + (-1)n^\mu n^\nu \right\}. \tag{5.40}$$

This equation exhibits the contributions to the photon propagator from transverse, longitudinal and scalar photons.

By analogy with the meson case, we interpret the first term in Eq. (5.40),

$${}_{\mathrm{T}}D_{\mathrm{F}}^{\mu\nu}(k) \equiv \frac{1}{k^2 + \mathrm{i}\varepsilon} \sum_{r=1}^{2} \varepsilon_r^\mu(\mathbf{k})\varepsilon_r^\nu(\mathbf{k}), \tag{5.41a}$$

as the exchange of transverse photons. In the language of Chapter 1, it corresponds to the interaction of charges via the transverse radiation field.

The interpretation of the remaining two terms in Eq. (5.40) follows not from considering longitudinal and scalar photons separately, but from combining them into a term proportional to $n^\mu n^\nu$ plus the remainder. Eq. (5.40) then becomes

$$D_{\mathrm{F}}^{\mu\nu}(k) = {}_{\mathrm{T}}D_{\mathrm{F}}^{\mu\nu}(k) + {}_{\mathrm{C}}D_{\mathrm{F}}^{\mu\nu}(k) + {}_{\mathrm{R}}D_{\mathrm{F}}^{\mu\nu}(k), \tag{5.42}$$

where

$${}_{\mathrm{C}}D_{\mathrm{F}}^{\mu\nu}(k) \equiv \frac{n^\mu n^\nu}{(kn)^2 - k^2}, \tag{5.41b}$$

$${}_{\mathrm{R}}D_{\mathrm{F}}^{\mu\nu}(k) \equiv \frac{1}{k^2 + \mathrm{i}\varepsilon} \left[\frac{k^\mu k^\nu - (kn)(k^\mu n^\nu + k^\nu n^\mu)}{(kn)^2 - k^2} \right], \tag{5.41c}$$

and it is these linear combinations, both of which involve longitudinal and scalar photons, which allow a simple interpretation.

We first consider Eq. (5.41b) in configuration space. From Eqs. (5.38) and (5.21a) we obtain

$$\begin{aligned} {}_{\mathrm{C}}D_{\mathrm{F}}^{\mu\nu}(x) &= \frac{g^{\mu 0} g^{\nu 0}}{(2\pi)^4} \int \frac{\mathrm{d}^3\mathbf{k}\, \mathrm{e}^{\mathrm{i}\mathbf{k}\cdot\mathbf{x}}}{|\mathbf{k}|^2} \int \mathrm{d}k^0\, \mathrm{e}^{-\mathrm{i}k^0 x^0} \\ &= g^{\mu 0} g^{\nu 0} \frac{1}{4\pi|\mathbf{x}|} \delta(x^0). \end{aligned} \tag{5.43}$$

This expression has the time dependence $[\delta(x^0)]$ and the space dependence $[1/|\mathbf{x}|]$ characteristic of an instantaneous Coulomb potential. Thus we see that the exchange of longitudinal and scalar photons, represented by the term (5.43), corresponds to the instantaneous Coulomb interaction between charges. In Chapter 1, we quantized the transverse radiation field only and treated the instantaneous Coulomb interaction as a classical potential, corresponding to the fact that the instantaneous Coulomb field does not represent independent dynamical degrees of freedom but is fully determined by the charges. In the present treatment, the longitudinal and scalar field components are also quantized and the instantaneous Coulomb interaction emerges as an exchange of longitudinal and scalar photons.

Finally, we must discuss the remainder term (5.41c). In Chapter 1 the complete electromagnetic interaction between charges was represented in terms of the interactions via the transverse radiation field and the instantaneous Coulomb fields. Both these have been accounted for in the present treatment, and for the two treatments to be equivalent the contribution of the remainder term (5.41c) to all observable quantities must vanish. This is indeed the case, the basic reason being that the electromagnetic field only interacts with the conserved charge–current density $s^\mu(x)$, Eqs. (5.2) and (5.4). We shall illustrate this for a simple example.

We shall see in Section 7.1, Eq. (7.14), that the scattering of charges by each other is, in lowest order of perturbation theory, given by the matrix element of the operator

$$\int \mathrm{d}^4x \int \mathrm{d}^4y s_1^\mu(x) D_{\mathrm{F}\mu\nu}(x-y) s_2^\nu(y). \tag{5.44}$$

Here $s_1^\mu(x)$ and $s_2^\nu(y)$ are the two interacting charge–current densities. It is clear from Eq. (5.43) that the contribution of ${}_{\mathrm{C}}D_{\mathrm{F}\mu\nu}(x-y)$ to (5.44) corresponds to the instantaneous Coulomb interaction between the charge densities $\rho_1(\mathbf{x}, x^0) = s_1^0(\mathbf{x}, x^0)/c$ and $\rho_2(\mathbf{y}, x^0) = s_2^0(\mathbf{y}, x^0)/c$. Similarly, the transverse propagator ${}_{\mathrm{T}}D_{\mathrm{F}\mu\nu}(x-y)$ accounts for the electromagnetic interaction between the current densities $\mathbf{j}_1(x) = \mathbf{s}_1(x)$ and $\mathbf{j}_2(y) = \mathbf{s}_2(y)$.

The contribution to (5.44) of the remainder term ${}_{\mathrm{R}}D_{\mathrm{F}\mu\nu}(x-y)$ is easily shown to vanish, on account of current conservation. Transforming this contribution to expression (5.44) into momentum space, one obtains

$$\frac{1}{(2\pi)^4} \int \mathrm{d}^4k s_1^\mu(-k) {}_{\mathrm{R}}D_{\mathrm{F}\mu\nu}(k) s_2^\nu(k), \tag{5.45}$$

where the momentum transforms $s_r^\mu(k)$, $r = 1, 2$, are defined, analogously to Eq. (5.38), by

$$s_r^\mu(x) = \frac{1}{(2\pi)^4} \int \mathrm{d}^4k s_r^\mu(k) \, \mathrm{e}^{-ikx}, \qquad r = 1, 2. \tag{5.46}$$

The current conservation equations, $\partial_\mu s_r^\mu(x) = 0$, translated into momentum space, become

$$k_\mu s_r^\mu(k) = 0, \qquad r = 1, 2. \tag{5.47}$$

We see from the explicit form (5.41c) that each term in ${}_{\mathrm{R}}D_{\mathrm{F}\mu\nu}(k)$ is proportional to either k_μ or k_ν or both. Hence it follows from Eq. (5.47) that the expression (5.45) vanishes.

This completes our discussion of the equivalence of the two formulations of quantum electrodynamics. In doing this, we employed a special frame of reference leading to a division of the fields into transverse, longitudinal and scalar parts. In general, such a division is not required, and we shall work with manifestly covariant expressions involving summations over all four polarization states. In particular, the photon propagator, Eqs. (5.38) and (5.39), which will be very important in the development of quantum electrodynamics, has this property.

PROBLEMS

5.1 Show that the Lagrangian density obtained from

$$\mathscr{L} = -\tfrac{1}{4}F_{\mu\nu}(x)F^{\mu\nu}(x)$$

by adding the term $-\frac{1}{2}(\partial_\mu A^\mu(x))(\partial_\nu A^\nu(x))$, i.e.

$$\mathscr{L} = -\tfrac{1}{4}F_{\mu\nu}(x)F^{\mu\nu}(x) - \tfrac{1}{2}(\partial_\mu A^\mu(x))(\partial_\nu A^\nu(x)),$$

is equivalent to the Lagrangian density, proposed by Fermi:

$$\mathscr{L} = -\tfrac{1}{2}(\partial_\nu A_\mu(x))(\partial^\nu A^\mu(x)).$$

5.2 From the commutation relations (5.28) show that

$$[a_3(\mathbf{k}) - a_0(\mathbf{k}), a_3^\dagger(\mathbf{k}) - a_0^\dagger(\mathbf{k})] = 0.$$

Show that the most general state representing the physical vacuum, i.e. the state in which there are no transverse photons present but which contains the most general allowed admixture of scalar and longitudinal photons, is given by

$$|\Psi_{\mathrm{SL}}\rangle = \sum_{n_1=0}^{\infty}\sum_{n_2=0}^{\infty}\ldots c(n_1, n_2, \ldots)\prod_{i=1}^{\infty}(\alpha_i^\dagger)^{n_i}|0\rangle$$

where

$$\alpha_i^\dagger \equiv a_3^\dagger(\mathbf{k}_i) - a_0^\dagger(\mathbf{k}_i),$$

$\mathbf{k}_i$ are the allowed wave vectors [see Eq. (1.13)], and $|0\rangle$ is the vacuum state in which there are no photons of any kind present. Show that the norm of this state is given by

$$\langle\Psi_{\mathrm{SL}}|\Psi_{\mathrm{SL}}\rangle = |c(0, 0, \ldots)|^2.$$

What is the most general state in which there are a definite number of transverse photons, with definite momenta and polarization vectors, present?

5.3 $|\Psi_T\rangle$ is a state which contains transverse photons only. Let

$$|\Psi'_T\rangle = \{1 + c[a_3^\dagger(\mathbf{k}) - a_0^\dagger(\mathbf{k})]\}|\Psi_T\rangle,$$

where c is a constant. Show that replacing $|\Psi_T\rangle$ by $|\Psi'_T\rangle$ corresponds to a gauge transformation, i.e.

$$\langle\Psi'_T|A^\mu(x)|\Psi'_T\rangle = \langle\Psi_T|A^\mu(x) + \partial^\mu\Lambda(x)|\Psi_T\rangle,$$

where

$$\Lambda(x) = \left(\frac{2\hbar c^2}{V\omega_\mathbf{k}^3}\right)^{1/2} \mathrm{Re}(ic\,\mathrm{e}^{-ikx}).$$

5.4 By making the minimal substitution

$$\partial_\alpha\phi(x) \to D_\alpha\phi(x) = [\partial_\alpha + \frac{ie}{\hbar c} A_\alpha(x)]\phi(x)$$

$$\partial_\alpha\phi^\dagger(x) \to [D_\alpha\phi(x)]^\dagger = [\partial_\alpha - \frac{ie}{\hbar c} A_\alpha(x)]\phi^\dagger(x)$$

in the Lagrangian density (3.23) of the complex Klein–Gordon field $\phi(x)$, derive the Lagrangian density $\mathscr{L}_I(x)$ for the interaction of the charged bosons, described by the field $\phi(x)$, with the electromagnetic field $A^\alpha(x)$.

Assuming that this interaction is invariant under the charge conjugation transformation $\mathscr{C}$, show that

$$\mathscr{C}A^\alpha(x)\mathscr{C}^{-1} = -A^\alpha(x).$$

[The transformation properties of $\phi(x)$ and of $s^\alpha(x)$, Eq. (3.33), under charge conjugation were discussed in Problem 3.5.]

Hence show that a single-photon state $|\mathbf{k}, r\rangle$ is an eigenstate of $\mathscr{C}$ with eigenvalue -1.

CHAPTER 6

The S-matrix expansion

We shall now progress from the discussion of the free fields to the realistic and much more interesting case of fields in interaction, in which particles can be scattered, created and destroyed. In essence this requires solving the coupled non-linear field equations for given conditions. In quantum electrodynamics, for example, one must solve the inhomogeneous wave equation (5.12) with the Dirac current density (4.28) as source term. This is an extremely difficult problem which has only been solved in perturbation theory, i.e. the Hamiltonian of the system is divided into that of the free fields plus an interaction term. The latter is treated as a perturbation which is justifiable if the interaction is sufficiently weak. For quantum electrodynamics, where the coupling of photons and electrons is measured by the small dimensionless fine structure constant $\alpha \approx 1/137$, this approach is outstandingly successful, not only in calculating processes in lowest order of perturbation theory but also in calculating higher-order corrections.

In the Heisenberg picture, which we have so far been using, this programme is still very complex, and it was decisive for the successful development of the theory to work instead in the interaction picture. In Section 6.2 we shall study the equations of motion of the interacting fields in the interaction picture and we shall obtain a perturbation series solution suitable for collision processes. This solution, known as the S-matrix expansion, is due to Dyson. The Dyson expansion of the S-matrix is of great importance since it contains the complete information about all collision processes in a form suitable for extracting the transition amplitude for a specific process to any order of perturbation theory. A systematic procedure for doing this will be developed in Section 6.3.

Before proceeding with these topics, we shall in Section 6.1 introduce natural units which considerably simplify details of the following calculations.

6.1 NATURAL DIMENSIONS AND UNITS

We have so far used c.g.s. units in which the fundamental dimensions, in terms of which quantities are expressed, are mass (M), length (L) and time (T). In relativistic quantum field theory, expressions and calculations are much simplified if one uses natural units (n.u.). In natural units one takes mass, action (A) and velocity (V) as fundamental dimensions and chooses $\hbar$ as unit of action and the velocity of light c as unit of velocity. Hence $\hbar = c = 1$ in natural units, and c.g.s. expressions are transformed into natural units by putting $\hbar = c = 1$. In such n.u. expressions all quantities have the dimensions of a power of M. Since

$$L = \frac{A}{MV} \quad \text{and} \quad T = \frac{A}{MV^2} \tag{6.1}$$

one has the general result that a quantity which has the c.g.s. dimensions

$$M^p L^q T^r = M^{p-q-r} A^{q+r} V^{-q-2r}, \tag{6.2}$$

has the n.u. dimensions M^{p-q-r}. In natural units, many quantities have the same dimension. For example, the momentum–energy relation for a particle of mass m becomes, in natural units,

$$E^2 = m^2 + \mathbf{p}^2 = m^2 + \mathbf{k}^2, \tag{6.3}$$

so that mass, momentum, energy and wave number all have the same natural dimension M. The c.g.s. expression for the dimensionless fine structure constant

$$\alpha = \frac{e^2}{4\pi\hbar c} = \frac{1}{137.04} \quad \text{(c.g.s.)} \tag{6.4a}$$

becomes

$$\alpha = \frac{e^2}{4\pi} = \frac{1}{137.04} \quad \text{(n.u.)} \tag{6.4b}$$

so that in natural units electric charge is dimensionless (M^0).

From the general relation (6.2) or by using particular equations one easily derives the n.u. dimensions of all quantities, and some of the more important ones are listed in Table 6.1.

Working in natural units it is very easy to obtain numerical results in any system of units. A quantity in natural units will have a dimension M^n. To convert this quantity to whatever c.g.s. units are convenient, one merely multiplies it by such powers of $\hbar$ and c, expressed in the appropriate units, as to

Table 6.1 The c.g.s. dimensions $M^pL^qT^r$ and the n.u. dimensions $M^n = M^{p-q-r}$ of some quantities.

Quantity	c.g.s. p	q	r	n.u. n
Action	1	2	-1	0
Velocity	0	1	-1	0
Mass	1	0	0	1
Length	0	1	0	-1
Time	0	0	1	-1
Lagrangian or Hamiltonian densities	1	-1	-2	4
Fine structure constant α	0	0	0	0
Electric charge	$\frac{1}{2}$	$\frac{3}{2}$	-1	0
Klein–Gordon field $\phi(x)$ *	$\frac{1}{2}$	$\frac{1}{2}$	-1	1
Electromagnetic field $A^{\mu}(x)$ *	$\frac{1}{2}$	$\frac{1}{2}$	-1	1
Dirac fields $\psi(x)$ and $\bar{\psi}(x)$ *	0	$-\frac{3}{2}$	0	$\frac{3}{2}$

* The dimensions of the fields can, for example, be obtained from the Lagrangian densities, Eqs. (3.5), (5.10) and (4.20).

give it the correct c.g.s. dimensions. One frequently interprets M as an energy and measures it in MeV. The conversion factors

$$\hbar = 6.58 \times 10^{-22}\ \text{MeV}\cdot\text{sec} \tag{6.5a}$$

$$\hbar c = 1.973 \times 10^{-11}\ \text{MeV}\cdot\text{cm} \tag{6.5b}$$

then enable one easily to express quantities in terms of MeV, centimetres and seconds. Two examples will illustrate this.

The Thomson cross-section (1.72) becomes, in natural units,

$$\sigma = \frac{8\pi}{3}\frac{\alpha^2}{m^2}. \tag{6.6}$$

With $m = 0.511$ MeV, we convert the right-hand side of this equation to cm^2 by multiplying by ($\hbar c$ in MeV·cm)^2 which, from Eq. (6.5b), gives

$$\sigma = \frac{8\pi}{3}\alpha^2\frac{(1.973 \times 10^{-11}\ \text{MeV}\cdot\text{cm})^2}{(0.511\ \text{MeV})^2} = 6.65 \times 10^{-25}\ \text{cm}^2.$$

Secondly, we quote the n.u. expression for the lifetime τ of the positronium ground state $1\,{}^1S_0$.‡ It is given by

$$\tau = \frac{2}{\alpha^5}\frac{1}{m}, \tag{6.7}$$

‡ See J. M. Jauch and F. Rohrlich, *The Theory of Photons and Electrons*, 2nd edn, Springer, New York, 1976, p. 286, Eq. (12-108).

where m is the mass of the electron. With m in MeV, we must multiply Eq. (6.7) by ($\hbar$ in MeV·sec), Eq. (6.5a), to obtain

$$\tau = \frac{2}{\alpha^5} \frac{(6.58 \times 10^{-22}\ \text{MeV}\cdot\text{sec})}{(0.511\ \text{MeV})} = 1.24 \times 10^{-10}\ \text{sec}.$$

This conversion factor is of course the same for converting any lifetime τ from natural units to seconds, the essential points being that in natural units τ has the dimension M^{-1} and must be expressed in $(\text{MeV})^{-1}$.

These examples illustrate how very easy it is to obtain numerical results in any c.g.s. units from equations expressed in natural units. No advantage is gained by tediously retaining factors of $\hbar$ and c throughout a calculation or by converting a n.u. equation into c.g.s. form by inserting the appropriate factors of $\hbar$ and c prior to substituting numerical values.

Although rarely required, the c.g.s. form of an equation is easily obtained from its n.u. form. In a sum of terms, one must multiply each term by appropriate powers of $\hbar$ and c to make all the terms have the same c.g.s. dimensions. [E.g. a factor $(E + k)$, with E interpreted as an energy and k a wave number, could be turned into $(E + c\hbar k)$ or into $(E/c + \hbar k)$, etc.] To obtain the correct c.g.s. dimensions for the whole expression, it must be multiplied by a factor $\hbar^a c^b$ with the exponents a and b determined from dimensional arguments. Usually they are easily guessed.

From now on we shall in general work in natural units.

6.2 THE S-MATRIX EXPANSION

So far we have mainly considered the free, i.e. non-interacting, fields, using the Heisenberg picture (H.P.) in which state vectors are constant in time and the operators carry the full time dependence.

We now turn to the study of the interacting fields. For example in quantum electrodynamics (QED), the interacting electron–positron and electromagnetic fields are described by the Lagrangian density

$$\mathscr{L} = \mathscr{L}_0 + \mathscr{L}_\text{I} \tag{6.8}$$

with the free-field Lagrangian density

$$\mathscr{L}_0 = \text{N}[\bar{\psi}(x)(\text{i}\gamma^\mu \partial_\mu - m)\psi(x) - \tfrac{1}{2}(\partial_\nu A_\mu(x))(\partial^\nu A^\mu(x))] \tag{6.9}$$

and the interaction Lagrangian density

$$\mathscr{L}_\text{I} = \text{N}[-s^\mu(x)A_\mu(x)] = \text{N}[e\bar{\psi}(x)\not{A}(x)\psi(x)] \tag{6.10}$$

[see Eqs. (4.66)–(4.68) and (5.10)]. In Eqs. (6.9) and (6.10) we have written the free-field and the interaction Lagrangian densities as normal products. This ensures, as for the free-field cases considered earlier, that the vacuum expectation values of all observables, e.g. energy or charge, vanish. Corresponding to the division (6.8), the complete Hamiltonian H of the system is split into the free-field Hamiltonian H_0 and the interaction Hamiltonian H_{I}:

$$H = H_0 + H_I. \tag{6.11}$$

As discussed at the beginning of this chapter, we shall employ the interaction picture (I.P.) which leads to two essential simplifications.[‡]

Firstly, in the I.P. the operators satisfy the Heisenberg-like equations of motion (1.87) but involving the free Hamiltonian H_0 only, not the complete Hamiltonian H.

Secondly, if the interaction Lagrangian density $\mathscr{L}_{\mathrm{I}}$ does not involve derivatives (and we shall restrict ourselves to this case until Chapter 14), the fields canonically conjugate to the interacting fields and to the free fields are identical. (For example in QED $\partial\mathscr{L}/\partial\dot{\psi}_\alpha = \partial\mathscr{L}_0/\partial\dot{\psi}_\alpha$, etc.) Since the I.P. and the H.P. are related by a unitary transformation, it follows that in the I.P. the interacting fields satisfy the same commutation relations as the free fields.

Thus in the I.P. the interacting fields satisfy the same equations of motion and the same commutation relations as the free-field operators. Consequently, we can take over the many results derived for free fields (in Chapters 3–5) as also true for the interacting fields in the I.P. In particular, the complete sets of plane wave states which we obtained continue to be solutions of the equations of motion, resulting in the same plane wave expansions of the field operators as before, the same number representations and the same explicit forms for the Feynman propagators.

In the I.P., the system is described by a time-dependent state vector $|\Phi(t)\rangle$. According to Eqs. (1.88) and (1.89), $|\Phi(t)\rangle$ satisfies the equation of motion

$$\mathrm{i}\frac{\mathrm{d}}{\mathrm{d}t}|\Phi(t)\rangle = H_{\mathrm{I}}(t)|\Phi(t)\rangle, \tag{6.12}$$

where

$$H_{\mathrm{I}}(t) = \mathrm{e}^{\mathrm{i}H_0(t-t_0)}H_{\mathrm{I}}^{\mathrm{S}}\,\mathrm{e}^{-\mathrm{i}H_0(t-t_0)} \tag{6.13}$$

is the interaction Hamiltonian in the I.P., with $H_{\mathrm{I}}^{\mathrm{S}}$ and $H_0 = H_0^{\mathrm{S}}$ being the interaction and free-field Hamiltonians in the Schrödinger picture (S.P.). $H_{\mathrm{I}}(t)$ is obtained by replacing, in $H_{\mathrm{I}}^{\mathrm{S}}$, the S.P. field operators by the time-dependent free-field operators. In Eqs. (6.12) and (6.13) we have omitted the labels I, used in Eqs. (1.88) and (1.89) to distinguish the I.P., as we shall be working exclusively in the I.P. in what follows.

Eq. (6.12) is a Schrödinger-like equation with the time-dependent Hamiltonian $H_{\mathrm{I}}(t)$. With the interaction 'switched off' (i.e. we put $H_{\mathrm{I}} \equiv 0$), the state

[‡] The interaction picture, and its relation to the Heisenberg and Schrödinger pictures, is discussed in the appendix to Chapter 1 (Section 1.5). The reader who is not intimately familiar with this material is advised to study this appendix in depth at this stage.

vector is constant in time. The interaction leads to the state $|\Phi(t)\rangle$ changing with time. Given that the system is in a state $|i\rangle$ at an initial time $t = t_i$, i.e.

$$|\Phi(t_i)\rangle = |i\rangle, \tag{6.14}$$

the solution of Eq. (6.12) with this initial condition gives the state $|\Phi(t)\rangle$ of the system at any other time t. It follows from the Hermiticity of the operator $H_I(t)$ that the time development of the state $|\Phi(t)\rangle$ according to Eq. (6.12) is a unitary transformation. Accordingly it preserves the normalization of states,

$$\langle\Phi(t)|\Phi(t)\rangle = \text{const.}, \tag{6.15}$$

and, more generally, the scalar product.

Clearly the formalism which we are here developing is not appropriate for the description of bound states but it is particularly suitable for scattering processes. In a collision process the state vector $|i\rangle$ will define an initial state, long before the scattering occurs ($t_i = -\infty$), by specifying a definite number of particles, with definite properties and far apart from each other so that they do not interact. (For example in QED $|i\rangle$ would specify a definite number of electrons, positrons and photons with given momenta, spins and polarizations.) In the scattering process, the particles will come close together, collide (i.e. interact) and fly apart again. Eq. (6.12) determines the state $|\Phi(\infty)\rangle$ into which the initial state

$$|\Phi(-\infty)\rangle = |i\rangle, \tag{6.14a}$$

evolves at $t = \infty$, long after the scattering is over and all particles are far apart again. The S-matrix relates $|\Phi(\infty)\rangle$ to $|\Phi(-\infty)\rangle$ and is defined by

$$|\Phi(\infty)\rangle = S|\Phi(-\infty)\rangle = S|i\rangle. \tag{6.16}$$

A collision can lead to many different final states $|f\rangle$, and all these possibilities are contained within $|\Phi(\infty)\rangle$. (For example, an electron–positron collision may result in elastic scattering, Bremsstrahlung (i.e. emission of photons), pair annihilation, etc.) Each of these final states $|f\rangle$ is specified in a way analogous to $|i\rangle$.

The transition probability that after the collision (i.e. at $t = \infty$) the system is in the state $|f\rangle$ is given by

$$|\langle f|\Phi(\infty)\rangle|^2. \tag{6.17}$$

($|\Phi(\infty)\rangle$ and $|i\rangle$ are assumed normed to unity.) The corresponding probability amplitude is

$$\langle f|\Phi(\infty)\rangle = \langle f|S|i\rangle \equiv S_{fi}. \tag{6.18}$$

With the state $|\Phi(\infty)\rangle$ expanded in terms of a complete orthonormal set of states,

$$|\Phi(\infty)\rangle = \sum_f |f\rangle\langle f|\Phi(\infty)\rangle = \sum_f |f\rangle S_{fi}, \tag{6.19}$$

the unitarity of the S-matrix can be written

$$\sum_f |S_{fi}|^2 = 1. \tag{6.20}$$

Eq. (6.20) expresses the conservation of probability. It is more general than the corresponding conservation of particles in non-relativistic quantum mechanics, since now particles can be created or destroyed.

In order to calculate the S-matrix we must solve Eq. (6.12) for the initial condition (6.14a). These equations can be combined into the integral equation

$$|\Phi(t)\rangle = |i\rangle + (-\mathrm{i})\int_{-\infty}^{t} \mathrm{d}t_1 H_{\mathrm{I}}(t_1)|\Phi(t_1)\rangle. \tag{6.21}$$

This equation can only be solved iteratively. The resulting perturbation solution, as a series in powers of H_{I}, will only be useful if the interaction energy H_{I} is small. This is the case for QED where the dimensionless coupling constant characterizing the photon–electron interaction is the fine structure constant $\alpha \approx 1/137$.

Solving Eq. (6.21) by iteration

$$|\Phi(t)\rangle = |i\rangle + (-\mathrm{i})\int_{-\infty}^{t} \mathrm{d}t_1 H_{\mathrm{I}}(t_1)|i\rangle$$
$$+ (-\mathrm{i})^2\int_{-\infty}^{t} \mathrm{d}t_1 \int_{-\infty}^{t_1} \mathrm{d}t_2 H_{\mathrm{I}}(t_1)H_{\mathrm{I}}(t_2)|\Phi(t_2)\rangle,$$

and so on, we obtain, in the limit $t \to \infty$, the S-matrix

$$S = \sum_{n=0}^{\infty} (-\mathrm{i})^n \int_{-\infty}^{\infty} \mathrm{d}t_1 \int_{-\infty}^{t_1} \mathrm{d}t_2 \ldots \int_{-\infty}^{t_{n-1}} \mathrm{d}t_n H_{\mathrm{I}}(t_1)H_{\mathrm{I}}(t_2)\ldots H_{\mathrm{I}}(t_n) \tag{6.22a}$$

$$= \sum_{n=0}^{\infty} \frac{(-\mathrm{i})^n}{n!} \int_{-\infty}^{\infty} \mathrm{d}t_1 \int_{-\infty}^{\infty} \mathrm{d}t_2 \ldots \int_{-\infty}^{\infty} \mathrm{d}t_n \mathrm{T}\{H_{\mathrm{I}}(t_1)H_{\mathrm{I}}(t_2)\ldots H_{\mathrm{I}}(t_n)\}. \tag{6.22b}$$

Here the time-ordered product $\mathrm{T}\{\ldots\}$ of n factors is the natural generalization of the definitions (3.53) and (4.59) for two factors, i.e. the factors are ordered so that later times stand to the left of earlier times, and all boson (fermion) fields are treated as though their commutators (anticommutators) vanish. The equivalence of the two forms (6.22a) and (6.22b) only holds if H_1 contains an even number of fermion factors (as in QED) so that the reordering process introduces no extra factors (-1). The equivalence of the two forms holds separately for each term of the series. Its verification is left as an exercise for the reader. Finally we rewrite Eq. (6.22b) in terms of the interaction Hamiltonian density $\mathscr{H}_{\mathrm{I}}(x)$ to obtain the explicitly covariant result

$$S = \sum_{n=0}^{\infty} \frac{(-\mathrm{i})^n}{n!} \int \ldots \int \mathrm{d}^4x_1\, \mathrm{d}^4x_2 \ldots \mathrm{d}^4x_n\, \mathrm{T}\{\mathscr{H}_{\mathrm{I}}(x_1)\mathscr{H}_{\mathrm{I}}(x_2)\ldots \mathscr{H}_{\mathrm{I}}(x_n)\}, \tag{6.23}$$

the integrations being over all space–time. This equation is the Dyson expansion of the S-matrix. It forms the starting point for the approach to perturbation theory used in this book.

We have seen that the amplitude for a particular transition $|i\rangle \to |f\rangle$ is given by $\langle f|S|i\rangle$. To pick out from the expansion (6.23) the parts which contribute to this matrix element is a complex problem to which we shall return in the next section, but we must first discuss the specification of the initial and final states $|i\rangle$ and $|f\rangle$.

In the above perturbation formalism the states $|i\rangle$ and $|f\rangle$ are, as usual, eigenstates of the unperturbed free-field Hamiltonian H_0, i.e. with the interaction switched off ($H_{\mathrm{I}} = 0$). This description appears wrong since the particles we are dealing with are real physical particles even when far apart. An electron, even when far away from other electrons, is surrounded by its photon cloud; it is a real electron, not a bare electron without its own electromagnetic field. Hence, the use of bare particle states $|i\rangle$ and $|f\rangle$ requires justification. One possible procedure is to appeal to the adiabatic hypothesis in which the interaction $H_{\mathrm{I}}(t)$ is replaced by $H_{\mathrm{I}}(t)f(t)$. The function $f(t)$ is chosen so that $f(t) = 1$ for a sufficiently long interval $-T \leqslant t \leqslant T$, and $f(t) \to 0$ monotonically as $t \to \pm\infty$. [In QED, for example, we could replace the elementary charge e by the time-dependent coupling constant $ef(t)$.] In this way the initial and final states are described by bare particles. During the interval $-\infty < t \leqslant -T$ the equation of motion (6.12), with $H_{\mathrm{I}}(t)$ replaced by $H_{\mathrm{I}}(t)f(t)$, generates the real physical particles from the bare particles, and during the interval $|t| \leqslant T$ we are dealing with the physical particles and the full interaction $H_{\mathrm{I}}(t)$. In particular, the full interaction is effective during the interval $-\tau \leqslant t \leqslant \tau$ while the particles are sufficiently close together to interact (i.e. we must choose $T \gg \tau$). The essence of the adiabatic hypothesis is that the scattering, which occurs during the interval $|t| \leqslant \tau$, cannot depend on our description of the system a long time before the scattering ($t \ll -\tau$) or a long time after the scattering ($t \gg \tau$). Only at the end of a calculation do we take the limit $T \to \infty$. Of course, if we calculate a process in lowest order perturbation theory [i.e. we use only the term of lowest order n in Eq. (6.23) which gives a non-vanishing result] then the interaction is exclusively used to cause the transition and not also to convert bare into real particles. We may then take the limit $T \to \infty$ from the start of the calculation and work with the full interaction $H_{\mathrm{I}}(t)$.

6.3 WICK'S THEOREM

We must now see how to obtain from the S-matrix expansion (6.23) the transition amplitude $\langle f|S|i\rangle$ for a particular transition $|i\rangle \to |f\rangle$ in a given order of perturbation theory. The Hamiltonian density $\mathscr{H}_{\mathrm{I}}(x)$ in Eq. (6.23) involves the interacting fields, each linear in creation and absorption operators. Hence the expansion (6.23) will describe a large number of different processes. However, only certain terms of the S-matrix will contribute to a given transition $|i\rangle \to |f\rangle$. For these terms must contain just the right absorption operators to destroy the particles present in $|i\rangle$, and they must contain the right creation operators to emit the particles present in $|f\rangle$. They may also contain additional creation and absorption operators which create particles which are subsequently reabsorbed. These particles are only present in intermediate states and are called virtual particles.

Calculations can be greatly simplified by avoiding the explicit introduction of virtual intermediate particles. This can be achieved by writing the S-matrix expansion as a sum of normal products, since in a normal product *all* absorption operators stand to the *right* of *all* creation operators. Such an operator first absorbs a certain number of particles and then emits some particles. It does not cause emission and reabsorption of intermediate particles. Each of these normal products will effect a particular transition $|i\rangle \to |f\rangle$ which can be represented by a Feynman graph, similar to those introduced in Chapters 3 and 4.

Consider, for example, Compton scattering ($e^- + \gamma \to e^- + \gamma$). The QED interaction Hamiltonian density is, from Eq. (6.10),

$$\mathscr{H}_I(x) = -\mathscr{L}_I(x) = -e\mathrm{N}[\bar{\psi}(x)\not{A}(x)\psi(x)]. \tag{6.24}$$

Since the negative (positive) frequency parts A^-, $\bar{\psi}^-$, ψ^- (A^+, ψ^+, $\bar{\psi}^+$) are linear in creation (absorption) operators for photons, electrons and positrons respectively, the only normal product which contributes to Compton scattering is

$$\bar{\psi}^- A^- \psi^+ A^+ .$$

The method for expanding the S-matrix as a sum of normal products which we shall now describe is due to Dyson and Wick.

We first of all summarize the general definition of a normal product. Let Q, $R, \ldots, W$ be operators of the type $\psi^\pm$, $A^\pm$, etc., i.e. each is linear in either creation or absorption operators, then

$$\mathrm{N}(QR\ldots W) = (-1)^P(Q'R'\ldots W'). \tag{6.25a}$$

Here $Q', R', \ldots, W'$ are the operators $Q, R, \ldots, W$ reordered so that all absorption operators (i.e. positive frequency parts) stand to the right of all creation operators (i.e. negative frequency parts). The exponent P is the number of interchanges of neighbouring fermion operators required to change the order $(QR\ldots W)$ into $(Q'R'\ldots W')$. We generalize the definition (6.25a) by requiring the normal product to obey the distributive law

$$\mathrm{N}(RS\ldots + VW\ldots) = \mathrm{N}(RS\ldots) + \mathrm{N}(VW\ldots). \tag{6.25b}$$

The QED interaction (6.24) is a normal product of field operators. We shall find that in other cases too the interaction Hamiltonian density can be written as a normal product, i.e.

$$\mathscr{H}_{\mathrm{I}}(x) = \mathrm{N}\{A(x)B(x)\ldots\}, \tag{6.26}$$

where each of the fields $A(x)$, $B(x)$, ..., is linear in creation and absorption operators. Hence we must consider the expansion into a sum of normal products of a 'mixed' T-product (i.e. a T-product whose factors are normal products), such as occurs in the S-matrix expansion (6.23).

From the definition of the normal product we have for two field operators $A \equiv A(x_1)$ and $B \equiv B(x_2)$ that

$$AB - \mathrm{N}(AB) = \begin{Bmatrix} [A^+, B^-]_+, & \text{for two fermion fields} \\ [A^+, B^-], & \text{otherwise} \end{Bmatrix}. \tag{6.27}$$

For two fermion fields the anticommutators, and in all other cases the commutators, are c-numbers, i.e. they do not involve creation or annihilation operators. [We had examples in Eqs. (3.41) and (4.53a).] Hence, the right-hand side of Eq. (6.27) is always a c-number. It is given by $\langle 0|AB|0\rangle$, as follows by taking the vacuum expectation value of Eq. (6.27). Hence Eq. (6.27) becomes:

$$AB = \mathrm{N}(AB) + \langle 0|AB|0\rangle. \tag{6.28}$$

Since

$$\mathrm{N}(AB) = \pm\mathrm{N}(BA), \tag{6.29}$$

the minus sign applying in the case of two fermion fields, the plus sign in all other cases, it follows from Eq. (6.28) that for $x_1^0 \neq x_2^0$

$$\mathrm{T}\{A(x_1)B(x_2)\} = \mathrm{N}\{A(x_1)B(x_2)\} + \langle 0|\mathrm{T}\{A(x_1)B(x_2)\}|0\rangle. \tag{6.30}$$

The case of equal times, $x_1^0 = x_2^0$, will be considered below.

The special notation

$$\underbrace{A(x_1)B}(x_2) \equiv \langle 0|\mathrm{T}\{A(x_1)B(x_2)\}|0\rangle \tag{6.31}$$

will be convenient for this vacuum expectation value which will be called the *contraction* of $A(x_1)$ and $B(x_2)$. Being a vacuum expectation value, it will vanish unless one of the field operators A and B creates particles which the other absorbs. The non-vanishing contractions are of course just the Feynman propagators, e.g. Eqs. (3.56), (3.60), (4.61) and (5.26):

$$\underbrace{\phi(x_1)\phi}(x_2) = \mathrm{i}\Delta_{\mathrm{F}}(x_1 - x_2) \tag{6.32a}$$

$$\underbrace{\phi(x_1)\phi^\dagger}(x_2) = \underbrace{\phi^\dagger(x_2)\phi}(x_1) = \mathrm{i}\Delta_{\mathrm{F}}(x_1 - x_2) \tag{6.32b}$$

$$\underbrace{\psi_\alpha(x_1)\bar{\psi}_\beta}(x_2) = -\underbrace{\bar{\psi}_\beta(x_2)\psi_\alpha}(x_1) = \mathrm{i}S_{\mathrm{F}\alpha\beta}(x_1 - x_2) \tag{6.32c}$$

$$\underbrace{A^\mu(x_1)A^\nu}(x_2) = \mathrm{i}D_{\mathrm{F}}^{\mu\nu}(x_1 - x_2). \tag{6.32d}$$

To generalize Eq. (6.30) to several operators $A \equiv A(x_1), \ldots, M \equiv M(x_m), \ldots$, the generalized normal product is defined by

$$\begin{aligned} &\mathrm{N}(ABCDEF \ldots JKLM \ldots) \\ &\qquad = (-1)^P \underbracket{AK}\,\underbracket{BC}\,\underbracket{EL} \ldots \mathrm{N}(DF \ldots JM \ldots) \end{aligned} \tag{6.33}$$

(on the left-hand side A is contracted with K, B with C, and E with L)

where P is the number of interchanges of neighbouring fermion operators required to change the order $(ABC \ldots)$ to $(AKB \ldots)$; for example

$$\begin{aligned} &\mathrm{N}(\psi_\alpha(x_1)\underbracket{\psi_\beta(x_2)A^\mu(x_3)\bar{\psi}_\gamma(x_4)\bar{\psi}_\delta(x_5)}) \\ &\qquad = (-1)\underbracket{\psi_\beta(x_2)\bar{\psi}_\delta(x_5)}\mathrm{N}(\psi_\alpha(x_1)A^\mu(x_3)\bar{\psi}_\gamma(x_4)). \end{aligned} \tag{6.34}$$

For the case of unequal times (i.e. $x_i^0 \neq x_j^0$, for $i \neq j$), Wick has proved the following generalization of Eq. (6.30):

$$\begin{aligned} \mathrm{T}(ABCD \ldots WXYZ) &= \mathrm{N}(ABCD \ldots WXYZ) \\ &+ \mathrm{N}(\underbracket{AB}C \ldots YZ) + \mathrm{N}(\underbracket{ABC} \ldots YZ) + \cdots + \mathrm{N}(ABC \ldots \underbracket{YZ}) \\ &+ \mathrm{N}(\underbracket{AB}\,\underbracket{CD} \ldots YZ) + \cdots + \mathrm{N}(AB \ldots \underbracket{WX}\,\underbracket{YZ}) \\ &+ \cdots. \end{aligned} \tag{6.35}$$

On the right-hand side of this equation appears the sum of all possible generalized normal products that can be formed from $(ABCD \ldots WXYZ)$, the first, second and third lines representing all terms with no, one and two contractions, and so on. Each term on the right-hand side of this equation contains all the factors in the same order in which they occur in the T-product on the left-hand side.

Eq. (6.35) states Wick's theorem. We shall not reproduce its proof which is by induction, and so not very illuminating.‡

With the interaction (6.26), the S-matrix expansion (6.23) contains the mixed T-products

$$\mathrm{T}\{\mathscr{H}_\mathrm{I}(x_1) \ldots \mathscr{H}_\mathrm{I}(x_n)\} = \mathrm{T}\{\mathrm{N}(AB \ldots)_{x_1} \ldots \mathrm{N}(AB \ldots)_{x_n}\}. \tag{6.36}$$

Wick extended the theorem (6.35) to include such mixed T-products. In each factor $\mathrm{N}(AB \ldots)_{x_r}$ we replace $x_r = (x_r^0, \mathbf{x}_r)$ by $\xi_r = (x_r^0 \pm \varepsilon, \mathbf{x}_r)$, $(\varepsilon > 0)$, depending on whether the substitution is made in the creation or absorption part of the field. Hence

$$\mathrm{T}\{\mathrm{N}(AB \ldots)_{x_1} \ldots \mathrm{N}(AB \ldots)_{x_n}\} = \lim_{\varepsilon \to 0} \mathrm{T}\{(AB \ldots)_{\xi_1} \ldots (AB \ldots)_{\xi_n}\}, \tag{6.37}$$

‡ G. C. Wick, *Phys. Rev.* **80** (1950) 268.

the normal and chronological orderings within each group $(AB\ldots)_{\xi_r}$ being the same on account of the $\pm\varepsilon$ in ξ_r^0. On expanding the right-hand side of Eq. (6.37) by Wick's theorem *before* going to the limit $\varepsilon \to 0$, contractions within one group $(AB\ldots)_{\xi_r}$ (i.e. over equal-times operators when $\varepsilon \to 0$) vanish as the group is already in normal order. We thus have the desired result: the mixed T-product (6.36) can be expanded according to Eq. (6.35), provided contractions over equal times are omitted:

$$\mathrm{T}\{\mathrm{N}(AB\ldots)_{x_1}\ldots\mathrm{N}(AB\ldots)_{x_n}\} = \mathrm{T}\{(AB\ldots)_{x_1}\ldots(AB\ldots)_{x_n}\}_{\text{no e.t.c.}} \quad (6.38)$$

where 'no e.t.c.' stands for 'no equal-times contractions'.

Eqs. (6.35) and (6.38) represent the desired result, enabling us to expand each term in the S-matrix expansion (6.23) into a sum of generalized normal products. Each of these normal products corresponds to a definite process, characterized by the operators not contracted which absorb and create the particles present in the initial and final states respectively. The non-vanishing contractions which occur in these generalized normal products are the Feynman propagators (6.32) corresponding to virtual particles being emitted and reabsorbed in intermediate states. In the next chapter we shall see how to evaluate these individual contributions to $\langle f|S|i\rangle$ which result from the application of Wick's theorem.

CHAPTER 7

Feynman diagrams and rules in QED

In the last chapter we obtained the S-matrix expansion (6.23) and Wick's theorem for writing the terms in this expansion as a sum of normal products. In this chapter we shall show how to calculate the matrix element $\langle f|S|i\rangle$ for a transition from an initial state $|i\rangle$ to a final state $|f\rangle$ in a given order of perturbation theory. For definiteness we shall give this development for the important case of QED. Once this case is understood, the corresponding formalism for others is easily derived.

In Section 7.1 we shall show how to pick out from the S-matrix expansion the terms which contribute to $\langle f|S|i\rangle$ in a given order of perturbation theory. These terms are easily identified. They are those normal products which contain the appropriate destruction and creation operators to destroy the particles present in the initial state $|i\rangle$ and create those present in the final state $|f\rangle$.

In Section 7.2 we shall evaluate the transition amplitude $\langle f|S|i\rangle$ in momentum space. This leads to Feynman diagrams as a way of interpreting the terms in the Wick expansion. There exists a one-to-one correspondence between the diagrams and the terms which can be summarized in simple rules. These enable one to write down transition amplitudes directly from the Feynman graphs, rather than proceed *ab initio* from Wick's theorem. In Section 7.3 we shall state these rules, known as Feynman rules, for QED. We shall have obtained these rules from the Dyson–Wick formalism, but historically they were first derived by Feynman using a strongly intuitive approach.

In the first three sections of this chapter we shall consider QED as the interaction of the electron–positron field with the electromagnetic field. In the last section (Section 7.4) we shall extend QED to include, in addition to the electron–positron field, other leptons, such as the muon and tauon.

7.1 FEYNMAN DIAGRAMS IN CONFIGURATION SPACE

The processes to which the individual terms in the S-matrix expansion (6.23), i.e. in

$$S = \sum_{n=0}^{\infty} S^{(n)} \equiv \sum_{n=0}^{\infty} \frac{(-i)^n}{n!} \int \ldots \int d^4x_1 \ldots d^4x_n \mathrm{T}\{\mathscr{H}_I(x_1) \ldots \mathscr{H}_I(x_n)\}, \quad (7.1)$$

contribute are of course determined by the nature of the interaction $\mathscr{H}_I(x)$. For QED this is given by Eq. (6.24):

$$\left.\begin{aligned} \mathscr{H}_I(x) &= -e\mathrm{N}\{\bar{\psi}(x)\not{A}(x)\psi(x)\} \\ &= -e\mathrm{N}\{(\bar{\psi}^+ + \bar{\psi}^-)(\not{A}^+ + \not{A}^-)(\psi^+ + \psi^-)\}_x. \end{aligned}\right\} \quad (7.2)$$

With $\psi^+(\bar{\psi}^-)$, $\bar{\psi}^+(\psi^-)$ and $A^+(A^-)$ being linear in absorption (creation) operators of electrons, positrons and photons, respectively, the interaction (7.2) gives rise to eight basic processes, e.g. the term $-e\mathrm{N}(\bar{\psi}^+ \not{A}^- \psi^+)_x$ corresponds to the annihilation of an electron–positron pair with the creation of a photon.

Using the conventions for Feynman diagrams explained at the end of Section 4.4, we can represent these eight processes by the Feynman graphs of Fig. 7.1, which have been grouped into pairs. The graphs in each pair correspond to absorption or emission of a photon, together with: (a) the scattering of an electron, (b) the scattering of a positron, (c) pair annihilation, or (d) pair creation. These diagrams illustrate the basic processes to which the QED interaction gives rise and will be referred to as the basic vertex part. All other QED Feynman diagrams are built up by combining such basic vertex parts.

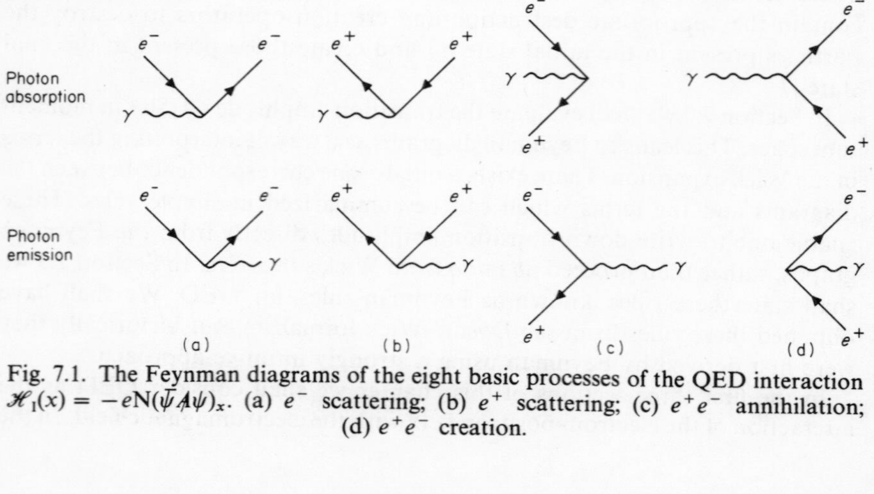

Fig. 7.1. The Feynman diagrams of the eight basic processes of the QED interaction $\mathscr{H}_I(x) = -e\mathrm{N}(\bar{\psi}\not{A}\psi)_x$. (a) e^- scattering; (b) e^+ scattering; (c) e^+e^- annihilation; (d) e^+e^- creation.

The diagrams of Fig. 7.1 also represent the processes arising from the first-order term $S^{(1)}$ in Eq. (7.1). However these are not real physical processes, i.e. for none of them can energy and momentum be conserved for real physical particles for which we must have $k^2 = 0$ for photons, and $p^2 = m^2$ for fermions. Consequently

$$\langle f|S^{(1)}|i\rangle = 0 \tag{7.3a}$$

for these transitions, as will be shown explicitly in the next section. More generally

$$\langle f|S^{(n)}|i\rangle = 0 \tag{7.3b}$$

for any unphysical process, i.e. for a transition between real physical states which violates a conservation law of the theory. This follows since S generates a solution of the equations of motion, so that

$$\langle f|S|i\rangle = 0 \tag{7.3c}$$

for an unphysical process, and since Eq. (7.1) is a power series in the coupling constant e.

To obtain real processes, we must go at least to the second-order term $S^{(2)}$ in Eq. (7.1). This term contains two factors $\mathscr{H}_{\mathrm{I}}$. Its expansion by Wick's theorem into a sum of normal products corresponds to all meaningful ways of joining two basic vertex parts into a Feynman diagram, as we shall now see.

Application of Wick's theorem, Eqs. (6.35) and (6.38), to $S^{(2)}$ leads to

$$S^{(2)} = \sum_{i=\mathrm{A}}^{F} S_i^{(2)} \tag{7.4}$$

where

$$S_{\mathrm{A}}^{(2)} = -\frac{e^2}{2!}\int \mathrm{d}^4x_1\,\mathrm{d}^4x_2\mathrm{N}[(\bar\psi\not{A}\psi)_{x_1}(\bar\psi\not{A}\psi)_{x_2}] \tag{7.5a}$$

$$S_{\mathrm{B}}^{(2)} = -\frac{e^2}{2!}\int \mathrm{d}^4x_1\,\mathrm{d}^4x_2\{\mathrm{N}[(\bar\psi\not{A}\underbrace{\psi)_{x_1}(\bar\psi}\not{A}\psi)_{x_2}] + \mathrm{N}[(\underbrace{\bar\psi\not{A}\psi)_{x_1}(\bar\psi\not{A}\psi}_{\ })_{x_2}]\} \tag{7.5b}$$

$$S_{\mathrm{C}}^{(2)} = -\frac{e^2}{2!}\int \mathrm{d}^4x_1\,\mathrm{d}^4x_2\mathrm{N}[(\bar\psi\gamma^\alpha \underbrace{A_\alpha\psi)_{x_1}(\bar\psi\gamma^\beta A_\beta}\psi)_{x_2}] \tag{7.5c}$$

$$S_{\mathrm{D}}^{(2)} = -\frac{e^2}{2!}\int \mathrm{d}^4x_1\,\mathrm{d}^4x_2\{\mathrm{N}[(\bar\psi\gamma^\alpha \underbrace{A_\alpha\underbrace{\psi)_{x_1}(\bar\psi}\gamma^\beta A_\beta}\psi)_{x_2}] + \mathrm{N}[(\underbrace{\bar\psi\gamma^\alpha \underbrace{A_\alpha\psi)_{x_1}(\bar\psi\gamma^\beta A_\beta}\psi}_{\ })_{x_2}]\} \tag{7.5d}$$

$$S_{\mathrm{E}}^{(2)} = -\frac{e^2}{2!}\int \mathrm{d}^4x_1\,\mathrm{d}^4x_2\,\mathrm{N}[(\bar{\psi}\not{A}\psi)_{x_1}(\bar{\psi}\not{A}\psi)_{x_2}] \tag{7.5e}$$

$$S_{\mathrm{F}}^{(2)} = -\frac{e^2}{2!}\int \mathrm{d}^4x_1\,\mathrm{d}^4x_2(\bar{\psi}\gamma^{\alpha}A_{\alpha}\psi)_{x_1}(\bar{\psi}\gamma^{\beta}A_{\beta}\psi)_{x_2}. \tag{7.5f}$$

The first of these terms, $S_{\mathrm{A}}^{(2)}$, Eq. (7.5a), is not very interesting. It corresponds to two processes of the kind illustrated in Fig. 7.1 going on independently of each other. Like $S^{(1)}$, this term does not lead to any real transitions.

The two terms in $S_{\mathrm{B}}^{(2)}$, Eq. (7.5b), are identically equal to each other as is seen by permuting the operators. This requires care since the fermion fields are anticommuting operators and four-component spinors. Permuting the two groups $(\bar{\psi}\not{A}\psi)$ involves an even permutation of fermion operators and the spinor indices of each group are self-contained.‡ Hence

$$\mathrm{N}[(\bar{\psi}\not{A}\psi)_{x_1}(\bar{\psi}\not{A}\psi)_{x_2}] = \mathrm{N}[(\bar{\psi}\not{A}\psi)_{x_2}(\bar{\psi}\not{A}\psi)_{x_1}]. \tag{7.6}$$

Using this result and interchanging the integration variables $x_1 \leftrightarrow x_2$ in the second term of Eq. (7.5b), one obtains

$$S_{\mathrm{B}}^{(2)} = -e^2\int \mathrm{d}^4x_1\,\mathrm{d}^4x_2\,\mathrm{N}[(\bar{\psi}\not{A}\psi)_{x_1}(\bar{\psi}\not{A}\psi)_{x_2}]. \tag{7.7}$$

This expression contains one fermion contraction. This is given by the fermion propagator (6.32c) which is a c-number and corresponds to a virtual intermediate fermion. For $t_2 < t_1$, we can think of it as a virtual electron propagating from x_2 to x_1, for $t_1 < t_2$ as a virtual positron propagating from x_1 to x_2. As explained in Section 4.4, no time-ordering is implied in the present formalism—indeed, all space–time points x_1 and x_2 are summed over—and these two cases are combined and jointly referred to as a virtual fermion propagating from x_2 (associated with $\bar{\psi}$) to x_1 (associated with ψ). In addition to this propagator, expression (7.7) contains two uncontracted fermion and two uncontracted photon operators. These absorb or create particles present initially or finally, so-called *external particles.* The operator $S_{\mathrm{B}}^{(2)}$ contributes to many real processes. (To conserve energy and momentum, the initial and final states must each contain two particles.) Since the operators in $S_{\mathrm{B}}^{(2)}$ are in normal order, it is easy to pick out the terms which contribute to a given process.

One of these processes is Compton scattering

$$\gamma + e^- \to \gamma + e^-, \tag{7.8}$$

‡ The reader can always resolve any cases of doubt by writing out explicitly the spinor indices.

already mentioned in Section 4.4. This process corresponds to selecting the positive frequency part $\psi^+(x_2)$ of $\psi(x_2)$ to absorb the initial electron, and the negative frequency part $\bar{\psi}^-(x_1)$ of $\bar{\psi}(x_1)$ to create the final electron. But either $A\!\!\!/^+(x_1)$ or $A\!\!\!/^+(x_2)$ can absorb the initial photon and correspondingly $A\!\!\!/^-(x_2)$ or $A\!\!\!/^-(x_1)$ must emit the final photon. Thus the part of Eq. (7.7) which causes Compton scattering is

$$S^{(2)}(\gamma e^- \to \gamma e^-) = S_a + S_b \tag{7.9}$$

where

$$S_a = -e^2 \int d^4x_1\, d^4x_2 \bar{\psi}^-(x_1)\gamma^\alpha i S_F(x_1 - x_2)\gamma^\beta A_\alpha^-(x_1) A_\beta^+(x_2)\psi^+(x_2) \tag{7.10a}$$

$$S_b = -e^2 \int d^4x_1\, d^4x_2 \bar{\psi}^-(x_1)\gamma^\alpha i S_F(x_1 - x_2)\gamma^\beta A_\beta^-(x_2) A_\alpha^+(x_1)\psi^+(x_2). \tag{7.10b}$$

In Eqs. (7.10), the operators have been put in a normal order and we have substituted Eq. (6.32c) for the fermion contraction.

The contributions S_a and S_b to Compton scattering are represented by the Feynman graphs in Figs. 7.2(a) and (b). The latter is the same as Fig. 4.3. (Remember that, except for the conventions about initial and final lines, there is no time ordering in Feynman graphs, so that the same graph can be drawn in many different ways.) In Fig. 7.2 we have attached the appropriate Lorentz indices (α, β) to vertices, and particle labels (γ, e^-) to external lines. We shall often omit these as redundant.

The other real processes described by Eq. (7.7) are Compton scattering by positrons, and the two-photon pair annihilation and creation processes, i.e.

$$\text{(i)}\ \gamma + e^+ \to \gamma + e^+, \qquad \text{(ii)}\ e^+ + e^- \to \gamma + \gamma, \qquad \text{(iii)}\ \gamma + \gamma \to e^+ + e^-. \tag{7.11}$$

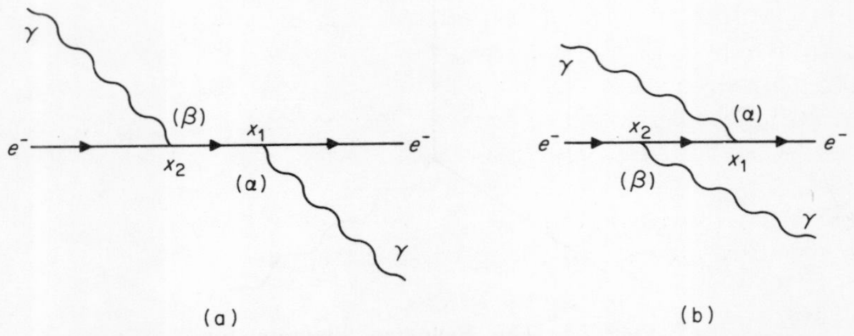

Fig. 7.2. The contributions S_a, S_b, Eqs. (7.10), to Compton scattering.

The corresponding Feynman diagrams are shown in Figs. 7.3–7.5. We leave it to the reader to write down the operators for the first two of these processes. For the pair creation process one obtains

$$S^{(2)}(2\gamma \to e^+e^-)$$

$$= -e^2 \int d^4x_1\, d^4x_2 \bar{\psi}^-(x_1)\gamma^\alpha iS_F(x_1 - x_2)\gamma^\beta\psi^-(x_2)A_\alpha^+(x_1)A_\beta^+(x_2). \quad (7.12)$$

Although we have only shown one diagram in Fig. 7.5, Eq. (7.12) actually gives two contributions since the operator $A_\beta^+(x_2)$ can absorb either of the initially present photons, with $A_\alpha^+(x_1)$ absorbing the other. A similar situation exists for the pair annihilation process.

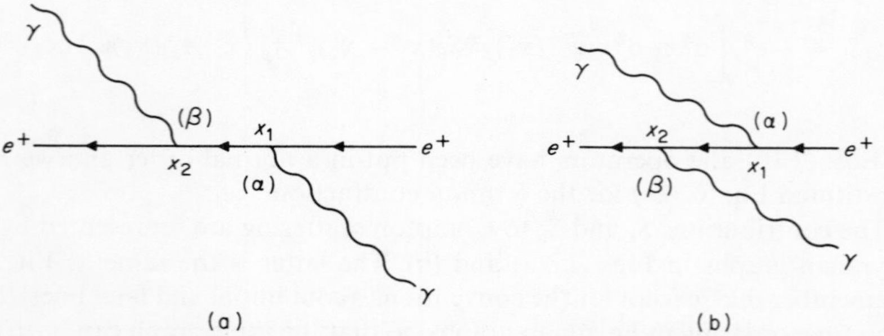

Fig. 7.3. The Feynman diagrams for Compton scattering by positrons.

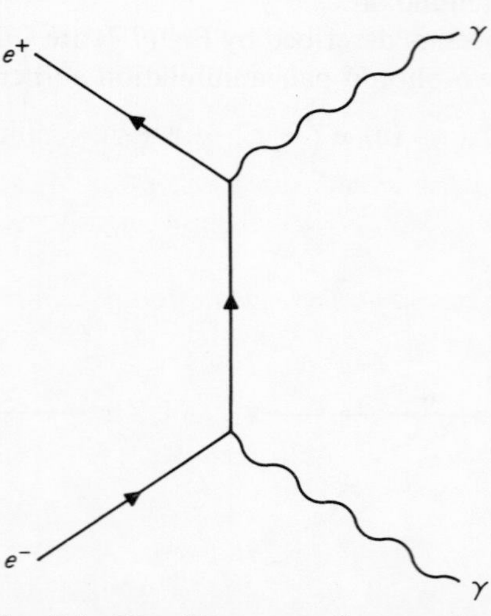

Fig. 7.4. The Feynman diagram for $e^+ + e^- \to \gamma + \gamma$.

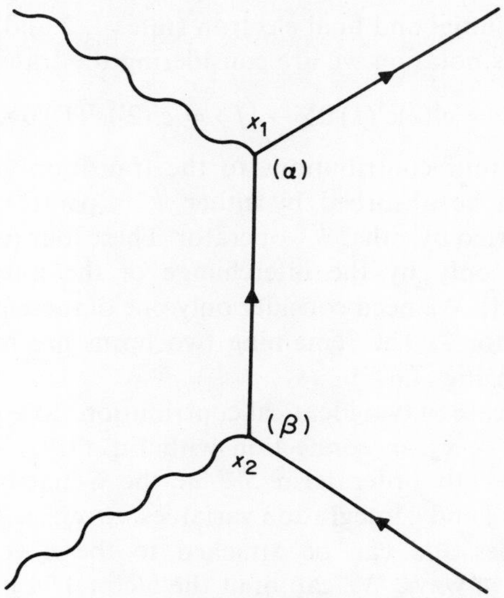

Fig. 7.5. The Feynman diagram for $\gamma + \gamma \rightarrow e^+ + e^-$, Eq. (7.12).

We next consider Eq. (7.5c). This term contains four uncontracted fermion operators. Accordingly, the real processes to which this term gives rise are fermion–fermion scattering: $e^- - e^-$, $e^+ - e^+$ or $e^- - e^+$ scattering, according to which positive and negative frequency parts are selected from the external fermion fields. The photon–photon contraction in Eq. (7.5c) describes the interaction between the charges as the exchange of transverse, longitudinal and scalar photons. This photon propagator occurs associated with two conserved current operators $s^\mu(x) = (\bar{\psi}\gamma^\mu\psi)_x$. As discussed in Section 5.3 [particularly the discussion of Eq. (5.44)], this covariant formulation is equivalent to the usual description of the interaction in terms of the instantaneous Coulomb interaction together with the exchange of transverse photons.

We next consider electron–electron scattering,

$$e^- + e^- \rightarrow e^- + e^-, \tag{7.13}$$

known as Møller scattering, in more detail. The part of the operator (7.5c) describing this process is

$$S^{(2)}(2e^- \rightarrow 2e^-)$$

$$= \frac{-e^2}{2!} \int d^4x_1\, d^4x_2 N[(\bar{\psi}^- \gamma^\alpha \psi^+)_{x_1} (\bar{\psi}^- \gamma^\beta \psi^+)_{x_2}] iD_{F\alpha\beta}(x_1 - x_2), \tag{7.14}$$

where we substituted Eq. (6.32d) for the photon contraction.

Let us label the initial and final electron states 1, 2 and 1′, 2′ respectively, i.e. with an obvious notation we are considering the transition

$$|i\rangle = c^{\dagger}(2)c^{\dagger}(1)|0\rangle \rightarrow |f\rangle = c^{\dagger}(2')c^{\dagger}(1')|0\rangle. \tag{7.15}$$

Eq. (7.14) gives four contributions to the transition (7.15), since either initial electron can be absorbed by either ψ^{+} operator, and either final electron can be emitted by either $\bar{\psi}^{-}$ operator. These four terms comprise two pairs which differ only by the interchange of the integration variables $x_1 \leftrightarrow x_2$ in Eq. (7.14). We need consider only one of these pairs and multiply the result by a factor 2. The remaining two terms are represented by the Feynman graphs in Fig. 7.6.

We had another case of two identical contributions to a process, related to the interchange $x_1 \leftrightarrow x_2$, in connection with Eq. (7.5b). This represents a general result. The nth order term $S^{(n)}$ in the S-matrix expansion (7.1) contains a factor $1/n!$ and n integration variables $x_1, x_2, \ldots, x_n$. These are only summation variables and can be attached to the n vertices of a given Feynman graph in $n!$ ways. We can omit the factor $1/n!$ if we consider only topologically different Feynman diagrams, i.e. diagrams which differ only in the labelling of vertices are considered the same. Some care is required in interpreting this statement. For example, the two diagrams of Fig. 7.6 are topologically different from each other because the two final electrons have different properties. (These were labelled 1′ and 2′. In practice they are the momenta and spins.) Permuting x_1 and x_2 does not interchange the two graphs of Fig. 7.6. As we shall see, their contributions occur with a relative minus sign and correspond to the 'direct minus exchange scattering' which the reader should recognize, from non-relativistic quantum mechanics, as characteristic of two identical fermions.

In order to obtain explicit expressions for these two contributions, let

$$\psi_j^{+}(x) = c(j)f_j(x), \qquad \bar{\psi}_j^{-}(x) = c^{\dagger}(j)g_j(x) \tag{7.16}$$

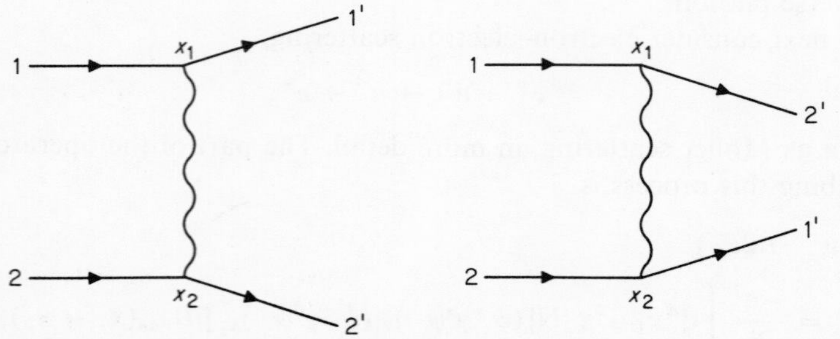

Fig. 7.6. The two diagrams for electron–electron scattering (Møller scattering).

be the parts of the operators $\psi^+(x)$ and $\bar{\psi}^-(x)$ proportional to $c(j)$ and $c^\dagger(j)$ respectively. $j = 1, 2, 1', 2'$ labels the electron states involved in the process. The part of the S-matrix operator (7.14) which effects the transition $|i\rangle \to |f\rangle$, Eq. (7.15), is then given by

$$S^{(2)}(e^-(1) + e^-(2) \to e^-(1') + e^-(2')) = S_a + S_b \tag{7.17a}$$

where S_a and S_b correspond to Figs. 7.6(a) and (b), and are given by

$$S_a = -e^2 \int d^4x_1\, d^4x_2 N[(\bar{\psi}_{1'}^- \gamma^\alpha \psi_1^+)_{x_1} (\bar{\psi}_{2'}^- \gamma^\beta \psi_2^+)_{x_2}] iD_{F\alpha\beta}(x_1 - x_2) \tag{7.17b}$$

$$S_b = -e^2 \int d^4x_1\, d^4x_2 N[(\bar{\psi}_{2'}^- \gamma^\alpha \psi_1^+)_{x_1} (\bar{\psi}_{1'}^- \gamma^\beta \psi_2^+)_{x_2}] iD_{F\alpha\beta}(x_1 - x_2). \tag{7.17c}$$

The relative minus sign of the two contributions is implied by the normal products in these equations. To arrange the creation and annihilation operators in both cases in the same order, e.g. as $c^\dagger(1')c^\dagger(2')c(1)c(2)$, requires the normal products in Eqs. (7.17b) and (7.17c) to be reordered equal to

$$-\bar{\psi}_{1'}^-(x_1)\bar{\psi}_{2'}^-(x_2)\psi_1^+(x_1)\psi_2^+(x_2) \quad \text{and} \quad +\bar{\psi}_{1'}^-(x_2)\bar{\psi}_{2'}^-(x_1)\psi_1^+(x_1)\psi_2^+(x_2)$$

respectively. Using Eqs. (7.16), we obtain from Eqs. (7.17) the transition amplitude

$$\begin{aligned}
&\langle f|S^{(2)}(2e^- \to 2e^-)|i\rangle \\
&\quad = \left\{-e^2 \int d^4x_1\, d^4x_2 g_{1'}(x_1)\gamma^\alpha f_1(x_1) g_{2'}(x_2)\gamma^\beta f_2(x_2) iD_{F\alpha\beta}(x_1 - x_2)\right\} \\
&\qquad - \{1' \leftrightarrow 2'\},
\end{aligned} \tag{7.18}$$

where the term $\{1' \leftrightarrow 2'\}$ is just the first expression in braces with the labels $1'$ and $2'$ of the two final electron states interchanged. Our final result (7.18) has the desired form of a 'direct' amplitude minus an 'exchange' amplitude, the two amplitudes being transformed into each other by exchanging the single-particle states of the two electrons in the final state. In non-relativistic quantum mechanics this result follows through the use of antisymmetric wavefunctions according to Pauli's principle. In the above field-theoretic derivation the anticommutativity of the fermion field operators is the crucial element.

These arguments generalize. Whenever the initial or final state contains several identical fermions, one obtains a completely antisymmetric transition amplitude $\langle f|S|i\rangle$. For example, if the initial state $|i\rangle$ contains s positrons in states $1, 2, \ldots, s$, the corresponding S-matrix operator will contain s uncontracted operators $N(\bar{\psi}(x_1)\bar{\psi}(x_2)\ldots\bar{\psi}(x_s))$. Any one of these operators

$\bar{\psi}(x_1), \ldots, \bar{\psi}(x_s)$ can absorb the positron in state 1, and so on, giving $s!$ terms whose sum is completely antisymmetric in the labels $1, 2, \ldots, s$, since the operators $\bar{\psi}(x_1), \ldots, \bar{\psi}(x_s)$ anticommute. An analogous argument holds for several identical final state fermions.

More curiously, the fact that the operator $\psi(x)$ can absorb an electron or create a positron implies that transition amplitudes are antisymmetric with respect to initial electron and final positron states. (A similar argument applies of course to $\bar{\psi}(x)$ and initial positrons and final electrons.) We have an example of this in electron–positron scattering,

$$e^+ + e^- \to e^+ + e^-,$$

known as Bhabha scattering. The part of the operator (7.5c) describing this process must contain the uncontracted operators $\bar{\psi}^+$, ψ^+, ψ^- and $\bar{\psi}^-$ to absorb and create the particles present initially and finally. As in the case of Møller scattering, four terms contribute which again reduce to two by the general argument given above. It is left as an exercise to the reader to derive from Eq. (7.5c) the following expression for the S-matrix operator for Bhabha scattering:

$$S^{(2)}(e^+e^- \to e^+e^-) = S_a + S_b \tag{7.19a}$$

where

$$S_a = -e^2 \int d^4x_1\, d^4x_2 N[(\bar{\psi}^- \gamma^\alpha \psi^+)_{x_1} (\bar{\psi}^+ \gamma^\beta \psi^-)_{x_2}] iD_{F\alpha\beta}(x_1 - x_2), \tag{7.19b}$$

$$S_b = -e^2 \int d^4x_1\, d^4x_2 N[(\bar{\psi}^- \gamma^\alpha \psi^-)_{x_1} (\bar{\psi}^+ \gamma^\beta \psi^+)_{x_2}] iD_{F\alpha\beta}(x_1 - x_2). \tag{7.19c}$$

The Feynman graph for S_a is shown in Fig. 7.7(a). It represents the scattering by photon exchange, as occurred for electron–electron scattering (Fig. 7.6). However, in the term S_b both initial particles are annihilated at x_2 and the final electron–positron pair is created at x_1. It corresponds to the *annihilation diagram* of Fig. 7.7(b). As in electron–electron scattering, there is a relative sign factor (-1) between the two contributions implicit in the normal products which becomes explicit if the creation and annihilation operators are brought into the same normal order in both cases. That the diagrams of Figs. 7.7(a) and (b) are related by the interchange of an initial electron state and a final positron state is brought out by 'deforming' diagram 7.7(b) into diagram 7.7(c). Comparing diagrams 7.7(a) and (c) one sees that the latter is obtained from the former by interchanging the initial electron line at x_1 and the final positron line at x_2.

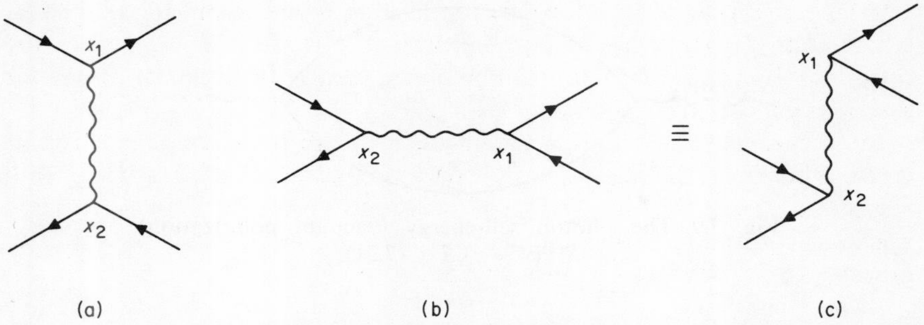

Fig. 7.7. The contributions S_a and S_b to electron–positron scattering (Bhabha scattering): (a) represents photon exchange; (b) and (c) are equivalent ways of representing the pair annihilation process.

We shall now discuss briefly the remaining second-order terms $S_D^{(2)}$ to $S_F^{(2)}$, Eqs. (7.5d)–(7.5f).

Eq. (7.5d) contains two uncontracted fermion fields and gives rise to two processes according to whether the fermion present initially and finally is an electron or a positron. The two terms in Eq. (7.5d) are again equal to each other. For the electron case this equation reduces to

$$S^{(2)}(e^- \to e^-)$$

$$= -e^2 \int d^4x_1\, d^4x_2 \bar{\psi}^-(x_1)\gamma^\alpha iS_F(x_1 - x_2)\gamma^\beta \psi^+(x_2) iD_{F\alpha\beta}(x_1 - x_2) \qquad (7.20)$$

which corresponds to the diagram of Fig. 7.8. It represents a modification of the properties of a *bare* electron due to its interaction with the radiation field. It is one of the processes—in fact the simplest—which converts a bare electron into a *physical* electron, i.e. one surrounded by its photon cloud. This interaction changes the energy of the system, that is, the mass of the physical electron as compared with that of the bare electron. This is known as the *self-energy* of the electron, and Fig. 7.8 is called a self-energy diagram. Its evaluation leads to a divergent integral. These divergent self-energy effects can be eliminated by incorporating them in the properties of the physical electron. This is the process of *renormalization* which will be studied in Chapter 9.

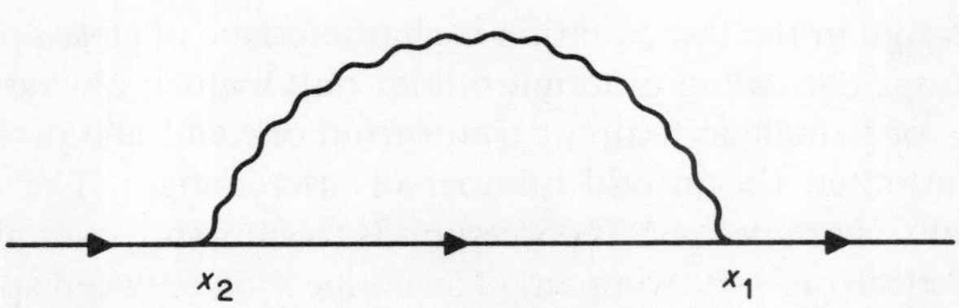

Fig. 7.8. The electron self-energy $S^{(2)}(e^- \to e^-)$, Eq. (7.20).

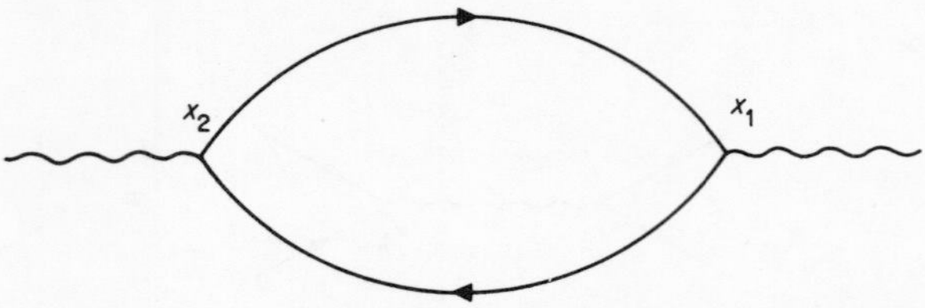

Fig. 7.9. The photon self-energy (vacuum polarization) $S^{(2)}(\gamma \to \gamma)$, Eq. (7.21).

Fig. 7.9 similarly describes a *photon self-energy* arising from the term $S_E^{(2)}$, Eq. (7.5e). The interaction between the electromagnetic and the electron–positron fields enables the photon to create a virtual electron–positron pair which subsequently annihilates again. An external electromagnetic field (for example the field of a heavy nucleus) will modify the distribution of these virtual electron–positron pairs, i.e. it will 'polarize the vacuum' in much the same way in which it would polarize a dielectric. For this reason such photon self-energy graphs are called *vacuum polarization* diagrams. Like the electron self-energy, they lead to infinities which are again eliminated by renormalization (see Chapter 9).

Eq. (7.5e) for the photon self-energy can be written

$$S^{(2)}(\gamma \to \gamma) = -e^2 \int \mathrm{d}^4x_1\, \mathrm{d}^4x_2 \mathrm{N}[(\overline{\psi} \not{A}^- \psi)_{x_1} (\overline{\psi} \not{A}^+ \psi)_{x_2}]. \tag{7.21}$$

Writing the spinor indices out explicitly, we can re-express the normal product in Eq. (7.21) as

$$\begin{aligned} &\mathrm{N}[(\overline{\psi}_\lambda \not{A}^-_{\lambda\mu} \psi_\mu)_{x_1} (\overline{\psi}_\sigma \not{A}^+_{\sigma\tau} \psi_\tau)_{x_2}] \\ &\qquad = (-1) \psi_\tau(x_2) \overline{\psi}_\lambda(x_1) \not{A}^-_{\lambda\mu}(x_1) \psi_\mu(x_1) \overline{\psi}_\sigma(x_2) \not{A}^+_{\sigma\tau}(x_2) \\ &\qquad = (-1) \mathrm{Tr}[\mathrm{i}S_\mathrm{F}(x_2 - x_1) \not{A}^-(x_1) \mathrm{i}S_\mathrm{F}(x_1 - x_2) \not{A}^+(x_2)]. \end{aligned} \tag{7.22}$$

(Here $\not{A}^-_{\lambda\mu}(x) \equiv \gamma^\alpha_{\lambda\mu} A^-_\alpha(x)$, etc.)

The minus sign in the last equation is characteristic of *closed fermion loops* (i.e. closed loops consisting of fermion lines only) which always involve the transposition of a single fermion operator from one end of a product of such factors to the other, i.e. an odd number of interchanges. The trace in Eq. (7.22) is equally characteristic. It corresponds to summing over all spin states of the virtual electron–positron pair. (The connection between spin sums and traces will be discussed in Section 8.2.)

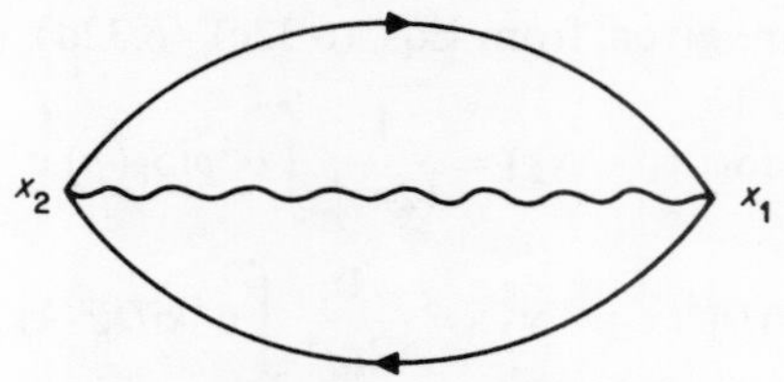

Fig. 7.10. The simplest vacuum diagram, Eq. (7.5f).

Finally, Fig. 7.10 shows the graph representing Eq. (7.5f). This diagram has no external lines and consequently does not cause any transitions. One can show that such *vacuum diagrams* (i.e. diagrams without external lines) may be omitted altogether, at any rate in elementary applications.

This completes our initial analysis of the various terms which occur on decomposing $S^{(2)}$ into normal products. We have seen that the terms obtained correspond to specific processes and how Feynman diagrams greatly aid their interpretation. No new features occur for the higher-order terms $S^{(3)}, \ldots$. The methods developed are sufficient to deal with such higher-order processes although the complexity of the mathematics increases rapidly with order.

7.2 FEYNMAN DIAGRAMS IN MOMENTUM SPACE

In the last section we developed a technique for deriving the S-matrix operator which generates a particular transition $|i\rangle \to |f\rangle$ in a given order. In practice one is usually interested in the corresponding matrix element $\langle f|S^{(n)}|i\rangle$. The states $|i\rangle$ and $|f\rangle$ are usually specified by the particles of known momenta and spin and polarization properties present initially and finally. Explicit calculations of the matrix elements lead to a reinterpretation of the Feynman graphs as diagrams in momentum space. By studying some specific cases we shall see that these diagrams are closely related to the mathematical expressions they represent. It is possible to formulate a set of rules which enable one to write down the matrix elements directly from the Feynman diagrams without detailed calculations. These Feynman rules, which will be given in the next section, are the linchpin of practical calculations in perturbation theory.

Calculation of the matrix elements, with $|i\rangle$ and $|f\rangle$ specified as momentum eigenstates of the particles present, essentially corresponds to Fourier transforming the fields into momentum space in order to pick out the appropriate absorption and creation operators. For the propagators, these

Fourier transforms are given, from Eqs. (6.32c), (6.32d), (4.63) and (5.27), by

$$\underbrace{\psi(x_1)\bar{\psi}(x_2)} = \mathrm{i}S_{\mathrm{F}}(x_1 - x_2) = \frac{1}{(2\pi)^4}\int \mathrm{d}^4 p \,\mathrm{i}S_{\mathrm{F}}(p)\,\mathrm{e}^{-\mathrm{i}p(x_1 - x_2)} \tag{7.23a}$$

$$\underbrace{A^{\alpha}(x_1)A^{\beta}(x_2)} = \mathrm{i}D_{\mathrm{F}}^{\alpha\beta}(x_1 - x_2) = \frac{1}{(2\pi)^4}\int \mathrm{d}^4 k \,\mathrm{i}D_{\mathrm{F}}^{\alpha\beta}(k)\,\mathrm{e}^{-\mathrm{i}k(x_1 - x_2)} \tag{7.23b}$$

where

$$S_{\mathrm{F}}(p) = \frac{\not{p} + m}{p^2 - m^2 + \mathrm{i}\varepsilon} \equiv \frac{1}{\not{p} - m + \mathrm{i}\varepsilon} \tag{7.24a}$$

$$D_{\mathrm{F}}^{\alpha\beta}(k) = \frac{-g^{\alpha\beta}}{k^2 + \mathrm{i}\varepsilon}. \tag{7.24b}$$

The Fourier expansions of the uncontracted fields ψ, $\bar{\psi}$ and A_{α} are given by Eqs. (4.38) and (5.16). The effect of the uncontracted operators ψ^{+}, $\bar{\psi}^{+}$ and A_{α}^{+}, which occur in a term of the S-matrix expansion, acting on $|i\rangle$, is to give the vacuum state $|0\rangle$. For example, it follows from Eqs. (4.38) and (5.16) that

$$\psi^{+}(x)|e^{-}\mathbf{p}\rangle = |0\rangle\left(\frac{m}{VE_{\mathbf{p}}}\right)^{1/2} u(\mathbf{p})\,\mathrm{e}^{-\mathrm{i}px} \tag{7.25a}$$

$$\bar{\psi}^{+}(x)|e^{+}\mathbf{p}\rangle = |0\rangle\left(\frac{m}{VE_{\mathbf{p}}}\right)^{1/2} \bar{v}(\mathbf{p})\,\mathrm{e}^{-\mathrm{i}px} \tag{7.25b}$$

$$A_{\alpha}^{+}(x)|\gamma\mathbf{k}\rangle = |0\rangle\left(\frac{1}{2V\omega_{\mathbf{k}}}\right)^{1/2} \varepsilon_{\alpha}(\mathbf{k})\,\mathrm{e}^{-\mathrm{i}kx} \tag{7.25c}$$

Here we have suppressed the spin and polarization labels. For example, $|e^{-}\mathbf{p}\rangle$ and $|\gamma\mathbf{k}\rangle$ stand for the one-electron and one-photon states

$$|e^{-}\mathbf{p}\rangle \equiv |e^{-}\mathbf{p}r\rangle = c_r^{\dagger}(\mathbf{p})|0\rangle, \qquad |\gamma\mathbf{k}\rangle \equiv |\gamma\mathbf{k}r\rangle = a_r^{\dagger}(\mathbf{k})|0\rangle, \qquad r = 1, 2,$$

and $u(\mathbf{p})$ and $\varepsilon_{\alpha}(\mathbf{k})$ are short for $u_r(\mathbf{p})$ and $\varepsilon_{r\alpha}(\mathbf{k})$. In the following we shall frequently simplify the notation in this way, writing $c(\mathbf{p})$ for $c_r(\mathbf{p})$, etc.

The effect of the uncontracted operators $\bar{\psi}^{-}$, ψ^{-} and A_{α}^{-}, which occur in a term of the S-matrix expansion, acting on $|0\rangle$, is to produce the final state $|f\rangle$. In particular, we find from Eqs. (4.38) and (5.16) that

$$\bar{\psi}^{-}(x)|0\rangle = \sum |e^{-}\mathbf{p}\rangle\left(\frac{m}{VE_{\mathbf{p}}}\right)^{1/2} \bar{u}(\mathbf{p})\,\mathrm{e}^{\mathrm{i}px} \tag{7.26a}$$

$$\psi^{-}(x)|0\rangle = \sum |e^{+}\mathbf{p}\rangle\left(\frac{m}{VE_{\mathbf{p}}}\right)^{1/2} v(\mathbf{p})\,\mathrm{e}^{\mathrm{i}px} \tag{7.26b}$$

$$A_{\alpha}^{-}(x)|0\rangle = \sum |\gamma\mathbf{k}\rangle\left(\frac{1}{2V\omega_{\mathbf{k}}}\right)^{1/2} \varepsilon_{\alpha}(\mathbf{k})\,\mathrm{e}^{\mathrm{i}kx} \tag{7.26c}$$

where the summations are over spin and polarization states, as well as momenta. It is straightforward to generalize the results (7.25) and (7.26) to states involving several particles.

Using Eqs. (7.23)–(7.26) it is easy to calculate S-matrix elements, as the following examples will show.

7.2.1 The first-order terms $S^{(1)}$

The Feynman graphs resulting from the first-order term

$$S^{(1)} = \mathrm{i}e \int \mathrm{d}^4x \mathrm{N}(\bar{\psi}A\!\!\!/\psi)_x \tag{7.27}$$

are just the basic vertex diagrams of Fig. 7.1. Let us calculate the matrix element $\langle f|S^{(1)}|i\rangle$ for one of these processes, namely for electron scattering with emission of a photon, illustrated in Fig. 7.11. In this figure we state the energy–momentum four-vectors of the particles involved but their spin and polarization labels have been suppressed, as discussed above. Fig. 7.11 represents the transition

$$|i\rangle = |e^-\mathbf{p}\rangle = c^\dagger(\mathbf{p})|0\rangle \to |f\rangle = |e^-\mathbf{p}';\gamma\mathbf{k}'\rangle = c^\dagger(\mathbf{p}')a^\dagger(\mathbf{k}')|0\rangle, \tag{7.28}$$

i.e. $|i\rangle$ consists of an electron of momentum $\mathbf{p}$ (and spin state $s = 1, 2$), and $|f\rangle$ of an electron of momentum $\mathbf{p}'$ (and spin state $s' = 1, 2$) plus a photon of momentum $\mathbf{k}'$ (and polarization state $r' = 1, 2$). From Eqs. (7.25)–(7.28) we obtain

$$\begin{aligned}\langle f|S^{(1)}|i\rangle &= \langle e^-\mathbf{p}';\gamma\mathbf{k}'|\mathrm{i}e\int \mathrm{d}^4x\bar{\psi}^-(x)\gamma^\alpha A_\alpha^-(x)\psi^+(x)|e^-\mathbf{p}\rangle \\ &= \mathrm{i}e\int \mathrm{d}^4x\left[\left(\frac{m}{VE_{\mathbf{p}'}}\right)^{1/2}\bar{u}(\mathbf{p}')\,\mathrm{e}^{\mathrm{i}p'x}\right]\gamma^\alpha\left[\left(\frac{1}{2V\omega_{\mathbf{k}'}}\right)^{1/2}\varepsilon_\alpha(\mathbf{k}')\,\mathrm{e}^{\mathrm{i}k'x}\right] \\ &\quad\times\left[\left(\frac{m}{VE_{\mathbf{p}}}\right)^{1/2}u(\mathbf{p})\,\mathrm{e}^{-\mathrm{i}px}\right].\end{aligned} \tag{7.29}$$

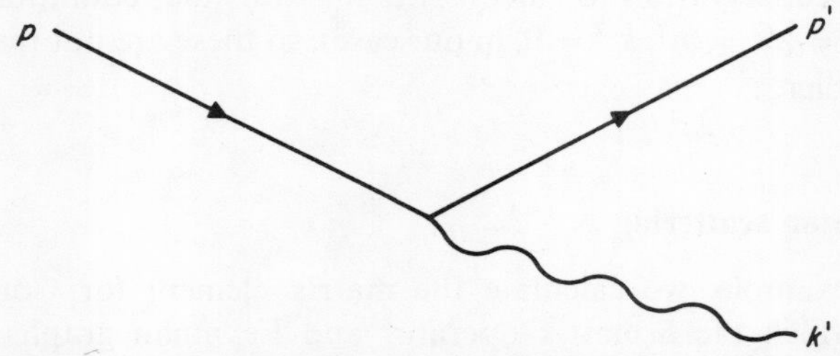

Fig. 7.11. The process $e^- \to e^- + \gamma$. The four-momenta of the particles are shown. The spin and polarization labels (s, s' and r) have been suppressed, as explained in the text.

The x-dependent terms in this expression give

$$\int d^4x \exp[ix(p' + k' - p)] = (2\pi)^4\delta^{(4)}(p' + k' - p), \tag{7.30}$$

where we have anticipated going to the limits of an infinite volume, $V \to \infty$, and an infinite time interval during which the transition may occur. From Eqs. (7.29) and (7.30) we obtain

$$\langle f|S^{(1)}|i\rangle = \left[(2\pi)^4\delta^{(4)}(p' + k' - p)\left(\frac{m}{VE_{\mathbf{p}}}\right)^{1/2}\left(\frac{m}{VE_{\mathbf{p}'}}\right)^{1/2}\left(\frac{1}{2V\omega_{\mathbf{k}'}}\right)^{1/2}\right]\mathcal{M} \tag{7.31}$$

where

$$\mathcal{M} = ie\bar{u}(\mathbf{p}')\not{\varepsilon}(\mathbf{k}' = \mathbf{p} - \mathbf{p}')u(\mathbf{p}). \tag{7.32}$$

Eqs. (7.31) and (7.32) are our final result. $\mathcal{M}$ is called the *Feynman amplitude* for the process represented by the Feynman graph in Fig. 7.11. Since this diagram is labelled by the momenta (and the implied spin and polarization labels) of the particles involved, it is called a Feynman diagram in momentum space, in contrast to the configuration space diagrams of the last section, e.g. Fig. 7.1(a).

The δ-function in Eq. (7.31) arose from the x-integration in Eq. (7.29) over the three exponential functions associated with the two fermion lines and the photon line which meet at the vertex x. This δ-function ensures conservation of energy and momentum for this process: $p = p' + k'$. [Correspondingly, the argument of the polarization vector $\varepsilon_\alpha(\mathbf{k}')$ in Eq. (7.32) was written $\mathbf{k}' = \mathbf{p} - \mathbf{p}'$.] We shall see that for more complicated Feynman diagrams such a δ-function is obtained in this way for each vertex, ensuring energy–momentum conservation at each vertex and consequently for the process as a whole.

For the process $e^- \to e^- + \gamma$ and the other first-order processes energy–momentum conservation is incompatible with the conditions for real particles ($p^2 = p'^2 = m^2$, $k'^2 = 0$, in our case), so these are not real processes, as stated earlier.

7.2.2 Compton scattering

As second example, we calculate the matrix element for Compton scattering, for which the S-matrix operator and Feynman graphs were given in Eqs. (7.9) and (7.10) and Fig. 7.2. Their counterparts in momentum space are shown in Fig. 7.12, corresponding to the transition

$$|i\rangle = c^\dagger(\mathbf{p})a^\dagger(\mathbf{k})|0\rangle \to |f\rangle = c^\dagger(\mathbf{p}')a^\dagger(\mathbf{k}')|0\rangle. \tag{7.33}$$

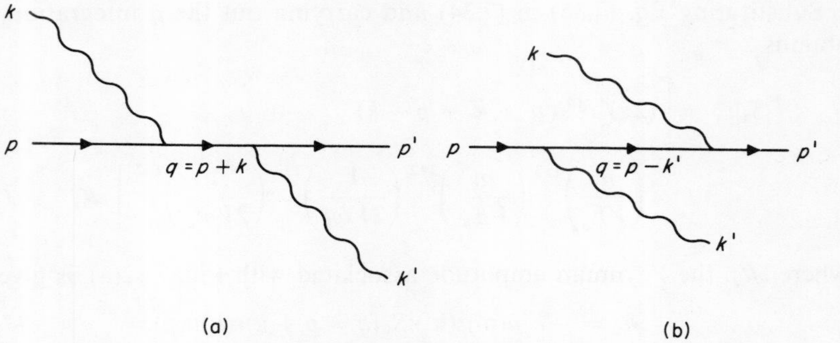

Fig. 7.12. Compton scattering by electrons.

The S-matrix element for this transition is derived from Eqs. (7.9) and (7.10). Using Eqs. (7.23a), (7.25) and (7.26), one obtains

$$\langle f|S_{\mathrm{a}}|i\rangle = -e^2 \int \mathrm{d}^4x_1\,\mathrm{d}^4x_2 \left[\left(\frac{m}{VE_{\mathbf{p}'}}\right)^{1/2} \bar{u}(\mathbf{p}')\,\mathrm{e}^{\mathrm{i}p'x_1}\right]\left[\left(\frac{1}{2V\omega_{\mathbf{k}'}}\right)^{1/2} \not{\varepsilon}(\mathbf{k}')\,\mathrm{e}^{\mathrm{i}k'x_1}\right]$$

$$\times \frac{1}{(2\pi)^4}\int \mathrm{d}^4q\,\mathrm{i}S_{\mathrm{F}}(q)\,\mathrm{e}^{-\mathrm{i}q(x_1-x_2)}$$

$$\times \left[\left(\frac{1}{2V\omega_{\mathbf{k}}}\right)^{1/2} \not{\varepsilon}(\mathbf{k})\,\mathrm{e}^{-\mathrm{i}kx_2}\right]\left[\left(\frac{m}{VE_{\mathbf{p}}}\right)^{1/2} u(\mathbf{p})\,\mathrm{e}^{-\mathrm{i}px_2}\right]. \tag{7.34}$$

Note that u and $\bar{u}$ are four-component spinors and that S_{F} and the factors $\not{\varepsilon}$ are 4×4 matrices. The spinor indices are suppressed, but these quantities must always be written in the correct order of matrix algebra.

The x_1 and x_2 integrations in Eq. (7.34) give

$$\int \mathrm{d}^4x_1 \exp[\mathrm{i}x_1(p' + k' - q)] \int \mathrm{d}^4x_2 \exp[\mathrm{i}x_2(q - p - k)]$$

$$= (2\pi)^4\delta^{(4)}(p' + k' - q)(2\pi)^4\delta^{(4)}(q - p - k)$$

$$= (2\pi)^4\delta^{(4)}(p' + k' - p - k)(2\pi)^4\delta^{(4)}(q - p - k). \tag{7.35}$$

Hence energy and momentum are conserved at each vertex and overall for the process. In particular, the energy–momentum q of the virtual intermediate electron is fixed:

$$q = p + k = p' + k'. \tag{7.36}$$

Substituting Eq. (7.35) in (7.34) and carrying out the q integration, one obtains

$$\langle f|S_{\mathrm{a}}|i\rangle = \left[(2\pi)^4\delta^{(4)}(p' + k' - p - k) \times \left(\frac{m}{VE_{\mathbf{p}}}\right)^{1/2}\left(\frac{m}{VE_{\mathbf{p}'}}\right)^{1/2}\left(\frac{1}{2V\omega_{\mathbf{k}}}\right)^{1/2}\left(\frac{1}{2V\omega_{\mathbf{k}'}}\right)^{1/2}\right]\mathcal{M}_{\mathrm{a}} \qquad (7.37)$$

where $\mathcal{M}_{\mathrm{a}}$, the Feynman amplitude associated with Fig. 7.12(a), is given by

$$\mathcal{M}_{\mathrm{a}} = -e^2\bar{u}(\mathbf{p}')\not{\varepsilon}(\mathbf{k}')\mathrm{i}S_{\mathrm{F}}(q = p + k)\not{\varepsilon}(\mathbf{k})u(\mathbf{p}). \qquad (7.38\mathrm{a})$$

It is left as an exercise for the reader to show that the second contribution to Compton scattering, $\langle f|S_{\mathrm{b}}|i\rangle$, is given by the same equation (7.37) with $\mathcal{M}_{\mathrm{a}}$ replaced by the Feynman amplitude for Fig. 7.12(b):

$$\mathcal{M}_{\mathrm{b}} = -e^2\bar{u}(\mathbf{p}')\not{\varepsilon}(\mathbf{k})\mathrm{i}S_{\mathrm{F}}(q = p - k')\not{\varepsilon}(\mathbf{k}')u(\mathbf{p}). \qquad (7.38\mathrm{b})$$

Our result, Eqs. (7.37) and (7.38), displays some general features which always occur in calculating S-matrix elements by these methods.

Firstly, the factors in Eqs. (7.38a) and (7.38b) are in the correct spinor order. Comparing these expressions with the Feynman graphs, Figs. 7.12(a) and (b), we can describe this order as: Following a fermion line *in the sense of its arrows*, corresponds to writing the spinor factors *from right to left*.

Secondly, comparing these results with Eqs. (7.31) and (7.32), we note many common features. The square brackets in Eqs. (7.31) and (7.37) each contain a δ-function for overall energy–momentum conservation (multiplied by $(2\pi)^4$), and factors $(1/2V\omega_{\mathbf{k}})^{1/2}$ and $(m/VE_{\mathbf{p}})^{1/2}$ for each external photon and fermion line respectively. The Feynman amplitudes (7.32) and (7.38) contain a factor (ie), associated with each vertex in the related Feynman graphs, and factors $\bar{u}$, u and $\not{\varepsilon}$, associated in an obvious manner with external electron and photon lines. The one additional feature in Eqs. (7.38) is the presence of the factors $\mathrm{i}S_{\mathrm{F}}(q)$ which correspond to the intermediate fermion lines in diagrams 7.12(a) and (b). These common features are examples of Feynman rules which will be fully discussed in the next section.

Finally, we see that, for both Figs. 7.12(a) and (b), the intermediate particle cannot be a real particle: $q^2 \neq m^2$, since we cannot have energy–momentum conservation for three real particles at a vertex. This is in contrast to the non-covariant perturbation theory of non-relativistic quantum mechanics, where time and space coordinates (and consequently energy and three-momentum) are treated on different footings: particles in intermediate states satisfy the energy–momentum conditions of real particles (i.e. $p^2 = m^2$, $k^2 = 0$) but energy is not conserved in intermediate states although three-momentum is.

We briefly consider Compton scattering by positrons in order to establish some differences of detail which occur for positrons. The Feynman graphs in

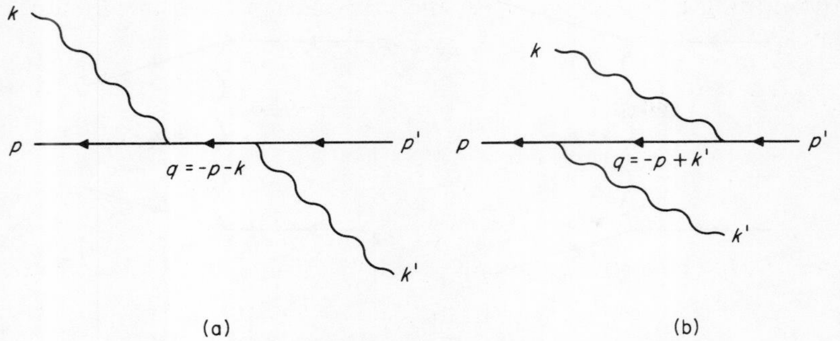

Fig. 7.13. Compton scattering by positrons.

momentum space for this process are shown in Fig. 7.13. We leave it as exercises for the reader to show from first principles that the Feynman diagram 7.13(a) again leads to Eq. (7.37), with $\mathscr{M}_{\mathrm{a}}$ replaced by

$$\mathscr{M}'_{\mathrm{a}} = e^2\bar{v}(\mathbf{p})\not{\varepsilon}(\mathbf{k})\mathrm{i}S_{\mathrm{F}}(q = -p - k)\not{\varepsilon}(\mathbf{k}')v(\mathbf{p}') \tag{7.39}$$

and to obtain the corresponding result for diagram 7.13(b).

In Eq. (7.39), the spinor $v(\mathbf{p}')$ relates to the final-state positron, and the spinor $\bar{v}(\mathbf{p})$ to the initial-state positron. The order of the spinor factors in this equation corresponds to writing these factors *from right to left* as one follows the fermion line *in the sense of its arrows*. This is the same prescription as for electrons. Care is also needed in interpreting the momentum labels on Feynman diagrams. For *external* lines the momenta shown are the *actual* four-momenta of the particles present initially and finally. This applies to electrons, positrons and photons. This means that on external *electron* lines, the flow of four-momentum is in the *same* sense as that of the arrows on the lines; on external *positron* lines it is in the sense *opposite* to that of the arrows. On *internal* fermion lines, on the other hand, the four-momentum labels on Feynman graphs *always* represent energy–momentum flow in the *same* direction as the arrows.

This completes our detailed analysis of Compton scattering. In the following examples, the detailed derivations will be left as exercises for the reader, and we shall concentrate on the remaining features of Feynman graphs not yet encountered.

7.2.3 Electron–electron scattering

The Feynman diagrams in configuration space for Møller scattering were shown in Fig. 7.6. The corresponding momentum space graphs are shown in

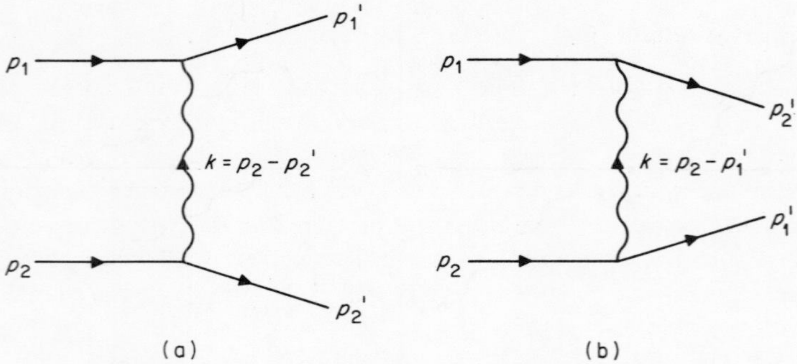

7.14. Electron–electron scattering (Møller scattering).

Fig. 7.14. The S-matrix element for the transition

$$|i\rangle = c^{\dagger}(\mathbf{p}_2)c^{\dagger}(\mathbf{p}_1)|0\rangle \rightarrow |f\rangle = c^{\dagger}(\mathbf{p}_2')c^{\dagger}(\mathbf{p}_1')|0\rangle \tag{7.40}$$

is obtained from Eqs. (7.17). One finds

$$\langle f|S^{(2)}(2e^- \rightarrow 2e^-)|i\rangle$$

$$= \left[(2\pi)^4\delta^{(4)}(p_1' + p_2' - p_1 - p_2)\Pi\left(\frac{m}{VE_{\mathbf{p}}}\right)^{1/2}\right](\mathcal{M}_{\mathrm{a}} + \mathcal{M}_{\mathrm{b}}) \tag{7.41a}$$

where the meaning of $\Pi(m/VE_{\mathbf{p}})^{1/2}$ should be clear to the reader,‡ and where the Feynman amplitudes corresponding to Figs. 7.14(a) and (b) are given by

$$\mathcal{M}_{\mathrm{a}} = -e^2\bar{u}(\mathbf{p}_1')\gamma^{\alpha}u(\mathbf{p}_1)\mathrm{i}D_{\mathrm{F}\alpha\beta}(k = p_2 - p_2')\bar{u}(\mathbf{p}_2')\gamma^{\beta}u(\mathbf{p}_2) \tag{7.41b}$$

$$\mathcal{M}_{\mathrm{b}} = +e^2\bar{u}(\mathbf{p}_2')\gamma^{\alpha}u(\mathbf{p}_1)\mathrm{i}D_{\mathrm{F}\alpha\beta}(k = p_2 - p_1')\bar{u}(\mathbf{p}_1')\gamma^{\beta}u(\mathbf{p}_2). \tag{7.41c}$$

The last two equations exhibit explicitly the relative minus sign of the direct and exchange amplitudes, which reflects the exclusion principle, as discussed in the last section. The new feature in these equations is the appearance of the factors $\mathrm{i}D_{\mathrm{F}\alpha\beta}(k)$, corresponding to the internal photon lines in the Feynman graphs 7.14. Since from Eq. (7.24b) $D_{\mathrm{F}\alpha\beta}(k) = D_{\mathrm{F}\alpha\beta}(-k)$, the sense of k along an internal photon line is arbitrary. However, a definite direction must be chosen for k in order to assign consistent signs to k in the δ-functions associated with the vertices at the two ends of an internal photon line. For example, with the choice for k in Fig. 7.14(a), i.e. from the bottom to the top vertex, $p_2 = p_2' + k$ and $p_1 + k = p_1'$, giving the correct overall energy–momentum conservation for the process.

‡ i.e. even *without* detailed derivation, but Eqs. (7.41) should of course be derived by the reader.

7.2.4 Closed loops

A new feature occurs for Feynman diagrams containing closed loops of internal lines, such as the electron and photon self-energy diagrams, Figs. 7.8 and 7.9. For diagrams without loops, such as we have been considering in this section so far, energy–momentum conservation at the vertices determines the four-momenta of all internal lines completely. For loop diagrams this is not the case. Consider, as a typical example, the electron self-energy. Its Feynman graph in momentum space is shown in Fig. 7.15. Conservation of energy and momentum at the two vertices gives

$$p = q + k = p' \tag{7.42}$$

but this does not determine the internal momenta k and q separately. The intuitive response to this is that one must sum over all allowed values of k and q to find the total amplitude.

To see that this conjecture is correct, we require the matrix element of the electron self-energy operator $S^{(2)}(e^- \to e^-)$, Eq. (7.20), for the transition

$$|i\rangle = c^\dagger(\mathbf{p})|0\rangle \to |f\rangle = c^\dagger(\mathbf{p}')|0\rangle. \tag{7.43}$$

We leave it to the reader to obtain the result

$$\begin{aligned}
&\langle f|S^{(2)}(e^- \to e^-)|i\rangle \\
&= -e^2\left(\frac{m}{VE_{\mathbf{p}}}\right)^{1/2}\left(\frac{m}{VE_{\mathbf{p}'}}\right)^{1/2}\int \mathrm{d}^4q\,\mathrm{d}^4k\,\delta^{(4)}(p'-k-q)\delta^{(4)}(k+q-p) \\
&\quad\times \mathrm{i}D_{\mathrm{F}\alpha\beta}(k)\bar{u}(\mathbf{p}')\gamma^\alpha \mathrm{i}S_{\mathrm{F}}(q)\gamma^\beta u(\mathbf{p}) \\
&= \left[(2\pi)^4\delta^{(4)}(p'-p)\left(\frac{m}{VE_{\mathbf{p}}}\right)^{1/2}\left(\frac{m}{VE_{\mathbf{p}'}}\right)^{1/2}\right]\mathscr{M}
\end{aligned} \tag{7.44a}$$

where

$$\mathscr{M} = \frac{-e^2}{(2\pi)^4}\int \mathrm{d}^4k\,\mathrm{i}D_{\mathrm{F}\alpha\beta}(k)\bar{u}(\mathbf{p})\gamma^\alpha \mathrm{i}S_{\mathrm{F}}(p-k)\gamma^\beta u(\mathbf{p}). \tag{7.44b}$$

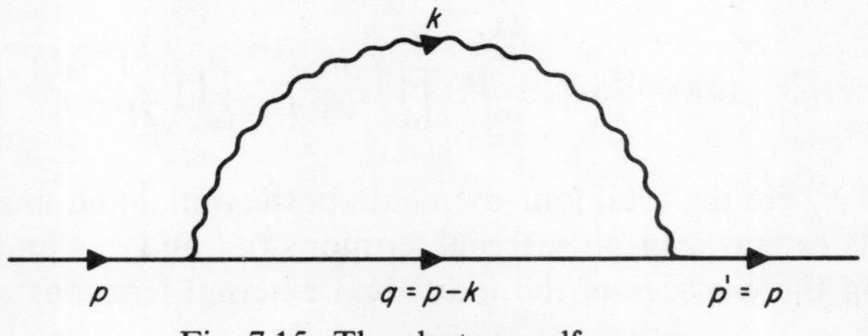

Fig. 7.15. The electron self-energy.

As expected, this Feynman amplitude contains an integration over all internal photon momenta k, while for each value of k the internal fermion momentum has the value $q = p - k$ which corresponds to energy–momentum conservation at the vertices. This integration over an internal momentum is typical of a closed loop. (For another example, see Problem 7.2 on the photon self-energy.)

Eqs. (7.44) display the same structure which occurred in all our other examples. The origins of the individual factors in Eq. (7.44a) should be clear to the reader. The Feynman amplitude $\mathscr{M}$ contains factors $\mathrm{i}S_\mathrm{F}$ and $\mathrm{i}D_\mathrm{F}$ for fermion and photon propagators, spinors u and $\bar{u}$ for the initial and final external electron lines, and a γ-factor for each vertex. The spinor quantities are in the expected order as one follows the fermion line in the direction of its arrows. The remaining factor $(-e^2) = (\mathrm{i}e)^2$ has its origin in the form of $S^{(2)}$ for QED, Eqs. (7.1) and (7.2).

7.3 FEYNMAN RULES FOR QED

The S-matrix elements $\langle f|S|i\rangle$ which we have calculated for various processes exhibit a definite structure, which allows one to identify individual factors and features with different aspects of the corresponding Feynman graphs. The same identification between the mathematical expressions and Feynman graphs is possible for all processes. Furthermore no new features occur for other processes. This enables one to construct a set of rules for writing $\langle f|S|i\rangle$ down directly from the Feynman graphs.

In this section we shall state these rules for QED. They represent a generalization and a tidying-up of earlier results. Their origins should be clear to the reader. Where appropriate, we shall give explanations and cross-references, but we do not repeat everything from scratch; for example, we are assuming the conventions about arrows and momentum labels for Feynman graphs.

The expression for the S-matrix element for a transition can at once be written down by generalizing our earlier results. For the transition $|i\rangle \to |f\rangle$, where the initial and final states are specified by the momenta (and spin and polarization variables) of the particles present, the S-matrix element is given by

$$\langle f|S|i\rangle = \delta_{fi} + \left[(2\pi)^4\delta^{(4)}(P_f - P_i)\prod_{\text{ext.}}\left(\frac{m}{VE}\right)^{1/2}\prod_{\text{ext.}}\left(\frac{1}{2V\omega}\right)^{1/2}\right]\mathscr{M}. \quad (7.45)$$

Here P_i and P_f are the total four-momenta in the initial and final states, and the products extend over all external fermions (e^- and e^+) and photons, E and ω being the energies of the individual external fermions and photons respectively.

The Feynman amplitude $\mathscr{M}$ is given by

$$\mathscr{M} = \sum_{n=1}^{\infty} \mathscr{M}^{(n)} \tag{7.46}$$

where the contribution $\mathscr{M}^{(n)}$ comes from the nth order perturbation term $S^{(n)}$. The Feynman amplitude $\mathscr{M}^{(n)}$ is obtained by drawing all topologically different, connected Feynman graphs in momentum space which contain n vertices and the correct external lines. The contribution to $\mathscr{M}^{(n)}$ from each graph is obtained from the following Feynman rules.

1. For each vertex, write a factor $\mathrm{i}e\gamma^{\alpha}$ [see Eqs. (7.1) and (7.2)].

2. For each internal photon line, labelled by the momentum k, write a factor [see Eq. (7.24b)]

$$\mathrm{i}D_{\mathrm{F}\alpha\beta}(k) = \mathrm{i}\,\frac{-g_{\alpha\beta}}{k^2 + \mathrm{i}\varepsilon}. \tag{7.47}$$

3. For each internal fermion line, labelled by the momentum p, write a factor [see Eq. (7.24a)]

$$\mathrm{i}S_{\mathrm{F}}(p) = \mathrm{i}\,\frac{1}{\not{p} - m + \mathrm{i}\varepsilon}. \tag{7.48}$$

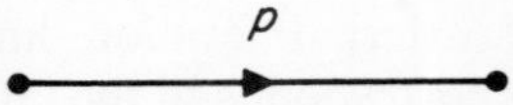

4. For each external line, write one of the following factors [see Eqs. (7.25) and (7.26)]:

(a) for each initial electron: $u_r(\mathbf{p})$ (7.49a)

(b) for each final electron: $\bar{u}_r(\mathbf{p})$ (7.49b)

(c) for each initial positron: $\bar{v}_r(\mathbf{p})$ (7.49c)

(d) for each final positron: $v_r(\mathbf{p})$ (7.49d)

(e) for each initial photon: $\varepsilon_{r\alpha}(\mathbf{k})$ (7.49e)

(f) for each final photon‡: $\varepsilon_{r\alpha}(\mathbf{k})$ (7.49f)

In Eqs. (7.49) $\mathbf{p}$ and $\mathbf{k}$ denote the three-momenta of the external particles, and $r(=1, 2)$ labels their spin and polarization states.

5. The spinor factors (γ-matrices, S_{F}-functions, four-spinors) for each fermion line are ordered so that, reading from right to left, they occur in the same sequence as following the fermion line in the direction of its arrows.

6. For each closed fermion loop, take the trace and multiply by a factor (-1).

‡ For linear polarization states, which we are using, $\varepsilon_{r\alpha}(\mathbf{k})$ is real. In general it is complex (e.g. for circular polarization), and we must then replace $\varepsilon_{r\alpha}(\mathbf{k})$ by $\varepsilon^{*}_{r\alpha}(\mathbf{k})$ for a final-state photon.

This rule follows directly from the corresponding result in configuration space, derived in Section 7.1 [see Eq. (7.22)].

7. The four-momenta associated with the three lines meeting at each vertex satisfy energy–momentum conservation. For each four-momentum q which is not fixed by energy–momentum conservation carry out the integration $(2\pi)^{-4} \int d^4q$. One such integration with respect to an internal momentum variable q occurs for each closed loop.

We had an example of this rule in Eq. (7.44b) for the electron self-energy. Inclusion of the factors $(2\pi)^{-4}$ in this rule is convenient, since all numerical factors (except for the phase factor of rule 8) are accounted for in this way, as will be shown below.

8. Multiply the expression by a phase factor δ_P which is equal to $+1$ (-1) if an even (odd) number of interchanges of neighbouring fermion operators is required to write the fermion operators in the correct normal order.

In general this phase factor is only of significance when the contributions of several Feynman graphs are added, and only the relative signs matter. The situation most frequently met is the one we discussed for (e^-e^-)- and (e^-e^+)-scattering, involving contributions from diagrams which differ only by the interchange of external fermion lines associated with identical fermion operators. This corresponds to the interchange of either (i) two initial $e^-(e^+)$ lines, or (ii) two final $e^-(e^+)$ lines, or (iii) an initial $e^-(e^+)$ line with a final $e^+(e^-)$ line.

It remains to justify our assertion, made when discussing rule 7, that the above rules allow for all numerical factors. The only factors not taken into account so far are factors $(2\pi)^4$ which occur together with δ-functions or result from propagators. The x-integration at each vertex gives a factor $(2\pi)^4$ [see Eq. (7.30)], and the Fourier transform of each propagator gives a factor $(2\pi)^{-4}$ [see Eqs. (7.23)]. For a Feynman diagram containing n vertices and $f_i(b_i)$ internal fermion (photon) lines, the Feynman amplitude contains a factor

$$[(2\pi)^4]^{n-f_i-b_i-1} \tag{7.50}$$

where the exponent -1 allows for the factor $(2\pi)^4$ which was separated out in Eq. (7.45). It is left as a problem for the reader to show that for a Feynman diagram containing l closed loops

$$n - f_i - b_i - 1 = -l. \tag{7.51}$$

Since one momentum integral $\int d^4q$ occurs for each loop, we may omit the factor (7.50) from the Feynman amplitude expression provided we replace each loop integral $\int d^4q$ by $(2\pi)^{-4} \int d^4q$.

This completes our discussion of Feynman rules for QED. The reader is recommended to use Feynman rules to re-derive the matrix elements $\langle f|S|i\rangle$ which we obtained from first principles earlier in this chapter. After a little practice, Feynman rules provide an extraordinarily simple method for obtaining even very complicated matrix elements, and for this reason they form the basis of most practical calculations. Similar diagrammatic techniques are also of great importance in many other fields, e.g. weak interactions (to be studied later in this book) and condensed matter physics, where analogous rules can be developed. In Appendix B, at the end of this book, we give a summary of these rules for QED, as well as those for the standard electro–weak theory which will be derived later.

7.4 LEPTONS

So far we have treated QED as the interaction of electrons and positrons with the electromagnetic field. But there are many electrically charged particles in nature. These can be divided into two categories: *hadrons* which also interact via strong forces (often called nuclear forces), and *leptons* which do not. Both types interact via weak as well as electromagnetic interactions.

Electrons and positrons are leptons, as are muons ($\mu^{\pm}$) and tauons ($\tau^{\pm}$), the latter discovered in 1975 only.[‡] Both muons and tauons have spin $\frac{1}{2}$ and charge $\pm e$. Furthermore, within experimental accuracy (which is high in the case of muon experiments), they exhibit all the properties of particles whose interactions are identical with those of electrons, except for their different masses: $m_\mu = 206.8\, m_e = 105.7$ MeV, $m_\tau = (1784 \pm 3)$ MeV. This is referred to as *$e - \mu - \tau$ universality*. Whether other particles exist in this series of so-called sequential leptons is an open question. Present experiments indicate only that none exist with a mass of less than about 15 GeV.

QED is usually understood to include the interaction of all kinds of leptons ($e, \mu, \tau, \ldots$) with the electromagnetic field. The reason why hadrons are excluded will become clear shortly. This extended QED which we shall now study displays a new richness: processes involving more than one kind of lepton.

Assuming universality, the extension of the theory is almost trivial. Like the electron, we describe each kind of lepton by a Dirac spinor field: $\psi_l(x)$, where l labels the kind of lepton: $l = e, \mu, \tau, \ldots$. The generalization of the free-field Lagrangian density for electrons, Eq. (4.67), is

$$\mathscr{L}_0 = \sum_l \bar{\psi}_l(x)(\mathrm{i}\gamma^\alpha \partial_\alpha - m_l)\psi_l(x), \tag{7.52a}$$

and making the minimal substitution (4.64b) (wtih $q = -e$) leads to the interaction Hamiltonian density

$$\mathscr{H}_{\mathrm{I}}(x) = -\mathscr{L}_{\mathrm{I}}(x) = -e \sum_l \mathrm{N}[\bar{\psi}_l(x) A\!\!\!/(x)\psi_l(x)]. \tag{7.52b}$$

[‡] Just as the names muon and tauon refer to both positively and negatively charged particles, so it is convenient to have a single word for electron and positron. It is usual to use electron for this purpose as well as for the negatively charged member of this pair. We shall follow this practice, adding the appropriate qualification when it is required to avoid ambiguity.

This equation describes a *local* interaction since all field operators are evaluated at a single space–time point. This is appropriate for the interaction of the electromagnetic field with a point particle. While, within the limits of current experiment, leptons are point-like, hadrons have finite size. For example, the experimental value of the proton radius is of the order 0.8×10^{-15} m.[‡] For this reason the electromagnetic interactions of charged hadrons cannot be described by expressions like (7.52b).

The second point to note about the interaction (7.52b) is that it consists of a sum of terms each of which involves *one* kind of lepton only. Hence, the interaction is described by basic vertex parts like those of Fig. 7.1, with *both* fermion lines at a vertex referring to the *same* kind of lepton. Instead of two electrons, as in Fig. 7.1, they could both be muons or both tauons. But we could not, for example, have one electron and one muon. The vertex part in Fig. 7.16 conserves charge but it does not occur with the interaction (7.52b) since it would require an interaction term of the form $-e\bar{\psi}_\mu \not{A} \psi_e$. Consequently, for any non-vanishing matrix element $\langle j|\mathscr{H}_I|i\rangle$ the *electron number* $N(e)$, defined by

$$N(e) = N(e^-) - N(e^+) \tag{7.53a}$$

in an obvious notation, is conserved, as are the *muon and tauon numbers*[§]

$$N(\mu) = N(\mu^-) - N(\mu^+) \tag{7.53b}$$

$$N(\tau) = N(\tau^-) - N(\tau^+). \tag{7.53c}$$

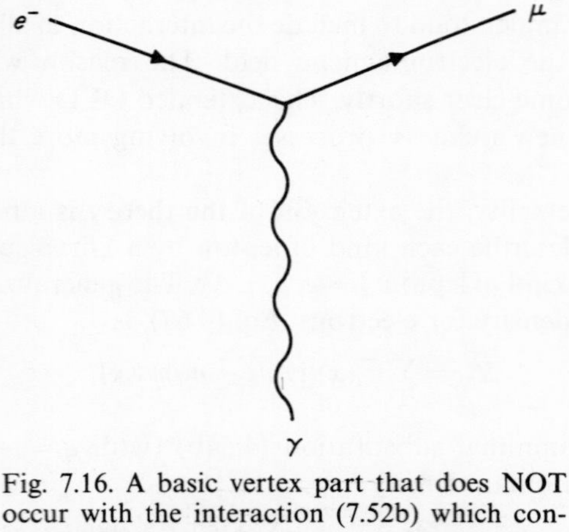

Fig. 7.16. A basic vertex part that does NOT occur with the interaction (7.52b) which conserves electron and muon numbers at each vertex.

[‡] The point-like nature of leptons has been tested to much shorter distances. (See Sections 8.4 and 8.5.)

[§] These definitions will be modified when we consider weak interactions.

Consequently processes like

$$e^- + \mu^+ \to e^+ + \mu^-, \tag{7.54}$$

although they conserve charge, are forbidden and indeed are not observed.

The extension of the S-matrix formalism and of the Feynman rules to the QED interaction (7.52b) is now straightforward. Each term $(\bar{\psi}A\!\!\!/\psi)$ in our original interaction (7.2), which allowed for electrons only, is replaced by a sum $\sum_l (\bar{\psi}_l A\!\!\!/\psi_l)$, and the S-matrix expansion (7.1) leads to

$$S = \sum_{n=0}^{\infty} \frac{(\mathrm{i}e)^n}{n!} \int \ldots \int \mathrm{d}^4x_1 \ldots \mathrm{d}^4x_n \sum_{l_1} \ldots \sum_{l_n} \times \mathrm{T}\{\mathrm{N}(\bar{\psi}_{l_1} A\!\!\!/\psi_{l_1})_{x_1} \ldots \mathrm{N}(\bar{\psi}_{l_n} A\!\!\!/\psi_{l_n})_{x_n}\}. \tag{7.55}$$

This expansion first of all contains terms involving one kind of lepton only. These are the terms we considered in the last two sections, but we could now be considering muons or tauons instead of electrons. $S^{(1)}$ is of this type and so are the terms in $S^{(2)}$ with $l_1 = l_2$ $(=e, \mu, \ldots)$. The more interesting terms, for which new processes occur, are those involving more than one kind of lepton.

Consider, for example, the $l_1 = \mu$, $l_2 = e$ term in $S^{(2)}$, given by

$$S^{(2)}_{\mu e} = -e^2 \int \mathrm{d}^4x_1\, \mathrm{d}^4x_2 \mathrm{T}\{\mathrm{N}(\bar{\psi}_\mu A\!\!\!/\psi_\mu)_{x_1} \mathrm{N}(\bar{\psi}_e A\!\!\!/\psi_e)_{x_2}\}. \tag{7.56}$$

Using Wick's theorem [Eqs. (6.35) and (6.38)] to expand the T-product in terms of normal products, we obtain

$$S^{(2)}_{\mu e} = -e^2 \int \mathrm{d}^4x_1\, \mathrm{d}^4x_2 \mathrm{N}[(\bar{\psi}_\mu A\!\!\!/\psi_\mu)_{x_1} (\bar{\psi}_e A\!\!\!/\psi_e)_{x_2}] - e^2 \int \mathrm{d}^4x_1\, \mathrm{d}^4x_2 \mathrm{N}[(\bar{\psi}_\mu \underbrace{A\!\!\!/\psi_\mu)_{x_1} (\bar{\psi}_e A\!\!\!/}\psi_e)_{x_2}]. \tag{7.57}$$

All other unequal-time contractions vanish as follows from the definition (6.31) of a contraction as a vacuum expectation value.‡

The first term in Eq. (7.57), like the term $S^{(2)}_{\mathrm{A}}$, Eq. (7.5a), corresponds to two independent unphysical processes of the kind shown in Fig. 7.1, except that now one refers to a muon instead of an electron.

The second term in Eq. (7.57) gives rise to processes which involve two external muons and two external electrons, and which must conserve charge, electron number and muon number. These include electron–muon scattering

‡ When dealing with several fermion fields, we must assume that the field operators for different fermion fields anticommute. We continue to assume that fermion and boson field operators always commute with each other. (See J. D. Bjorken and S. D. Drell, *Relativistic Quantum Fields*, McGraw-Hill, New York, 1965, p. 98.)

and, more interestingly, the process

$$e^+ + e^- \to \mu^+ + \mu^-, \tag{7.58}$$

i.e. the annihilation of an (e^+e^-) pair leading to the creation of a $(\mu^+\mu^-)$ pair. The term in $S_{\mu e}^{(2)}$ responsible for this process is

$$S^{(2)}(e^+e^- \to \mu^+\mu^-)$$
$$= -e^2 \int \mathrm{d}^4x_1\, \mathrm{d}^4x_2 \mathrm{N}[(\bar{\psi}_\mu^- \gamma^\alpha \psi_\mu^-)_{x_1} (\bar{\psi}_e^+ \gamma^\beta \psi_e^+)_{x_2}] \mathrm{i}D_{\mathrm{F}\alpha\beta}(x_1 - x_2), \tag{7.59}$$

and from this operator one can calculate the transition matrix elements. For the transition shown in the Feynman graph in Fig. 7.17, i.e.

$$|i\rangle = |e^-\mathbf{p}_2; e^+\mathbf{p}_1\rangle = c_e^\dagger(\mathbf{p}_2)d_e^\dagger(\mathbf{p}_1)|0\rangle$$
$$\to |f\rangle = |\mu^-\mathbf{p}_2'; \mu^+\mathbf{p}_1'\rangle = c_\mu^\dagger(\mathbf{p}_2')d_\mu^\dagger(\mathbf{p}_1')|0\rangle, \tag{7.60}$$

one would in this way obtain the Feynman amplitude

$$\mathscr{M}^{(2)}(e^+e^- \to \mu^+\mu^-) = -\mathrm{i}e^2\bar{u}_\mu(\mathbf{p}_2')\gamma^\alpha v_\mu(\mathbf{p}_1')D_{\mathrm{F}\alpha\beta}(p_1 + p_2)\bar{v}_e(\mathbf{p}_1)\gamma^\beta u_e(\mathbf{p}_2). \tag{7.61}$$

Here the labels e and μ, attached to the fermion lines in Fig. 7.17, to creation and absorption operators in Eq. (7.60), and to spinors in Eq. (7.61), distinguish electrons and muons.

We do not advocate deriving the Feynman amplitude (7.61) from first principles (other than as an exercise), since it is trivial to extend the rules for calculating amplitudes, which were given in the last section, to QED involving several leptons. The S-matrix operator (7.59) differs from the operator S_b, Eq. (7.19c), for the annihilation diagram, Fig. 7.7(b), for (e^+e^-) scattering in that in the final state the electrons are replaced by muons. Hence the amplitude (7.61) can be derived in two steps.

Firstly, obtain the Feynman amplitude $\mathscr{M}_b$ corresponding to the operator S_b, Eq. (7.19c), for the transition analogous to (7.60) but with all particles electrons. $\mathscr{M}_b$ can be written down directly, using the Feynman rules of Section 7.3.

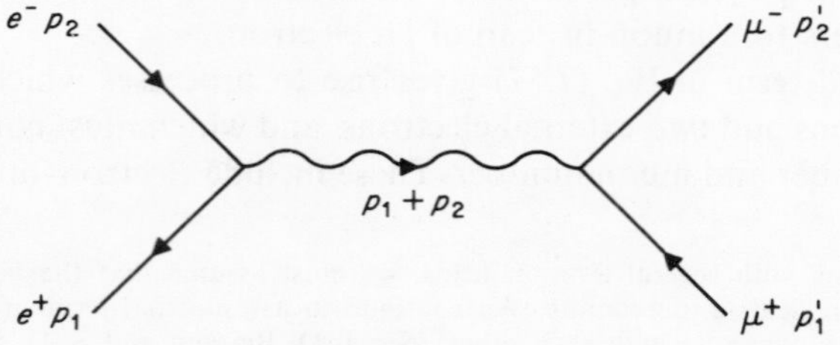

Fig. 7.17. The process $e^+ + e^- \to \mu^+ + \mu^-$.

Secondly, in the expression for $\mathscr{M}_b$, replace all quantities referring to the final state electrons by the corresponding quantities for muons.

It is left as an exercise for the reader to verify that this procedure leads to Eq. (7.61).

There is one important difference between the $e^+e^- \rightarrow \mu^+\mu^-$ process and (e^+e^-) scattering. For the latter, a second contribution stems from S_a, Eq. (7.19b), corresponding to Fig. 7.7(a). For the former, as we have seen, Eq. (7.57) gives no such contribution. It would correspond to replacing the final electron lines in Fig. 7.7(a) by muon lines, so that each vertex would involve one electron line and one muon line (e.g. the vertex at the top would look like Fig. 7.16), violating the conservation of both electron number $N(e)$ and of muon number $N(\mu)$ at each vertex.

From this example, it is easy to see how to extend the rules of Section 7.3. For any process, one must draw all relevant Feynman diagrams which conserve $N(e)$, $N(\mu)$, ... at each vertex, i.e. the two lepton lines entering and leaving a vertex must be of the same kind (both e or both μ, etc.). The Feynman amplitude corresponding to each of these diagrams is then written down directly using the Feynman rules of the last section.

PROBLEMS

7.1 Derive the lowest-order non-vanishing S-matrix element (7.19) and hence the corresponding Feynman amplitude for Bhabha scattering, i.e. the process

$$e^+(\mathbf{p}_1, r_1) + e^-(\mathbf{p}_2, r_2) \rightarrow e^+(\mathbf{p}'_1, s_1) + e^-(\mathbf{p}'_2, s_2).$$

7.2 Show that the Feynman amplitude for the photon self-energy diagram in Fig. 7.9 is given by

$$\mathscr{M} = \frac{-e^2}{(2\pi)^4}\int \mathrm{d}^4p\,\mathrm{Tr}\,[\not\varepsilon_r(\mathbf{k})S_F(p+k)\not\varepsilon_r(\mathbf{k})S_F(p)]$$

where $\mathbf{k}$ and $\varepsilon_r(\mathbf{k})$ are the momentum and polarization vectors of the photon.

7.3 A real scalar field $\phi(x)$, associated with a spin-zero boson B, is described by the Lagrangian density

$$\mathscr{L}(x) = \mathscr{L}_0(x) + \mathscr{L}_I(x)$$

where $\mathscr{L}_0$ is the free-field density (3.5), and

$$\mathscr{L}_I(x) = g[\phi(x)]^4/4!$$

describes an interaction of the field with itself, with g a real coupling constant. (Normal ordering of operators is assumed throughout.)

Write down the S-matrix expansion, and pick out the normal ordered term that gives rise to the BB scattering process

$$B(\mathbf{k}_1) + B(\mathbf{k}_2) \rightarrow B(\mathbf{k}_3) + B(\mathbf{k}_4)$$

in first-order perturbation theory. Draw the Feynman diagram representing this term, and show that the corresponding S-matrix element is given by

$$\langle k_3, k_4|S^{(1)}|k_1, k_2\rangle = (2\pi)^4\delta^{(4)}(k_3 + k_4 - k_1 - k_2)\prod_i\left(\frac{1}{2V\omega_i}\right)^{1/2}\mathcal{M}$$

with the Feynman amplitude $\mathcal{M} = ig$. [Note that $\mathcal{M}$ is independent of the boson four-momenta $k_i^\alpha \equiv (\omega_i, \mathbf{k}_i)$.]

7.4 Pseudo-scalar meson theory is defined by the Lagrangian density

$$\mathcal{L}(x) = \mathcal{L}_0(x) + \mathcal{L}_I(x)$$

where

$$\mathcal{L}_0(x) = \tfrac{1}{2}[\partial_\alpha\phi(x)\partial^\alpha\phi(x) - \mu^2\phi^2(x)] + \bar{\psi}(x)(i\gamma^\alpha\partial_\alpha - m)\psi(x)$$

represents a free real spin 0 field $\phi(x)$ and a free fermion field $\psi(x)$, and

$$\mathcal{L}_I(x) = -ig\bar{\psi}(x)\gamma_5\psi(x)\phi(x)$$

describes their interaction.

The interaction Lagrangian density $\mathcal{L}_I(x)$ is similar to that of QED, except that $e\gamma^\alpha$ is replaced by $(-ig\gamma_5)$, and the photon field $A_\alpha(x)$ is replaced by the meson field $\phi(x)$. Exploit this similarity to write down the Feynman rules for pseudo-scalar meson theory.

7.5 A real scalar field $\phi(x)$ is described by the Lagrangian density

$$\mathcal{L}(x) = \mathcal{L}_0(x) + \mu U(\mathbf{x})\phi^2(x),$$

where $\mathcal{L}_0$ is the free-field Lagrangian density (3.5), and $U(\mathbf{x})$ is a static external potential.

Derive the equation of motion

$$(\Box + \mu^2)\phi(x) = 2\mu U(\mathbf{x})\phi(x).$$

Show that, in lowest order, the S-matrix element for an incoming boson, with momentum $k_i = (\omega_i, \mathbf{k}_i)$, to be scattered to a state with momentum $k_f = (\omega_f, \mathbf{k}_f)$, is given by

$$\langle \mathbf{k}_f|S^{(1)}|\mathbf{k}_i\rangle = \frac{i2\pi\,\delta(\omega_f - \omega_i)}{(2V\omega_i)^{1/2}(2V\omega_f)^{1/2}}\,2\mu\tilde{U}(\mathbf{k}_f - \mathbf{k}_i)$$

where

$$\tilde{U}(\mathbf{q}) = \int d^3\mathbf{x}\,U(\mathbf{x})\,e^{-i\mathbf{q}\cdot\mathbf{x}}.$$

This type of problem, with a static external potential, will be considered further in Chapter 8.

CHAPTER 8

QED Processes in lowest order

In the last chapter we established the Feynman rules for obtaining the matrix element S_{fi} for any collision process in QED. In this chapter we shall start by deriving from S_{fi} the experimentally observable quantities, i.e. the cross-sections. This is a straightforward generalization of the corresponding kinematical and phase–space arguments of non-relativistic collision theory.

The cross-sections obtained in this way are fully polarized, i.e. the photons and leptons present initially and finally are in definite polarization states. (As is customary, we use the term 'polarization state' for both photons and fermions, meaning a spin state in the latter case.) In most practical situations, the beams of colliding particles are unpolarized, and the polarizations of the particles produced in the collision are not observed. It then becomes necessary to average and sum over polarization states of initial and final particles respectively. The very powerful and elegant techniques for performing these spin and polarization sums are developed in Sections 8.2 and 8.3. The corresponding formalism for analysing polarization properties is more complex, and we shall consider a simple example only.

In Sections 8.4–8.6, we shall illustrate our results by deriving the cross-sections, in lowest non-vanishing order of perturbation theory, for some of the processes considered in the previous chapter. By the end of this chapter, the reader should be able to deal in a similar way with any collision problem in QED. (A reader who tires of these applications should not be tempted also to omit Sections 8.7–8.9 which introduce some fundamental new ideas.)

We shall extend the S-matrix formalism to allow for the presence of an external electromagnetic field, i.e. of a field whose quantum fluctuations are negligible, so that it can be described by an unquantized classical field. As an application of these ideas we shall consider the scattering of electrons by the Coulomb field of a nucleus, both elastic scattering (Section 8.7) and inelastic scattering accompanied by emission of radiation, i.e. bremsstrahlung (Section 8.8).

In studying these Coulomb scattering processes, we shall encounter a new feature. There exists the possibility of the emission by a charged particle of one or more *very soft* photons (i.e. with very little energy). Experimentally, because of finite energy resolution, the distinction between elastic and inelastic scattering becomes blurred. This unrealistic separation into elastic and inelastic scattering events leads to the infrared divergence. In the last section of this chapter we shall see how this difficulty is resolved.

8.1 THE CROSS-SECTION

We consider a scattering process in which two particles, they may be leptons or photons, with four-momenta $p_i = (E_i, \mathbf{p}_i)$, $i = 1, 2$, collide and produce N final particles with momenta $p'_f = (E'_f, \mathbf{p}'_f)$, $f = 1, \ldots, N$. Initial and final particles are assumed to be in definite polarization states. As in Chapter 7, the indices labelling these states will in general be suppressed. Eq. (7.45), defining the Feynman amplitude $\mathscr{M}$ for this process, can now be written

$$S_{fi} = \delta_{fi} + (2\pi)^4\,\delta^{(4)}\Big(\sum p'_f - \sum p_i\Big)\prod_i\left(\frac{1}{2VE_i}\right)^{1/2} \times \prod_f\left(\frac{1}{2VE'_f}\right)^{1/2}\prod_l (2m_l)^{1/2}\mathscr{M} \tag{8.1}$$

where the index l runs over all external leptons in the process.

Eq. (8.1) corresponds to the limit of an infinite time interval, $T \to \infty$, and an infinite volume, $V \to \infty$. For finite T and V, we would have obtained the same expression (8.1) with

$$(2\pi)^4\,\delta^{(4)}\Big(\sum p'_f - \sum p_i\Big) = \lim_{\substack{T\to\infty\\ V\to\infty}} \delta_{TV}\Big(\sum p'_f - \sum p_i\Big) \equiv \lim_{\substack{T\to\infty\\ V\to\infty}} \int_{-T/2}^{T/2} \mathrm{d}t \int_V \mathrm{d}^3\mathbf{x}\, \exp\Big[\mathrm{i}x\Big(\sum p'_f - \sum p_i\Big)\Big] \tag{8.2}$$

replaced by $\delta_{TV}(\sum p'_f - \sum p_i)$. In deriving the cross-section, it will help to take T and V finite, to begin with. In this case the transition probability per unit time

$$w = |S_{fi}|^2/T \tag{8.3}$$

involves the factor $[\delta_{TV}(\sum p'_f - \sum p_i)]^2$. For large values of T and V we can then take

$$\delta_{TV}\left(\sum p'_f - \sum p_i\right) = (2\pi)^4\,\delta^{(4)}\left(\sum p'_f - \sum p_i\right) \tag{8.4}$$

and

$$\left[\delta_{TV}\left(\sum p'_f - \sum p_i\right)\right]^2 = TV(2\pi)^4\,\delta^{(4)}\left(\sum p'_f - \sum p_i\right) \tag{8.5}$$

with errors which tend to zero as $T \to \infty$ and $V \to \infty$. Hence Eq. (8.3) becomes

$$w = V(2\pi)^4\,\delta^{(4)}\left(\sum p'_f - \sum p_i\right)\left(\prod_i \frac{1}{2VE_i}\right)\left(\prod_f \frac{1}{2VE'_f}\right)\left(\prod_l (2m_l)\right)|\mathscr{M}|^2. \tag{8.6}$$

Eq. (8.6) is the transition rate to one definite final state. To obtain the transition rate to a group of final states with momenta in the intevals $(\mathbf{p}'_f, \mathbf{p}'_f + \mathrm{d}\mathbf{p}'_f), f = 1, \ldots, N$, we must multiply w by the number of these states which is

$$\prod_f \frac{V\,\mathrm{d}^3\mathbf{p}'_f}{(2\pi)^3}. \tag{8.7}$$

The differential cross-section is the transition rate into this group of final states for one scattering centre and unit incident flux. With our choice of normalization for the states, the volume V which we are considering contains one scattering centre, and the incident flux is v_{rel}/V where v_{rel} is the relative velocity of the colliding particles.

Combining these results with Eq. (8.6), we obtain the required expression for the differential cross-section

$$\begin{aligned} \mathrm{d}\sigma &= w\frac{V}{v_{\text{rel}}}\prod_f \frac{V\,\mathrm{d}^3\mathbf{p}'_f}{(2\pi)^3} \\ &= (2\pi)^4\,\delta^{(4)}\left(\sum p'_f - \sum p_i\right)\frac{1}{4E_1E_2v_{\text{rel}}}\left(\prod_l (2m_l)\right)\left(\prod_f \frac{\mathrm{d}^3\mathbf{p}'_f}{(2\pi)^3 2E'_f}\right)|\mathscr{M}|^2. \end{aligned} \tag{8.8}$$

Eq. (8.8) holds in any Lorentz frame in which the colliding particles move collinearly. In such a frame the relative velocity v_{rel} is given by the expression

$$E_1E_2v_{\text{rel}} = [(p_1p_2)^2 - m_1^2m_2^2]^{1/2} \tag{8.9}$$

where m_1 and m_2 are the rest masses of the colliding particles. Two important examples of such frames are the centre-of-mass (CoM) system, and the laboratory (Lab) system. In the CoM system we have $\mathbf{p}_1 = -\mathbf{p}_2$, and hence

$$v_{\text{rel}} = \frac{|\mathbf{p}_1|}{E_1} + \frac{|\mathbf{p}_2|}{E_2} = |\mathbf{p}_1| \frac{E_1 + E_2}{E_1 E_2} \quad \text{(CoM).} \tag{8.10a}$$

In the laboratory system, the target particle (particle 2, say) is at rest, $\mathbf{p}_2 = 0$, and

$$v_{\text{rel}} = \frac{|\mathbf{p}_1|}{E_1} \quad \text{(Lab).} \tag{8.10b}$$

Eqs. (8.10) of course also follow from the general result (8.9).

The relativistic invariance of the cross-section formula (8.8) follows from Eq. (8.9) and from the Lorentz invariance of $\mathrm{d}^3\mathbf{p}/2E$ for any four-vector $p = (E, \mathbf{p})$.‡

Because of conservation of energy and momentum, the final-state momenta $\mathbf{p}'_1, \ldots, \mathbf{p}'_N$ are not all independent variables. In order to obtain a differential cross-section in the independent variables appropriate to a given situation, we integrate Eq. (8.8) with respect to the remaining variables. We illustrate this for the frequently occurring case of a process leading to a two-body final state. Eq. (8.8) now becomes

$$\mathrm{d}\sigma = f(p'_1, p'_2)\, \delta^{(4)}(p'_1 + p'_2 - p_1 - p_2)\, \mathrm{d}^3\mathbf{p}'_1\, \mathrm{d}^3\mathbf{p}'_2 \tag{8.12a}$$

where

$$f(p'_1, p'_2) \equiv \frac{1}{64\pi^2 v_{\text{rel}} E_1 E_2 E'_1 E'_2} \left(\prod_l (2m_l) \right) |\mathcal{M}|^2. \tag{8.12b}$$

Integration of Eq. (8.12a) with respect to $\mathbf{p}'_2$ gives

$$\mathrm{d}\sigma = f(p'_1, p'_2)\, \delta(E'_1 + E'_2 - E_1 - E_2) |\mathbf{p}'_1|^2 \mathrm{d}|\mathbf{p}'_1|\, \mathrm{d}\Omega'_1 \tag{8.13}$$

where $\mathbf{p}'_2 = \mathbf{p}_1 + \mathbf{p}_2 - \mathbf{p}'_1$, and integrating Eq. (8.13) over $|\mathbf{p}'_1|$ we obtain§

$$\mathrm{d}\sigma = f(p'_1, p'_2) |\mathbf{p}'_1|^2\, \mathrm{d}\Omega'_1 \left[\frac{\partial (E'_1 + E'_2)}{\partial |\mathbf{p}'_1|} \right]^{-1} \tag{8.15}$$

‡ $\mathrm{d}^3\mathbf{p}/2E$ can be written in the explicitly invariant form

$$\frac{\mathrm{d}^3\mathbf{p}}{2E} = \int \mathrm{d}^4 p\, \delta(p^2 - m^2)\theta(p^0) \tag{8.11}$$

where $m^2 = E^2 - \mathbf{p}^2$, $\theta(p^0)$ is the step function (3.54), and the integration is with respect to p^0 over the range $-\infty < p^0 < \infty$.

§ We are here using the general relation

$$\int f(x, y)\, \delta[g(x, y)]\, \mathrm{d}x = \int f(x, y)\, \delta[g(x, y)] \left(\frac{\partial x}{\partial g} \right)_y \mathrm{d}g$$
$$= \left[\frac{f(x, y)}{(\partial g/\partial x)_y} \right]_{g=0} \tag{8.14}$$

where $p_2' = p_1 + p_2 - p_1'$ and the partial derivative is evaluated with the polar angles θ_1', ϕ_1' of the vector $\mathbf{p}_1'$ constant.

To obtain the differential cross-section in the CoM system, we note that in the CoM system $\mathbf{p}_1' = -\mathbf{p}_2'$. From

$$(E_f')^2 = (m_f')^2 + |\mathbf{p}_f'|^2, \qquad f = 1, 2, \tag{8.16}$$

we find

$$\frac{\partial(E_1' + E_2')}{\partial|\mathbf{p}_1'|} = |\mathbf{p}_1'| \frac{E_1 + E_2}{E_1' E_2'}, \tag{8.17}$$

and combining Eqs. (8.15), (8.12b), (8.10a) and (8.17) we obtain the CoM differential cross-section

$$\left(\frac{d\sigma}{d\Omega_1'}\right)_{\text{CoM}} = \frac{1}{64\pi^2(E_1 + E_2)^2} \frac{|\mathbf{p}_1'|}{|\mathbf{p}_1|} \left(\prod_l (2m_l)\right) |\mathscr{M}|^2. \tag{8.18}$$

Finally, we note that all the cross-section formulae which we have derived apply irrespective of whether identical particles are present or not. However, on calculating total cross-sections in cases where two or more final-state particles are identical, one must integrate only over those ranges of angles which correspond to physically distinguishable events. For example, if the CoM cross-section (8.18) refers to a process with two identical particles in the final state, then the scattering angles $(\theta_1', \phi_1') = (\alpha, \beta)$ and $(\theta_1', \phi_1') = (\pi - \alpha, \pi + \beta)$ describe the same process. Hence the total CoM cross-section is obtained by integrating Eq. (8.18) only over the forward hemisphere $0 \leqslant \theta_1' \leqslant \frac{1}{2}\pi$, i.e.

$$\sigma_{\text{CoM}}^{\text{tot}} = \int_0^1 d(\cos\theta_1') \int_0^{2\pi} d\phi_1' \left(\frac{d\sigma}{d\Omega_1'}\right)_{\text{CoM}} = \frac{1}{2} \int_{4\pi} d\Omega_1' \left(\frac{d\sigma}{d\Omega_1'}\right)_{\text{CoM}}, \tag{8.19}$$

where the last integral is over the complete solid angle 4π, as indicated.

8.2 SPIN SUMS

In the last section we considered a reaction in which the initial and final states are completely specified, including the polarization states of the leptons and photons present initially and finally. In many experiments, the colliding particles are unpolarized and the polarizations of the final-state particles are not detected. To obtain the corresponding unpolarized cross-section from Eq. (8.8), we must *average* $|\mathscr{M}|^2$ over all initial polarization states, and we must *sum* it over all final polarization states. In this section we shall show how to obtain these averages and sums over initial and final lepton spins. We shall find that the unpolarized cross-section can always be expressed in terms of traces of products of γ-matrices.

Consider a Feynman amplitude of the form

$$\mathcal{M} = \bar{u}_s(\mathbf{p}')\Gamma u_r(\mathbf{p}). \tag{8.20}$$

This occurs, for example, for Compton scattering [see Fig. 7.12 and Eqs. (7.38)]. Here the spinors $u_r(\mathbf{p})$ and $\bar{u}_s(\mathbf{p}')$ completely specify the momenta and spins of the electron in the initial and final states, and the operator Γ is a 4×4 matrix built up out of γ-matrices. Eq. (8.20) gives rise to an unpolarized cross-section proportional to

$$X \equiv \tfrac{1}{2} \sum_{r=1}^{2} \sum_{s=1}^{2} |\mathcal{M}|^2 \tag{8.21}$$

where we have averaged over initial spins ($\frac{1}{2}\sum_r$) and summed over final spins ($\sum_s$). Defining

$$\tilde{\Gamma} \equiv \gamma^0 \Gamma^\dagger \gamma^0, \tag{8.22}$$

we can write Eq. (8.21) as

$$X = \tfrac{1}{2} \sum_r \sum_s (\bar{u}_s(\mathbf{p}')\Gamma u_r(\mathbf{p}))(\bar{u}_r(\mathbf{p})\tilde{\Gamma} u_s(\mathbf{p}')). \tag{8.23}$$

Writing out the spinor indices explicitly, this can be written

$$X = \tfrac{1}{2}\left(\sum_s u_{s\delta}(\mathbf{p}')\bar{u}_{s\alpha}(\mathbf{p}')\right)\Gamma_{\alpha\beta}\left(\sum_r u_{r\beta}(\mathbf{p})\bar{u}_{r\gamma}(\mathbf{p})\right)\tilde{\Gamma}_{\gamma\delta}.$$

We introduce the positive energy projection operator [Eqs. (A.31) and (A.35)],

$$\Lambda^+_{\alpha\beta}(\mathbf{p}) = \left(\frac{\not{p} + m}{2m}\right)_{\alpha\beta} = \sum_{r=1}^{2} u_{r\alpha}(\mathbf{p})\bar{u}_{r\beta}(\mathbf{p}) \tag{8.24a}$$

in order to eliminate the sums over positive energy states.‡ This leads to our final result

$$\begin{aligned} X &= \tfrac{1}{2}\Lambda^+_{\delta\alpha}(\mathbf{p}')\Gamma_{\alpha\beta}\Lambda^+_{\beta\gamma}(\mathbf{p})\tilde{\Gamma}_{\gamma\delta} \\ &= \tfrac{1}{2}\,\mathrm{Tr}\,[\Lambda^+(\mathbf{p}')\Gamma\Lambda^+(\mathbf{p})\tilde{\Gamma}] \\ &= \tfrac{1}{2}\,\mathrm{Tr}\left[\frac{\not{p}' + m}{2m}\Gamma\frac{\not{p} + m}{2m}\tilde{\Gamma}\right]. \end{aligned} \tag{8.25}$$

In addition to the amplitude (8.20), involving the absorption and emission of an external negative lepton, there are also Feynman amplitudes

‡ Equation numbers (A.·) refer to equations in Appendix A at the end of the book. This appendix gives a self-contained account of the properties of Dirac spinors, etc. which we here require, and a reader not familiar with these is advised to study the appendix.

of the form

$$\mathcal{M} = \bar{v}_s(\mathbf{p}')\Gamma v_r(\mathbf{p}) \tag{8.26a}$$

$$\mathcal{M} = \bar{u}_s(\mathbf{p}')\Gamma v_r(\mathbf{p}) \tag{8.26b}$$

$$\mathcal{M} = \bar{v}_s(\mathbf{p}')\Gamma u_r(\mathbf{p}). \tag{8.26c}$$

These represent: (a) absorption and emission of a positive lepton, as in Compton scattering by positrons [Eq. (7.39) and Fig. 7.13]; (b) creation of a lepton pair, as in $2\gamma \to e^+e^-$ (Fig. 7.5); and (c) annihilation of a lepton pair, as in $e^+e^- \to 2\gamma$ (Fig. 7.4).

The spin sums for these cases are performed as for the case which we considered in detail, but using the negative energy projection operator [Eqs. (A.31) and (A.35)]

$$\Lambda^-_{\alpha\beta}(\mathbf{p}) = -\left(\frac{\not{p} - m}{2m}\right)_{\alpha\beta} = -\sum_{r=1}^{2} v_{r\alpha}(\mathbf{p})\bar{v}_{r\beta}(\mathbf{p}) \tag{8.24b}$$

to eliminate sums over negative energy states. For example, Eq. (8.26b) leads to

$$\begin{aligned} \tfrac{1}{2}\sum_r\sum_s |\mathcal{M}|^2 &= -\tfrac{1}{2}\,\mathrm{Tr}\,[\Lambda^+(\mathbf{p}')\Gamma\Lambda^-(\mathbf{p})\tilde{\Gamma}] \\ &= \tfrac{1}{2}\,\mathrm{Tr}\left[\frac{\not{p}' + m}{2m}\Gamma\frac{\not{p} - m}{2m}\tilde{\Gamma}\right]. \end{aligned} \tag{8.27}$$

Spin sums, and consequently traces like Eqs. (8.25) and (8.27), frequently occur in practice. There exist simple techniques for calculating such traces. These use algebraic identities for γ-matrices (see Appendix A, Section A.2) and some general rules for calculating the traces of products of γ-matrices (see Section A.3). Later in this chapter we shall repeatedly apply these results and methods in calculating the unpolarized cross-sections for various processes in QED.

To conclude this section we briefly discuss how to calculate the spin polarization properties of a process. This involves evaluating $|\mathcal{M}|^2$ for specific initial and final spin states. This can be done either by using a specific matrix representation for the spinors or by employing helicity or spin projection operators to select the appropriate spin states. The latter technique again leads to traces and is usually the more convenient one.

We shall illustrate this method for the particular process resulting from the Feynman amplitude (8.20) in which the incident electron has positive helicity and the outgoing electron has negative helicity. The cross-section for this helicity flip process is proportional to

$$\begin{aligned} X &= |\bar{u}_2(\mathbf{p}')\Gamma u_1(\mathbf{p})|^2 \\ &= (\bar{u}_2(\mathbf{p}')\Gamma u_1(\mathbf{p}))(\bar{u}_1(\mathbf{p})\tilde{\Gamma} u_2(\mathbf{p}')). \end{aligned} \tag{8.28}$$

We introduce the helicity projection operators

$$\Pi^{\pm}(\mathbf{p}) = \tfrac{1}{2}(1 \pm \sigma_{\mathbf{p}}) \tag{A.37}$$

which have the properties

$$\Pi^{+}(\mathbf{p})u_r(\mathbf{p}) = \delta_{1r}u_r(\mathbf{p}), \qquad \Pi^{-}(\mathbf{p})u_r(\mathbf{p}) = \delta_{2r}u_r(\mathbf{p}). \tag{A.40}$$

Eq. (8.28) then becomes

$$\begin{aligned} X &= (\bar{u}_2(\mathbf{p}')\Gamma\Pi^{+}(\mathbf{p})u_1(\mathbf{p}))(\bar{u}_1(\mathbf{p})\tilde{\Gamma}\Pi^{-}(\mathbf{p}')u_2(\mathbf{p}')) \\ &= \sum_r \sum_s (\bar{u}_s(\mathbf{p}')\Gamma\Pi^{+}(\mathbf{p})u_r(\mathbf{p}))(\bar{u}_r(\mathbf{p})\tilde{\Gamma}\Pi^{-}(\mathbf{p}')u_s(\mathbf{p}')) \\ &= \mathrm{Tr}\,[\Lambda^{+}(\mathbf{p}')\Gamma\Pi^{+}(\mathbf{p})\Lambda^{+}(\mathbf{p})\tilde{\Gamma}\Pi^{-}(\mathbf{p}')], \end{aligned} \tag{8.29}$$

where the last line follows from Eqs. (8.23) and (8.25) with Γ and $\tilde{\Gamma}$ replaced by $\Gamma\Pi^{+}(\mathbf{p})$ and $\tilde{\Gamma}\Pi^{-}(\mathbf{p}')$.

In the relativistic limit $E \gg m$, the helicity projection operators (A.40) simplify to

$$\Pi^{\pm}(\mathbf{p}) = \tfrac{1}{2}(1 \pm \gamma^5) \qquad (E \gg m) \tag{A.43}$$

which also leads to a considerable simplification of Eq. (8.29) in the relativistic limit $E \gg m$, $E' \gg m$.

8.3 PHOTON POLARIZATION SUMS

In the last section we showed how to perform spin sums in order to obtain unpolarized cross-sections. We now consider the corresponding photon polarization sums. We met an example of this for Thomson scattering in Section 1.4.4, where we first obtained the fully polarized cross-section, Eq. (1.69), and then explicitly performed the summing and averaging over final and initial polarizations by means of Eq. (1.71). An alternative covariant formalism exists for obtaining the unpolarized cross-section directly. This formalism depends on the gauge invariance of the theory, the consequences of which we shall now consider in more detail.

Gauge invariance of the theory implies the gauge invariance of the matrix elements, i.e. of the Feynman amplitudes. It is of course only the matrix element itself, corresponding to the sum of all possible Feynman graphs in a given order of perturbation theory, which must be gauge-invariant. The contributions to the amplitude from individual Feynman graphs are in general not gauge-invariant. For example, for Compton scattering the individual amplitudes $\mathscr{M}_a$ and $\mathscr{M}_b$, Eqs. (7.38a) and (7.38b), are not gauge invariant, but their sum, $(\mathscr{M}_a + \mathscr{M}_b)$, is. (The verification of this statement, using the method to be developed in this section, is left as a problem for the reader. See Problem 8.7.)

For any process involving external photons, the Feynman amplitude $\mathcal{M}$ is of the form

$$\mathcal{M} = \varepsilon^{\alpha}_{r_1}(\mathbf{k}_1)\varepsilon^{\beta}_{r_2}(\mathbf{k}_2)\ldots\mathcal{M}_{\alpha\beta\ldots}(\mathbf{k}_1, \mathbf{k}_2, \ldots) \tag{8.30}$$

with one polarization vector $\varepsilon(\mathbf{k})$ for each external photon, and the tensor amplitude $\mathcal{M}_{\alpha\beta\ldots}(\mathbf{k}_1, \mathbf{k}_2, \ldots)$ independent of these polarization vectors. [This follows from our fourth Feynman rule, Eqs. (7.49e) and (7.49f). We are again using real polarization vectors.]

The polarization vectors are of course gauge-dependent. For example, for a free photon, described in a Lorentz gauge by the plane wave

$$A^{\mu}(x) = \text{const.}\ \varepsilon^{\mu}_{r}(\mathbf{k})\, \mathrm{e}^{\pm ikx}$$

the gauge transformation

$$A^{\mu}(x) \to A^{\mu}(x) + \partial^{\mu}f(x), \qquad \text{with } f(x) = \tilde{f}(k)\, \mathrm{e}^{\pm ikx},$$

implies

$$\varepsilon^{\mu}_{r}(\mathbf{k})\, \mathrm{e}^{\pm ikx} \to [\varepsilon^{\mu}_{r}(\mathbf{k}) \pm \mathrm{i}k^{\mu}\tilde{f}(k)]\, \mathrm{e}^{\pm ikx}. \tag{8.31}$$

Invariance of the amplitude (8.30) under this transformation requires

$$k^{\alpha}_{1}\mathcal{M}_{\alpha\beta\ldots}(\mathbf{k}_1, \mathbf{k}_2, \ldots) = k^{\beta}_{2}\mathcal{M}_{\alpha\beta\ldots}(\mathbf{k}_1, \mathbf{k}_2, \ldots) = \cdots = 0, \tag{8.32}$$

i.e. when any external photon polarization vector is replaced by the corresponding four-momentum, the amplitude must vanish.

To illustrate how Eq. (8.32) is used to calculate photon polarization sums, we consider, as simple example, the matrix element

$$\mathcal{M}_{r}(\mathbf{k}) = \varepsilon^{\alpha}_{r}(\mathbf{k})\mathcal{M}_{\alpha}(\mathbf{k})$$

corresponding to a process involving one external photon. The gauge invariance now implies

$$k^{\alpha}\mathcal{M}_{\alpha}(\mathbf{k}) = 0. \tag{8.33}$$

The unpolarized cross-section for the process is proportional to

$$X = \sum_{r=1}^{2} |\mathcal{M}_{r}(\mathbf{k})|^2 = \mathcal{M}_{\alpha}(\mathbf{k})\mathcal{M}^{*}_{\beta}(\mathbf{k}) \sum_{r=1}^{2} \varepsilon^{\alpha}_{r}(\mathbf{k})\varepsilon^{\beta}_{r}(\mathbf{k}). \tag{8.34}$$

Using the relation

$$\sum_{r=1}^{2} \varepsilon^{\alpha}_{r}(\mathbf{k})\varepsilon^{\beta}_{r}(\mathbf{k}) = -g^{\alpha\beta} - \frac{1}{(kn)^2}[k^{\alpha}k^{\beta} - (kn)(k^{\alpha}n^{\beta} + k^{\beta}n^{\alpha})], \tag{8.35}$$

which follows from Eqs. (5.39) and (5.40) for a real photon ($k^2 = 0$), and the gauge condition (8.33), we at once obtain from Eq. (8.34)

$$\sum_{r=1}^{2} |\mathcal{M}_{r}(\mathbf{k})|^2 = -\mathcal{M}^{\alpha}(\mathbf{k})\mathcal{M}^{*}_{\alpha}(\mathbf{k}). \tag{8.36}$$

Eq. (8.36) is our desired result, and it is easily extended to transitions involving several external photons. This formalism necessitates working in a general Lorentz gauge, as the explicit gauge invariance of the matrix element may be lost in a particular gauge. (We shall meet an example of this when discussing Compton scattering in Section 8.6.) However, in practice, it may be advantageous to choose a particular gauge, which simplifies the algebra of the trace sums, and to carry out the photon polarization sums explicitly, as was done for Thomson scattering in Section 1.4.4. The use of both techniques will be illustrated later in this chapter (see Sections 8.6 and 8.8 on Compton scattering and on bremsstrahlung).

8.4 LEPTON PAIR PRODUCTION IN (e^+e^-) COLLISIONS

As a first illustration of the use of the above methods in calculating processes to lowest non-vanishing order of perturbation theory, we shall consider the processes in which an electron–positron pair annihilates in collision, producing a charged lepton pair (l^+l^-). These processes are of considerable interest and have been studied experimentally over a wide range of energies. In this section we shall take the final lepton pair to be muons or tauons, but not electrons. The case of Bhabha scattering (i.e. $e^+e^- \to e^+e^-$) will be considered in the next section.

We already considered the process

$$e^+(\mathbf{p}_1, r_1) + e^-(\mathbf{p}_2, r_2) \to l^+(\mathbf{p}'_1, s_1) + l^-(\mathbf{p}'_2, s_2) \tag{8.37}$$

(where $l = \mu, \tau, \ldots$) in Section 7.4. Its Feynman amplitude, corresponding to the Feynman graph of Fig. 7.17, is given by Eq. (7.61) which we now write in slightly modified notation as

$$\mathcal{M}(r_1, r_2, s_1, s_2) = \mathrm{i}e^2[\bar{u}_{s_2}(\mathbf{p}'_2)\gamma_\alpha v_{s_1}(\mathbf{p}'_1)]_{(l)} \frac{1}{(p_1 + p_2)^2} [\bar{v}_{r_1}(\mathbf{p}_1)\gamma^\alpha u_{r_2}(\mathbf{p}_2)]_{(e)}. \tag{8.38}$$

The labels (l) and (e) distinguish quantities referring to leptons and to electrons. In Eq. (8.38) we have dropped the term $(+\mathrm{i}\varepsilon)$ in the photon propagator. This term is only of significance at the pole of the propagator, and in the present case $(p_1 + p_2)^2 \geqslant 4m_e^2$ cannot vanish.

For the unpolarized cross-section we require

$$X = \tfrac{1}{4} \sum_{r_1}\sum_{r_2}\sum_{s_1}\sum_{s_2} |\mathcal{M}(r_1, r_2, s_1, s_2)|^2. \tag{8.39}$$

Using the hermiticity condition $\gamma^{\alpha\dagger} = \gamma^0\gamma^\alpha\gamma^0$ [Eq. (A.6)], Eq. (8.38) gives

$$\mathcal{M}^*(r_1, r_2, s_1, s_2) = -\mathrm{i}e^2[\bar{v}_{s_1}(\mathbf{p}'_1)\gamma_\beta u_{s_2}(\mathbf{p}'_2)]_{(l)} \frac{1}{(p_1 + p_2)^2} [\bar{u}_{r_2}(\mathbf{p}_2)\gamma^\beta v_{r_1}(\mathbf{p}_1)]_{(e)} \tag{8.40}$$

and Eq. (8.39) becomes

$$X = \frac{e^4}{4[(p_1 + p_2)^2]^2} A_{(l)\alpha\beta} B_{(e)}{}^{\alpha\beta}. \tag{8.41}$$

Here $A_{(l)\alpha\beta}$ is given by

$$\begin{aligned} A_{(l)\alpha\beta} &= \sum_{s_1}\sum_{s_2} [(\bar{u}_{s_2}(\mathbf{p}'_2)\gamma_\alpha v_{s_1}(\mathbf{p}'_1))(\bar{v}_{s_1}(\mathbf{p}'_1)\gamma_\beta u_{s_2}(\mathbf{p}'_2))]_{(l)} \\ &= \mathrm{Tr}\left[\frac{\not{p}'_2 + m_l}{2m_l}\gamma_\alpha \frac{\not{p}'_1 - m_l}{2m_l}\gamma_\beta\right], \end{aligned} \tag{8.41a}$$

where we used the energy projection operators (8.24a) and (8.24b). Similarly one obtains

$$B_{(e)}{}^{\alpha\beta} = \mathrm{Tr}\left[\frac{\not{p}_1 - m_e}{2m_e}\gamma^\alpha \frac{\not{p}_2 + m_e}{2m_e}\gamma^\beta\right]. \tag{8.41b}$$

The traces (8.41a) and (8.41b) are easily evaluated using the results of Appendix A, Sections A.2 and A.3. Since the trace of a product of an odd number of γ-matrices vanishes [Eq. (A.16)], Eq. (8.41a) becomes

$$A_{(l)\alpha\beta} = \frac{1}{4m_l^2}[\mathrm{Tr}(\not{p}'_2\gamma_\alpha\not{p}'_1\gamma_\beta) - m_l^2\,\mathrm{Tr}(\gamma_\alpha\gamma_\beta)],$$

and from Eqs. (A.17) this gives

$$A_{(l)\alpha\beta} = \frac{1}{m_l^2}[p'_{1\alpha}p'_{2\beta} + p'_{2\alpha}p'_{1\beta} - (m_l^2 + p'_1p'_2)g_{\alpha\beta}]. \tag{8.42a}$$

Similarly one finds

$$B_{(e)}{}^{\alpha\beta} = \frac{1}{m_e^2}[p_1^\alpha p_2^\beta + p_2^\alpha p_1^\beta - (m_e^2 + p_1p_2)g^{\alpha\beta}]. \tag{8.42b}$$

Substituting Eqs. (8.42) into Eq. (8.41), one obtains

$$\begin{aligned} X = \frac{e^4}{2m_e^2m_l^2[(p_1+p_2)^2]^2}\{&(p_1p'_1)(p_2p'_2) + (p_1p'_2)(p_2p'_1) \\ &+ m_e^2(p'_1p'_2) + m_l^2(p_1p_2) + 2m_e^2m_l^2\}. \end{aligned} \tag{8.43}$$

So far we have worked in an arbitrary reference frame. We now specialize to the CoM frame, as specified in Fig. 8.1. For experiments carried out with colliding electron and positron beams this is the same as the laboratory system.‡

‡ This is, of course, the exceptional situation. More usually, the 'target' particle is at rest in the laboratory frame; see Eq. (8.10b).

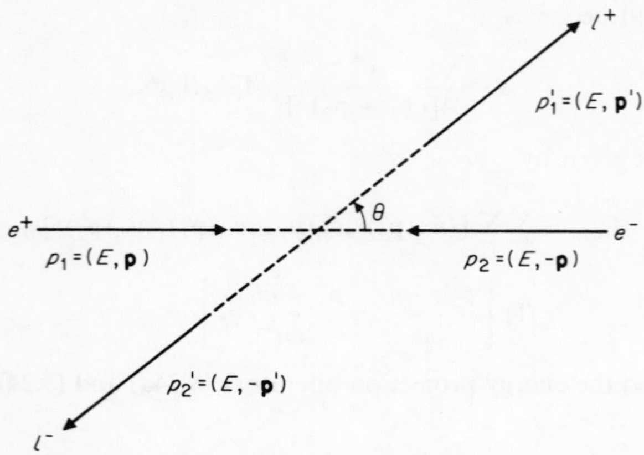

Fig. 8.1. Kinematics for the process $e^+e^- \to l^+l^-$ in the CoM system.

The kinematic factors occuring in Eq. (8.43) now take the form

$$\left.\begin{aligned} p_1p_1' = p_2p_2' = E^2 - pp'\cos\theta, \quad & p_1p_2' = p_2p_1' = E^2 + pp'\cos\theta \\ p_1p_2 = E^2 + p^2, \quad & p_1'p_2' = E^2 + p'^2 \\ (p_1 + p_2)^2 &= 4E^2 \end{aligned}\right\} \tag{8.44a}$$

where

$$p \equiv |\mathbf{p}|, \qquad p' \equiv |\mathbf{p}'|. \tag{8.44b}$$

Furthermore, since $E \geqslant m_\mu \approx 207m_e$, it is a very good approximation to take $p \equiv |\mathbf{p}| = E$, and to neglect terms proportional to m_e^2 inside the curly brackets in Eq. (8.43). On making these approximations and substituting Eqs. (8.43) and (8.44) in the CoM cross-section formula (8.18), we finally obtain

$$\left(\frac{\mathrm{d}\sigma}{\mathrm{d}\Omega}\right)_{\mathrm{CoM}} = \frac{\alpha^2}{16E^4}\left(\frac{p'}{E}\right)(E^2 + m_l^2 + p'^2\cos^2\theta) \tag{8.45a}$$

for the differential cross-section, and

$$\sigma_{\mathrm{tot}} = \frac{\pi\alpha^2}{4E^4}\left(\frac{p'}{E}\right)[E^2 + m_l^2 + \tfrac{1}{3}p'^2] \tag{8.45b}$$

for the total cross-section. In the extreme relativistic limit, $E \gg m_l$, these formulae reduce to the much quoted

$$\left.\begin{aligned} \left(\frac{\mathrm{d}\sigma}{\mathrm{d}\Omega}\right)_{\mathrm{CoM}} &= \frac{\alpha^2}{16E^2}(1 + \cos^2\theta) \\ \sigma_{\mathrm{tot}} &= \frac{\pi\alpha^2}{3E^2} \end{aligned}\right\} \quad (E \gg m_l). \tag{8.46}$$

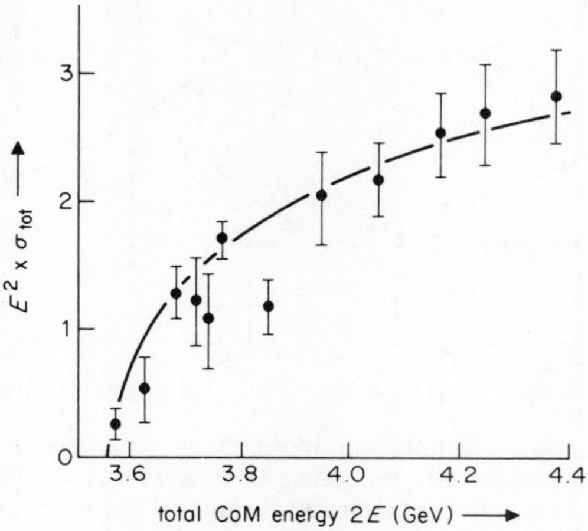

Fig. 8.2. $E^2 \times \sigma_{tot}$ (in arbitrary units) for the process $e^+e^- \to \tau^+\tau^-$ near threshold $2E = 2m_\tau$. [After W. Bacino *et al.*, *Phys. Rev. Lett.* **41** (1978), 13.] ⧫: experimental data; curve: fit of cross-section formula (8.45b) with $m_\tau = 1782$ MeV.

Both the processes $e^+e^- \to \mu^+\mu^-$ and $e^+e^- \to \tau^+\tau^-$ have been extensively studied over a wide range of energies. For the τ, which was first discovered in this reaction, the threshold region is of particular interest. The τ is a highly unstable particle.‡ It is detected by its decay products so that its precise mass is not easy to determine. The best values are obtained by fitting the total cross-section (8.45b) to the experimental data in the threshold region. Fig. 8.2 shows the excellent agreement obtained in such a fit to the experiments of W. Bacino *et al.*, *Phys. Rev. Lett.*, **41** (1978), 13. The tauon mass obtained in this way is $m_\tau = 1782^{+2}_{-7}$ MeV. Other experiments are compatible with this value but with somewhat larger errors.

At relativistic energies the predictions of Eq. (8.46) are in excellent agreement with experiment. Typical results shown in Fig. 8.3 which is based on the work of D. P. Barber *et al.*, *Phys. Rev. Lett.*, **43** (1979), 1915. These high-energy results are of great interest since they probe the interaction down to very small distances and so represent a severe test of QED. In the CoM system, the energy of the virtual photon in the process is $2E$, implying a timescale of order $\hbar/2E$ and a corresponding distance scale of order $\hbar c/2E$. For $E \approx 15$ GeV, this corresponds to a distance of the order 7×10^{-3} f. The agreement between theory and experiment implies that even at these very

‡ The decay mechanism and lifetime of the τ will be discussed in Section 11.6.1.

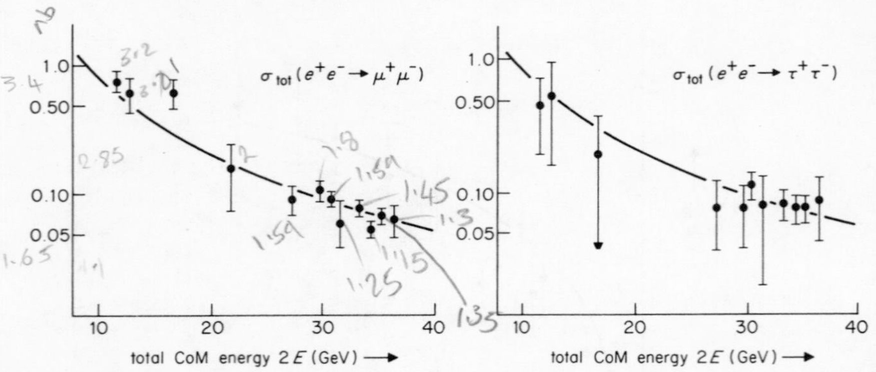

Fig. 8.3. The total cross-sections (in nb) for the processes $e^+e^- \to \mu^+\mu^-$ and $e^+e^- \to \tau^+\tau^-$ at relativistic energies. [After D. P. Barber *et al.*, *Phys. Rev. Lett.* **43**, (1979) 1915.] ⧳: experimental data; curves: theoretical cross-section formula (8.46).

small distances the electron, muon and tauon are adequately described as point charges. This small value should be compared with the experimental r.m.s. charge radius of the proton which is of order 0.8 f.

Eqs. (8.45) also predict substantial cross-sections for the (l^+l^-) production of any other charged leptons which may exist in the sequence $e, \mu, \tau, \ldots$. Consequently, these should be comparatively easy to detect in the (e^+e^-) annihilation process. Current experiments show that no such particle exists with mass less than about 15 GeV.

8.5 BHABHA SCATTERING

We now consider elastic e^+e^- scattering. This process is a little more complicated than those considered in the last section since, in addition to the annihilation diagram [Figs. 7.17 or 7.7(b)], the scattering diagram [Fig. 7.7(a)] also contributes. The Feynman amplitude for the process

$$e^+(\mathbf{p}_1, r_1) + e^-(\mathbf{p}_2, r_2) \to e^+(\mathbf{p}_1', s_1) + e^-(\mathbf{p}_2', s_2) \tag{8.47}$$

is

$$\mathscr{M} = \mathscr{M}_a + \mathscr{M}_b$$

where $\mathscr{M}_a$ and $\mathscr{M}_b$ correspond to the scattering and annihilation diagrams and are given by

$$\mathscr{M}_a = -ie^2[\bar{u}(\mathbf{p}_2')\gamma_\alpha u(\mathbf{p}_2)]\frac{1}{(p_1 - p_1')^2}[\bar{v}(\mathbf{p}_1)\gamma^\alpha v(\mathbf{p}_1')] \tag{8.48a}$$

$$\mathscr{M}_b = ie^2[\bar{u}(\mathbf{p}_2')\gamma_\alpha v(\mathbf{p}_1')]\frac{1}{(p_1 + p_2)^2}[\bar{v}(\mathbf{p}_1)\gamma^\alpha u(\mathbf{p}_2)]. \tag{8.48b}$$

We have here suppressed the spin indices again. $\mathcal{M}_b$ is of course just the amplitude (8.38) with e^+e^- for the final state lepton pair. We again note the relative minus sign between these two terms corresponding to Feynman rule 8 (Section 7.3).

We shall evaluate the cross-section for this process in the CoM system, restricting ourselves for simplicity to the important case of the relativistic high-energy limit. The kinematics are now defined by Eqs. (8.44) with $p' = p$ and $E \gg m(\equiv m_e)$. The cross-section formula (8.18) now leads to

$$\left(\frac{d\sigma}{d\Omega}\right)_{CoM} = \frac{m^4}{16\pi^2 E^2}(X_{aa} + X_{bb} + X_{ab} + X_{ab}^*), \tag{8.49}$$

where

$$X_{aa} = \tfrac{1}{4}\sum_{spins} |\mathcal{M}_a|^2 \tag{8.50a}$$

$$X_{bb} = \tfrac{1}{4}\sum_{spins} |\mathcal{M}_b|^2 \tag{8.50b}$$

$$X_{ab} = \tfrac{1}{4}\sum_{spins} \mathcal{M}_a\mathcal{M}_b^*, \tag{8.50c}$$

the summations being over the spins of all four fermions.

The term X_{bb} follows at once as a special case of X, Eq. (8.43), with $m_l = m$ and $E \gg m$:

$$X_{bb} = \frac{e^4}{16m^4}\left[1 + \cos^2\theta + O\left(\frac{m^2}{E^2}\right)\right]. \tag{8.51}$$

The evaluation of X_{aa} is essentially similar to that of X and is left as an exercise for the reader. The result is

$$X_{aa} = \frac{e^4}{2m^4[(p_1 - p_1')^2]^2}\{(p_1p_2)(p_1'p_2') + (p_1p_2')(p_2p_1') + O(E^2m^2)\}$$

$$= \frac{e^4}{8m^4\sin^4(\theta/2)}\left[1 + \cos^4\frac{\theta}{2} + O\left(\frac{m^2}{E^2}\right)\right]. \tag{8.52}$$

The interference term X_{ab}, Eq. (8.50c), is more complicated, and we shall give its evaluation in some detail in order to illustrate how such more complicated spin sums can be handled simply. From Eqs. (8.50c) and (8.48),

$$X_{ab} = \frac{-e^4}{4(p_1 - p_1')^2(p_1 + p_2)^2}\sum_{spins}\{[\bar{u}(\mathbf{p}_2')\gamma_\alpha u(\mathbf{p}_2)][\bar{u}(\mathbf{p}_2)\gamma_\beta v(\mathbf{p}_1)]$$

$$\times [\bar{v}(\mathbf{p}_1)\gamma^\alpha v(\mathbf{p}_1')][\bar{v}(\mathbf{p}_1')\gamma^\beta u(\mathbf{p}_2')]\}$$

$$= \frac{-e^4}{4(p_1 - p_1')^2(p_1 + p_2)^2}\operatorname{Tr}\left\{\frac{\not{p}_2' + m}{2m}\gamma_\alpha\frac{\not{p}_2 + m}{2m}\gamma_\beta\frac{\not{p}_1 - m}{2m}\gamma^\alpha\frac{\not{p}_1' - m}{2m}\gamma^\beta\right\}$$

$$= \frac{-e^4}{64m^4(p_1 - p_1')^2(p_1 + p_2)^2}[\operatorname{Tr}(\not{p}_2'\gamma_\alpha\not{p}_2\gamma_\beta\not{p}_1\gamma^\alpha\not{p}_1'\gamma^\beta) + O(E^2m^2)].$$

We must therefore evaluate the trace of a product of eight γ-matrices. It is usually possible and highly desirable to simplify such a product before taking the trace, rather than blindly use Eq. (A.18c). The contraction identities (Appendix A, Section A.2) are particularly useful in this respect. By means of

$$\gamma_\lambda \gamma_\alpha \gamma_\beta \gamma_\gamma \gamma^\lambda = -2\gamma_\gamma \gamma_\beta \gamma_\alpha, \qquad \gamma_\lambda \gamma_\alpha \gamma_\beta \gamma^\lambda = 4g_{\alpha\beta} \tag{A.14a}$$

we find that the trace in X_{ab} equals

$$-2\,\text{Tr}\,(\not{p}_2' \not{p}_1 \gamma_\beta \not{p}_2 \not{p}_1' \gamma^\beta) = -8(p_2 p_1')\,\text{Tr}\,(\not{p}_2' \not{p}_1) = -32(p_2 p_1')(p_2' p_1),$$

whence X_{ab} becomes

$$\begin{aligned} X_{\text{ab}} &= \frac{e^4}{2m^4(p_1 - p_1')^2(p_1 + p_2)^2}[(p_1 p_2')(p_2 p_1') + \text{O}(E^2 m^2)] \\ &= \frac{-e^4}{8m^4 \sin^2(\theta/2)}\left[\cos^4\frac{\theta}{2} + \text{O}\left(\frac{m^2}{E^2}\right)\right]. \end{aligned} \tag{8.53}$$

We see from this equation that X_{ab} is real. Hence substituting Eqs. (8.51)–(8.53) in Eq. (8.49), we obtain for the CoM differential cross-section in the high-energy limit ($E \gg m$)

$$\left(\frac{\mathrm{d}\sigma}{\mathrm{d}\Omega}\right)_{\text{CoM}} = \frac{\alpha^2}{8E^2}\left[\frac{1 + \cos^4(\theta/2)}{\sin^4(\theta/2)} + \frac{1 + \cos^2\theta}{2} - \frac{2\cos^4(\theta/2)}{\sin^2(\theta/2)}\right]. \tag{8.54}$$

The three terms in this equation correspond to the photon exchange diagram, Fig. 7.7(a), the annihilation diagram 7.7(b), and the interference term between them. It should be compared with the corresponding result (8.46) for the process $e^+e^- \to l^+l^-$ with $l \neq e$, when only the annihilation diagram is present.

At small angles, the exchange term dominates, giving rise to an infinite cross-section in the forward direction, $\theta = 0$, and an infinite total cross-section. These features are a consequence of the infinite range of the electromagnetic forces or, equivalently, of the zero mass of the photon. As $\theta \to 0$, the four-momentum $k^\alpha = (p_1 - p_1')^\alpha$ of the exchanged photon tends to zero, and the factor

$$\frac{1}{k^2 + \mathrm{i}\varepsilon} = \frac{1}{(p_1 - p_1')^2 + \mathrm{i}\varepsilon}$$

in the photon propagator diverges, from which the divergence of the amplitude (8.48a) and of the cross-section (8.54) follows.‡

At large angles the photon exchange term and the annihilation term are of comparable importance, and sensitive to the short distance behaviour. For

‡ As is often done, we suppressed the term $+\mathrm{i}\varepsilon$ in the photon propagator in Eq. (8.48a). This term is only relevant at the pole, $(p_1 - p_1')^2 = 0$.

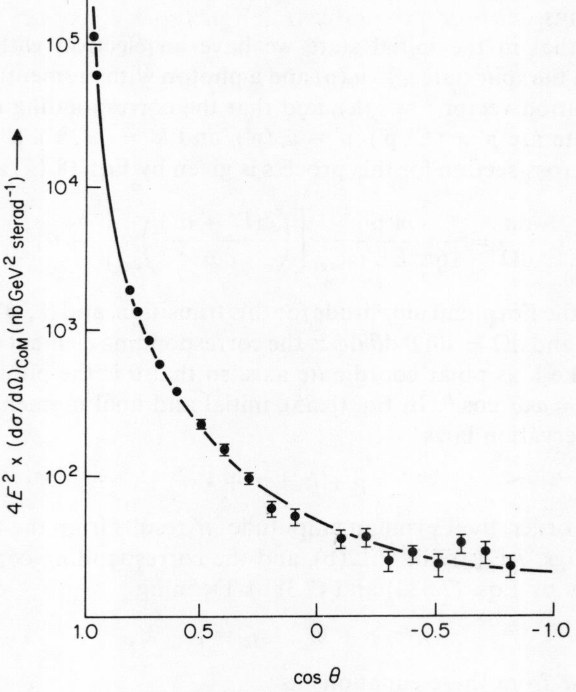

Fig. 8.4. The differential cross-section $(d\sigma/d\Omega)_{CoM}$ for Bhabha scattering, $e^+e^- \to e^+e^-$, at the total CoM energy $2E = 34$ GeV, [After H. J. Behrend *et al.*, *Phys. Lett*, **103B** (1981), 148.] ⧳: experimental data; curve: QED cross-section formula (8.54).

the annihilation term this was discussed in the last section. For the exchange diagram, the exchanged photon has the wave number $|\mathbf{k}| = |\mathbf{p}_1 - \mathbf{p}'_1| = 2E \sin(\theta/2)$, with an associated wavelength $\lambda = 2\pi/|\mathbf{k}|$.

Experimentally, the predicted behaviour has been confirmed over a wide range of energies and angles, and the interaction has been tested down to very short distances, comparable to those probed in the $e^+e^- \to \mu^+\mu^-$ and $e^+e^- \to \tau^+\tau^-$ experiments which were discussed in the last section. Typical results for Bhabha scattering are shown in Fig. 8.4.

8.6 COMPTON SCATTERING

We shall now derive the cross-section for Compton scattering. In this process a photon is present in both the initial and final states, and we shall apply our earlier results to carry out the photon polarization sums, as well as the electron spin sums.

Suppose that in the initial state we have an electron with momentum $p = (E, \mathbf{p})$ in the spin state $u \equiv u_r(\mathbf{p})$ and a photon with momentum $k = (\omega, \mathbf{k})$ and polarization vector $\varepsilon \equiv \varepsilon_s(\mathbf{k})$, and that the corresponding quantities for the final state are $p' = (E', \mathbf{p}')$, $u' \equiv u_{r'}(\mathbf{p}')$, and $k' = (\omega', \mathbf{k}')$, $\varepsilon' \equiv \varepsilon_{s'}(\mathbf{k}')$. The differential cross-section for this process is given by Eqs. (8.15) and (8.12b) as

$$\frac{d\sigma}{d\Omega} = \frac{m^2\omega'}{16\pi^2 EE'\omega v_{\text{rel}}}\left[\left(\frac{\partial(E'+\omega')}{\partial\omega'}\right)_{\theta\phi}\right]^{-1}|\mathscr{M}|^2 \tag{8.55}$$

where $\mathscr{M}$ is the Feynman amplitude for this transition, and (θ, ϕ) are the polar angles of $\mathbf{k}'$, and $d\Omega = \sin\theta \, d\theta \, d\phi$ is the corresponding element of solid angle. We shall take $\mathbf{k}$ as polar coordinate axis, so that θ is the photon scattering angle: $\mathbf{k}\cdot\mathbf{k}' = \omega\omega'\cos\theta$. In Eq. (8.55), initial and final momenta are related by the conservation laws

$$p + k = p' + k'. \tag{8.56}$$

In lowest order, the Feynman amplitude $\mathscr{M}$ results from the two Feynman graphs in Figs. 7.12(a) and 7.12(b), and the corresponding contributions to $\mathscr{M}$ are given by Eqs. (7.38a) and (7.38b). Defining

$$f_1 \equiv p + k, \qquad f_2 \equiv p - k', \tag{8.57}$$

we obtain $\mathscr{M}$ from these equations as

$$\mathscr{M} = \mathscr{M}_a + \mathscr{M}_b \tag{8.58}$$

where

$$\mathscr{M}_a = -ie^2\frac{\bar{u}'\not{\varepsilon}'(\not{f}_1 + m)\not{\varepsilon}u}{2(pk)}, \qquad \mathscr{M}_b = ie^2\frac{\bar{u}'\not{\varepsilon}(\not{f}_2 + m)\not{\varepsilon}'u}{2(pk')}. \tag{8.59}$$

These results refer to a general reference frame. In most experiments, the photon beam is incident on a target of nearly stationary electrons. We shall now specialize Eq. (8.55) to the laboratory frame in which $p = (m, 0, 0, 0)$ and

$$\mathbf{p}' = \mathbf{k} - \mathbf{k}', \tag{8.60a}$$

$$E' = [m^2 + (\mathbf{k} - \mathbf{k}')^2]^{1/2} = [m^2 + \omega^2 + \omega'^2 - 2\omega\omega'\cos\theta]^{1/2}. \tag{8.60b}$$

From Eq. (8.56) we have generally

$$pk = p'k + k'k = pk' + k'k$$

which in the laboratory system reduces to

$$\omega' = \frac{m\omega}{m + \omega(1 - \cos\theta)}. \tag{8.61}$$

This equation gives the energy shift of the scattered photon due to the recoil of the target electron. From Eq. (8.60b) we find

$$\left(\frac{\partial(E' + \omega')}{\partial\omega'}\right)_{\theta\phi} = \frac{m\omega}{E'\omega'}, \tag{8.62}$$

so that Eq. (8.55) gives for the differential cross-section in the laboratory frame

$$\left(\frac{d\sigma}{d\Omega}\right)_{\text{Lab}} = \frac{1}{(4\pi)^2}\left(\frac{\omega'}{\omega}\right)^2 |\mathcal{M}|^2. \tag{8.63}$$

The cross-sections (8.55) and (8.63) are fully polarized, i.e. both initial and final electrons and photons are in definite polarization states. To obtain the cross-section for Compton scattering by an unpolarized electron target, and with the spin of the final electron undetected, we must sum and average the above cross-section formulas over final and initial electron spins. To obtain the unpolarized cross-section, we must also sum and average over final and initial photon polarizations. We shall illustrate both the methods of Section 1.4.4 and of Section 8.3 for handling photon polarization.

To obtain the unpolarized cross-section directly, we use the covariant method of Section 8.3. Writing

$$\mathcal{M} \equiv \varepsilon_\alpha \varepsilon'_\beta \mathcal{M}^{\alpha\beta}, \tag{8.64}$$

we obtain

$$\tfrac{1}{4}\sum_{\text{pol}}\sum_{\text{spin}} |\mathcal{M}|^2 = \tfrac{1}{4}\sum_{\text{spin}} \mathcal{M}^{\alpha\beta}\mathcal{M}^*_{\alpha\beta} \tag{8.65}$$

where the summations are over initial and final electron spins and photon polarizations. Eq. (8.65) is the analogue of Eq. (8.36) for the case of two external photons. Carrying out the spin summations, we obtain from Eqs. (8.65) and (8.58)–(8.59)

$$\begin{aligned}\tfrac{1}{4}\sum_{\text{pol}}\sum_{\text{spin}} |\mathcal{M}|^2 &= \tfrac{1}{4}\sum_{\text{pol}}\sum_{\text{spin}} \{|\mathcal{M}_\text{a}|^2 + |\mathcal{M}_\text{b}|^2 + \mathcal{M}_\text{a}\mathcal{M}^*_\text{b} + \mathcal{M}_\text{b}\mathcal{M}^*_\text{a}\}\\ &= \frac{e^4}{64m^2}\left\{\frac{X_{\text{aa}}}{(pk)^2} + \frac{X_{\text{bb}}}{(pk')^2} - \frac{X_{\text{ab}} + X_{\text{ba}}}{(pk)(pk')}\right\},\end{aligned} \tag{8.66}$$

where

$$X_{\text{aa}} = \text{Tr}\,\{\gamma^\beta(\not{f}_1 + m)\gamma^\alpha(\not{p} + m)\gamma_\alpha(\not{f}_1 + m)\gamma_\beta(\not{p}' + m)\} \tag{8.67a}$$

$$X_{\text{bb}} = \text{Tr}\,\{\gamma^\alpha(\not{f}_2 + m)\gamma^\beta(\not{p} + m)\gamma_\beta(\not{f}_2 + m)\gamma_\alpha(\not{p}' + m)\} \tag{8.67b}$$

$$X_{\text{ab}} = \text{Tr}\,\{\gamma^\beta(\not{f}_1 + m)\gamma^\alpha(\not{p} + m)\gamma_\beta(\not{f}_2 + m)\gamma_\alpha(\not{p}' + m)\} \tag{8.67c}$$

$$X_{\text{ba}} = \text{Tr}\,\{\gamma^\alpha(\not{f}_2 + m)\gamma^\beta(\not{p} + m)\gamma_\alpha(\not{f}_1 + m)\gamma_\beta(\not{p}' + m)\}. \tag{8.67d}$$

Note that the effect of the substitutions

$$k \leftrightarrow -k', \qquad \varepsilon \leftrightarrow \varepsilon', \tag{8.68a}$$

is to induce the transformations

$$f_1 \leftrightarrow f_2, \qquad \mathcal{M}_a \leftrightarrow \mathcal{M}_b \tag{8.68b}$$

and hence

$$X_{aa} \leftrightarrow X_{bb}, \qquad X_{ab} \leftrightarrow X_{ba}, \tag{8.68c}$$

We need therefore only calculate X_{aa} and X_{ab} from first principles. From Eq. (8.66), $X_{ba} = X_{ab}^*$. Furthermore, it follows from Eqs. (8.67c) and (8.67d) and the general property (A.20a) of γ-matrices, that $X_{ab} = X_{ba}$. Hence X_{ab} is real, and it is symmetric with respect to the transformation (8.68a), which provides two useful checks in its calculation.

The traces in Eqs. (8.67) involve products of up to eight γ-matrices. Their computation is much simplified by the use of the contraction identities (Appendix A, Section A.2) which eliminate four γ-matrices. We illustrate this for X_{aa}. The trace in Eq. (8.67a) contains the factor

$$\begin{aligned} Y &\equiv \gamma^\beta(\not{f}_1 + m)\gamma^\alpha(\not{p} + m)\gamma_\alpha(\not{f}_1 + m)\gamma_\beta \\ &= \gamma^\beta(\not{f}_1 + m)(-2\not{p} + 4m)(\not{f}_1 + m)\gamma_\beta \\ &= 4\not{f}_1\not{p}\not{f}_1 + m[-16(pf_1) + 16f_1^2] + m^2(4\not{p} - 16\not{f}_1) + 16m^3. \end{aligned}$$

Hence, using Eqs. (A.16) and (A.18a), (A.18b), one obtains directly

$$\begin{aligned} X_{aa} &= \mathrm{Tr}\{Y(\not{p}' + m)\} \\ &= 16\{2(f_1p)(f_1p') - f_1^2(pp') + m^2[-4(pf_1) + 4f_1^2] \\ &\quad + m^2[(pp') - 4(f_1p')] + 4m^4\}. \end{aligned} \tag{8.69}$$

If we express all quantities in terms of the three linearly independent scalars

$$p^2 = p'^2 = m^2, \qquad pk = p'k', \qquad pk' = p'k, \tag{8.70}$$

X_{aa} simplifies to

$$X_{aa} = 32[m^4 + m^2(pk) + (pk)(pk')]. \tag{8.71a}$$

From Eqs. (8.68) we have at once

$$X_{bb} = 32[m^4 - m^2(pk') + (pk)(pk')]. \tag{8.71b}$$

The interference term X_{ab}, Eq. (8.67c), is similarly computed with the result

$$X_{ab} = 16m^2[2m^2 + (pk) - (pk')]. \tag{8.71c}$$

As expected, $X_{\rm ab}$ is real and symmetric, i.e. the substitution $k \leftrightarrow -k'$ transforms $X_{\rm ab}$ into

$$X_{\rm ba} = X_{\rm ab}. \tag{8.71d}$$

Substituting Eqs. (8.71a)–(8.71d) into Eq. (8.66), we obtain

$$\frac{1}{4}\sum_{\rm pol}\sum_{\rm spin} |\mathscr{M}|^2 = \frac{e^4}{2m^2}\left\{\left(\frac{pk}{pk'} + \frac{pk'}{pk}\right)\right.$$
$$\left. + 2m^2\left(\frac{1}{pk} - \frac{1}{pk'}\right) + m^4\left(\frac{1}{pk} - \frac{1}{pk'}\right)^2\right\}. \tag{8.72}$$

In the laboratory system, $pk = m\omega$, $pk' = m\omega'$, and from Eq. (8.61)

$$\frac{1}{\omega} - \frac{1}{\omega'} = \frac{1}{m}(\cos\theta - 1).$$

Hence Eq. (8.72) reduces to

$$\left[\frac{1}{4}\sum_{\rm pol}\sum_{\rm spin} |\mathscr{M}|^2\right]_{\rm Lab} = \frac{e^4}{2m^2}\left\{\frac{\omega}{\omega'} + \frac{\omega'}{\omega} - \sin^2\theta\right\}, \tag{8.73}$$

and Eq. (8.63) gives the unpolarized cross-section

$$\left(\frac{d\sigma}{d\Omega}\right)_{\rm Lab} = \frac{\alpha^2}{2m^2}\left(\frac{\omega'}{\omega}\right)^2\left\{\frac{\omega}{\omega'} + \frac{\omega'}{\omega} - \sin^2\theta\right\}. \tag{8.74}$$

By means of Eq. (8.61), ω' can of course be eliminated altogether from this equation. In the low-energy limit $\omega \ll m$, we have $\omega' \approx \omega$, i.e. the kinetic energy of the recoil electron is negligible, and Eq. (8.74) reduces to the Thomson scattering cross-section, Eq. (1.69a).

We shall now derive the cross-section for initial and final photons in states of definite polarization, i.e. summing and averaging over electron spins only. On also summing and averaging over photon polarizations, using the method of Section 1.4.4, we shall regain the result (8.74). In this case, the trace calculations cannot be simplified through use of the contraction identities. However, they are greatly facilitated by a suitable choice of gauge. In any reference frame it is possible to find a Lorentz gauge in which the vacuum contains no longitudinal or scalar photons, and free photons are transverse (see Section 5.2). In this gauge, the polarization vectors of the external photons are of the form $\varepsilon = (0, \boldsymbol{\varepsilon})$, $\varepsilon' = (0, \boldsymbol{\varepsilon}')$, with

$$\varepsilon k = -\boldsymbol{\varepsilon}\cdot\mathbf{k} = 0, \qquad \varepsilon' k' = -\boldsymbol{\varepsilon}'\cdot\mathbf{k}' = 0. \tag{8.75a}$$

The analysis is further simplified if we work in the laboratory frame, $p = (m, 0, 0, 0)$, in which we also have

$$p\varepsilon = p\varepsilon' = 0. \tag{8.75b}$$

It follows from the anticommutation relations $[\gamma^\alpha, \gamma^\beta]_+ = 2g^{\alpha\beta}$ and the Dirac equation $(\not{p} - m)u(p) = 0$, that

$$\not{p}\not{\varepsilon}u = -m\not{\varepsilon}u, \qquad \not{p}\not{\varepsilon}'u = -m\not{\varepsilon}'u,$$

so that the matrix elements (8.59) simplify to

$$\mathscr{M}_{\mathrm{a}} = -\mathrm{i}e^2 \frac{\bar{u}'\not{\varepsilon}'\not{k}\not{\varepsilon}u}{2(pk)}, \qquad \mathscr{M}_{\mathrm{b}} = -\mathrm{i}e^2 \frac{\bar{u}'\not{\varepsilon}\not{k}'\not{\varepsilon}'u}{2(pk')}. \tag{8.76}$$

Note that Eq. (8.76) does not give a gauge-invariant expression for the matrix element $\mathscr{M} = \mathscr{M}_{\mathrm{a}} + \mathscr{M}_{\mathrm{b}}$. For example, the gauge transformation $\varepsilon \to \varepsilon + \lambda k$, where λ is a constant, leads to $\mathscr{M}_{\mathrm{a}} \to \mathscr{M}_{\mathrm{a}}$ (since $\not{k}\not{k} = k^2 = 0$) but $\mathscr{M}_{\mathrm{b}} \nleftrightarrow \mathscr{M}_{\mathrm{b}}$. This is of course due to the fact that we dropped the terms $p\varepsilon$ and $p\varepsilon'$ which are zero in our gauge.

Summing and averaging over electron spins now gives

$$\frac{1}{2}\sum_{\text{spin}} |\mathscr{M}|^2 = \frac{e^4}{32m^2}\left\{\frac{Y_{\mathrm{aa}}}{(pk)^2} + \frac{Y_{\mathrm{bb}}}{(pk')^2} + \frac{Y_{\mathrm{ab}} + Y_{\mathrm{ba}}}{(pk)(pk')}\right\} \tag{8.77}$$

where

$$Y_{\mathrm{aa}} = \mathrm{Tr}\,\{\not{\varepsilon}'\not{k}\not{\varepsilon}(\not{p} + m)\not{\varepsilon}\not{k}\not{\varepsilon}'(\not{p}' + m)\} \tag{8.78a}$$

$$Y_{\mathrm{bb}} = \mathrm{Tr}\,\{\not{\varepsilon}\not{k}'\not{\varepsilon}'(\not{p} + m)\not{\varepsilon}'\not{k}'\not{\varepsilon}(\not{p}' + m)\} \tag{8.78b}$$

$$Y_{\mathrm{ab}} = \mathrm{Tr}\,\{\not{\varepsilon}'\not{k}\not{\varepsilon}(\not{p} + m)\not{\varepsilon}'\not{k}'\not{\varepsilon}(\not{p}' + m)\} \tag{8.78c}$$

$$Y_{\mathrm{ba}} = \mathrm{Tr}\,\{\not{\varepsilon}\not{k}'\not{\varepsilon}'(\not{p} + m)\not{\varepsilon}\not{k}\not{\varepsilon}'(\not{p}' + m)\}. \tag{8.78d}$$

Substituting $k \leftrightarrow -k'$, $\varepsilon \leftrightarrow \varepsilon'$, again leads to $\mathscr{M}_{\mathrm{a}} \leftrightarrow \mathscr{M}_{\mathrm{b}}$, and

$$Y_{\mathrm{aa}} \leftrightarrow Y_{\mathrm{bb}}, \qquad Y_{\mathrm{ab}} \leftrightarrow Y_{\mathrm{ba}}, \tag{8.79}$$

and $Y_{\mathrm{ab}} = Y_{\mathrm{ba}} = Y_{\mathrm{ab}}^*$.

The traces in Eqs. (8.78) contain products of up to eight γ-matrices. We reduce this number using

$$\not{A}\not{B} = -\not{B}\not{A} + 2AB. \tag{8.80a}$$

For $A = B$, we have

$$\not{A}\not{A} = A^2, \tag{8.80b}$$

and, in particular,

$$\not{p}\not{p} = m^2, \qquad \not{k}\not{k} = 0, \qquad \not{\varepsilon}\not{\varepsilon} = \not{\varepsilon}'\not{\varepsilon}' = -1. \tag{8.80c}$$

For $AB = 0$, we have

$$\not{A}\not{B} = -\not{B}\not{A}, \tag{8.80d}$$

which will be particularly useful on account of Eqs. (8.75).

We illustrate the use of these tricks by computing Y_{aa}. Since $\not{k}\not{\varepsilon}\not{\varepsilon}\not{k} = -\not{k}\not{k} = 0$, Eq. (8.78a) reduces to

$$Y_{aa} = \mathrm{Tr}\,\{\not{\varepsilon}'\not{k}\not{\varepsilon}\not{p}\not{\varepsilon}\not{k}\not{\varepsilon}'\not{p}'\} = \mathrm{Tr}\,\{\not{\varepsilon}'\not{k}\not{p}\not{k}\not{\varepsilon}'\not{p}'\}$$

since $\not{\varepsilon}\not{p}\not{\varepsilon} = -\not{p}\not{\varepsilon}\not{\varepsilon} = \not{p}$. Permuting $\not{p}$ and $\not{k}$, and using $\not{k}\not{k} = 0$, we obtain

$$\begin{aligned} Y_{aa} = 2(pk)\,\mathrm{Tr}\,\{\not{\varepsilon}'\not{k}\not{\varepsilon}'\not{p}'\} &= 8(pk)[2(\varepsilon'k)(\varepsilon'p') + (kp')] \\ &= 8(pk)[2(\varepsilon'k)^2 + (pk')], \end{aligned} \tag{8.81a}$$

since $p' - k = p - k'$ implies $\varepsilon'p' = \varepsilon'k$ and $kp' = pk'$. From Eq. (8.79) we have

$$Y_{bb} = -8(pk')[2(\varepsilon k')^2 - (pk)]. \tag{8.81b}$$

The interference term Y_{ab}, Eq. (8.78c), is more complicated to evaluate. Its simplification depends essentially on writing $p' = p + k - k'$, so that the orthogonality relations (8.75) and Eqs. (8.80) can be used to the full. In this way one finds

$$Y_{ab} = 8(pk)(pk')[2(\varepsilon\varepsilon')^2 - 1] - 8(k\varepsilon')^2(pk') + 8(k'\varepsilon)^2(pk), \tag{8.81c}$$

which is real and symmetric (i.e. $Y_{ab} = Y_{ba}$). From Eqs. (8.81), (8.77) and (8.63), one obtains the differential cross-section for Compton scattering of polarized photons

$$\left(\frac{d\sigma}{d\Omega}\right)_{\text{Lab,pol}} = \frac{\alpha^2}{4m^2}\left(\frac{\omega'}{\omega}\right)^2\left\{\frac{\omega}{\omega'} + \frac{\omega'}{\omega} + 4(\varepsilon\varepsilon')^2 - 2\right\}. \tag{8.82}$$

Eq. (8.82) is known as the Klein–Nishina formula. From it one obtains the unpolarized cross-section by summing and averaging over final and initial photon polarizations. Since $\varepsilon\varepsilon' = -\boldsymbol{\varepsilon}\cdot\boldsymbol{\varepsilon}'$, one can write Eq. (1.71)

$$\tfrac{1}{2}\sum_{\text{pol}} (\varepsilon\varepsilon')^2 = \tfrac{1}{2}(1 + \cos^2\theta), \tag{8.83}$$

and applying this equation to Eq. (8.82) one at once regains the unpolarized cross-section formula (8.74).

8.7 SCATTERING BY AN EXTERNAL FIELD

So far, the electromagnetic field has been described by a quantized field, involving photon creation and annihilation operators. In some problems, where the quantum fluctuations of the field are unimportant, it may be adequate to describe the field as a purely classical function of the space–time coordinates. An example would be the scattering of electrons or positrons by an applied 'external' electromagnetic field $A_e^{\alpha}(x)$, such as the Coulomb field of

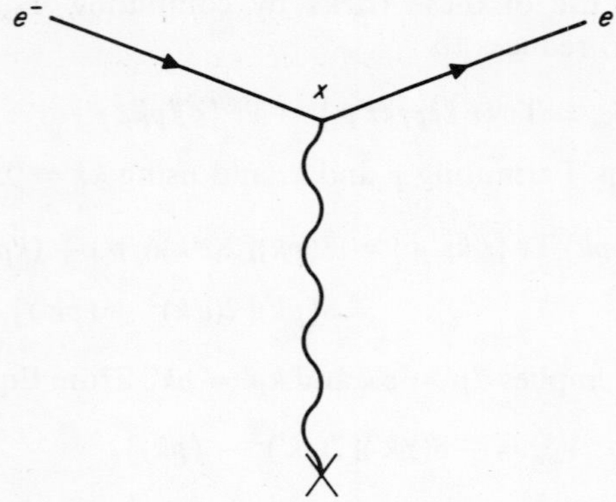

Fig. 8.5. Feynman graph in configuration space for electron scattering by an external source, marked by the cross.

a heavy nucleus.‡ More generally, one may have to consider both types of field, replacing $A^\alpha(x)$ by the sum of the quantized and the classical fields, $A^\alpha(x) + A^\alpha_e(x)$. The S-matrix expansion of QED, Eqs. (7.1) and (7.2), then becomes

$$S = \sum_{n=0}^{\infty} \frac{(ie)^n}{n!} \int \dots \int d^4x_1 \dots d^4x_n \mathrm{T}\{\mathrm{N}[\bar{\psi}(A\!\!\!/ + A\!\!\!/_e)\psi]_{x_1} \dots \mathrm{N}[\bar{\psi}(A\!\!\!/ + A\!\!\!/_e)\psi]_{x_n}\}. \tag{8.84}$$

As a simple example, we consider the scattering of an electron by a static external field

$$A^\alpha_e(x) = A^\alpha_e(\mathbf{x}) = \frac{1}{(2\pi)^3} \int d^3\mathbf{q}\, e^{i\mathbf{q}\cdot\mathbf{x}} A^\alpha_e(\mathbf{q}) \tag{8.85}$$

where the Fourier transform of the field in momentum space, $A^\alpha_e(\mathbf{q})$, has been introduced for later convenience. In lowest order, the scattering arises from the first-order term in Eq. (8.84):

$$S^{(1)}_e = ie \int d^4x \bar{\psi}^-(x) A\!\!\!/_e(x) \psi^+(x). \tag{8.86}$$

This is represented by the Feynman diagram in Fig. 8.5, in which the source of the classical field is represented by a cross. Consider the scattering of an electron from a state $|i\rangle$, with momentum $p = (E, \mathbf{p})$ and spinor $u_r(\mathbf{p})$, to a

‡ The meaning of 'external' in the present context and in the description of Feynman graphs (where we talk of external lines, particles, etc.) should not be confused.

state $|f\rangle$, with momentum $p' = (E', \mathbf{p}')$ and spinor $u_s(\mathbf{p}')$. The evaluation of the matrix element $\langle f|S_e^{(1)}|i\rangle$ for this transition is similar to that in Section 7.2.1 for electron scattering by the quantized field. Going over to momentum space, one easily obtains

$$\langle f|S_e^{(1)}|i\rangle = \left[(2\pi)\delta(E'-E)\left(\frac{m}{VE}\right)^{1/2}\left(\frac{m}{VE'}\right)^{1/2}\right]\mathscr{M}, \tag{8.87}$$

where

$$\mathscr{M} = ie\bar{u}_s(\mathbf{p}')\not{A}_e(\mathbf{q} = \mathbf{p}' - \mathbf{p})u_r(\mathbf{p}). \tag{8.88}$$

These equations should be compared with Eqs. (7.31) and (7.32) for the case of the quantized field. Unlike Eq. (7.31), Eq. (8.87) does not contain a momentum conserving δ-function, since we are ignoring the momentum of the source of the field, which experiences a recoil, $\mathbf{q} = \mathbf{p}' - \mathbf{p}$ being the momentum transferred to the particle by the field. On the other hand, the δ-function in Eq. (8.87) leads to conservation of the electron's energy, i.e. to elastic scattering ($|\mathbf{p}'| = |\mathbf{p}|$). It implies that the recoil energy of the source is negligible, and is a consequence of the static field assumption (8.85). The Feynman amplitude (8.88) is represented by the momentum space Feynman diagram of Fig. 8.6, in which it must be remembered that $|\mathbf{p}'| = |\mathbf{p}|$.

In addition to the replacement

$$(2\pi)^4\,\delta^{(4)}(p' + k' - p) \rightarrow (2\pi)\,\delta(E' - E)$$

Eqs. (8.87) and (8.88) differ from Eqs. (7.31) and (7.32) in that the photon factors $(1/2V\omega_{\mathbf{k}'})^{1/2}\varepsilon_\alpha(\mathbf{k}')$ are replaced by the external field factor $A_{e\alpha}(\mathbf{q})$.

These results are easily generalized and lead to the following two changes

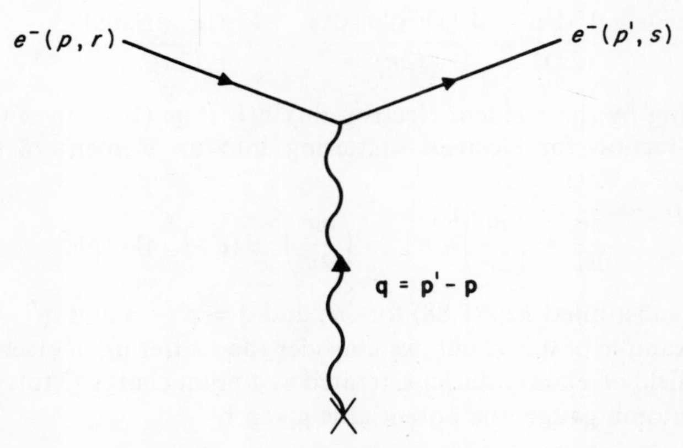

Fig. 8.6. Feynman graph in momentum space for electron scattering by an external source; $|\mathbf{p}'| = |\mathbf{p}|$.

in the Feynman rules, discussed in Section 7.3, in order to allow for the interaction with an external static field.

(i) In Eq. (7.45), relating the Feynman amplitude $\mathscr{M}$ of a process to its S-matrix element $\langle f|S|i\rangle$, we must make the replacement

$$(2\pi)^4\, \delta^{(4)}(P_f - P_i) \rightarrow (2\pi)\, \delta(E_f - E_i) \tag{8.89}$$

where E_i and E_f are the total energies of all particles present in the initial and final states.

(ii) Corresponding to rule 4, Eqs. (7.49), for external lines (i.e. particles present initially or finally), we must add the following rule to the Feynman rules 1–8 of Section 7.3:

9. For each interaction of a charged particle with an external static field $A_e(x)$, write a factor

$$A_{e\alpha}(\mathbf{q}) = \int \mathrm{d}^3\mathbf{x}\, \mathrm{e}^{-i\mathbf{q}\cdot\mathbf{x}} A_{e\alpha}(\mathbf{x}) \tag{8.90}$$

(α)

Here $\mathbf{q}$ labels the momentum transferred from the field source (marked ×) to the particle. (The energy of the particle is conserved at the vertex.)

We next derive the cross-section for electron scattering by an arbitrary static external field in terms of the corresponding Feynman amplitude $\mathscr{M}$. The argument is very similar to that of Section 8.1 and we shall state it only briefly. Eq. (8.87) leads to the transition probability per unit time

$$w = \frac{1}{T}|\langle f|S_e^{(1)}|i\rangle|^2 = 2\pi\delta(E' - E)\left(\frac{m}{VE}\right)^2 |\mathscr{M}|^2,$$

where, as in Eq. (8.6), we are considering a long but finite time interval T. Multiplying w by the density of final states

$$\frac{V\,\mathrm{d}^3\mathbf{p}'}{(2\pi)^3} = \frac{V|\mathbf{p}'|^2\,\mathrm{d}|\mathbf{p}'|\,\mathrm{d}\Omega'}{(2\pi)^3} = \frac{V|\mathbf{p}'|E'\,\mathrm{d}E'\,\mathrm{d}\Omega'}{(2\pi)^3},$$

and dividing by the incident electron flux $v/V = |\mathbf{p}|/(VE)$ gives the differential cross-section for electron scattering into an element of solid angle $\mathrm{d}\Omega'$

$$\frac{\mathrm{d}\sigma}{\mathrm{d}\Omega'} = \left(\frac{m}{2\pi}\right)^2 |\mathscr{M}|^2 = \left(\frac{me}{2\pi}\right)^2 |\bar{u}_s(\mathbf{p}')\not{A}_e(\mathbf{q})u_r(\mathbf{p})|^2, \tag{8.91}$$

where we substituted Eq. (8.88) for $\mathscr{M}$, and $\mathbf{q} = \mathbf{p}' - \mathbf{p}$ and $|\mathbf{p}'| = |\mathbf{p}|$.

As an example of this result, we consider the scattering of electrons by the Coulomb field of a heavy nucleus, treated as a point charge (Mott scattering). In the Coulomb gauge, the potential is given by

$$A_e^{\alpha}(x) = \left(\frac{Ze}{4\pi|\mathbf{x}|}, 0, 0, 0\right), \tag{8.92a}$$

with the momentum space potential

$$A_e^{\alpha}(\mathbf{q}) = \left(\frac{Ze}{|\mathbf{q}|^2}, 0, 0, 0\right). \tag{8.92b}$$

On substituting Eq. (8.92b) in Eq. (8.91), and summing and averaging over electron spins, one obtains the unpolarized cross-section for Coulomb scattering

$$\begin{aligned}\frac{d\sigma}{d\Omega'} &= \frac{(2m\alpha Z)^2}{|\mathbf{q}|^4} \cdot \tfrac{1}{2}\sum_{r,s} |\bar{u}_s(\mathbf{p}')\gamma^0 u_r(\mathbf{p})|^2 \\ &= \frac{(\alpha Z)^2}{2|\mathbf{q}|^4} \operatorname{Tr}\{(\not{p}' + m)\gamma^0(\not{p} + m)\gamma^0\} \\ &= \frac{2(\alpha Z)^2}{|\mathbf{q}|^4}(E^2 + \mathbf{p}\cdot\mathbf{p}' + m^2). \end{aligned} \tag{8.93a}$$

Introducing the scattering angle θ, we have

$$\mathbf{p}\cdot\mathbf{p}' = |\mathbf{p}|^2 \cos\theta, \qquad |\mathbf{q}|^2 = |\mathbf{p}' - \mathbf{p}|^2 = 4|\mathbf{p}|^2 \sin^2(\theta/2), \tag{8.94}$$

and, with $|\mathbf{p}| = Ev$, Eq. (8.93a) reduces to the Mott scattering formula

$$\frac{d\sigma}{d\Omega'} = \frac{(\alpha Z)^2}{4E^2v^4 \sin^4(\theta/2)}[1 - v^2 \sin^2(\theta/2)] \tag{8.93b}$$

for the scattering of relativistic electrons by a Coulomb field. In the non-relativistic limit, this reduces to the Rutherford scattering formula

$$\frac{d\sigma}{d\Omega'} = \frac{(\alpha Z)^2}{4m^2v^4 \sin^4(\theta/2)}. \tag{8.95}$$

We have here considered the nucleus as a point charge. We only mention that the treatment is easily modified to apply to the realistic case of a nucleus whose charge is distributed over a finite volume. For high-energy electrons, this leads to an important method of investigating the nuclear charge distribution. In particular, one obtains in this way the r.m.s. radius of the charge distribution which has previously been referred to in connection with the finite size of the proton. (See Problem 8.1.)

The above analysis is easily extended to give the polarization properties of the electrons in Coulomb scattering.

At non-relativistic energies, the answer is of course well known from non-relativistic quantum mechanics. The Coulomb interaction (8.92) and the scattering amplitude are spin-independent, i.e. spin is conserved in Rutherford scattering. Interpreted in terms of helicity, this means that if the incident electron has positive helicity, then the outgoing electron has positive helicity for forward scattering, and negative helicity for backward scattering. At intermediate angles, it follows from the rotational properties of spinors that, for scattering through an angle θ, the probability of positive helicity (i.e. no helicity flip) is $\cos^2(\theta/2)$, and of negative helicity (i.e. helicity flip) is $\sin^2(\theta/2)$. These results of course also follow from the matrix element in the first line of Eq. (8.93a). For the non-relativistic limit, it is most natural to use the Dirac–Pauli spinor representation [Eqs. (A.72) and (A.73) in Appendix A] for the spinors $u_r(\mathbf{p})$ and $u_s(\mathbf{p}')$. The indices r and s then label the spin components referred to a fixed axis in space. It follows that in the non-relativistic limit the leading term of the matrix element in Eq. (8.93a) is proportional to

$$\lim_{\substack{\mathbf{p}\to 0\\ \mathbf{p}'\to 0}} (\bar{u}_s(\mathbf{p}')\gamma^0 u_r(\mathbf{p})) = u_s^\dagger(\mathbf{0})u_r(\mathbf{0}) = \delta_{sr},$$

i.e. spin is conserved.

At relativistic energies, the scattering is spin-dependent, due to the interaction of the electron's magnetic moment with the magnetic field which the electron sees in its own rest-frame. [The velocity-dependent term $-v^2\sin^2(\theta/2)$ in the Mott formula (8.93b) similarly results from this magnetic scattering.] To obtain the polarization properties, it is now most convenient to use the helicity states and the helicity projection operator formalism of Section 8.2. We consider a transition in which the incident electron has positive helicity [i.e. it is in the spin state $u_1(\mathbf{p})$]. The probability that the scattered electron has positive or negative helicity is, from Eq. (8.93a), proportional to

$$X_s = |\bar{u}_s(\mathbf{p}')\gamma^0 u_1(\mathbf{p})|^2, \quad \begin{cases} s = 1, & \text{positive helicity} \\ s = 2, & \text{negative helicity.} \end{cases} \tag{8.96}$$

From Eqs. (8.28) and (8.29) we see that this can be written

$$X_s = \mathrm{Tr}\,\{\Lambda^+(\mathbf{p}')\gamma^0\Pi^+(\mathbf{p})\Lambda^+(\mathbf{p})\gamma^0\Pi^\pm(\mathbf{p}')\}, \tag{8.97}$$

where the plus and minus signs on $\Pi^\pm(\mathbf{p}')$ correspond to $s = 1$ (no helicity flip) and $s = 2$ (helicity flip) respectively.

In the extreme relativistic limit $E = E' \gg m$, the helicity projection operators simplify to

$$\Pi^\pm(\mathbf{p}) = \tfrac{1}{2}(1 \pm \gamma^5). \tag{A.43}$$

Correspondingly, Eq. (8.97) reduces to

$$X_s = \frac{1}{16m^2}\,\mathrm{Tr}\,\{(\not{p}' + m)\gamma^0(1 + \gamma^5)(\not{p} + m)\gamma^0(1 \pm \gamma^5)\}. \tag{8.98}$$

Since p and p' are of order E, we would expect the leading contributions to X_s to come from the part of the trace in Eq. (8.98) which contains both $\not{p}$ and $\not{p}'$ as factors, so that X_s would be of order $(E/m)^2$. It is easy to show that for

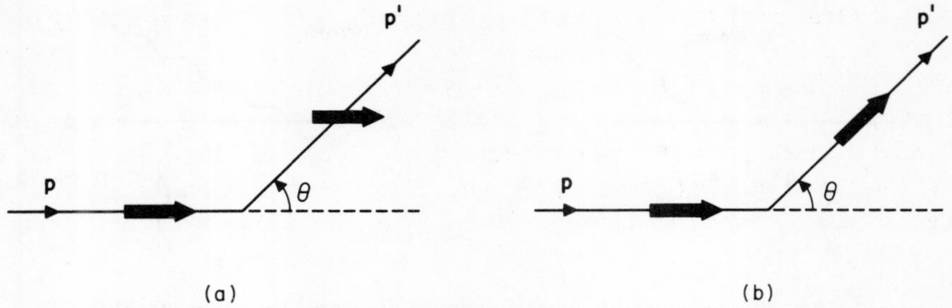

Fig. 8.7. The two extreme regimes for Coulomb scattering. (The broad arrows represent the electron spins.) (a) Non-relativistic energies, $|\mathbf{p}| \ll E$: *spin* is conserved. (b) Extreme relativistic energies, $E \gg m$: *helicity* is conserved.

X_1, i.e. no helicity flip, this is indeed the case, but that for X_2, i.e. helicity flip, the only non-vanishing term is the part of the trace in Eq. (8.98) which is proportional to m^2 (i.e. independent of p and p'), so that X_2 is of order unity.‡ Thus in the extreme relativistic limit $E \gg m$, the helicity flip amplitude vanishes, and helicity is conserved. This contrasts with the non-relativistic limit $|\mathbf{p}| \ll m$, where spin is conserved and the probability of helicity flip is $\sin^2(\theta/2)$. These two regimes of non-relativistic and extreme relativistic energies are schematically illustrated in Fig. 8.7. At intermediate energies, the traces in Eq. (8.97) must be evaluated exactly.

8.8 BREMSSTRAHLUNG

The deflection of an electron by the Coulomb field of a heavy nucleus, which we studied in the last section, implies that the electron must emit radiation and consequently be slowed down in its motion. More generally, the scattering of any charged particle leads to the emission of radiation. Both this process and the radiation are referred to as bremsstrahlung (literally translated, braking radiation). In this section we shall consider bremsstrahlung resulting from the scattering of electrons by the Coulomb field of a heavy nucleus. This process is of considerable practical importance. It is primarily responsible for the slowing down of fast electrons in their passage through matter, and it is widely used in electron accelerators to produce photon beams.

‡ Using $[\gamma^5, \gamma^\alpha]_+ = 0$, $(\gamma^5)^2 = 1$, and

$$\mathrm{Tr}\,\gamma^5 = \mathrm{Tr}\,(\gamma^5\gamma^\lambda) = \mathrm{Tr}\,(\gamma^5\gamma^\lambda\gamma^\mu) = \mathrm{Tr}\,(\gamma^5\gamma^\lambda\gamma^\mu\gamma^\nu) = 0, \tag{A.21}$$

one easily proves these points and calculates the traces in Eq. (8.98). (One finds that only the term containing p and p' makes a non-vanishing contribution to X_1.) The detailed proofs are left as exercises for the reader.

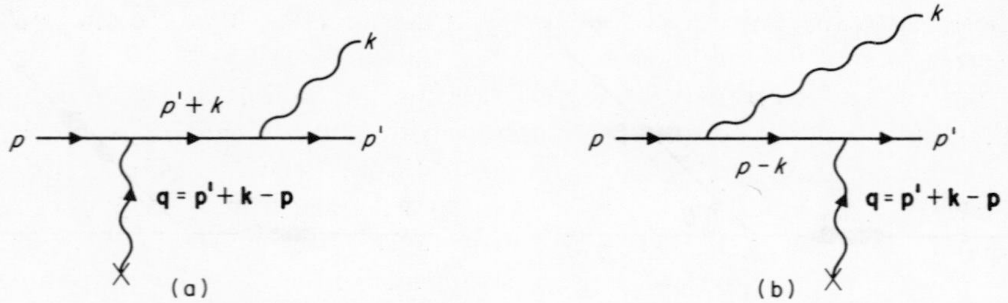

Fig. 8.8. The momentum space Feynman diagrams for bremsstrahlung.

Unlike elastic Coulomb scattering, we are now considering a process in which both the quantized field and the external field (i.e. the nuclear Coulomb field) play a role; the former to emit the radiation, the latter to ensure conservation of energy and momentum. In lowest order, the process is brought about by the second-order term ($n = 2$) in the S-matrix expansion (8.84) but, by this stage, the reader should have no difficulty in writing down the corresponding S-matrix element directly in momentum space. The two Feynman diagrams responsible for the transition are shown in Fig. 8.8, which also specifies the momenta of the particles. (The spin and polarization labels are again suppressed.) From the Feynman rules of Sections 7.3 and 8.7, the S-matrix element is given by

$$\langle f|S|i\rangle = 2\pi\delta(E' + \omega - E)\left(\frac{m}{VE}\right)^{1/2}\left(\frac{m}{VE'}\right)^{1/2}\left(\frac{1}{2V\omega}\right)^{1/2}\mathscr{M} \tag{8.99}$$

where

$$\begin{aligned}\mathscr{M} &= -e^2\bar{u}(\mathbf{p}')[\not{\varepsilon}(\mathbf{k})\mathrm{i}S_\mathrm{F}(p' + k)\not{A}_\mathrm{e}(\mathbf{q}) + \not{A}_\mathrm{e}(\mathbf{q})\mathrm{i}S_\mathrm{F}(p - k)\not{\varepsilon}(\mathbf{k})]u(\mathbf{p}) \\ &= -\mathrm{i}e^2\bar{u}(\mathbf{p}')\left[\not{\varepsilon}(\mathbf{k})\frac{\not{p}' + \not{k} + m}{2p'k}\not{A}_\mathrm{e}(\mathbf{q}) + \not{A}_\mathrm{e}(\mathbf{q})\frac{\not{p} - \not{k} + m}{-2pk}\not{\varepsilon}(\mathbf{k})\right]u(\mathbf{p})\end{aligned} \tag{8.100}$$

and $A_{\mathrm{e}\alpha}(\mathbf{q} = \mathbf{p}' + \mathbf{k} - \mathbf{p})$ is the momentum space Coulomb potential, given in Eq. (8.92b).

The derivation of the cross-section formula for bremsstrahlung from Eq. (8.99) is very similar to that given in the last section for Coulomb scattering. Multiplying the transition rate $|\langle f|S|i\rangle|^2/T$ by the density of final states

$$V\,\mathrm{d}^3\mathbf{p}'\,V\,\mathrm{d}^3\mathbf{k}/(2\pi)^6$$

and dividing by the incident electron flux $|\mathbf{p}|/(VE)$ leads to

$$\mathrm{d}\sigma = \frac{m^2}{(2\pi)^5 2\omega}\frac{|\mathbf{p}'|}{|\mathbf{p}|}|\mathscr{M}|^2\,\mathrm{d}^3\mathbf{k}\,\mathrm{d}\Omega', \tag{8.101}$$

where $d\Omega'$ is the element of solid angle in the direction $\mathbf{p}'$ of the scattered electron.

Summing and averaging Eq. (8.101) over final and initial electron spins is a straightforward but lengthy calculation which leads to the Bethe–Heitler cross-section formula for bremsstrahlung in a Coulomb field.‡ We shall restrict ourselves to the simpler but interesting situation in which the emitted photon has very low-energy, the so-called 'soft photon' limit $\omega \approx 0$. In this limit $\mathbf{q} = \mathbf{p}' - \mathbf{p}$ and $|\mathbf{p}'| = |\mathbf{p}|$, as for elastic scattering. In the Feynman amplitude (8.100), we neglect the $\not{k}$ terms in the numerators of the electron propagators and, using the Dirac equation, easily obtain

$$\begin{aligned} \mathcal{M} &= -\mathrm{i}e^2\bar{u}(\mathbf{p}')\not{A}_e(\mathbf{q})u(\mathbf{p})\left[\frac{p'\varepsilon}{p'k} - \frac{p\varepsilon}{pk}\right] \qquad (\omega \approx 0) \\ &= -e\mathcal{M}_0\left[\frac{p'\varepsilon}{p'k} - \frac{p\varepsilon}{pk}\right] \qquad (\omega \approx 0) \end{aligned} \tag{8.102}$$

where $\varepsilon \equiv \varepsilon(\mathbf{k})$ and $\mathcal{M}_0$ is the Feynman amplitude, Eq. (8.88), for elastic scattering without photon emission. Substituting Eq. (8.102) in Eq. (8.101) and comparing with Eq. (8.91), we can write the cross-section for soft bremsstrahlung in the form

$$\left(\frac{d\sigma}{d\Omega'}\right)_B = \left(\frac{d\sigma}{d\Omega'}\right)_0 \frac{\alpha}{(2\pi)^2}\left[\frac{p'\varepsilon}{p'k} - \frac{p\varepsilon}{pk}\right]^2 \frac{d^3\mathbf{k}}{\omega} \qquad (\omega \approx 0) \tag{8.103}$$

where $(d\sigma/d\Omega')_0$ is the cross-section for elastic scattering without photon emission, Eq. (8.91). In deriving Eqs. (8.102) and (8.103), we have not used the explicit form of $A_{e\alpha}(\mathbf{q})$, so that these equations hold for soft bremsstrahlung in an arbitrary static external field, not just for a Coulomb field.

Eqs. (8.102) and (8.103) contain two interesting features. Firstly, they each factorize into the corresponding quantity for elastic scattering without photon emission, multiplied by a factor which relates to the soft photon. Secondly, both the amplitude $\mathcal{M}$ and the cross-section (8.103) are singular in the infra-red limit $\omega \to 0$. This infra-red singularity arises because for $k = 0$ we have $p^2 = p'^2 = m^2$, so that the intermediate-state electrons in the Feynman diagrams in Figs. 8.8(a) and (b) possess the four-momenta of a *real* electron, and correspondingly the propagators in Eq. (8.100) diverge.§ Both the factorization property and the infra-red singularity are characteristic features of soft photon emission processes. We shall discuss them further in the next section.

‡ The formula is derived in, for example, C. Itzykson and J. B. Zuber, *Quantum Field Theory*, McGraw-Hill, New York, 1980, Section 5-2-4, and in J. M. Jauch and F. Rohrlich, *The Theory of Photons and Electrons*, 2nd edn, Springer, New York, 1976, Section 15-6.

§ In a concise but horrible jargon, this situation is often described by saying that, as $k \to 0$, the internal electron lines in the Feynman diagrams 8.8(a) and (b) go on the mass shell; or, even more concisely and horribly, that they go on shell.

It remains to sum over the polarization of the emitted photons, assuming that this is not observed. We shall do this by the gauge-invariant method of Section 8.3. Applying Eq. (8.36) to Eq. (8.103), we at once obtain the cross-section for the emission of soft bremsstrahlung in electron scattering by a static external field

$$\left(\frac{d\sigma}{d\Omega'}\right)_B = \left(\frac{d\sigma}{d\Omega'}\right)_0 \frac{(-\alpha)}{(2\pi)^2}\left[\frac{p'}{p'k} - \frac{p}{pk}\right]^2 \frac{d^3\mathbf{k}}{\omega} \qquad (\omega \approx 0). \tag{8.104}$$

8.9 THE INFRA-RED DIVERGENCE

The above results on soft photon emission in electron scattering have important implications for experiments on elastic electron scattering. In such an experiment a photon may be emitted which is too soft to be detected, and it is the energy resolution ΔE of the apparatus which determines whether such a photon emission event is recorded as elastic or inelastic scattering. Consequently, the *experimental* cross-section is the sum of the elastic cross-section and the cross-section for bremsstrahlung of energy less than ΔE, i.e. it should be written

$$\left(\frac{d\sigma}{d\Omega'}\right)_{\text{Exp}} = \left(\frac{d\sigma}{d\Omega'}\right)_{\text{El}} + \left(\frac{d\sigma}{d\Omega'}\right)_B . \tag{8.105}$$

Here $(d\sigma/d\Omega')_{\text{El}}$ is the elastic cross-section and $(d\sigma/d\Omega')_B$ is the soft bremsstrahlung cross-section (8.104) integrated over the range of photon energy $0 \leqslant \omega \leqslant \Delta E$:

$$\left(\frac{d\sigma}{d\Omega'}\right)_B = \left(\frac{d\sigma}{d\Omega'}\right)_0 \alpha B, \tag{8.106}$$

where $(d\sigma/d\Omega')_0$ is the elastic cross-section in lowest order of perturbation theory and

$$B = \frac{-1}{(2\pi)^2}\int_{0\leqslant|\mathbf{k}|\leqslant\Delta E} \frac{d^3\mathbf{k}}{\omega}\left[\frac{p'}{p'k} - \frac{p}{pk}\right]^2 . \tag{8.107}$$

We are assuming that ΔE is sufficiently small so that the soft photon result (8.104) is valid for $\omega \leqslant \Delta E$.

Unfortunately the integrand in Eq. (8.107) behaves like $1/\omega$ for small ω, so that the integral is logarithmically divergent at the lower limit of integration. This divergence is known as the *infra-red catastrophe*. It is a consequence of the zero mass of the photon, and one way of dealing with this problem, due to Feynman, is to assign a fictitious small mass $\lambda(\neq 0)$ to the photon and at the end of the calculation take the limit $\lambda \to 0$ to regain QED.

Introducing a non-zero photon mass leads to modifications of Eq. (8.107).

The amplitude $\mathscr{M}$, Eq. (8.100), is modified since in the denominators of the electron propagators we must now put $k^2 = \lambda^2$, instead of $k^2 = 0$. It is easy to show that Eq. (8.103) is replaced by

$$\frac{d\sigma}{d\Omega'} = \left(\frac{d\sigma}{d\Omega'}\right)_0 \frac{\alpha}{(2\pi)^2} \left[\frac{2p'\varepsilon}{2p'k + \lambda^2} + \frac{2p\varepsilon}{-2pk + \lambda^2}\right]^2 \frac{d^3\mathbf{k}}{\omega}. \tag{8.108}$$

With non-zero mass, photons can be longitudinally as well as transversely polarized. The photon polarization sum is now effected by means of the formula for massive spin 1 bosons‡

$$\sum_{r=1}^{3} \varepsilon_{r\alpha}\varepsilon_{r\beta} = -g_{\alpha\beta} + \frac{k_\alpha k_\beta}{\lambda^2}.‡ \tag{8.109}$$

The term $k_\alpha k_\beta/\lambda^2$ gives no contribution to the unpolarized cross-section in the limit when $\lambda \to 0$, Hence the unpolarized soft bremsstrahlung cross-section, integrated over the photon energy range $\lambda \leqslant \omega \leqslant \Delta E$, is given by Eq. (8.106) with B, Eq. (8.107), now replaced by

$$B(\lambda) = \frac{-1}{(2\pi)^2} \int \frac{d^3\mathbf{k}}{\omega_\lambda} \left[\frac{2p'}{2p'k + \lambda^2} + \frac{2p}{-2pk + \lambda^2}\right]^2 \tag{8.110}$$

where $\omega_\lambda \equiv (\lambda^2 + \mathbf{k}^2)^{1/2}$ and the integration is over wave vectors $\mathbf{k}$ such that $\lambda \leqslant \omega_\lambda \leqslant \Delta E$. For $\lambda > 0$, $B(\lambda)$ is finite and well-defined. For $\lambda \to 0$, $B(\lambda)$ goes over into Eq. (8.107) and diverges again.

Returning to Eqs. (8.105) and (8.106), we see that the bremsstrahlung cross-section (8.106) is of order α relative to the lowest-order elastic cross-section $(d\sigma/d\Omega')_0$. Consequently it would be inconsistent to use the latter for $(d\sigma/d\Omega')_{El}$ in Eq. (8.105). Instead, we must include corrections of order α to the elastic cross-section, which arise from the next order in perturbation theory. Fig. 8.9 shows a Feynman diagram responsible for such a correction to the lowest-order graph, Fig. 8.6. We shall consider these corrections in detail in Section 9.7. For the moment we shall write

$$\left(\frac{d\sigma}{d\Omega'}\right)_{El} = \left(\frac{d\sigma}{d\Omega'}\right)_0 [1 + \alpha R(\lambda)] \tag{8.111}$$

with the correction term $R(\lambda)$ also a function of the photon mass λ. Combining Eqs. (8.105), (8.106) and (8.111), we obtain

$$\left(\frac{d\sigma}{d\Omega'}\right)_{Exp} = \left(\frac{d\sigma}{d\Omega'}\right)_0 \{1 + \alpha[B(\lambda) + R(\lambda)] + O(\alpha^2)\} \tag{8.112}$$

where the term $O(\alpha^2)$ is to remind the reader that there exist higher-order corrections.

‡ For a derivation of this formula, see Section 11.3, Eqs. (11.24)–(11.27).

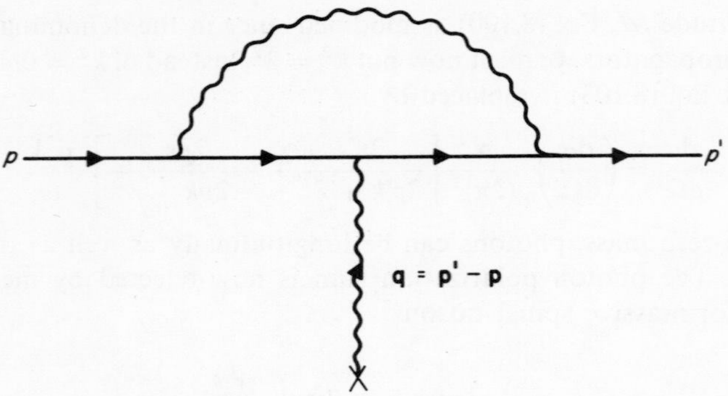

Fig. 8.9. A correction term to the elastic electron scattering by an external field.

We have seen that $B(\lambda) \to \infty$ as $\lambda \to 0$. When calculating $R(\lambda)$ in Section 9.7, we shall find that $R(\lambda)$ is also singular at $\lambda = 0$, with $R(\lambda) \to -\infty$ as $\lambda \to 0$ in such a way that the singularities in $B(\lambda)$ and $R(\lambda)$ exactly cancel and $[B(\lambda) + R(\lambda)]$ is well-defined and finite. Hence we can easily take the limit $\lambda \to 0$ in Eq. (8.112), obtaining a finite correction of order α to the lowest-order elastic cross-section $(\mathrm{d}\sigma/\mathrm{d}\Omega')_0$. Such corrections are called *radiative corrections.*

The experimental cross-section (8.112) depends on the energy resolution ΔE which occurs in Eqs. (8.107) and (8.110), and so depends on the details of the experimental set-up. Because of this, the radiative corrections are sometimes calculated and subtracted from the experimental data to give 'radiatively corrected cross-sections'. In this way the results of different experiments can be compared with each other and with theoretical predictions for $(\mathrm{d}\sigma/\mathrm{d}\Omega')_0$.

We can now understand the origin of the infra-red catastrophe better. It arises through treating soft bremsstrahlung and elastic scattering as separate processes in perturbation theory. This separation is artificial as one always observes elastic scattering together with some soft bremsstrahlung, and the cross-section for this is finite.

We have so far considered the emission of one soft photon and the infra-red divergence to which this gives rise. For high-resolution experiments, the emission of many soft photons may become important, resulting in infra-red divergences of higher order in α. We must then also modify Eq. (8.111) to take into account higher-order corrections to the elastic scattering. The statement that the infra-red divergences cancel exactly to all orders of perturbation theory, leaving a finite radiative correction of order α, is the famous Bloch–Nordsieck theorem. In high-energy experiments, where radiative corrections can become as large as 50 per cent in some kinematic regions, these multiphoton contributions must be incorporated. It can be shown‡ that the most important radiative correction is given by the generalization of Eq. (8.112) to all orders and obtained from it by exponentiation, i.e.

$$\left(\frac{d\sigma}{d\Omega'}\right)_{\text{Exp}} = \left(\frac{d\sigma}{d\Omega'}\right)_0 e^{\alpha[B(\lambda)+R(\lambda)]}. \tag{8.113}$$

Correspondingly, the elastic cross-section, calculated to all orders of perturbation theory, is given by

$$\left(\frac{d\sigma}{d\Omega'}\right)_{\text{El}} = \left(\frac{d\sigma}{d\Omega'}\right)_0 e^{\alpha R(\lambda)}. \tag{8.114}$$

Since $R(\lambda) \to -\infty$ as $\lambda \to 0$, it follows that the pure elastic cross-section, without photon emission, vanishes: there is no truly elastic scattering; all observed scattering is accompanied by emission of radiation. This corresponds to the result of classical electrodynamics that an accelerated charge must radiate.

Any process involving charged particles initially or finally can be accompanied by the emission of soft photons, and infra-red divergences occur in all these processes. Our conclusions, reached for the case of elastic electron scattering by an external field, hold generally. In particular, the Bloch–Nordsieck theorem applies, i.e. for all processes in QED the infra-red divergences cancel exactly to all orders of perturbation theory, leaving finite radiative corrections of order α. In the next two chapters we shall go on to calculate these corrections.

PROBLEMS

8.1 In our discussion of electron scattering by an infinitely heavy nucleus, Eqs. (8.92)–(8.95), we treated the nucleus as a point charge. More realistically, we could treat the nucleus as a spherical charge distribution $Ze\rho(r)$, where

$$\int d^3\mathbf{r}\rho(r) = 1.$$

Show that the elastic scattering cross-section is now given by

$$\frac{d\sigma}{d\Omega'} = \left(\frac{d\sigma}{d\Omega'}\right)_M |F(\mathbf{q})|^2$$

where $(d\sigma/d\Omega')_M$ is the Mott cross-section (8.93b), $ZeF(\mathbf{q})$ is the Fourier transform of the charge distribution and $\mathbf{q} = \mathbf{p}' - \mathbf{p}$ ($\mathbf{p}$ and $\mathbf{p}'$ are the initial and final momenta of the electron).

‡ See the comprehensive article on the infra-red problem by D. R. Yennie, S. C. Frautschi and H. Suura, *Ann. Phys.* (N.Y.), **13** (1961), 379.

$F(\mathbf{q})$ is called the form factor of the charge distribution. Show that it is a function of $|\mathbf{q}|^2$ only, and that the root-mean-square radius r_m of the charge distribution is given by

$$r_m^2 = -6 \left. \frac{dF(\mathbf{q})}{d(|\mathbf{q}|^2)} \right|_{|\mathbf{q}|=0}.$$

8.2 Consider elastic $e^- - \mu^-$ scattering at energies sufficiently high so that the mass of the electron can be neglected throughout. Show that the differential cross-section for this process, in the frame of reference in which the muon is initially at rest, is given by

$$\frac{d\sigma}{d\Omega'} = \left(\frac{d\sigma}{d\Omega'}\right)_M \left[1 + \frac{2E}{m_\mu}\sin^2\frac{\theta}{2}\right]^{-1} \left[1 - \frac{q^2}{2m_\mu^2}\tan^2\frac{\theta}{2}\right].$$

Here E and E' are the initial and final energies of the electron, and θ is the angle through which the electron is scattered. q is the four-momentum of the exchanged photon, whence

$$q^2 = -4EE'\sin^2(\theta/2),$$

and $(d\sigma/d\Omega')_M$ is the Mott cross-section (8.93b) in the extreme relativistic limit $v \to 1$ (and, of course, $Z = 1$).

8.3 In the last problem the interaction of the muon is characterized by the vertex factor $ie\gamma^\alpha$ (Feynman rule 1). The method of the last problem can be applied to electron–proton scattering by replacing the factor $ie\gamma^\alpha$ of the muon vertex by the factor

$$-ie\left[\gamma^\alpha F_1(q^2) + \frac{\kappa_p}{2m_p} F_2(q^2) i\sigma^{\alpha\beta} q_\beta\right],$$

where q is the momentum transfer four-vector of the electron, and κ_p is the anomalous magnetic moment of the proton in units of the Bohr magneton. $F_1(q^2)$ and $F_2(q^2)$ are form factors representing the internal structure of the proton, with $F_1(0) = F_2(0) = 1$, so that for $\mathbf{q} \to 0$, i.e. a stationary proton interacting with static electric and magnetic fields, the proton has the correct electrostatic and magnetostatic interactions.

Show that in the laboratory frame in which the proton is initially at rest the differential cross-section for elastic scattering of electrons of energy $E(\gg m_e)$ is given by the Rosenbluth cross-section

$$\frac{d\sigma}{d\Omega'} = \left(\frac{d\sigma}{d\Omega'}\right)_M \left[1 + \frac{2E}{m_p}\sin^2\frac{\theta}{2}\right]^{-1}$$

$$\times \left\{\left[F_1^2(q^2) - \frac{\kappa_p^2}{4m_p^2} q^2 F_2^2(q^2)\right] - \frac{q^2}{2m_p^2}[F_1(q^2) + \kappa_p F_2(q^2)]^2 \tan^2\frac{\theta}{2}\right\}.$$

Here θ is the angle of scattering, and $(d\sigma/d\Omega')_M$ is the Mott cross-section (8.93b) with $Z = 1$ and $v = 1$.

8.4 Show that the probability of helicity flip occurring in Mott scattering is given by

$$\frac{m^2\sin^2(\theta/2)}{E^2\cos^2(\theta/2) + m^2\sin^2(\theta/2)}.$$

8.5 Show that the Mott scattering formula (8.93b) also gives the differential cross-section for the scattering of positrons by a heavy nucleus, treated as a point charge. (This equality of electron and positron scattering holds for the lowest-order calculations only.)

8.6 Show that the differential cross-section in the centre-of-mass system for electron–electron scattering in the high-energy limit ($E \gg m$) is given by

$$\left(\frac{\mathrm{d}\sigma}{\mathrm{d}\Omega}\right)_{\mathrm{CoM}} = \frac{\alpha^2}{8E^2}\left[\frac{1+\cos^4(\theta/2)}{\sin^4(\theta/2)} + \frac{2}{\sin^2(\theta/2)\cos^2(\theta/2)} + \frac{1+\sin^4(\theta/2)}{\cos^4(\theta/2)}\right]$$

where θ is the scattering angle and E is the energy of either electron in the CoM system.

8.7 Show that the Feynman amplitude for Compton scattering

$$\mathscr{M} = \mathscr{M}_{\mathrm{a}} + \mathscr{M}_{\mathrm{b}} \qquad (8.58\text{–}59)$$

is gauge-invariant, although the individual contributions $\mathscr{M}_{\mathrm{a}}$ and $\mathscr{M}_{\mathrm{b}}$ are not, by considering the gauge transformation [compare Eq. (8.31)]

$$\varepsilon(\mathbf{k}) \to \varepsilon(\mathbf{k}) + \lambda k, \qquad \varepsilon'(\mathbf{k}') \to \varepsilon'(\mathbf{k}') + \lambda' k'.$$

CHAPTER 9

Radiative corrections

In the last chapter we applied QED to calculate processes in lowest order of perturbation theory. On taking higher orders into account, one expects corrections of the order of the fine structure constant α to the lowest-order results, known as radiative corrections. However, on doing such a calculation one encounters divergent integrals. The divergent electron self-energy term, Eq. (7.44b), corresponding to the Feynman diagram, Fig. 7.15, is a typical example.

In this chapter we shall show how to overcome these difficulties. This involves three steps. Firstly, one regularizes the theory, that is, one modifies it so that it remains finite and well-defined in all orders of perturbation theory. The second step originates from the recognition that the non-interacting leptons and photons from which perturbation theory starts are not the same thing as the real physical particles which interact. The interaction modifies the properties of the particles, e.g. the charge and mass of the electron, and the predictions of the theory must be expressed in terms of the properties of the physical particles, not of the non-interacting (or bare) particles. The second step, called renormalization, consists of relating the properties of the physical particles to those of the bare particles and expressing the predictions of the theory in terms of the masses and charges of the physical particles. The third step consists of reverting from the regularized theory back to QED. The original infinities of QED now appear in the relations between bare and physical particles. These relations, like the bare particles themselves, are totally unobservable. In contrast, the observable predictions of the theory, expressed in terms of the measured charges and masses of the particles, remain finite as QED is restored. In particular, the radiative corrections are finite and of order α.

The programme we have outlined can be carried through to all orders of perturbation theory, so that the radiative corrections can be calculated to extraordinarily high accuracy. The complete agreement of these predictions with equally precise experiments, for example for the anomalous magnetic moments of leptons and for the Lamb shift, constitutes one of the great triumphs of physics.

In this chapter we shall almost exclusively consider the calculations of radiative corrections in lowest order of perturbation theory. We shall develop the general methods for this in Sections 9.1–9.5, and we shall consider applications and comparison with experiment in Section 9.6. The more technical details of regularization are relegated to the next chapter. As a further application, in Section 9.7, we shall once more consider the infra-red divergence completing the discussion which was started in Section 8.9. Finally, in Section 9.8, we shall briefly indicate how the considerations of this chapter can be generalized to yield finite radiative corrections to all orders of perturbation theory.

9.1 THE SECOND-ORDER RADIATIVE CORRECTIONS OF QED

The radiative corrections to any process in QED are obtained, like the lowest-order matrix elements themselves, from the S-matrix expansion, Eq. (8.84), using the Feynman rules of Sections 7.3 and 8.7. The Feynman diagrams representing the radiative corrections to a process contain additional vertices, compared with the diagrams describing the process in lowest order of perturbation theory, corresponding to the emission and reabsorption of virtual photons. Restricting oneself to the Feynman diagrams which contain two extra vertices corresponds to calculating the radiative corrections in lowest order of perturbation theory, involving only one virtual photon. Thus these corrections are of second order in the electronic charge, i.e. first order in the fine structure constant, relative to the lowest-order matrix element.

To introduce the basic ideas involved in calculating second-order radiative corrections, we shall consider the elastic scattering of electrons by a static external field $A_{\mathrm{e}}^{\mu}(\mathbf{x})$. As we saw in Section 8.7, the lowest order of perturbation theory describes this process by the Feynman diagram of Fig. 9.1 and by the Feynman amplitude

$$\mathscr{M}^{(0)} = \mathrm{i}e_0\bar{u}(\mathbf{p}')\not{A}_{\mathrm{e}}(\mathbf{p}' - \mathbf{p})u(\mathbf{p}). \tag{9.1}$$

Here and from now on the charge of the bare (i.e. non-interacting) electron is denoted by $(-e_0)$. Similarly, we shall denote the mass of the bare electron by m_0.

The radiative corrections to the amplitude (9.1) follow from the S-matrix

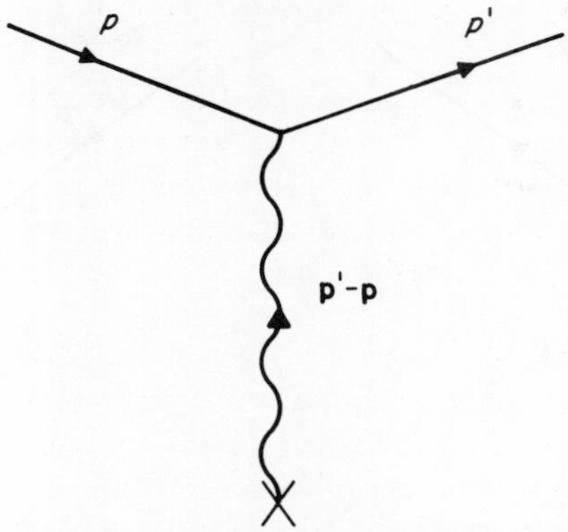

Fig. 9.1. The lowest-order contribution to the elastic scattering of electrons by a static external field.

expansion (8.84). We shall assume that the external field is weak, so that we need only retain terms linear in $A_e^\mu(\mathbf{x})$ in this expansion. It can then be written

$$S = 1 + \sum_{n=1}^{\infty} \frac{(ie_0)^n}{(n-1)!} \int \cdots \int d^4x_1 \ldots d^4x_n$$
$$\times \mathrm{T}\{\mathrm{N}(\bar{\psi}\not{A}_e\psi)_{x_1}\mathrm{N}(\bar{\psi}\not{A}\psi)_{x_2} \ldots \mathrm{N}(\bar{\psi}\not{A}\psi)_{x_n}\}. \tag{9.2}$$

The leading amplitude $\mathcal{M}^{(0)}$ stems from the $n = 1$ term in Eq. (9.2). The $n = 2$ term in Eq. (9.2) is linear in the quantized radiation field $A^\mu(x)$. Hence it necessarily involves emission or absorption of a photon and describes inelastic processes such as bremsstrahlung, discussed in Section 8.8. The second-order radiative correction follows from the $n = 3$ term in Eq. (9.2).

The four contributions to this second-order correction are shown in Fig. 9.2. Each of these can be regarded as a modification of the lowest-order diagram 9.1, corresponding to one of the substitutions shown in Fig. 9.3. For example, the Feynman graph 9.2(a) is obtained by making the substitution of Fig. 9.3(a) in the incoming electron line of diagram 9.1, and so on.

The loop diagrams 9.3(a) and (b) we have met before (see Figs. 7.8, 7.9 and 7.15). They represent the self-energy parts of an electron and of a photon due to the interaction of the electron–positron field with the photon field. The electron self-energy loop in diagram 9.3(a) represents the lowest-order process which turns a bare electron into a real physical (i.e. interacting) electron. Similarly, the loop of diagram 9.3(b) represents the photon self-energy in lowest, i.e. second order only. The fermion–photon interaction here manifests itself in the creation and annihilation of a virtual electron–positron pair, and the loop diagram 9.3(b) is referred to as a vacuum polarization diagram. Lastly, the substitution shown in Fig. 9.3(c) represents

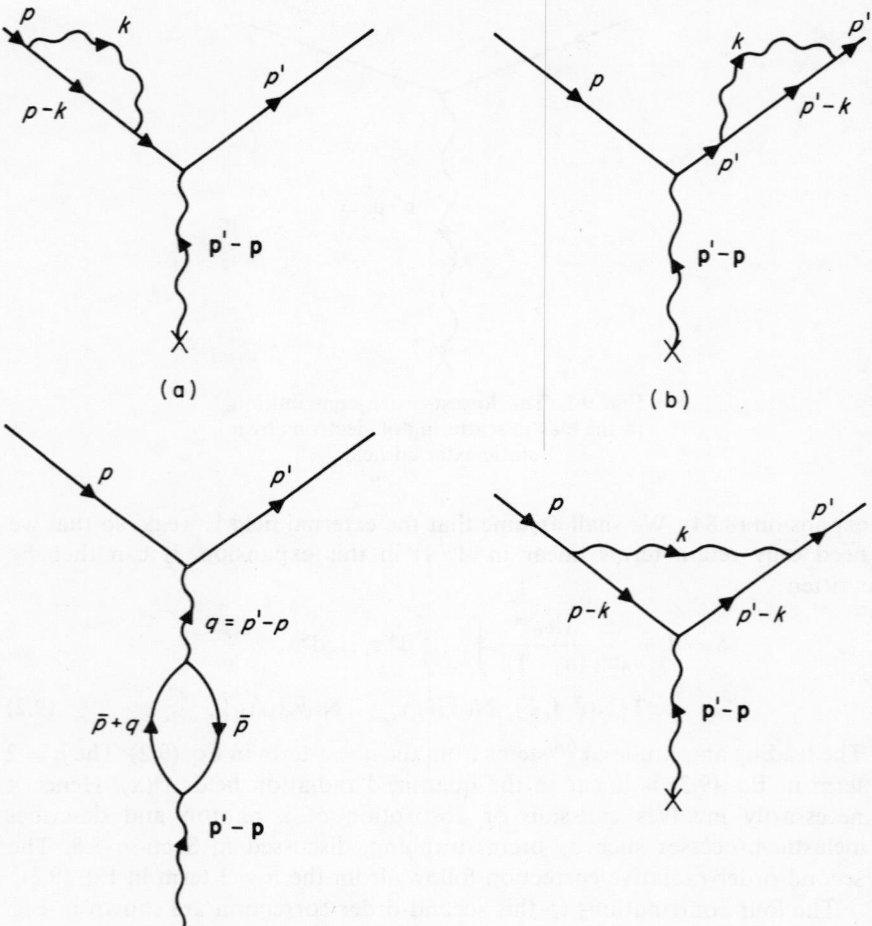

Fig. 9.2. The four contributions to the second-order radiative corrections to electron scattering.

the lowest-order modification of the basic vertex part, i.e. of the basic fermion–photon interaction $\mathrm{N}(\bar{\psi}\not{A}\psi)$, due to the emission and reabsorption of a virtual photon during the process of interaction.

The Feynman amplitudes for the diagrams 9.2(a)–(d) follow from the Feynman rules and are given by

$$\mathscr{M}_{\mathrm{a}}^{(2)} = \mathrm{i}e_0\bar{u}(\mathbf{p}')\not{A}_{\mathrm{e}}(\mathbf{p}' - \mathbf{p})\mathrm{i}S_{\mathrm{F}}(p)\mathrm{i}e_0^2\Sigma(p)u(\mathbf{p}) \tag{9.3a}$$

$$\mathscr{M}_{\mathrm{b}}^{(2)} = \mathrm{i}e_0\bar{u}(\mathbf{p}')\mathrm{i}e_0^2\Sigma(p')\mathrm{i}S_{\mathrm{F}}(p')\not{A}_{\mathrm{e}}(\mathbf{p}' - \mathbf{p})u(\mathbf{p}) \tag{9.3b}$$

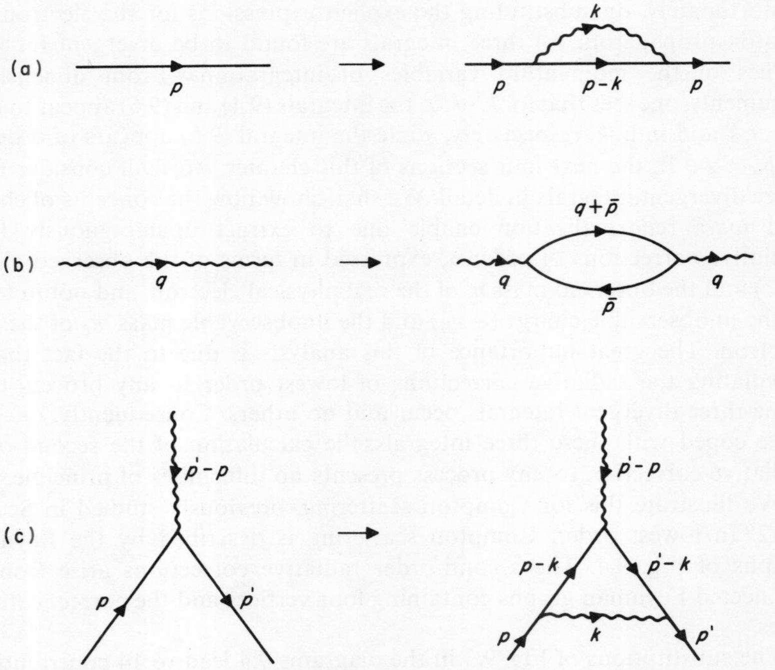

Fig. 9.3. The second-order corrections to fermion and photon lines and to the basic vertex part: (a) the fermion self-energy part $ie_0^2\Sigma(p)$, Eq. (9.4); (b) the photon self-energy part $ie_0^2\Pi^{\mu\nu}(q)$, Eq. (9.5); (c) the vertex part $e_0^2\Lambda^{\mu}(p', p)$, Eq. (9.6).

$$\mathscr{M}_{\mathrm{c}}^{(2)} = \mathrm{i}e_0\bar{u}(\mathbf{p}')\gamma^{\lambda}u(\mathbf{p})\mathrm{i}D_{\mathrm{F}\lambda\mu}(q)\mathrm{i}e_0^2\Pi^{\mu\nu}(q)A_{\mathrm{e}\nu}(\mathbf{p}'-\mathbf{p}) \tag{9.3c}$$

$$\mathscr{M}_{\mathrm{d}}^{(2)} = \mathrm{i}e_0\bar{u}(\mathbf{p}')e_0^2\Lambda^{\mu}(p', p)u(\mathbf{p})A_{\mathrm{e}\mu}(\mathbf{p}'-\mathbf{p}) \tag{9.3d}$$

where

$$\mathrm{i}e_0^2\Sigma(p) = \frac{(\mathrm{i}e_0)^2}{(2\pi)^4}\int \mathrm{d}^4k\,\mathrm{i}D_{\mathrm{F}\alpha\beta}(k)\gamma^{\alpha}\mathrm{i}S_{\mathrm{F}}(p-k)\gamma^{\beta} \tag{9.4}$$

$$\mathrm{i}e_0^2\Pi^{\mu\nu}(q) = \frac{(\mathrm{i}e_0)^2}{(2\pi)^4}(-1)\,\mathrm{Tr}\int \mathrm{d}^4\bar{p}\,\gamma^{\mu}\mathrm{i}S_{\mathrm{F}}(\bar{p}+q)\gamma^{\nu}\mathrm{i}S_{\mathrm{F}}(\bar{p}) \tag{9.5}$$

and

$$e_0^2\Lambda^{\mu}(p', p) = \frac{(\mathrm{i}e_0)^2}{(2\pi)^4}\int \mathrm{d}^4k\,\gamma^{\alpha}\mathrm{i}S_{\mathrm{F}}(p'-k)\gamma^{\mu}\mathrm{i}S_{\mathrm{F}}(p-k)\gamma^{\beta}\mathrm{i}D_{\mathrm{F}\alpha\beta}(k). \tag{9.6}$$

We see that in order to calculate the second-order radiative correction to electron scattering we must evaluate the three loop integrals (9.4)–(9.6).

Unfortunately, on substituting the explicit expressions for the electron and photon propagators, all three integrals are found to be divergent for large values of the momentum variables of integration.[‡] From dimensional arguments, one sees that for $k \to \infty$ the integrals (9.4) and (9.6) appear to be of order k and in $\ln k$, respectively, while the integral (9.5) appears of order $\bar{p}^2$ as $\bar{p} \to \infty$.[§] In the next four sections of this chapter, we shall consider these three divergent integrals in detail. We shall show how the concepts of charge and mass renormalization enable one to extract unambiguously finite radiative corrections of order α, expressed in terms of the observed charge $(-e)$ and the observed mass m of the real physical electron, and not in terms of the unobservable charge $(-e_0)$ and the unobservable mass m_0 of the bare electron. The great importance of this analysis is due to the fact that in calculating the radiative corrections of lowest order to any process these same three divergent integrals occur and no others. Consequently, once we have coped with these three integrals, the calculation of the second-order radiative correction to any process presents no difficulties of principle.

We illustrate this for Compton scattering, previously studied in Section 7.2.2. In lowest order, Compton scattering is described by the Feynman graphs of Fig. 9.4. The second-order radiative corrections arise from all connected Feynman graphs containing four vertices and the correct external lines.

The substitutions of Fig. 9.3 in the diagrams 9.4 lead to 14 contributions, with insertions of the electron self-energy part (in internal or external electron lines), of the photon self-energy part and of the vertex correction being responsible for 6, 4 and 4 graphs, respectively. Four of these graphs are shown in Fig. 9.5. The Feynman amplitudes for these 14 graphs will contain the three divergent loop integrals (9.4)–(9.6) as factors, similar to the way they occur in the electron scattering amplitudes (9.3). After renormalization, the evaluation of these amplitudes presents no difficulties and leads to finite radiative corrections of order α.

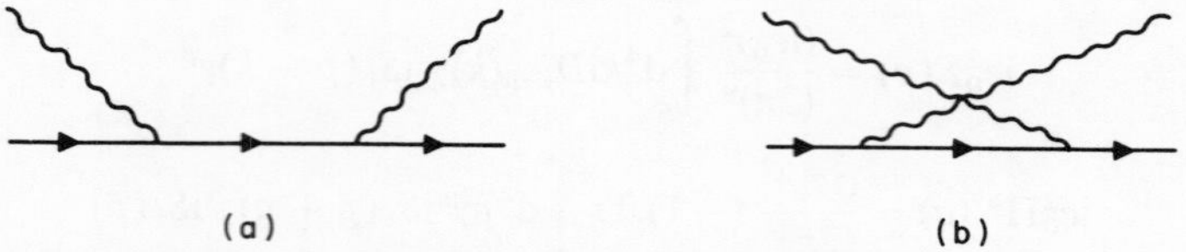

Fig. 9.4. The two lowest-order Feynman graphs for Compton scattering.

[‡] In addition, the integrals (9.4) and (9.6) also lead to infra-red divergences at the lower limit $k \to 0$. These infra-red divergences will also be dealt with later in this chapter.

[§] Of course, these dimensional arguments only give the maximum possible degree of divergence of each integral. If the coefficient of the leading divergence happens to vanish, the actual divergence is less severe.

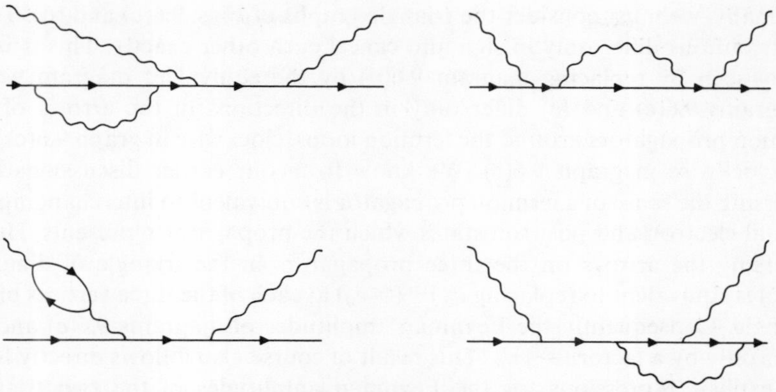

Fig. 9.5. Four of the fourth-order contributions to Compton scattering, obtained by inserting a self-energy or vertex modification in the Feynman graph 9.4(a).

In addition to these 14 Feynman diagrams, there are 4 fourth-order diagrams which are not obtained from the second-order diagrams 9.4 by inserting a self-energy or vertex correction. There are shown in Figs. 9.6(a)–(d).

Writing down the Feynman amplitudes for the diagrams 9.6(a) and (b), using the Feynman rules, one finds that these amplitudes are finite and well-defined, yielding radiative corrections of order α.

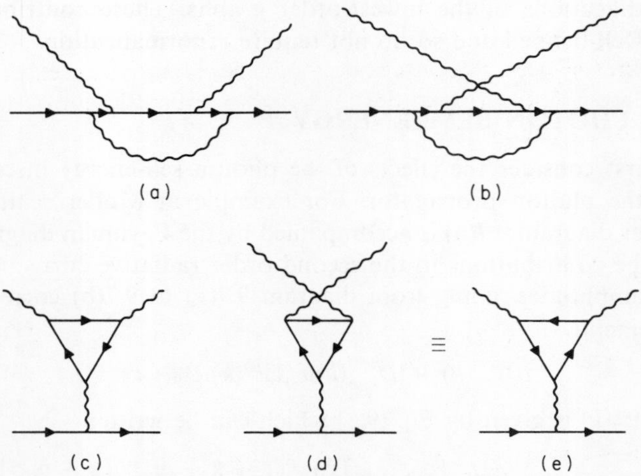

Fig. 9.6. The fourth-order contributions to Compton scattering which cannot be obtained by inserting a self-energy or vertex modification in the lowest-order Feynman graphs 9.4.

Finally, we must consider the triangle graphs of Figs. 9.6(c) and (d). Their contributions differ only in sign and cancel each other exactly. This is most easily seen by replacing diagram 9.6(d) by the equivalent diagram 9.6(e). Diagrams 9.6(c) and (e) differ only in the directions of the arrows of the fermion propagators around the fermion loops, clockwise in graph 9.6(c) and anticlockwise in graph 9.6(e). We know from our earlier discussions that reversing the sense of a fermion propagator is equivalent to interchanging the virtual electron and positron states which the propagator represents. Hence reversing the arrows on the three propagators in the triangle of diagram 9.6(c) is equivalent to replacing e_0 by $(-e_0)$ at each of the three vertices of the triangle. Consequently the Feynman amplitudes of diagrams 9.6(c) and (e) differ only by a factor $(-1)^3$. (This result of course also follows directly from the explicit expressions for the Feynman amplitudes of the two triangle graphs.) This result for the triangle diagrams is a particular case of Furry's theorem which states that the contributions of diagrams which contain a closed fermion loop with an odd number of vertices cancel. Such diagrams always occur in pairs differing only in the senses of the arrows on the fermion loops, and, by the same argument as above, their contributions will cancel.

The situation which has here been outlined for Compton scattering is characteristic of radiative corrections. The second-order radiative correction to any process is obtained by modifying the lowest-order graphs in all possible ways according to the substitutions of Fig. 9.3. After renormalization of the divergent loop integrals (9.4)–(9.6), these modified graphs give finite second-order radiative corrections. In general, there will be additional second-order radiative corrections from graphs which are not self-energy or vertex modifications of the lowest-order graphs. These contributions are finite and well-defined and so do not require renormalization.

9.2 THE PHOTON SELF-ENERGY

We shall first consider the effects of the photon self-energy insertion, Fig. 9.3(b), in the photon propagator. For example in Møller scattering, the lowest-order diagram 9.7(a) is accompanied by the Feynman diagram 9.7(b) as one of the contributions to the second-order radiative correction. In the Feynman amplitudes, going from diagram 9.7(a) to 9.7(b) corresponds to the replacement

$$\mathrm{i}D_{\mathrm{F}\alpha\beta}(k) \to \mathrm{i}D_{\mathrm{F}\alpha\mu}(k)\mathrm{i}e_0^2\Pi^{\mu\nu}(k)\mathrm{i}D_{\mathrm{F}\nu\beta}(k), \tag{9.7}$$

where $\mathrm{i}e_0^2\Pi^{\mu\nu}(k)$ is given by Eq. (9.5) which can be written

$$\mathrm{i}e_0^2\Pi^{\mu\nu}(k) = \frac{-e_0^2}{(2\pi)^4}\int \mathrm{d}^4p\,\frac{\mathrm{Tr}\,[\gamma^\mu(\not{p} + \not{k} + m_0)\gamma^\nu(\not{p} + m_0)]}{[(p+k)^2 - m_0^2 + \mathrm{i}\varepsilon][p^2 - m_0^2 + \mathrm{i}\varepsilon]}. \tag{9.8}$$

This integral is quadratically divergent for large p. In order to handle it, we

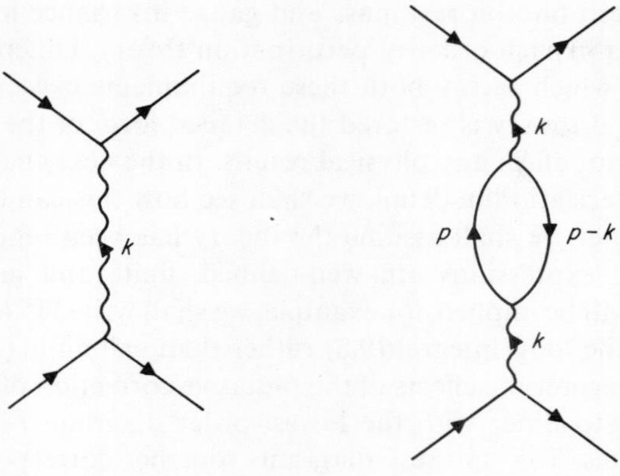

Fig. 9.7. Møller scattering: (b) represents the second-order photon self-energy correction to the lowest-order diagram (a).

must regularize it, that is, we must modify it so that it becomes a well-defined finite integral. For example, this could be achieved by multiplying the integrand in Eq. (9.8) by the convergence factor

$$f(p^2, \Lambda^2) = \left(\frac{-\Lambda^2}{p^2 - \Lambda^2}\right)^2. \tag{9.9}$$

Here Λ is a cut-off parameter. For large but finite values of Λ, the integral now behaves like $\int d^4p/p^6$ for large p, and is well-defined and convergent. For $\Lambda \to \infty$, the factor $f(p^2, \Lambda^2)$ tends to unity, and the original theory is restored. One can think of such convergence factors either as a mathematical device, introduced to overcome a very unsatisfactory feature of QED, or as a genuine modification of QED at very high energies, i.e. at very small distances, which should show up in experiments at sufficiently high energies. We shall return to this point later.

The convergence factor (9.9) was introduced in order to illustrate the idea of regularization in a simple way. However, this factor does not provide a suitable regularization procedure for QED, as it does not ensure zero rest mass for the real physical photon, nor the related gauge invariance of the theory.‡ It is most natural and desirable to employ a regularization procedure

‡ A non-zero photon rest mass μ would modify Maxwell's equations and, in particular, the characteristic long-range behaviour of electromagnetic phenomena. For example, the long-range $1/r$ Coulomb potential would turn into the Yukawa potential $e^{-\mu r}/r$ with the short range $1/\mu$. Laboratory tests of the Coulomb law provide the upper limit $\mu \leqslant 1 \times 10^{-14}$ eV. The smallest upper limit has been obtained from studies of the variation of the magnetic field of Jupiter over large distances which gives $\mu \leqslant 6 \times 10^{-16}$ eV. [L. Davis *et al.*, *Phys. Rev. Lett.* **35** (1975), 1402, which contains earlier references.]

which ensures zero photon rest mass and gauge invariance for all values of the cut-off Λ and in each order of perturbation theory. Different regularization formalisms which satisfy both these requirements exist. In the limit in which the original theory is restored the detailed form of the regularization procedure does not affect any physical results. In the next chapter, where we shall study regularization in detail, we shall see how this can be achieved. In the present chapter we shall assume the theory has been regularized in this way, so that all expressions are well-defined, finite and gauge-invariant. Regularization will be implied; for example, we shall write $\Pi^{\mu\nu}(k)$ for the regularized form of the loop integral (9.8) rather than calling it $\Pi^{\mu\nu}(k, \Lambda)$.

In order to interpret the effects of the radiative correction of Fig. 9.7(b), we shall consider it together with the lowest-order diagram, Fig. 9.7(a), from which it originates. Taking these diagrams together corresponds to the replacement shown in Fig. 9.8, i.e. the propagator modification

$$\mathrm{i}D_{\mathrm{F}\alpha\beta}(k) \to \mathrm{i}D_{\mathrm{F}\alpha\beta}(k) + \mathrm{i}D_{\mathrm{F}\alpha\mu}(k)\mathrm{i}e_0^2\Pi^{\mu\nu}(k)\mathrm{i}D_{\mathrm{F}\nu\beta}(k) \tag{9.10a}$$

or, more explicitly,

$$\frac{-\mathrm{i}g_{\alpha\beta}}{k^2 + \mathrm{i}\varepsilon} \to \frac{-\mathrm{i}g_{\alpha\beta}}{k^2 + \mathrm{i}\varepsilon} + \frac{-\mathrm{i}g_{\alpha\mu}}{k^2 + \mathrm{i}\varepsilon}\,\mathrm{i}e_0^2\Pi^{\mu\nu}(k)\,\frac{-\mathrm{i}g_{\nu\beta}}{k^2 + \mathrm{i}\varepsilon}. \tag{9.10b}$$

We can simplify this expression. It follows from Lorentz invariance that $\Pi^{\mu\nu}(k)$ must be of the form

$$\Pi^{\mu\nu}(k) = -g^{\mu\nu}A(k^2) + k^\mu k^\nu B(k^2), \tag{9.11}$$

since this is the most general second-rank tensor which can be formed using only the four-vector k^μ. Furthermore, the photon propagator always occurs coupled to conserved currents, leading to expressions, analogous to Eq. (5.45), of the form

$$\int \mathrm{d}^4k s_1^\alpha(-k)\mathrm{i}D_{\mathrm{F}\alpha\beta}(k)s_2^\beta(k).$$

If in this expression we make the replacement (9.10a) and substitute Eq. (9.11) for $\Pi^{\mu\nu}(k)$, it follows from current conservation, Eq. (5.47), that terms proportional to the photon momentum k give vanishing contributions. Hence, we can omit the term $k^\mu k^\nu B(k^2)$ from Eq. (9.11) on substituting this

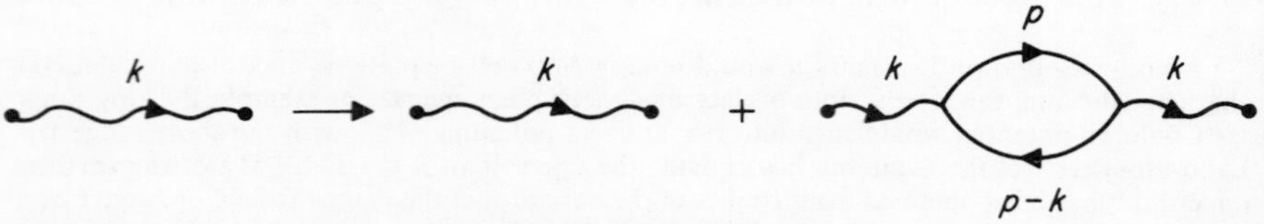

Fig. 9.8. The modified photon propagator.

equation in Eqs. (9.10), with the result

$$\frac{-\mathrm{i}g_{\alpha\beta}}{k^2+\mathrm{i}\varepsilon} \to \frac{-\mathrm{i}g_{\alpha\beta}}{k^2+\mathrm{i}\varepsilon}\left[1 - e_0^2 A(k^2)\frac{1}{k^2+\mathrm{i}\varepsilon}\right]. \tag{9.12}$$

The right-hand expression in Eq. (9.12) represents the photon propagator, modified to include the second-order photon self-energy effects. We rewrite Eq. (9.12) in the form

$$\frac{-\mathrm{i}g_{\alpha\beta}}{k^2+\mathrm{i}\varepsilon} \to \frac{-\mathrm{i}g_{\alpha\beta}}{k^2+\mathrm{i}\varepsilon + e_0^2 A(k^2)} + \mathrm{O}(e_0^4). \tag{9.13}$$

The equivalence of Eqs. (9.12) and (9.13) follows since after regularization all quantities, and in particular $A(k^2)$, are finite. Hence in the spirit of perturbation theory, we can, for sufficiently small e_0^2, expand the right-hand side of Eq. (9.13) in powers of e_0^2, thereby regaining Eq. (9.12).

The left-hand side of Eq. (9.13) represents the photon propagator in lowest order of perturbation theory, i.e. the propagator of the bare, non-interacting photon. This propagator possesses a pole at $k^2 = 0$, corresponding to the bare photon having zero rest mass. [Quite generally, the propagator of a particle of rest mass m possesses a pole for four-momenta p such that $p^2 = m^2$. Eqs. (3.59) and (7.48) illustrate this for the meson and fermion propagators.] The right-hand side of Eq. (9.13) represents the photon propagator including second-order self-energy corrections, i.e. it is the propagator of a real physical photon [albeit to $\mathrm{O}(e_0^2)$ only]. If, as discussed previously, the real photon rest mass also vanishes, then the real photon propagator must also have its pole at $k^2 = 0$. From Eq. (9.13) this implies

$$A(0) \equiv A(k^2 = 0) = 0. \tag{9.14}$$

Hence $A(k^2)$ can be written

$$A(k^2) = k^2 A'(0) + k^2 \Pi_{\mathrm{c}}(k^2) \tag{9.15}$$

where

$$A'(0) \equiv A'(k^2 = 0) = \left.\frac{\mathrm{d}A(k^2)}{\mathrm{d}(k^2)}\right|_{k^2=0}$$

and $\Pi_{\mathrm{c}}(k^2)$ vanishes linearly with k^2 at $k^2 = 0$. Substituting Eq. (9.15) into (9.12) and multiplying by e_0^2, we obtain

$$\frac{-\mathrm{i}g_{\alpha\beta}}{k^2+\mathrm{i}\varepsilon}\, e_0^2 \to \frac{-\mathrm{i}g_{\alpha\beta}}{k^2+\mathrm{i}\varepsilon}\, e_0^2[1 - e_0^2 A'(0)] + \frac{\mathrm{i}g_{\alpha\beta}}{k^2+\mathrm{i}\varepsilon}\, e_0^4 \Pi_{\mathrm{c}}(k^2). \tag{9.16}$$

In formulating the Feynman rules, we associated a charge e_0 with each vertex. In writing Eq. (9.16), we have effectively incorporated into it the two factors e_0 which occur at the vertices at the ends of a photon propagator.

We must now interpret Eq. (9.16). The first term on the right-hand side is just the left-hand term multiplied by the constant $[1 - e_0^2 A'(0)]$. It is as though the magnitude of the two charges interacting through the photon propagator [for example, in the Møller scattering diagram 9.7(b)] is not e_0 but e, given by

$$e^2 = e_0^2[1 - e_0^2 A'(0)]. \tag{9.17}$$

In this equation we have introduced the concept of *charge renormalization.* Eq. (9.17) defines the *renormalized electronic charge* $(-e)$, i.e. the charge of the physical, interacting electron, in contrast to the charge $(-e_0)$ of the bare non-interacting electron. We have considered the photon self-energy in second order only. There are of course higher-order corrections and we shall write Eq. (9.17) more completely as

$$e \equiv Z_3^{1/2} e_0 = e_0[1 - \tfrac{1}{2}e_0^2 A'(0) + O(e_0^4)]. \tag{9.18}$$

The constant Z_3 relating the bare charge e_0 to the charge e of the real physical particle (to all orders in e_0) is called a *renormalization constant.* The right-hand side of Eq. (9.18) gives the explicit expression for this constant correct to the second order in e_0.

It is of course the charge $(-e)$ of the real physical electron which is observable, not the charge $(-e_0)$ which was introduced as the coupling constant of the free fields. Hence, all observable quantities, such as cross-sections, should be expressed in terms of the real charge e, not in terms of the bare charge e_0. From Eq. (9.18) we have

$$e_0 = e[1 + O(e^2)]$$

so that Eq. (9.16) can be written

$$\frac{-ig_{\alpha\beta}}{k^2 + i\varepsilon} e_0^2 \rightarrow \frac{-ig_{\alpha\beta}}{k^2 + i\varepsilon} e^2 + \frac{ig_{\alpha\beta}}{k^2 + i\varepsilon} e^4 \Pi_c(k^2) + O(e^6). \tag{9.19}$$

Eq. (9.19) is our final result. It gives the photon propagator (times e^2), expressed in terms of the real charge e and accurate to terms in e^4. The first term on the right-hand side is the original photon propagator but multiplied by the square of the renormalized charge e, instead of the bare charge e_0. The second term, of order α $(\equiv e^2/4\pi)$ relative to the first term, will lead to an observable radiative correction of this order in any process involving the photon propagator $iD_F(k)$ in the lowest order of perturbation theory.

We have discussed renormalization of the regularized theory for a general value of the cut-off parameter Λ, i.e. before taking the appropriate limit [in our unrealistic example, Eq. (9.9), $\Lambda \rightarrow \infty$] which restores QED.‡ We must

‡ It should be stressed that even for well-behaved finite theories renormalization is necessary in order to express theoretical predictions for cross-sections, etc., in terms of observable quantities like the charge e of real particles, rather than in terms of unobservable quantities such as the bare charge e_0.

now consider what happens as we take this limit. In the next chapter we shall consider regularization in detail. Here we shall only state the results. Before proceeding to the limit which restores the original theory, all quantities are well-defined and finite. As the limit is taken, the regularized integral $\Pi_c(k^2)$ tends to a well-defined finite limit which is independent of the detailed form of the regularization procedure. On the other hand, as the limit is taken, divergences reappear in the relation between e and e_0, Eq. (9.18). ($A'(0)$ and the renormalization constant Z_3 become infinite.) But this is a relation between the observable charge e of a real physical particle and the bare charge e_0 of a non-interacting particle, which is a theoretical construct and completely unobservable. Thus Eq. (9.18) is itself not amenable to experimental tests, and it is only in such untestable equations that divergences appear. Looking back at our final result for the modified photon propagator, Eq. (9.19), we see that in the QED limit it only involves the measured elementary charge e and the well-defined and finite limit of the loop integral $\Pi_c(k^2)$. Thus, the modified propagator, and the Feynman amplitudes and the physical predictions to which it gives rise, are well-defined and finite, even in the QED limit of the regularized theory. An explicit expression for the loop integral $\Pi_c(k^2)$ will be derived in Section 10.4, and it will be employed later in this chapter in considering the radiative corrections to electron scattering by an external field.

9.3 THE ELECTRON SELF-ENERGY

We next consider the insertion of a fermion self-energy correction in a fermion propagator, shown in Fig. 9.3(a) and given by Eq. (9.4). After simplification by means of the contraction identities (A.14), this equation can be written

$$\mathrm{i}e_0^2\Sigma(p) = \frac{e_0^2}{(2\pi)^4}\int \mathrm{d}^4k \, \frac{1}{k^2 + \mathrm{i}\varepsilon} \, \frac{2\not{p} - 2\not{k} - 4m_0}{(p-k)^2 - m_0^2 + \mathrm{i}\varepsilon}. \tag{9.20}$$

This loop integral is ultra-violet divergent (i.e. in the limit $k \to \infty$). Its treatment is very similar to that of the photon self-energy in the last section, so that we can be brief except where differences occur.

One of the differences is that the regularization and renormalization of Eq. (9.20) leads to integrals which are not only ultra-violet divergent but also infra-red divergent, i.e. they also diverge in the limit of zero photon momentum k. We have met infra-red divergences in our analysis of elastic and inelastic electron scattering (Section 8.9) and shall discuss their significance further in Section 9.7. Here we conveniently remove infra-red and ultra-violet divergences from Eq. (9.20) by the replacement

$$\frac{1}{k^2 + \mathrm{i}\varepsilon} \to \frac{1}{k^2 - \lambda^2 + \mathrm{i}\varepsilon} - \frac{1}{k^2 - \Lambda^2 + \mathrm{i}\varepsilon} \tag{9.21}$$

where λ is a small infra-red cut-off parameter which ultimately is set equal to zero. In effect, this corresponds to temporarily assigning a small non-zero rest mass λ to the photon. Λ is again an ultra-violet cut-off parameter, and taking the limit $\Lambda \to \infty$ restores QED. In the following we shall assume that the loop integral (9.20) has been regularized through the replacement (9.21).

Proceeding as in the last section, we shall consider the fermion self-energy loop together with the propagator into which it is inserted, as shown in Fig. 9.9 and given by

$$\frac{\mathrm{i}}{\not p - m_0 + \mathrm{i}\varepsilon} \to \frac{\mathrm{i}}{\not p - m_0 + \mathrm{i}\varepsilon} + \frac{\mathrm{i}}{\not p - m_0 + \mathrm{i}\varepsilon}\, \mathrm{i}e_0^2\Sigma(p)\, \frac{\mathrm{i}}{\not p - m_0 + \mathrm{i}\varepsilon}. \tag{9.22}$$

We rewrite this expression using the identity

$$\frac{1}{A - B} = \frac{1}{A} + \frac{1}{A} B \frac{1}{A} + \frac{1}{A} B \frac{1}{A} B \frac{1}{A} + \cdots \tag{9.23}$$

which holds for any two operators A and B, not necessarily commuting. (The identity is easily verified by post-multiplying it by $A - B$.) By means of this identity, Eq. (9.22) can be written, correct to terms of order e_0^2,

$$\frac{\mathrm{i}}{\not p - m_0 + \mathrm{i}\varepsilon} \to \frac{\mathrm{i}}{\not p - m_0 + e_0^2\Sigma(p) + \mathrm{i}\varepsilon} + \mathrm{O}(e_0^4). \tag{9.24}$$

The left-hand side of this equation is the propagator of the non-interacting fermion. As expected, it has a pole at $\not p = m_0$, corresponding to the bare rest mass m_0 of the non-interacting particle.‡ The expression on the right-hand side of Eq. (9.24) represents the propagator of the interacting physical fermion. Hence, we require its pole to be at $\not p = m$, where

$$m = m_0 + \delta m \tag{9.25}$$

is the real fermion rest mass; due to the interaction of the fermion field and the electromagnetic field, the rest mass m of the real fermion differs from the rest mass m_0 of the non-interacting bare fermion. In our perturbation treatment, δm is a power series in e_0^2, and in the lowest order—to which we are working—$\delta m = \mathrm{O}(e_0^2)$. m is called the renormalized mass and the replacement of m_0 by m using Eq. (9.25) is known as *mass renormalization*. It is similar to and as essential as charge renormalization: the experimentally determined mass of the electron is m, not m_0, and predictions from theory must be expressed in terms of the observable properties of the real interacting particles.

‡ This is merely a concise way of saying that

$$\frac{\mathrm{i}}{\not p - m_0 + \mathrm{i}\varepsilon} \equiv \frac{\mathrm{i}(\not p + m_0)}{p^2 - m_0^2 + \mathrm{i}\varepsilon}$$

has a pole at $p^2 = m_0^2$.

Fig. 9.9. The modified fermion propagator.

To identify the pole of the modified propagator (9.24) at $\not{p} = m$, we rewrite its denominator. It follows from Lorentz invariance that $\Sigma(p)$ can depend on the momentum vector p through $\not{p}$ and $p^2(= \not{p}\not{p})$ only. It will be convenient to expand $\Sigma(p)$ in powers of $(\not{p} - m)$ in the form

$$\Sigma(p) = A + (\not{p} - m)B + (\not{p} - m)\Sigma_c(p), \tag{9.26}$$

where A and B are constants (i.e. independent of p) and $\Sigma_c(p)$ vanishes linearly with $(\not{p} - m)$ at $\not{p} = m$. In particular,

$$A = \Sigma(p)\,|_{\not{p}=m}. \tag{9.27}$$

Substituting Eqs. (9.25) and (9.26) in the modified propagator on the right-hand side of Eq. (9.24), we see that this propagator has a pole at $\not{p} = m$, provided

$$\delta m = -e_0^2 A. \tag{9.28}$$

Eq. (9.24) then reduces to

$$\frac{\mathrm{i}}{\not{p} - m_0 + \mathrm{i}\varepsilon} \to \frac{\mathrm{i}}{(\not{p} - m)(1 + e_0^2 B) + e_0^2(\not{p} - m)\Sigma_c(p) + \mathrm{i}\varepsilon} + \mathrm{O}(e_0^4), \tag{9.29}$$

or, retaining only terms to $\mathrm{O}(e_0^2)$,

$$\frac{\mathrm{i}}{\not{p} - m_0 + \mathrm{i}\varepsilon} \to \frac{\mathrm{i}}{\not{p} - m + \mathrm{i}\varepsilon}[(1 - e_0^2 B) - e_0^2\Sigma_c(p)] + \mathrm{O}(e_0^4). \tag{9.30}$$

An alternative way of carrying out the mass renormalization is to express the Hamiltonian of QED in terms of the real electron mass m, instead of the bare mass m_0,

$$\mathscr{H} = \mathscr{H}_0 + \mathscr{H}_\mathrm{I} \tag{9.31a}$$

where

$$\mathscr{H}_0 = -\dot{A}_\nu(x)\dot{A}^\nu(x) + \tfrac{1}{2}(\partial_\mu A_\nu(x))(\partial^\mu A^\nu(x)) + \mathrm{i}\psi^\dagger(x)\dot{\psi}(x) - \bar{\psi}(x)(\mathrm{i}\not{\partial} - m)\psi(x), \tag{9.31b}$$

$$\mathscr{H}_\mathrm{I} = -e_0\bar{\psi}(x)\not{A}(x)\psi(x) - \delta m\bar{\psi}(x)\psi(x). \tag{9.31c}$$

As implied by our notation, we shall treat $\mathscr{H}_0$ as the free-field Hamiltonian density and $\mathscr{H}_\mathrm{I}$ as the interaction Hamiltonian density. The non-interacting fermion now has the physical mass m and satisfies the Dirac equation

$$(\mathrm{i}\not{\partial} - m)\psi(x) = 0,$$

etc., and the fermion propagator becomes, in lowest order,

$$\frac{\mathrm{i}}{\not{p} - m + \mathrm{i}\varepsilon}.$$

The use of the real electron mass m for the non-interacting fermion is compensated for through the *mass counterterm* $-\delta m\bar{\psi}\psi$ appearing as an additional interaction term in Eq. (9.31c). Graphically this term is represented by Fig. 9.10 corresponding to a two-line vertex at which an electron (or positron) is destroyed and recreated.

If we use this division, Eqs. (9.31a)–(9.31c), of the Hamiltonian density in the S-matrix expansion (6.23), the Feynman rules which we obtained in Sections 7.3 and 8.7 must be modified in two respects.

Firstly, the bare mass m_0 is replaced by the real fermion mass m throughout; in particular, in the fermion propagator. [Eq. (7.48), as it stands, already states the modified rule 3, but we must now interpret m as the real mass; in Chapter 7 it denoted the bare mass.]

Secondly, using Eq. (9.31c) as the interaction, leads to extra contributions to the S-matrix expansion, represented by Feynman graphs containing the two-line vertex part, Fig. 9.10. We see from the S-matrix expansion, Eq. (6.23), that with Eq. (9.31c) as interaction, each two-line vertex in a graph leads to a factor $\mathrm{i}\delta m$ in the Feynman amplitude, just as each basic vertex part leads to a factor $\mathrm{i}e\gamma^{\alpha}$ (rule 1 of Section 7.3). For each two-line vertex, we must therefore write a factor

$$\mathrm{i}\delta m = -\mathrm{i}e_0^2 A = -\,\mathrm{i}e_0^2\Sigma(p)|_{\not{p}=m}. \qquad (9.32)$$

We see that as a consequence of the replacement

$$m_0 \to m \qquad (9.33a)$$

for the free fermion, each Feynman graph containing a fermion self-energy loop must be considered together with an identical graph in which the self-energy loop has been replaced by a two-line vertex, Fig. 9.10. (These two graphs are of the same order in e_0.) The net effect on the Feynman amplitude of taking into account both graphs is the replacement

$$\mathrm{i}e_0^2\Sigma(p) \to \mathrm{i}e_0^2\Sigma(p) + \mathrm{i}\delta m = \mathrm{i}e_0^2(\not{p} - m)B + \mathrm{i}e_0^2(\not{p} - m)\Sigma_{\mathrm{c}}(p), \quad (9.33b)$$

where we used Eqs. (9.26) and (9.32). We see that, quite generally, the mass counterterm cancels the constant term A arising from the loop term $\mathrm{i}e_0^2\Sigma(p)$.

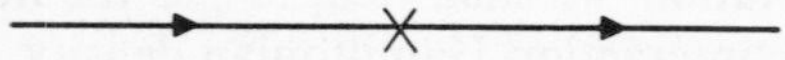

Fig. 9.10. The two-line vertex graph representing the mass counterterm $-\delta m\bar{\psi}\psi$.

Returning to Eq. (9.24), we see that if in this equation we make the replacements (9.33a) and (9.33b), we obtain

$$\frac{\mathrm{i}}{\not{p} - m + \mathrm{i}\varepsilon} \to \frac{\mathrm{i}}{(\not{p} - m)(1 + e_0^2 B) + e_0^2(\not{p} - m)\Sigma_c(p) + \mathrm{i}\varepsilon} + O(e_0^4). \quad (9.34)$$

This equation agrees with our earlier result (9.29), as it must, since the two derivations differ only in the way the same total Hamiltonian is split into free-field and interaction parts.

The modified fermion propagator has so far been expressed in terms of the bare charge e_0. We now define a renormalized charge e by the relation

$$e^2 \equiv Z_2 e_0^2 = e_0^2(1 - e_0^2 B) + O(e_0^6). \quad (9.35)$$

The general interpretation of this charge renormalization is the same as we met in the last section but it has of course a different origin; it is due to the fermion self-energy, not the photon self-energy. Multiplying Eq. (9.30) by e_0^2, in order to incorporate the charges e_0 which are associated with the two vertices at the ends of the propagator, and expressing e_0 in terms of e, we can write Eq. (9.30) as

$$\frac{\mathrm{i}e_0^2}{\not{p} - m_0 + \mathrm{i}\varepsilon} \to \frac{\mathrm{i}e^2}{\not{p} - m + \mathrm{i}\varepsilon}[1 - e^2\Sigma_c(p)] + O(e^6). \quad (9.36)$$

The right-hand side of Eq. (9.36) gives the renormalized fermion propagator (times e^2), correct to terms in e^4. The first term, $\mathrm{i}e^2(\not{p} - m + \mathrm{i}\varepsilon)^{-1}$, is simply the zeroth approximation, i.e. the bare fermion expression (in lowest order $e = e_0$ and $m = m_0$). The term containing Σ_c is a radiative correction of order α to the zeroth approximation.

Finally, we must take the limit $\Lambda \to \infty$ in order to regain QED. From Eq. (9.20), the loop integral $\Sigma(p)$ appears to be linearly divergent in this limit. As is shown explicitly in Section 10.2, it is actually only logarithmically divergent, with A given by

$$A = -\frac{3m}{8\pi^2} \ln \frac{\Lambda}{m}. \quad (9.37)$$

One finds that the constants B and Z_2 are also logarithmically divergent in the limit $\Lambda \to \infty$, while the correction term $\Sigma_c(p)$ remains well-defined and finite in this limit and independent of the details of the regularization procedure. It is from this term that measurable radiative corrections, of order α, to the lowest-order predictions are derived. In contrast, the divergent constants A, B and Z_2 occur only in untestable relations connecting physical and bare quantities.

9.4 EXTERNAL LINE RENORMALIZATION

In the last two sections we considered the second-order self-energy insertions in photon and fermion propagators. When considering radiative corrections, these insertions must of course also be made in external lines. Their only effect now is a charge renormalization but they do not lead to any finite radiative corrections. We shall derive these results for the case of an initially present electron.

Proceeding as in the last section, we consider the incident electron line together with its self-energy insertions, i.e. the replacement shown in Fig. 9.11. The corresponding replacement in the Feynman amplitudes is, from Eqs. (9.32) and (9.33), given by

$$u(\mathbf{p}) \to u(\mathbf{p}) + \frac{\mathrm{i}}{\not{p} - m + \mathrm{i}\varepsilon}\,[\mathrm{i}e_0^2(\not{p} - m)B + \mathrm{i}e_0^2(\not{p} - m)\Sigma_c(p)]u(\mathbf{p}). \tag{9.38}$$

Since $(\not{p} - m)u(\mathbf{p}) = 0$ and $(\not{p} - m)\Sigma_c(p)$ vanishes quadratically with $\not{p} - m$ as $\not{p} \to m$ [see Eq. (9.26)], we can drop the last term in Eq. (9.38) and write it

$$u(\mathbf{p}) \to \left[1 - \frac{e_0^2}{\not{p} - m + \mathrm{i}\varepsilon}(\not{p} - m)B\right]u(\mathbf{p}). \tag{9.39}$$

Unfortunately the term proportional to B in this equation is indeterminate, as it stands. This ambiguity is resolved by explicitly using the adiabatic hypothesis, discussed in Section 6.2, to describe how the self-energy effects convert the incident electron from a bare particle to a physical particle. In effect, the interaction is switched off as $t \to \pm\infty$ by multiplying the charge e_0 by a suitable factor $f(t)$, so that the interaction (9.31c) is replaced by

$$\mathscr{H}_{\mathrm{I}} = -e_0 f(t)\bar{\psi}(x)\not{A}(x)\psi(x) - \delta m[f(t)]^2\bar{\psi}(x)\psi(x). \tag{9.40}$$

The precise form of the function $f(t)$ is not important. We require that $f(t) \to 0$ as $t \to \pm\infty$, and that $f(t)$ does not differ significantly from unity during a time interval T which is long compared to the duration of the scattering process considered. In terms of the Fourier transform

$$f(t) = \int_{-\infty}^{\infty} F(E)\,\mathrm{e}^{\mathrm{i}Et}\,\mathrm{d}E = \int_{-\infty}^{\infty} F(E)\,\mathrm{e}^{\mathrm{i}qx}\,\mathrm{d}E, \tag{9.41}$$

where $q \equiv (E, \mathbf{0})$, we require the normalization

$$f(0) = \int_{-\infty}^{\infty} F(E)\,\mathrm{d}E = 1 \tag{9.42}$$

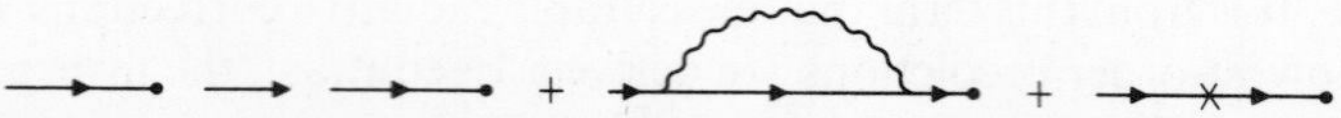

Fig. 9.11. The second-order modification of an external initial electron line.

and that $F(E)$ is almost a δ-function with a peak of width $1/T$ situated at $E = 0$. As $F(E) \to \delta(E)$, the original theory with $f(t) = 1$ is restored.

The modified interaction (9.40) behaves in many ways like an external field interaction. But whereas the static external field $A_e^\alpha(\mathbf{x})$, Eq. (8.85), conserves energy but not three-momentum, the interaction (9.40) conserves three-momentum but not energy. The effect of using the energy–non-conserving interaction (9.40) is to replace the original fermion self-energy insertion, Fig. 9.12(a), by the modified insertion shown in Fig. 9.12(b), where the vectors $q = (E, \mathbf{0})$ and $q' = (E', \mathbf{0})$ in the fermion propagators describe the non-conservation of energy at the vertices. Correspondingly, Eq. (9.39) is replaced by

$$u(\mathbf{p}) \to \left[1 - \int \mathrm{d}E\, \mathrm{d}E' F(E) F(E') \frac{e_0^2 B}{\not{p} - \not{q} - \not{q}' - m + i\varepsilon} (\not{p} - \not{q} - m)\right] u(\mathbf{p}). \tag{9.43}$$

The evaluation of this integral becomes trivial if in the numerator we make the replacement

$$(\not{p} - \not{q} - m) \to (\not{p} - \not{q} - m) - \tfrac{1}{2}(\not{p} - m) = \tfrac{1}{2}(\not{p} - 2\not{q} - m),$$

justified since $(\not{p} - m)u(\mathbf{p}) = 0$. In the resulting integral we can make the further replacement

$$\tfrac{1}{2}(\not{p} - 2\not{q} - m) \to \tfrac{1}{2}(\not{p} - \not{q} - \not{q}' - m),$$

since apart from this factor the integral is symmetric in q and q'. With these replacements and the normalization condition (9.42) for $F(E)$, Eq. (9.43) reduces to

$$u(\mathbf{p}) \to (1 - \tfrac{1}{2} e_0^2 B) u(\mathbf{p}). \tag{9.44}$$

The last expression is independent of the adiabatic switch-off function $F(E)$. Hence it already expresses our result in the limit $F(E) \to \delta(E)$ which restores our original theory with $f(t) = 1$.

Eq. (9.44) is a second-order result. We can rewrite it in the form

$$u(\mathbf{p}) \to Z_2^{1/2} u(\mathbf{p}) \tag{9.45a}$$

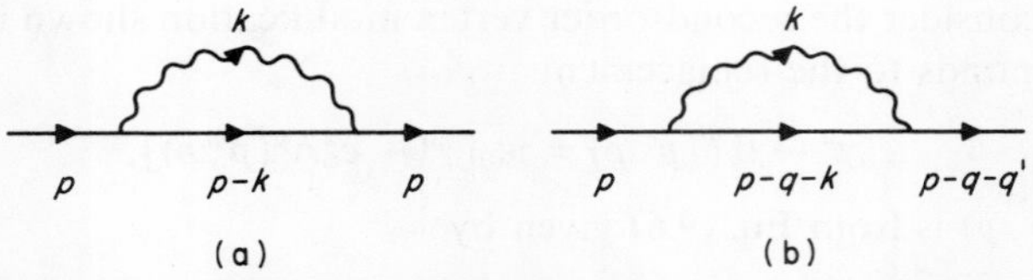

Fig. 9.12. The fermion self-energy loop: (a) for the QED interaction (9.31c); (b) for the modified interaction (9.40); $q = (E, \mathbf{0})$ and $q' = (E', \mathbf{0})$ represent the energy non-conservation at the vertices.

where the renormalization constant Z_2, defined in Eq. (9.35), relates the bare charge e_0 to the physical charge e. Although we have only derived Eq. (9.45a) in second-order perturbation theory, it holds to all orders.

Similar arguments applied to self-energy insertions in other external fermion lines lead to the analogous results

$$\bar{u}(\mathbf{p}) \to Z_2^{1/2}\bar{u}(\mathbf{p}), \qquad v(\mathbf{p}) \to Z_2^{1/2}v(\mathbf{p}), \qquad \bar{v}(\mathbf{p}) \to Z_2^{1/2}\bar{v}(\mathbf{p}). \tag{9.45b}$$

Similarly, the photon self-energy insertion of Fig. 9.3(b) in an external photon line leads to

$$\varepsilon^\mu(\mathbf{k}) \to Z_3^{1/2}\varepsilon^\mu(\mathbf{k}), \tag{9.45c}$$

where the charge renormalization constant Z_3 is defined by Eq. (9.18).

The modifications (9.45) of the wave functions of the external particles due to self-energy effects, are referred to as external line renormalization or wave function renormalization. When considering the modified photon and fermion propagators in the last two sections, we interpreted the parameters Z_3 and Z_2 as renormalization constants of the charges acting at the vertices at the ends of the photon and fermion propagators respectively [see Eqs. (9.18) and (9.35)]:

$$e_0 \to e = Z_3^{1/2}e_0, \qquad e_0 \to e = Z_2^{1/2}e_0. \tag{9.46}$$

We can equally interpret the wavefunction renormalizations (9.45) as charge renormalizations. For this purpose, we associate the factors $Z_3^{1/2}$ and $Z_2^{1/2}$ in Eqs. (9.45) with the charges acting at the vertices to which the external lines are attached. Eqs. (9.46) are now valid generally for each internal or external line attached to a vertex. For external lines, these charge renormalizations are the *only* self-energy effects (apart from the electron mass renormalization, $m_0 \to m$, allowed for automatically through the mass counterterm). This is in contrast to self-energy modifications of photon and fermion propagators [see Eqs. (9.19) and (9.36)] which lead to additional finite radiative corrections.

9.5 THE VERTEX MODIFICATION

We finally consider the second-order vertex modification shown in Fig. 9.13. This corresponds to the replacement

$$\mathrm{i}e_0\gamma^\mu \to \mathrm{i}\Gamma^\mu(p', p) = \mathrm{i}e_0[\gamma^\mu + e_0^2\Lambda^\mu(p', p)], \tag{9.47}$$

where $\Lambda^\mu(p', p)$ is from Eq. (9.6) given by

$$\Lambda^\mu(p', p) = \frac{-\mathrm{i}}{(2\pi)^4}\int \frac{\mathrm{d}^4k}{k^2 + \mathrm{i}\varepsilon}\gamma^\alpha \frac{1}{\not{p}' - \not{k} - m + \mathrm{i}\varepsilon}\gamma^\mu \frac{1}{\not{p} - \not{k} - m + \mathrm{i}\varepsilon}\gamma_\alpha. \tag{9.48}$$

(m now of course denotes the real fermion mass.) $\Lambda^\mu(p', p)$ is both ultra-violet

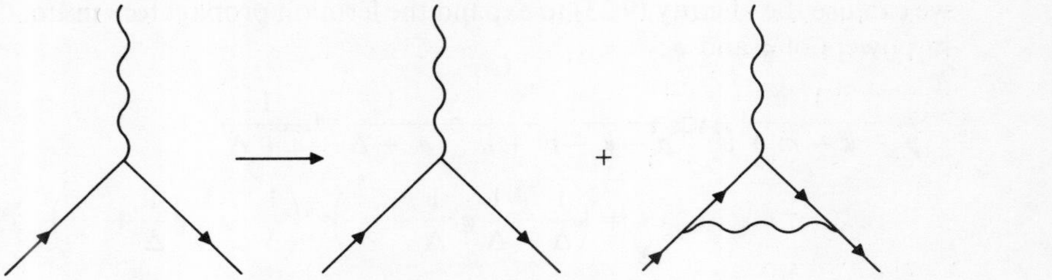

Fig. 9.13. The second-order vertex modification.

and infra-red divergent. We regularize it by the same replacement (9.21) of the photon propagator which we used for the fermion self-energy loop.

When considering the fermion self-energy, we separated off the free-particle part $\Sigma(p)|_{\not p=m}$, Eq. (9.27). The troublesome part of $\Lambda^{\mu}(p', p)$, which diverges logarithmically as the cut-off parameter Λ tends to infinity and which we want to separate off, is again given by the free-particle value

$$\bar{u}(\mathbf{P})\Lambda^{\mu}(P, P)u(\mathbf{P}). \tag{9.49}$$

Here $u(\mathbf{P})$ is a free-particle spinor and P a free-particle momentum vector, i.e. $P^2 = m^2$. From Lorentz invariance, expression (9.49) must be of the form

$$a\bar{u}(\mathbf{P})\gamma^{\mu}u(\mathbf{P}) + bP^{\mu}\bar{u}(\mathbf{P})u(\mathbf{P}), \tag{9.50}$$

where a and b are scalar constants. But from Gordon's identity (see Problem A.2)

$$P^{\mu}\bar{u}(\mathbf{P})u(\mathbf{P}) = m\bar{u}(\mathbf{P})\gamma^{\mu}u(\mathbf{P}), \tag{9.51}$$

and combining the last three equations we can write

$$\bar{u}(\mathbf{P})\Lambda^{\mu}(P, P)u(\mathbf{P}) = L\bar{u}(\mathbf{P})\gamma^{\mu}u(\mathbf{P}), \tag{9.52}$$

where L is a scalar constant.

For general four-vectors p and p', we define $\Lambda_c^{\mu}(p', p)$ by

$$\Lambda^{\mu}(p', p) = L\gamma^{\mu} + \Lambda_c^{\mu}(p', p). \tag{9.53}$$

We see from Eq. (9.52) that for a free-particle four-momentum P

$$\bar{u}(\mathbf{P})\Lambda_c^{\mu}(P, P)u(\mathbf{P}) = 0. \tag{9.54}$$

The motivation for writing $\Lambda^{\mu}(p', p)$ in the form (9.53) is that in the limit $\Lambda \to \infty$, in which QED is restored, L diverges but the second term $\Lambda_c^{\mu}(p', p)$ remains well-defined and finite. This can be seen from the expression (9.48) as it stands. With the abbreviations

$$\Delta \equiv \not P - \not k - m + \mathrm{i}\varepsilon, \qquad q \equiv p - P, \qquad q' \equiv p' - P,$$

we can use the identity (9.23) to expand the fermion propagators in Eq. (9.48) in powers of $\not{q}$ and $\not{q}'$:

$$\frac{1}{\not{p}' - \not{k} - m + i\varepsilon}\gamma^{\mu}\frac{1}{\not{p} - \not{k} - m + i\varepsilon} \equiv \frac{1}{\not{q}' + \Delta}\gamma^{\mu}\frac{1}{\not{q} + \Delta}$$
$$= \left(\frac{1}{\Delta} - \frac{1}{\Delta}\not{q}'\frac{1}{\Delta} + \cdots\right)\gamma^{\mu}\left(\frac{1}{\Delta} - \frac{1}{\Delta}\not{q}\frac{1}{\Delta} + \cdots\right). \quad (9.55)$$

Substituting this expansion in Eq. (9.48), we see that the leading term in $\Lambda^{\mu}(p', p)$ arises from $(1/\Delta)\gamma^{\mu}(1/\Delta)$ in Eq. (9.55) and is simply $\Lambda^{\mu}(P, P)$. This term may (and indeed does) diverge logarithmically as $k \to \infty$, since Δ is linear in k. All other terms necessarily converge as $k \to \infty$, since they contain additional factors Δ in the denominator.

Substituting Eq. (9.53) in Eq. (9.47), we obtain

$$ie_0\gamma^{\mu} \to i\Gamma^{\mu}(p', p) = ie_0[\gamma^{\mu}(1 + e_0^2 L) + e_0^2\Lambda_c^{\mu}(p', p)]. \quad (9.56)$$

The term proportional to γ^{μ} on the right-hand side of this equation is the original basic vertex part $ie_0\gamma^{\mu}$ but with a renormalized charge. We define a charge renormalization constant Z_1 by

$$e \equiv \frac{e_0}{Z_1} = e_0(1 + e_0^2 L) + O(e_0^5), \quad (9.57)$$

where $O(e_0^5)$ indicates that there are also higher-order contributions to the charge renormalization which results from all vertex modifications, whereas we have only considered the lowest, second-order correction. Expressing the right-hand side of Eq. (9.56) in terms of the renormalized charge e, we can write this equation

$$ie_0\gamma^{\mu} \to i\Gamma^{\mu}(p', p) = ie[\gamma^{\mu} + e^2\Lambda_c^{\mu}(p', p)] + O(e^5), \quad (9.58)$$

where $O(e^5)$ again indicates higher-order corrections.

Eq. (9.58) is our final result for the second-order vertex modification. It consists of the charge renormalization (9.57) and the correction term $\Lambda_c^{\mu}(p', p)$. In the regularized form, all quantities are well-defined and finite. In the limit $\Lambda \to \infty$, in which QED is restored, L (and Z_1) become infinite but this only affects the unobservable relation (9.57). On the other hand, $\Lambda_c^{\mu}(p', p)$ tends to a well-defined finite limit which is independent of the regularization procedure and which contributes to the lowest-order radiative correction of a process.

We now combine the charge renormalization (9.57) resulting from the vertex modification with the charge renormalizations (9.46) resulting from photon and fermion self-energy effects. Since each vertex has one photon line and two fermion lines attached, it follows from these equations that the net effect is the replacement of the bare charge e_0 at each vertex (i.e. everywhere in the theory) by the renormalized charge

$$e = e_0 Z_3^{1/2} Z_2 / Z_1. \tag{9.59}$$

This and all earlier results in this chapter have been derived in second-order perturbation theory only, but they can be shown to hold in all orders.

The result (9.59) allows a remarkable simplification. This is due to the fact that the fermion self-energy insertion $\Sigma(p)$, Eq. (9.4), and the vertex insertion $\Lambda^\mu(p', p)$, Eq. (9.6), are related by Ward's identity which is given by

$$\frac{\partial \Sigma(p)}{\partial p_\mu} = \Lambda^\mu(p, p). \tag{9.60}$$

(See Problem 9.2.) Ward's identity relates the fermion self-energy graph, Fig. 9.14(a), to the vertex modification obtained from it by insertion of a zero-energy photon in the intermediate fermion propagator, as shown in Fig. 9.14(b). Eq. (9.60) is a second-order result. However, Ward's identity can be generalized and holds in all orders of perturbation theory, allowing one to obtain higher-order vertex modifications by differentiation of higher-order fermion self-energy insertions. This greatly simplifies the calculation of higher-order radiative corrections.

Ward's identity also implies a relation between the charge renormalization constants Z_1 and Z_2, which we shall now derive. With $u(\mathbf{P})$ a free-particle spinor, we obtain from Eq. (9.60)

$$\bar{u}(\mathbf{P}) \frac{\partial \Sigma(P)}{\partial P_\mu} u(\mathbf{P}) = \bar{u}(\mathbf{P}) \Lambda^\mu(P, P) u(\mathbf{P}). \tag{9.61}$$

Using Eq. (9.26), we obtain for the left-hand side of Eq. (9.61)

$$B\bar{u}(\mathbf{P})\gamma^\mu u(\mathbf{P}),$$

since

$$\bar{u}(\mathbf{P})(\not{P} - m) = \Sigma_c(P) u(\mathbf{P}) = 0.$$

From Eqs. (9.53) and (9.54), the right-hand side of Eq. (9.61) equals

$$L\bar{u}(\mathbf{P})\gamma^\mu u(\mathbf{P}),$$

Fig. 19.14. (a) The second-order self-energy loop. (b) The vertex modification obtained by inserting a zero-energy photon in the fermion propagator of (a).

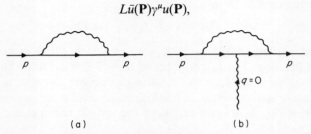

and hence

$$B = L. \tag{9.62}$$

From Eqs. (9.35) and (9.57), this relation can be expressed in terms of the charge-renormalization constants Z_1 and Z_2 as

$$Z_2 = Z_1. \tag{9.63}$$

Although we have only derived Eq. (9.63) in second-order perturbation theory, it is an exact relation holding in all orders of perturbation theory.

As a consequence of the equality (9.63), Eq. (9.59) reduces to

$$e = e_0 Z_3^{1/2}. \tag{9.64}$$

Thus the charge renormalization does not depend on fermion self-energy effects or vertex modifications, but originates solely from photon self-energy effects, i.e. from vacuum polarization. This has an interesting consequence when considering not only electrons and positrons, but also other kinds of leptons, $e^{\pm}$, $\mu^{\pm}$, ... It is easy to see that Eqs. (9.63) and (9.64) continue to hold for each type of lepton, with the same constant Z_3 in each case.‡ Consequently the observed equality of the physical charges of particles implies the equality of their bare charges.

We have now completed our analysis of the second-order modifications of photon and fermion lines and of vertices, and we can summarize our results as follows. If we ascribe the physical masses (m_e, m_μ, ...) to the leptons and throughout replace the bare charge e_0 by the physical charge $e = e_0 Z_3^{1/2}$, then the only modifications due to second-order self-energy and vertex corrections in QED are the propagator modifications

$$\frac{-\mathrm{i}g^{\alpha\beta}}{k^2 + \mathrm{i}\varepsilon} \to \frac{-\mathrm{i}g^{\alpha\beta}}{k^2 + \mathrm{i}\varepsilon}[1 - e^2\Pi_c(k^2)] + O(e^4) \tag{9.65a}$$

$$\frac{\mathrm{i}}{\not{p} - m + \mathrm{i}\varepsilon} \to \frac{\mathrm{i}}{\not{p} - m + \mathrm{i}\varepsilon}[1 - e^2\Sigma_c(p)] + O(e^4) \tag{9.65b}$$

and the vertex modification

$$\mathrm{i}e_0\gamma^\mu \to \mathrm{i}e[\gamma^\mu + e^2\Lambda_c^\mu(p', p)] + O(e^5). \tag{9.65c}$$

As stated earlier, in the limit $\Lambda \to \infty$ in which QED is restored, the regularized functions Π_c, Σ_c and Λ_c^μ tend to well-defined finite limits so that the modifications (9.65) lead to radiative corrections of order α. Instead of interpreting regularization as a mathematical device for coping with the divergences of the theory, we can keep the cut-off parameter Λ finite and

‡ In addition to the vacuum polarization loops formed from electron–positron pairs, the photon self-energy will now contain contributions from loops formed from μ^+–μ^- pairs, etc.

interpret the regularized theory as a genuine modification of QED. We must then ask what limits are set on the value of Λ and the validity of QED by experiment. The most restrictive limits are obtained from the lepton pair processes $e^+e^- \to l^+l^-$. As discussed in Section 8.4, these processes probe the behaviour of the photon propagator at $k^2 = (2E)^2$, where E is the electron energy in the centre-of-mass system and is about 15 GeV in the highest-energy experiments to date. If the modified propagator (9.21), with Λ kept finite, is used, then agreement with the experimental data is only obtained if $\Lambda \gtrsim 150$ GeV, corresponding to distances of order $\Lambda^{-1} \lesssim 2 \times 10^{-3}$ f. We conclude that, at presently accessible energies, the observable predictions of QED are insensitive to modifications of the theory at distances much shorter than 10^{-3} f.

In contrast to the regularized functions Π_c, Σ_c and Λ_c^μ, one finds that δm diverges like $\ln \Lambda$ as $\Lambda \to \infty$, and indeed all divergent quantities of QED diverge logarithmically. In our perturbation treatment, the leading divergent terms are of order α, so that appreciable differences between bare and physical quantities occur only for values of Λ which are enormously large. For example, we see from Eqs. (9.28) and (9.37) that for a significant electron mass correction δm to occur, i.e. $\delta m = \mathrm{O}(m)$, we require

$$\Lambda = \mathrm{O}(m\, \mathrm{e}^{2\pi/3\alpha}) \approx 10^{121} \text{ GeV}.$$

On the other hand, for $\Lambda \ll 10^{121}$ GeV, one obtains $\delta m \ll m$, so that it seems reasonable to treat the mass correction δm in perturbation theory. Thus the reader who is rightly worried about treating large or infinite quantities in perturbation theory, even in unphysical relations, should think of Λ as finite but much less than 10^{121} GeV. Furthermore, the physical predictions of the theory will not be measurably different from those obtained in the limit $\Lambda \to \infty$, provided Λ is much larger than 150 GeV.

9.6 APPLICATIONS

We have so far shown how to calculate well-defined finite radiative corrections of order α. Applications of these results lead to some of the most spectacular successes of modern physics. In particular, for the anomalous magnetic moment of the electron and the muon and for the energy levels of the hydrogen atom (the Lamb shift) comparison of theory with experiments leads to extraordinarily precise agreement. In this section we shall give these comparisons for both problems. For the magnetic moment we shall derive the radiative correction of order α to the Dirac value. The Lamb shift calculation is a bound-state problem and its proper treatment requires a fairly elaborate extension of the theory we have developed. We shall limit ourselves to giving a much simpler approximate non-relativistic derivation, due to Bethe, which calculates correctly the dominant contribution to the Lamb shift.

9.6.1 The anomalous magnetic moments

The magnetic moment of a particle shows up through the scattering of the particle by a magnetic field. For this reason we shall once more study the elastic scattering of an electron by a static external field. We considered this process in lowest order in Section 8.7 and found that it is represented by the Feynman graph in Fig. 9.15(a) and that its Feynman amplitude is given by Eq. (8.88):

$$ie\bar{u}(\mathbf{p}')\not{A}_{\mathrm{e}}(\mathbf{q} = \mathbf{p}' - \mathbf{p})u(\mathbf{p}). \tag{9.66}$$

The radiative corrections of order α to this process stem from the Feynman graphs in Figs. 9.15(b)–(g). After renormalization, only diagrams (b) and (c) give contributions, with the Feynman amplitude to order e^3 given, from Eqs. (9.65), by

$$ie\bar{u}(\mathbf{p}')\gamma^{\mu}u(\mathbf{p})A_{\mathrm{e}\mu}(\mathbf{q}) + ie\bar{u}(\mathbf{p}')\gamma^{\mu}u(\mathbf{p})[-e^2\Pi_{\mathrm{c}}(q^2)]A_{\mathrm{e}\mu}(q)$$
$$+ ie\bar{u}(\mathbf{p}')[e^2\Lambda^{\mu}_{\mathrm{c}}(p', p)]u(\mathbf{p})A_{\mathrm{e}\mu}(\mathbf{q}). \tag{9.67}$$

All other effects are absorbed into the mass and charge renormalizations. In particular, as we saw in Section 9.4, there are no observable radiative corrections associated with the external line diagrams (d)–(g). In connection with the charge renormalization resulting from the vacuum polarization

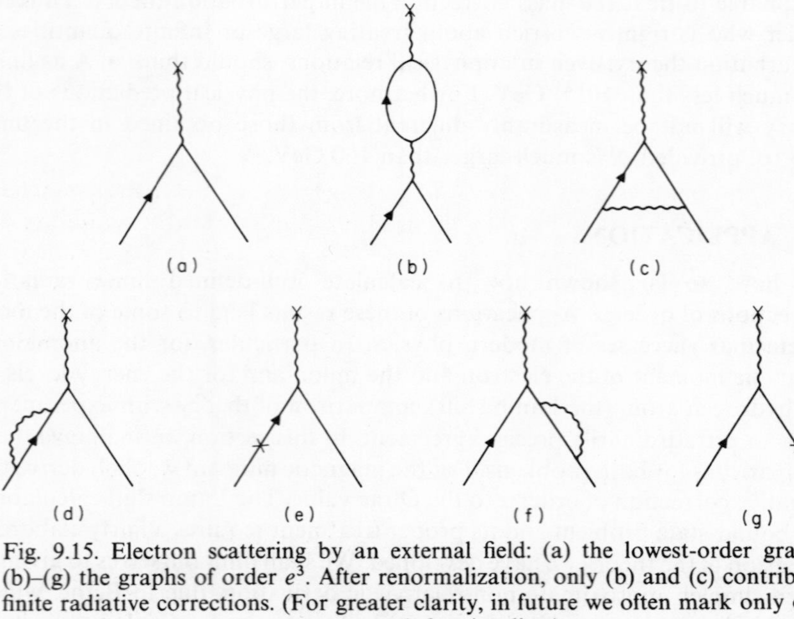

Fig. 9.15. Electron scattering by an external field: (a) the lowest-order graph; (b)–(g) the graphs of order e^3. After renormalization, only (b) and (c) contribute finite radiative corrections. (For greater clarity, in future we often mark only one arrow on each fermion line.)

graph (b), one point should be noted. When considering the vacuum polarization loop in Section 9.2, we associated a charge renormalization $e_0 \to e = e_0 Z_3^{1/2}$ with the charge acting at each end of the photon propagator. In the present context, i.e. for diagram (b), one of these factors $Z_3^{1/2}$ is absorbed into renormalizing the charges which act as source of the external field $A_{e\mu}(\mathbf{q})$.

To use Eq. (9.67) we need explicit expressions for $\Pi_c(q^2)$ and $\Lambda_c^\mu(p', p)$. In Section 10.4 we shall derive the result

$$e^2\Pi_c(q^2) = -\frac{2\alpha}{\pi}\int_0^1 dz z(1-z)\ln\left[1 - \frac{q^2 z(1-z)}{m^2}\right]. \tag{9.68}$$

For $q^2 \ll m^2$, the logarithm may be expanded to give

$$e^2\Pi_c(q^2) = \frac{\alpha}{15\pi}\left(\frac{q^2}{m^2}\right) + \cdots \quad (q^2 \ll m^2). \tag{9.69}$$

In the next chapter, Section 10.5, we shall also show how to evaluate $\Lambda_c^\mu(p', p)$. In particular, we shall see that the third term in Eq. (9.67) contains the term

$$ie\bar{u}(\mathbf{p}')\left[\frac{i\alpha}{4\pi m}\sigma^{\mu\nu}q_\nu\right]u(\mathbf{p})A_{e\mu}(\mathbf{q}). \tag{9.70}$$

It is this contribution to the Feynman amplitude (9.67) that we wish to interpret. To do so, we use the Gordon identity (see Problem A.2) to rewrite the lowest-order scattering amplitude (9.66) as

$$\frac{ie}{2m}\bar{u}(\mathbf{p}')[(p'+p)^\mu + i\sigma^{\mu\nu}q_\nu]u(\mathbf{p})A_{e\mu}(\mathbf{q}). \tag{9.71}$$

In the non-relativistic limit of slowly moving particles and a static magnetic field, the second term in Eq. (9.71) is just the amplitude for the scattering of a spin $\frac{1}{2}$ particle with magnetic moment $(-e/2m)$, i.e. with gyromagnetic ratio $g = 2$. The amplitude (9.70) is of the same form as the spin term in (9.71), i.e. it represents a correction to the value of the magnetic moment of the electron as given by the Dirac theory. This anomalous magnetic moment

$$-\frac{e}{2m}\left(1 + \frac{\alpha}{2\pi}\right)$$

corresponds to a shift in the g-factor, usually quoted in the form

$$a_e \equiv \frac{g-2}{2} = \frac{\alpha}{2\pi} = 0.00116. \tag{9.72}$$

This result, first derived by Schwinger in 1948, is in excellent agreement with the first measurements by Kusch and Foley (1947, 1948a) who obtained the value

$$a_e = 0.00119 \pm 0.00005.^{\ddagger}$$

Subsequently, both theory and experiment have been greatly refined. Theoretically, the higher-order contributions of order α^2 and of order α^3 to a_e have been derived. The result of these very heavy calculations (the α^3 term involves 72 very complicated Feynman graphs) is

$$10^9 a_e = 1159652.4 \pm 0.4.$$

The most recent experiments by Van Dyck *et al.* (1978) give

$$10^9 a_e = 1159652.41 \pm 0.20.$$

The last two numbers display a most remarkable agreement between theory and experiment to an accuracy greater than one part in a million.

A similar discussion applies to the anomalous magnetic moment of the muon. Since Eq. (9.72) is independent of the lepton mass, we obtain the same value for the muon in lowest order

$$a_\mu \equiv \frac{g_\mu - 2}{2} = \frac{\alpha}{2\pi} = 0.00116,$$

where we have written g_μ to distinguish the muon g-factor from that of the electron. In higher order, differences arise since vacuum polarization loops may involve any kind of lepton pair. This is illustrated in Fig. 9.16, where diagrams (a), (b) and (c), (d) show two e^4 contributions to a_e and a_μ respectively. Since the masses of the intermediate leptons occur in the denominators of the corresponding propagators and since $m_\mu/m_e \approx 207$, the contribution of the muon pair diagram (b) is completely negligible compared with that of the electron pair diagram (a). (Another way of putting this is to say that because of the much smaller electron mass, compared to m_μ, it is much easier to create a virtual electron–positron pair than a muon pair.) For the muon, on the other hand, the electron pair diagram (d) makes a large contribution to a_μ, in fact it is substantially bigger than that of the muon diagram (c). Similar conclusions hold generally for the contributions of vacuum polarization corrections.

High precision measurements of the muon magnetic moment have been performed, a recent experimental value (Bailey *et al.*, 1979) being

$$10^9 a_\mu = 1165924 \pm 8.5.$$

‡ The theoretical and experimental data quoted in this section and their sources are given in F. H. Combley, *Rep. Prog. Phys.*, **42** (1979), 1889. This article reviews the recent most ingenious measurements of $(g - 2)$ factors for electrons and muons and summarizes the current state of the theory.

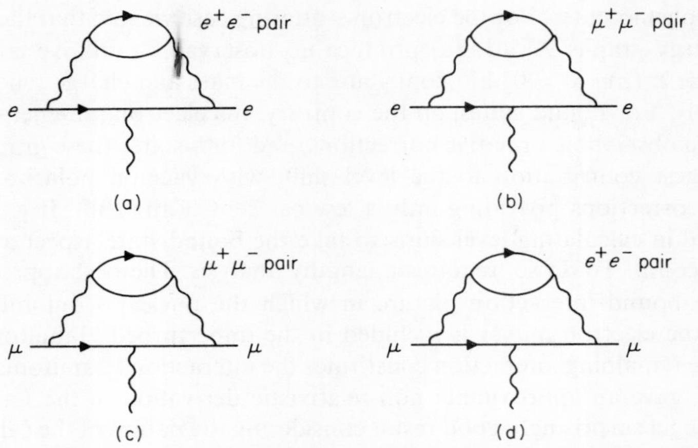

Fig. 9.16. e^4 vacuum polarization contributions to the g-factors of: (i) the electron: (a) and (b); (ii) the muon: (c) and (d).

The best current theoretical prediction is

$$10^9 a_\mu = 1165851.7 \pm 2.3.$$

The last two numbers show a small discrepancy between theory and experiment. This can be attributed to the effect of strong interactions, e.g. vacuum polarization graphs involving $\pi^+\pi^-$ pairs. The detailed analysis shows that these hadronic contributions are sufficiently large to close the gap between the theoretical and the experimental data for the muon. On the other hand, the hadronic effects are completely negligible for the electron and do not upset the excellent agreement between theory and experiment.

9.6.2 The Lamb shift

As a second important application we shall look at the radiative corrections to the energy levels of the hydrogen atom. Historically, the measurements by Lamb and Retherford in 1947 gave the main impetus to the development of modern QED. According to the Dirac theory, the $2s_{1/2}$ and $2p_{1/2}$ levels of hydrogen are degenerate. Lamb and Retherford's original experiment gave about 1000 MHz for the level splitting $E(2s_{1/2}) - E(2p_{1/2})$. This shift of the bound-state energy levels and the resulting splitting are known as the Lamb shift.

In the last sub-section we considered electron scattering by an external static potential, and we saw that the radiative corrections of order α stem from the Feynman graphs in Figs. 9.15(b)–(g). We can think of the same graphs as describing a bound-state level in hydrogen, if we interpret the electron lines and propagators not in terms of free-particle states but in terms of hydrogenic states. For the electron scattering case we saw that the electron self-energy graphs 9.15(d)–(g) produce no observable radiative corrections [because $\Sigma_c(p)u(\mathbf{p}) = 0$] but contribute to the mass and charge renormalization only. For bound states, on the contrary, the electron self-energy graphs produce observable radiative corrections, and for s-states these graphs make the largest contribution to the level shift, with vacuum polarization and vertex corrections providing only a few per cent of the shift. It is therefore essential in calculating level shifts to take the bound-state aspect accurately into account. To do so, requires a lengthy analysis. The best approach is to use the bound interaction picture in which the nuclear Coulomb field in which the electron moves is included in the unperturbed Hamiltonian and only the remaining interaction constitutes the interaction Hamiltonian. Bethe in 1947 gave an approximate non-relativistic derivation of the Lamb shift, obtaining a surprisingly good result considering the nature of the calculation. We shall restrict ourselves to the Bethe approach as it is much simpler and clearly exhibits the main origin of the Lamb shift.‡

Bethe attributes the shift of a bound-state energy level to the self-energy of the electron in that bound state. However, a part of this self-energy effect has already been allowed for in using the physical electron mass in the calculation, and not the bare mass. Hence the true level shift is the difference between the self-energies of the bound and the free electron.

In Bethe's calculation, the hydrogen atom is treated non-relativistically, and second-order perturbation theory is used to calculate the interaction between the electron and the transverse photons. This is the formulation of QED which we gave in Chapter 1. The interaction Hamiltonian is, from Eq. (1.62),

$$H_{\mathrm{I}} = -\frac{e}{m}\,\mathbf{A}(x)\cdot\mathbf{p}.^{\S} \tag{9.73}$$

The level shift of a hydrogenic state $|nl\rangle \equiv \phi_{nl}(\mathbf{x})$ (where n and l are the principal and angular momentum quantum numbers) is then given by

$$\delta E(nl) = -\sum_{\lambda}\sum_{\mathbf{k}}\sum_{r=1,2} \frac{|\langle \lambda, n_r(\mathbf{k}) = 1|H_{\mathrm{I}}|nl\rangle|^2}{E_\lambda + k - E_n} \tag{9.74}$$

‡ For a rigorous treatment, employing the bound interaction picture, the reader is referred to J. M. Jauch and F. Rohrlich, *The Theory of Photons and Electrons*, 2nd edn, Springer, New York, 1976, Sections 15-4 and S5-3. An alternative approach, which utilizes the Bethe derivation for the non-relativistic part of the calculation, is discussed in, for example, J. D. Bjorken and S. D. Drell, *Relativistic Quantum Mechanics*, McGraw-Hill, New York, 1964, Section 8.7, or C. Itzykson and J. B. Zuber, *Quantum Field Theory*, McGraw-Hill, New York, 1980, Section 7-3-2.

§ The $\mathbf{A}^2$ interaction term in Eq. (1.62) is independent of the electron momentum. Therefore it produces the same electron self-energy effects for the bound electron and the free electron, and so does not contribute to the level shift.

where the intermediate state $|\lambda, n_r(\mathbf{k}) = 1\rangle$ consists of the hydrogen atom in one of the complete set of states $|\lambda\rangle \equiv \phi_\lambda(\mathbf{x})$, together with one transverse photon. (E_λ and E_n are the energy eigenvalues of $|\lambda\rangle$ and $|nl\rangle$.)

The matrix elements in Eq. (9.74) are given by Eq. (1.65). Using the dipole approximation, we replace the exponential in the matrix element (1.65) by unity.‡ Substituting for the matrix elements in Eq. (9.74), one obtains

$$\delta E(nl) = -\sum_\lambda \sum_{\mathbf{k}} \sum_{r=1,2} \left(\frac{e}{m}\right)^2 \frac{1}{2Vk} \frac{|\langle\lambda|\boldsymbol{\varepsilon}_r(\mathbf{k})\cdot\mathbf{p}|nl\rangle|^2}{E_\lambda + k - E_n}. \tag{9.75}$$

As in Section 1.3, we sum over photon polarizations ($r = 1, 2$) and, after converting the momentum sum into an integral, integrate over photon directions [see the corresponding analysis leading to Eq. (1.53)]. In this way, we obtain

$$\delta E(nl) = -\frac{1}{6\pi^2}\left(\frac{e}{m}\right)^2 \int_0^\infty k\,\mathrm{d}k \sum_\lambda \frac{|\langle\lambda|\mathbf{p}|nl\rangle|^2}{E_\lambda + k - E_n} \tag{9.76}$$

where

$$\langle\lambda|\mathbf{p}|nl\rangle = \int \mathrm{d}^3\mathbf{x}\,\phi_\lambda^*(\mathbf{x})(-\mathrm{i}\boldsymbol{\nabla})\phi_{nl}(\mathbf{x}). \tag{9.77}$$

The integral in Eq. (9.76) is linearly divergent as $k \to \infty$.

The corresponding self-energy $\delta E_f(\mathbf{p})$ for a free electron with momentum $\mathbf{p}$ is given by the same expression (9.76) where the matrix element (9.77) is now between plane-wave states and is diagonal, so that

$$\delta E_f(\mathbf{p}) = -\frac{1}{6\pi^2}\left(\frac{e}{m}\right)^2 \mathbf{p}^2 \int_0^\infty \mathrm{d}k. \tag{9.78}$$

This self-energy is proportional to the kinetic energy of the electron and can be interpreted in terms of a correction to the electron mass. Since the electron in the state $|nl\rangle$ has a momentum distribution, the corresponding self-energy is given by

$$\delta E_f(nl) = -\frac{1}{6\pi^2}\left(\frac{e}{m}\right)^2 \langle nl|\mathbf{p}^2|nl\rangle \int_0^\infty \mathrm{d}k. \tag{9.79}$$

This integral, like that in Eq. (9.76), is linearly divergent as $k \to \infty$.

If one uses the physical mass of the electron in calculating the level shift of the state $|nl\rangle$, then the self-energy $\delta E_f(nl)$ has already been taken into account, and the observed level shift $\Delta E(nl)$ is given by

$$\Delta E(nl) = \delta E(nl) - \delta E_f(nl). \tag{9.80}$$

‡ This is justified since the virtual photons which contribute significantly to the sum (9.74) have wavelengths which are large compared to the Bohr radius. See Eq. (9.88) below.

Since

$$\langle nl|\mathbf{p}^2|nl\rangle = \sum_\lambda |\langle\lambda|\mathbf{p}|nl\rangle|^2, \tag{9.81}$$

we obtain from Eqs. (9.76), (9.79) and (9.80)

$$\Delta E(nl) = \frac{1}{6\pi^2}\left(\frac{e}{m}\right)^2 \sum_\lambda |\langle\lambda|\mathbf{p}|nl\rangle|^2 \int_0^\infty \mathrm{d}k \, \frac{E_\lambda - E_n}{E_\lambda - E_n + k}. \tag{9.82}$$

This integral is only logarithmically divergent as $k \to \infty$. To make it converge, we replace the infinite upper limit by a finite cut-off $k = K \sim m$, i.e. we suppress contributions to the self-energy from virtual photons with energy $k \gtrsim m$. We may try to justify this cut-off as follows. In emitting a virtual photon, the electron experiences a recoil. If the non-relativistic treatment of the electron is meaningful, this recoil and hence the virtual photon energy k must be small compared to the electron rest mass. In other words, only transitions to non-relativistic hydrogenic states and virtual photons with energy $k \ll m$ may be important. Hence taking the upper limit of the integral in Eq. (9.82) as $k = K \sim m$, and assuming that in the sum over λ in this equation

$$|E_\lambda - E_n| \ll K$$

holds for the terms which matter, we obtain from Eq. (9.82)

$$\Delta E(nl) = \frac{1}{6\pi^2}\left(\frac{e}{m}\right)^2 \sum_\lambda |\langle\lambda|\mathbf{p}|nl\rangle|^2 (E_\lambda - E_n) \ln \frac{K}{|E_\lambda - E_n|}. \tag{9.83}$$

From this equation, the shift of any bound-state enegy level is obtained. The fact that it depends only logarithmically on the cut-off K makes it insensitive to the exact value chosen for K.

To evaluate the λ-summation in Eq. (9.83), Bethe defines an average excitation energy $\langle E - E_n\rangle$ by the equation

$$\sum_\lambda |\langle\lambda|\mathbf{p}|nl\rangle|^2 (E_\lambda - E_n)\{\ln \langle E - E_n\rangle - \ln |E_\lambda - E_n|\} = 0. \tag{9.84}$$

Eq. (9.83) then becomes

$$\Delta E(nl) = \frac{1}{6\pi^2}\left(\frac{e}{m}\right)^2 \ln \frac{K}{\langle E - E_n\rangle} \sum_\lambda |\langle\lambda|\mathbf{p}|nl\rangle|^2 (E_\lambda - E_n). \tag{9.85}$$

The summation over λ in this equation can be performed, giving

$$\sum_\lambda |\langle\lambda|\mathbf{p}|nl\rangle|^2 (E_\lambda - E_n) = \tfrac{1}{2}e^2|\phi_{nl}(0)|^2$$

$$= \begin{cases} \dfrac{e^2}{2\pi a^3 n^3}, & \text{if } l = 0 \text{ (s-states)}, \\ 0, & \text{if } l \neq 0, \end{cases} \tag{9.86}$$

where $a = 4\pi/me^2$ is the Bohr radius.[‡] Substitution of this result into Eq. (9.85) gives

$$\Delta E(nl) = \frac{8}{3\pi}\frac{\alpha^3}{n^3}\,\text{Ry}\ln\frac{K}{\langle E - E_n\rangle}\,\delta_{l0}, \tag{9.87}$$

where $\text{Ry} \equiv e^2/(8\pi a) = 13.6$ eV is the Rydberg energy.

Eq. (9.87) is the final result of Bethe's calculation. According to it, only s-states experience a level shift due to electron self-energy effects. For the 2s states of hydrogen, Bethe uses the value

$$\langle E - E_{2s}\rangle = 17.8\ \text{Ry}, \tag{9.88}$$

obtained by computation.[§] Thus the important intermediate states of the hydrogen atom are indeed non-relativistic although they are highly excited continuum states. Using the value (9.88) and $K = m$, Bethe obtains from Eq. (9.87) the Lamb shift

$$E(2s_{1/2}) - E(2p_{1/2}) = 1040\ \text{MHz}, \tag{9.89}$$

in remarkable agreement with the experimental value of 1057.8 ± 0.1 MHz, obtained by Triebwasser, Dayhoff and Lamb in 1953 as the culmination of their measurements.

A proper relativistic calculation of the second-order radiative corrections leads to the value 1052.1 MHz for the $2s_{1/2} - 2p_{1/2}$ splitting.[¶] Such a calculation of course contains no arbitrary cut-off parameter K, and in addition to the electron self-energy it takes into account all e^2 radiative corrections, i.e. also the contributions from vacuum polarization and vertex corrections. States other than s-states now also experience level shifts although these are much smaller, e.g. the $2p_{1/2}$ level is shifted downwards by 12.9 MHz.

We shall only consider the vacuum polarization contribution. We see from Eqs. (9.67) and (9.69) that the effect of the vacuum polarization on the scattering of non-relativistic electrons by a Coulomb potential (8.92) corresponds to modifying the Coulomb potential through the replacement

$$\frac{Ze}{|\mathbf{q}|^2} \to \frac{Ze}{|\mathbf{q}|^2}\left(1 + \frac{\alpha}{15\pi}\frac{|\mathbf{q}|^2}{m^2}\right), \tag{9.90a}$$

since for the static external field $q = (0, \mathbf{q})$. The corresponding modification in configuration space is, from Eq. (8.85),

$$\frac{Ze}{4\pi|\mathbf{x}|} \to \frac{Ze}{4\pi|\mathbf{x}|} + \frac{Ze\alpha}{15\pi m^2}\,\delta(\mathbf{x}). \tag{9.90b}$$

‡ A simple derivation of this result is given in J. J. Sakurai, *Advanced Quantum Mechanics*, Addison-Wesley, Reading, Mass., 1967, pp. 70–71.

§ A later more accurate calculation gives the value 16.640 Ry.

¶ This and the other results we quote, together with detailed references, are given in the books by Jauch and Rohrlich and by Itzykson and Zuber which we list in the footnote on p. 204.

Since the bound states of the hydrogen atom are extremely non-relativistic, we can use the right-hand side of Eq. (9.90b) with $Z = 1$ as an effective potential to calculate the level shifts of the hydrogen atom due to vacuum polarization. In first-order perturbation theory, these shifts are given by

$$\Delta E_{nl}(\text{vac. pol.}) = \frac{-e^2\alpha}{15\pi m^2}\,|\phi_{nl}(0)|^2 = -\frac{8\alpha^3}{15\pi n^3}\,\text{Ry}\,\delta_{l0}. \tag{9.91}$$

For the 2s level of hydrogen this gives -27 MHz, a result which was first obtained by Uehling in 1935. We see from Eq. (9.91) that vacuum polarization only shifts s levels, the shift being downwards. Qualitatively one can understand this in terms of the polarization by the external Coulomb field of the virtual electron–positron pairs, the electrons being attracted towards the nucleus, the positrons repelled. Thus the virtual electrons screen the nuclear charge. However, an s-state atomic electron will penetrate inside this screening and see the full nuclear charge, i.e. it will experience a more attractive potential, leading to stronger binding.

Because of the great importance of the Lamb shift as a test of QED (the results we quoted differ by about 6 MHz between theory and experiment), both calculations and measurements of the Lamb shift have been greatly refined. Calculations have taken into account higher-order radiative corrections and other small effects such as the finite size and finite mass of the nucleus and the use of relativistic wavefunctions for the bound states of the hydrogen atom. The current most complete calculations give 1057.916 ± 0.010 MHz (Erickson, 1971) and 1057.864 ± 0.014 MHz (Mohr, 1975). These values are to be compared with the most recent experimental data of 1057.893 ± 0.020 MHz (Lundeen and Pipkin, 1975) and 1057.862 ± 0.020 MHz (Andrews and Newton, 1976). As in the case of the anomalous magnetic moments of the electron and the muon, the precision of both experiment and theory, and their agreement, can only be described as stunning. Lamb shift measurements and calculations also exist for other levels in hydrogen, for deuterium and for the He^+ ion, and in all these cases theory and experiment are in good agreement.

9.7 THE INFRA-RED DIVERGENCE

In Sections 8.8 and 8.9 we studied the scattering of electrons by an external field. Because of the finite energy resolution in any experiment, the observed elastic scattering cross-section always includes some bremsstrahlung, i.e. inelastic scattering with emission of a soft photon. We saw that the latter contribution is infra-red divergent and asserted that this divergence is exactly cancelled by an infra-red divergence in the radiative correction of order α to the elastic scattering. We shall now demonstrate this exact cancellation.

The Feynman amplitude, including the lowest-order radiative corrections, for elastic electron scattering is given by Eq. (9.67). The O(α) correction in the cross-section comes from the interference term of the lowest-order amplitude [the first term in Eq. (9.67)] with the radiative correction terms [the second and third terms in Eq. (9.67)]. Π_c is infra-red finite, and the infra-red divergence arises from the Λ_c^μ term which we shall now study.

We regularize Eq. (9.48) for $\Lambda^\mu(p', p)$ by means of the replacement in Eq. (9.21), obtaining

$$e^2\Lambda^\mu(p', p) = \frac{-ie^2}{(2\pi)^4}\int \frac{d^4k}{k^2 - \lambda^2 + i\varepsilon} f(k) \times \left\{ \frac{\gamma^\alpha(\not{p}' - \not{k} + m)\gamma^\mu(\not{p} - \not{k} + m)\gamma_\alpha}{[(p' - k)^2 - m^2 + i\varepsilon][(p - k)^2 - m^2 + i\varepsilon]} \right\} \tag{9.92}$$

where

$$f(k) \equiv \frac{\lambda^2 - \Lambda^2}{k^2 - \Lambda^2 + i\varepsilon}. \tag{9.93}$$

Since we are now interested in the infra-red divergence when $k \to 0$, and not in the ultra-violet divergence ($k \to \infty$), we shall omit the cut-off factor $f(k)$. We shall similarly drop terms linear in k and k^2 in the numerator and denominator in the expression in curly brackets in Eq. (9.92). Using $p^2 = p'^2 = m^2$ and the Dirac equation, we can simplify Eq. (9.92) to give

$$e^2\bar{u}(\mathbf{p}')\Lambda^\mu(p', p)u(\mathbf{p}) = \frac{-ie^2}{(2\pi)^4}\bar{u}(\mathbf{p}')\gamma^\mu u(\mathbf{p})\left[\int \frac{d^4k}{k^2 - \lambda^2 + i\varepsilon}\frac{(p'p)}{(p'k)(pk)} + \cdots\right] \tag{9.94}$$

where, as throughout the following, the dots indicate terms which are finite in the limit $\lambda \to 0$ and which we therefore neglect. We evaluate the integral (9.94) by means of the identity

$$\frac{1}{k^2 - \lambda^2 + i\varepsilon} \equiv P\frac{1}{k^2 - \lambda^2} - i\pi\delta(k^2 - \lambda^2) = P\frac{1}{k^2 - \lambda^2} - \frac{i\pi}{2\omega_\lambda}[\delta(k^0 - \omega_\lambda) + \delta(k^0 + \omega_\lambda)] \tag{9.95}$$

where $\omega_\lambda \equiv (\lambda^2 + \mathbf{k}^2)^{1/2}$. Performing the k^0-integration in (9.94) and omitting the infra-red finite contribution from the principal value part of Eq. (9.95), one obtains

$$e^2\bar{u}(\mathbf{p}')\Lambda^\mu(p', p)u(\mathbf{p}) = e^2\bar{u}(\mathbf{p}')\gamma^\mu u(\mathbf{p})A(p', p) + \cdots \tag{9.96a}$$

where

$$A(p', p) \equiv \frac{-1}{2(2\pi)^3} \int \frac{d^3\mathbf{k}}{\omega_\lambda} \frac{(p'p)}{(p'k)(pk)}. \tag{9.96b}$$

In Eq. (9.67) we require the renormalized part of Eq. (9.96) which is given by Eq. (9.53), i.e.

$$e^2\bar{u}(\mathbf{p}')\Lambda^\mu_c(p', p)u(\mathbf{p}) = e^2\bar{u}(\mathbf{p}')[\Lambda^\mu(p', p) - L\gamma^\mu]u(\mathbf{p}). \tag{9.97}$$

From Eqs. (9.53), (9.54) and (9.96) we obtain

$$e^2\bar{u}(\mathbf{p})\Lambda^\mu(p, p)u(\mathbf{p}) = e^2 L\bar{u}(\mathbf{p})\gamma^\mu u(\mathbf{p}) = e^2\bar{u}(\mathbf{p})\gamma^\mu u(\mathbf{p})A(p, p) + \cdots$$

and a similar equation with p replaced by p', whence

$$L = A(p, p) + \cdots = A(p', p') + \cdots. \tag{9.98}$$

Combining Eqs. (9.96)–(9.98) leads to

$$\begin{aligned} &e^2\bar{u}(\mathbf{p}')\Lambda^\mu_c(p', p)u(\mathbf{p}) \\ &\quad = e^2\bar{u}(\mathbf{p}')\gamma^\mu u(\mathbf{p})\{A(p', p) - \tfrac{1}{2}A(p', p') - \tfrac{1}{2}A(p, p)\} + \cdots \\ &\quad = e^2\bar{u}(\mathbf{p}')\gamma^\mu u(\mathbf{p})\left\{\frac{1}{4(2\pi)^3}\int \frac{d^3\mathbf{k}}{\omega_\lambda}\left[\frac{p'}{p'k} - \frac{p}{pk}\right]^2\right\} + \cdots. \end{aligned}$$

Substituting this expression in Eq. (9.67), we obtain the Feynman amplitude for elastic electron scattering

$$\mathscr{M} = \mathscr{M}_0\left\{1 + \frac{e^2}{4(2\pi)^3}\int \frac{d^3\mathbf{k}}{\omega_\lambda}\left[\frac{p'}{p'k} - \frac{p}{pk}\right]^2\right\} + \cdots \tag{9.99}$$

where $\mathscr{M}_0$ is the lowest-order elastic scattering amplitude

$$\mathscr{M}_0 = \mathrm{i}e\bar{u}(\mathbf{p}')\not{A}_e(\mathbf{p}' - \mathbf{p})u(\mathbf{p}).$$

Hence the elastic scattering cross-section (8.91) becomes

$$\left(\frac{d\sigma}{d\Omega}\right)_{El} = \left(\frac{d\sigma}{d\Omega}\right)_0\left\{1 + \frac{\alpha}{(2\pi)^2}\int \frac{d^3\mathbf{k}}{\omega_\lambda}\left[\frac{p'}{p'k} - \frac{p}{pk}\right]^2\right\} + \cdots \tag{9.100}$$

where $(d\sigma/d\Omega)_0$ is the lowest-order elastic scattering cross-section.

The cross-section (9.100) is infra-red divergent in the limit as $\lambda \to 0$ and $\omega_\lambda \to |\mathbf{k}|$. On forming the experimentally observed cross-section (8.105), we see that this divergence in the elastic scattering is exactly cancelled by the infra-red divergence which occurs in the soft bremsstrahlung cross-section, Eqs. (8.106) and (8.110), in the limit $\lambda \to 0$. Hence the experimental quantity remains finite as the limit $\lambda \to 0$ is taken, as asserted in Section 8.9. As stated in that section, this conclusion holds in all orders of perturbation theory: the infra-red divergences which occur in the higher-order radiative corrections exactly cancel those in the inelastic processes involving emission of several soft photons.

9.8 HIGHER-ORDER RADIATIVE CORRECTIONS: RENORMALIZABILITY

So far we have considered radiative corrections of order α only. The renormalization procedure we have developed can be extended and leads to finite radiative corrections in all orders of perturbation theory. The proof of this renormalizability of QED is of fundamental importance but because of its complexity we confine ourselves to a qualitative discussion only.‡

When studying radiative corrections of order α, we saw that these arise in two ways: from the e^2 modifications to propagators and basic vertex parts of the lowest-order Feynman graphs, and from higher-order graphs which cannot be obtained in this way. (This was illustrated for Compton scattering in Figs. 9.5 and 9.6.) This situation persists for higher-order corrections, and we shall first of all consider the higher-order modifications of propagators and vertices, starting with the electron propagator.

The e^2 correction to the electron propagator results from the insertion of the electron self-energy loop $ie_0^2\Sigma(p)$, Fig. 9.17, in the bare electron

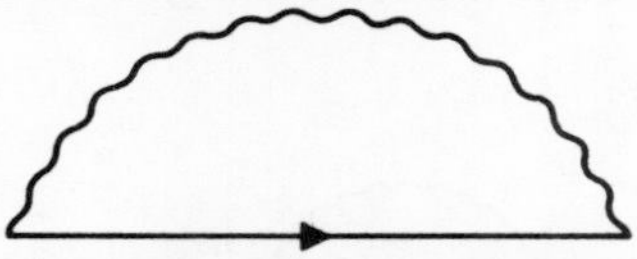

Fig. 9.17. The second-order electron self-energy insertion $ie_0^2\Sigma(p)$.

propagator, giving Figs. 9.9 and Eqs. (9.22). In higher orders, iterations of two, three or more such electron self-energy loops will occur. Their combined contributions produce the modification shown in Fig. 9.18 and given by

$$\frac{i}{\not{p} - m_0 + i\varepsilon} \to \frac{i}{\not{p} - m_0 + i\varepsilon} + \frac{i}{\not{p} - m_0 + i\varepsilon} ie_0^2\Sigma(p) \frac{i}{\not{p} - m_0 + i\varepsilon}$$

$$+ \frac{i}{\not{p} - m_0 + i\varepsilon} ie_0^2\Sigma(p) \frac{i}{\not{p} - m_0 + i\varepsilon} ie_0^2\Sigma(p) \frac{i}{\not{p} - m_0 + i\varepsilon} + \cdots \quad (9.101a)$$

$$= \frac{i}{\not{p} - m_0 + e_0^2\Sigma(p) + i\varepsilon}, \quad (9.101b)$$

where we used the identity (9.23) to obtain the last expression.

‡ For complete treatments, the reader is referred to the books listed in the footnote on p. 204 and to J. D. Bjorken and S. D. Drell, *Relativistic Quantum Fields*, McGraw-Hill, New York, 1965, Chapter 19.

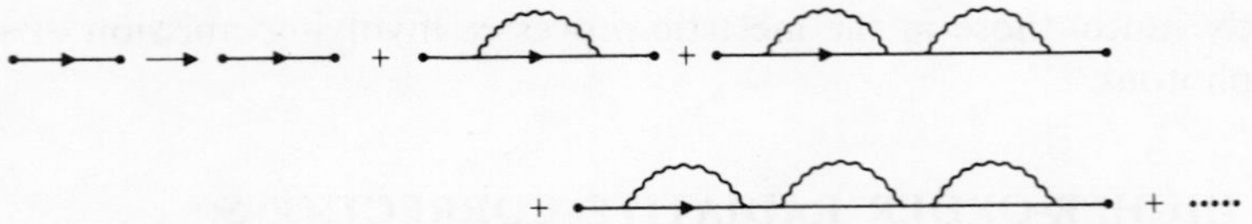

Fig. 9.18. The electron propagator obtained by including the electron self-energy insertion of order e^2 and its iterations.

There are of course many other electron self-energy insertions. To include them all, we define a *proper* Feynman graph as a graph which cannot be split into two graphs by cutting a single internal line. Figs. 9.19(a) and (b) show some proper and improper electron self-energy graphs. We now re-define $ie_0^2\Sigma(p)$ as the sum of all proper electron self-energy insertions, as indicated in Fig. 9.20. With this interpretation of $ie_0^2\Sigma(p)$, the infinite series of terms on the right-hand side of Eq. (9.101a) contains all electron self-energy graphs, so that the expression (9.101b) represents the complete (i.e. exact) electron propagator.

The expression (9.101b) for the complete electron propagator is of the same form as our earlier second-order result, Eq. (9.24), if in the latter equation we reinterpret $ie_0^2\Sigma(p)$ and omit the term $O(e_0^4)$. Hence, the discussion of mass

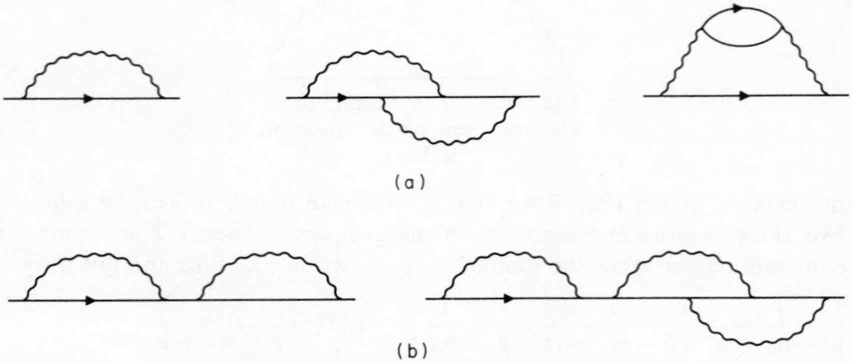

Fig. 9.19. Some electron self-energy graphs: (a) proper graphs; (b) improper graphs.

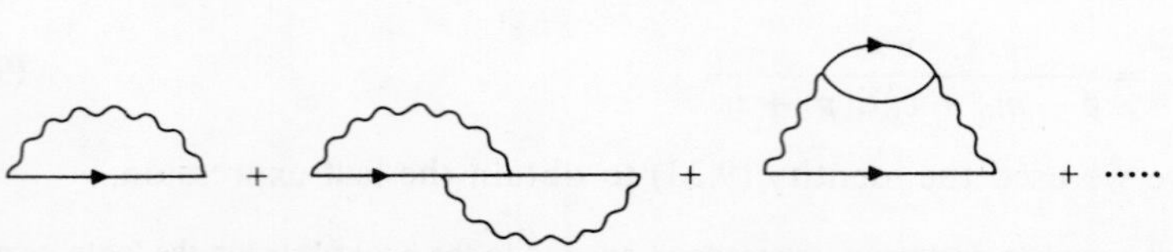

Fig. 9.20. The graphical representation of $ie_0^2\Sigma(p)$, redefined as the sum of all proper electron self-energy insertions.

renormalization, leading from Eq. (9.24) to Eq. (9.29), goes through exactly as before, giving

$$\frac{ie_0^2}{\not{p} - m_0 + i\varepsilon} \to \frac{ie_0^2}{(\not{p} - m)(1 + e_0^2 B) + e_0^2(\not{p} - m)\Sigma_c(p) + i\varepsilon}. \tag{9.102}$$

In Eq. (9.102) the bare and renormalized masses are again related by Eqs. (9.25), (9.27) and (9.28), which now hold to all orders, and B is again defined by Eq. (9.26). $ie_0^2\Sigma(p)$ in Eqs. (9.26) and (9.27) is the complete proper electron self-energy insertion, Fig. 9.20. Defining the renormalized charge e by

$$e^2 \equiv Z_2 e_0^2 = e_0^2/(1 + e_0^2 B), \tag{9.103}$$

we obtain from Eq. (9.102)

$$\frac{ie_0^2}{\not{p} - m_0 + i\varepsilon} \to \frac{ie^2}{(\not{p} - m) + e^2(\not{p} - m)\Sigma_c(p) + i\varepsilon} \tag{9.104a}$$

which holds to all orders in e^2. In lowest order, Eqs. (9.103) and (9.104a) reduce to our previous results, Eqs. (9.35) and (9.36), where $\Sigma_c(p)$ is now calculated from just the first graph of Fig. 9.20.

Alternatively, and equivalently, one can introduce mass counterterms as in Eqs. (9.31). The analysis goes through exactly as before, with the counter term (9.32) now defined in terms of the complete proper electron self-energy insertion. In this way one obtains, instead of Eq. (9.104a),

$$\frac{ie_0^2}{\not{p} - m + i\varepsilon} \to \frac{ie^2}{(\not{p} - m) + e^2(\not{p} - m)\Sigma_c(p) + i\varepsilon} \tag{9.104b}$$

in agreement with our earlier second-order results (9.34) and (9.35). We shall adopt this approach in what follows.

One can similarly deal with the photon propagator. We redefine $ie_0^2\Pi^{\mu\nu}(k)$ as the sum of all proper photon self-energy insertions, as indicated in Fig. 9.21. Iteration of this complete proper photon self-energy insertion leads, analogously to Eq. (9.101a), to the propagator modification

$$\frac{-ig_{\alpha\beta}}{k^2 + i\varepsilon} \to \frac{-ig_{\alpha\beta}}{k^2 + i\varepsilon} + \frac{-ig_{\alpha\mu}}{k^2 + i\varepsilon}\, ie_0^2\Pi^{\mu\nu}(k)\, \frac{-ig_{\nu\beta}}{k^2 + i\varepsilon}$$

$$+ \frac{-ig_{\alpha\mu}}{k^2 + i\varepsilon}\, ie_0^2\Pi^{\mu\nu}(k)\, \frac{-ig_{\nu\sigma}}{k^2 + i\varepsilon}\, ie_0^2\Pi^{\sigma\tau}(k)\, \frac{-ig_{\tau\beta}}{k^2 + i\varepsilon} + \cdots \tag{9.105a}$$

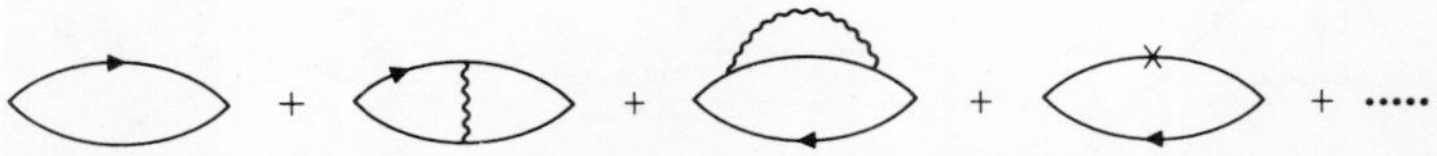

Fig. 9.21. The graphical representation of $ie_0^2\Pi^{\mu\nu}(k)$, redefined as the sum of all proper photon self-energy insertions.

which incorporates all photon self-energy terms, proper and improper. If we substitute Eq. (9.11) for $\Pi^{\mu\nu}(k)$ and omit terms proportional to $k^\mu k^\nu$, since the propagator is always coupled to conserved currents, we can sum the series in Eq. (9.105a) and obtain

$$\frac{-\mathrm{i}g_{\alpha\beta}}{k^2 + \mathrm{i}\varepsilon} \to \frac{-\mathrm{i}g_{\alpha\beta}}{k^2 + \mathrm{i}\varepsilon + e_0^2 A(k^2)}. \tag{9.105b}$$

This is just Eq. (9.13) which, with the redefinition of $\mathrm{i}e_0^2\Pi^{\mu\nu}(k)$, is now exact to all orders in e^2, rather than to $O(e_0^4)$ only. We again demand $A(0) = 0$, Eq. (9.14). If we multiply Eq. (9.105b) by e_0^2 and express e_0^2 in terms of the renormalized charge

$$e \equiv e_0 Z_3^{1/2} = e_0[1 + e_0^2 A'(0)]^{-1/2} \tag{9.106}$$

we finally obtain for the complete photon propagator

$$\frac{-\mathrm{i}g_{\alpha\beta}}{k^2 + \mathrm{i}\varepsilon} e_0^2 \to \frac{-\mathrm{i}g_{\alpha\beta}}{k^2 + \mathrm{i}\varepsilon + e^2\Pi_c(k^2)} e^2. \tag{9.107}$$

$A'(0)$ and $\Pi_c(k^2)$ are again defined in Eq. (9.15). In lowest order of perturbation theory, Eqs. (9.106) and (9.107) reduce to our previous results, Eqs. (9.18) and (9.19).

In Section 9.4 we considered external line renormalization in lowest order of perturbation theory. We found that for external lines the only self-energy effects are mass and charge renormalizations. This result can be shown to hold in all orders of perturbation theory.

Finally, we must consider the vertex function $\mathrm{i}\Gamma^\mu(p', p)$. The second-order treatment of Section 9.5 is easily generalized by including all proper vertex modifications and redefining $\mathrm{i}e_0^3\Lambda^\mu(p', p)$ as the sum of all such modifications, as indicated in Fig. 9.22. With this interpretation of $\mathrm{i}e_0^3\Lambda^\mu(p', p)$, the basic results, Eqs. (9.57) and (9.58), remain unchanged, except that the qualifying terms $[O(e_0^5)$ and $O(e^5)]$ are absent. Since the Ward identity (9.60) can be shown to hold to all orders, it follows that the relations $Z_1 = Z_2$ and $e = e_0 Z_3^{1/2}$, Eqs. (9.63) and (9.64), are also exact.

We have now generalized the relations between bare and physical masses and charges to all orders. Correspondingly, the modifications of the bare

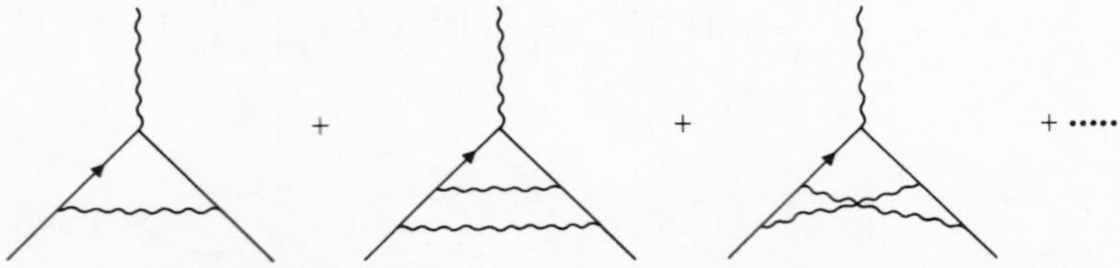

Fig. 9.22. The graphical representation of $\mathrm{i}e_0^3\Lambda^\mu(p', p)$, redefined as the sum of all proper vertex modifications.

propagators and the basic vertex part are given, to all orders, by

$$\frac{\mathrm{i}}{\not p - m + \mathrm{i}\varepsilon} \to \frac{\mathrm{i}}{(\not p - m) + e^2(\not p - m)\Sigma_c(p) + \mathrm{i}\varepsilon} \tag{9.108a}$$

$$\frac{-\mathrm{i}g_{\alpha\beta}}{k^2 + \mathrm{i}\varepsilon} \to \frac{-\mathrm{i}g_{\alpha\beta}}{k^2 + \mathrm{i}\varepsilon + e^2\Pi_c(k^2)} \tag{9.108b}$$

$$\mathrm{i}e_0\gamma^\mu \to \mathrm{i}e[\gamma^\mu + e^2\Lambda_c^\mu(p', p)]. \tag{9.108c}$$

Although the right-hand sides of these equations no longer depend explicitly on the bare charge e_0, they do depend implicitly on e_0 since Σ_c, Π_c and Λ_c^μ are expressed in terms of e_0. Dyson, Salam, Ward and others have shown that these quantities can be expressed in a consistent manner, order by order, in terms of the physical charge e. The right-hand sides of Eqs. (9.108) therefore represent the renormalized propagators and the renormalized vertex function.

Not all radiative corrections are due to self-energy and vertex modifications. To clarify the distinction, we define the process of *reduction* of a graph as removing self-energy and vertex modifications from it, i.e. as replacing them by bare propagators and basic vertex parts. Fig. 9.23 illustrates the process of reduction schematically. A graph from which all self-energy and vertex modifications have been removed, so that it cannot be reduced further, is called *irreducible* or alternatively a *skeleton graph.* For example, for Compton scattering the graphs in Figs. 9.4 and 9.6 are irreducible,

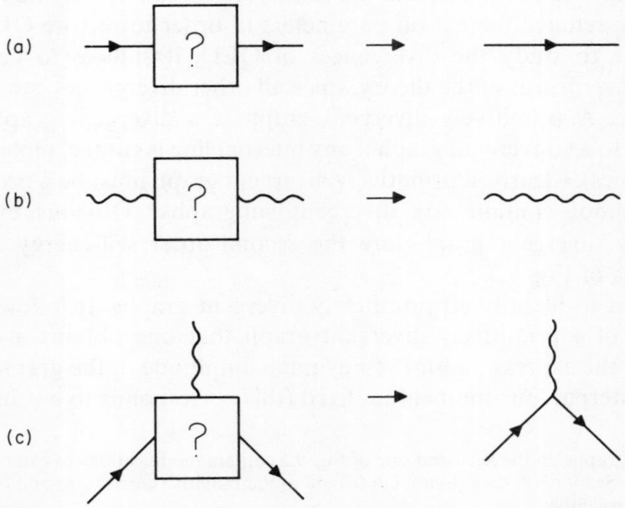

Fig. 9.23. Reduction of (a) an electron propagator; (b) a photon propagator; (c) a vertex part.

whereas the graphs in Fig. 9.5 are reducible to the same skeleton graph, Fig. 9.4(a).

To calculate the Feynman amplitude for an arbitrary process to order n, we combine the above results and proceed as follows. Firstly, we draw all skeleton diagrams which contribute to the process (i.e. have the correct external lines) and have not more than n vertices. From these skeleton graphs, all graphs contributing to the process up to nth order are obtained by replacing the bare propagators and the basic vertex parts by the renormalized propagators and vertex functions, Eqs. (9.108), expanding these equations up to the appropriate powers in e^2. Thus, to calculate the amplitude for Compton scattering to fourth order we require the second- and fourth-order skeleton graphs. These are shown in Figs. 9.4 and 9.6. (The triangle graphs 9.6(c) and (d) give zero contribution, as shown in Section 9.1, and should be omitted.) Substitution of Eqs. (9.108), expanded to $O(e^2)$, in the second-order skeleton graphs 9.4 generates the second-order amplitude as well as the contributions of all reducible fourth-order graphs; for example, those shown in Fig. 9.5.‡ The skeleton graphs in Fig. 9.6 are already of fourth order. Hence, no additional fourth-order graphs are generated by the replacements (9.108), and in the amplitudes for these skeleton graphs we merely replace the bare charge e_0 by the physical charge e.

We have now outlined a general method of calculating higher-order radiative corrections, expressed in terms of the mass and charge of the physical electron. In order to deal with well-defined finite quantities we had to regularize the theory through the introduction of suitable cut-off parameters. We must now consider whether the radiative corrections remain finite in the limit as we remove the cut-off parameters in order to restore QED.

In order to study the divergences of QED, it suffices to consider the *primitive divergences* of the theory, since all other divergences can be built up from these. A primitively divergent graph is a divergent graph which is converted to a convergent graph if any internal line is cut (i.e. replaced by two external lines). Clearly, a primitively divergent graph must be a proper graph, and it cannot contain any divergent subgraphs.§ Obvious examples of primitively divergent graphs are the second-order self-energy and vertex corrections of Fig. 9.3.

We want to identify all primitively divergent graphs. It follows from the definition of a primitively divergent graph that one obtains a convergent result if in the expression for the Feynman amplitude of the graph one keeps any one internal four-momentum fixed (this corresponds to cutting an inter-

‡ The two graphs on the left-hand side of Fig. 9.5 contain modifications to external lines only. As we saw in Section 9.4, they do not contribute to the radiative corrections and so need not be considered explicitly.

§ Any graph isolated from another graph G by cutting a finite number of internal lines is called a subgraph of G.

nal line) and integrates over all others. Hence the degree of divergence of this Feynman amplitude can be obtained by naive dimensional arguments, i.e. by counting powers of momentum variables of integration in the numerators and denominators.

Let the primitively divergent graph G have n vertices, $f_i(b_i)$ internal fermion (photon) lines and $f_e(b_e)$ external fermion (photon) lines. If d is the number of internal momenta not fixed by energy–momentum conservation at the vertices, then the dimensionality of the Feynman amplitude of G is

$$K = 4d - f_i - 2b_i. \tag{9.109}$$

There are n δ-functions associated with the vertices of G. One of these δ-functions ensures overall conservation of energy and momentum and only involves external momenta. Hence of the $(f_i + b_i)$ internal momenta, only

$$d = f_i + b_i - (n - 1) \tag{9.110}$$

are independent variables. We also have the relations

$$2n = f_e + 2f_i, \qquad n = b_e + 2b_i. \tag{9.111}$$

Combining Eqs. (9.109)–(9.111), we obtain

$$K = 4 - \tfrac{3}{2}f_e - b_e \geqslant 0 \tag{9.112}$$

as necessary condition for G to be a primitively divergent graph. $K = 0, 1, \ldots$ means at most a logarithmic, linear, ... divergence.

Eq. (9.112) is a most remarkable result because it depends only on the number of external fermion and photon lines of the graph and is independent of its internal structure. Furthermore it provides the vital information that the only graphs which may possibly be primitively divergent are those with $(f_e, b_e) = (0, 2), (0, 3), (0, 4), (2. 0)$ and $(2, 1)$, and that the divergences are at most quadratic. Two of these five types of graphs are in fact convergent. We met the simplest example of the type $(f_e, b_e) = (0, 3)$ in Section 9.1, Fig. 9.6, where we saw that the triangle graphs occur in pairs which exactly cancel. Since this result generalizes to all higher-order graphs of this type, one can omit such graphs altogether. Secondly, graphs with $(f_e, b_e) = (0, 4)$ do not lead to divergences. These graphs describe the scattering of light by light. The simplest Feynman graph for this process is shown in Fig. 9.24. From Eq. (9.112) such graphs could be logarithmically divergent. One can show that as a consequence of gauge invariance they are strongly convergent.

The remaining three types of graphs can be primitively divergent. They are just the electron and photon self-energy graphs and the vertex modifications. The only primitively divergent self-energy graphs are the second-order corrections, Figs. 9.3(a) and (b). The second-order vertex modification, Fig. 9.3(c), is also primitively divergent, but there exist infinitely many primitively

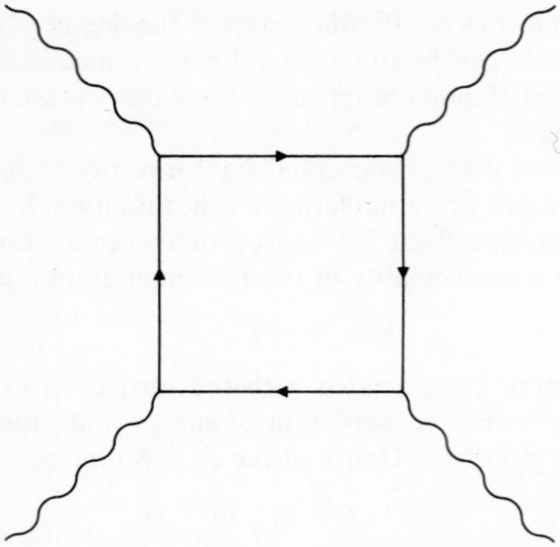

Fig. 9.24. The simplest light–light scattering diagram.

divergent higher-order vertex modifications; for example, the third graph on the right-hand side of Fig. 9.22 is primitively divergent.

This exhausts the enumeration of primitively divergent graphs. In particular, the irreducible diagrams of any physical process ($f_e + b_e \geqslant 4$) are finite, and infinities can only arise through insertions of self-energy and vertex modifications in these graphs. Hence if the residual modifications Σ_c, Π_c and Λ_c^μ, which result from renormalization and which occur in Eqs. (9.108), remain finite to all orders, then inserting them into the irreducible graphs cannot lead to divergences. The same approach which we used to analyse the primitively divergent second-order corrections (essentially a Taylor series expansion of the convergent integrands in the integrals representing the Feynman amplitudes) can be used to show that all primitively divergent contributions to Σ_c, Π_c and Λ_c^μ remain finite as the cut-off parameters are removed. This result can be extended to all proper self-energy and vertex modifications. Consequently, the predictions of QED, expressed in terms of the physical mass and charge of the electron, remain finite as all cut-off parameters are removed.

A field theory is called *renormalizable* if its predictions in terms of a finite number of parameters (i.e. masses and coupling constants) remain finite when all cut-offs are removed. QED is an example of a renormalizable theory. In such a theory, where the results are well-defined and finite in the limit as the momentum cut-off Λ tends to infinity, the results are insensitive to the form of the cut-off, provided only that Λ is much greater than the momentum scale of the process under consideration. In other words, for QED the discussion following Eqs. (9.65) goes through essentially unchanged to all orders, and theoretical predictions obtained with a finite Λ are not measurably different from those obtained in the limit $\Lambda \to \infty$, provided Λ is very much greater than about 150 GeV. In contrast, a theory which is not renormalizable can still be made well-defined and finite by the introduction of suitable cut-off parameters Λ. However, in such a non-renormalizable theory the physical predictions diverge in the limit $\Lambda \to \infty$ and hence are inevitably sensitive to the form and magnitude of the cut-offs, even for very large Λ.

PROBLEMS

9.1 In Section 9.1 we used general arguments to show that the Feynman amplitudes of the two triangle graphs, Figs. 9.6(c) and (d), differ only in sign and exactly cancel each other. Derive this result from the explicit forms of these amplitudes. [*Note:* This proof does not require the evaluation of the traces to which these diagrams give rise, but only the use of the general properties of traces (see Section A.3 in Appendix A) to relate the two expressions to each other.]

9.2 From $[S_F(p)]^{-1} = \not{p} - m$ and $[S_F(p)][S_F(p)]^{-1} = 1$, derive

$$\partial S_F(p)/\partial p_\mu = -S_F(p)\gamma^\mu S_F(p).$$

Hence, derive the Ward identity

$$\frac{\partial \Sigma(p)}{\partial p_\mu} = \Lambda^\mu(p, p). \tag{9.60}$$

CHAPTER 10

Regularization‡

In the last chapter we saw that the calculations of radiative corrections in QED lead to divergent loop integrals. These divergences are removed by regularization, i.e. suitable modification of these integrals. After renormalization, those integrals which enter physical predictions remain finite when the regularization is removed, i.e. when the original theory is restored. There exist several regularization formalisms, and the regularized integrals depend on the formalism employed. However, in the limit in which the original theory is restored, the physical predictions become independent of the method of regularization used. At any rate, whenever different methods have been used, they have led to the same results. In this chapter we shall consider the explicit evaluation of single loop integrals using two regularization procedures.

Historically the oldest procedure is the cut-off method which we discussed and used for illustration in Chapter 9. It has the advantage of relating the divergences to the short-distance and high-energy behaviour of the theory. To illustrate this method, we shall in Section 10.2 use it to calculate the electron mass shift δm. The cut-off method is difficult to apply in all but the simplest cases. In particular, using this method makes it difficult to ensure gauge invariance and the validity of the Ward identity to all orders of perturbation theory. In order to do so, one must adopt a clumsy and complicated cut-off procedure, such as the Pauli–Villars method.§

‡ This chapter deals with the technical details of regularization procedures. They will not be required elsewhere in this book, and readers not interested in them may omit this chapter.

§ The Pauli–Villars formalism is discussed in, for example, J. M. Jauch and F. Rohrlich, *The Theory of Photons and Electrons*, 2nd edn, Springer, New York, 1976, Section 10-9, and in C. Itzykson and J. B. Zuber, *Quantum Field Theory*, McGraw-Hill, New York, 1980, Sections 7-1-1 and 8-4-2.

More recently, an alternative method, known as dimensional regularization, has been developed. It is simpler to apply than the cut-off method, and it has the great advantage that it automatically ensures gauge invariance, and relatedly the validity of the Ward identity, to all orders of perturbation theory. Dimensional regularization is therefore of particular importance for QED and, even more so, for non-Abelian gauge theories, such as quantum chromodynamics and the Weinberg–Salam unified theory of weak and electromagnetic interactions. For non-Abelian gauge theories, gauge invariance implies several Ward identities. These are so restrictive that it is very difficult to satisfy them when employing a cut-off regularization procedure. In contrast, dimensional regularization automatically respects these identities and gauge invariance to all orders of perturbation theory, and for non-Abelian gauge theories this formalism has been almost exclusively used. In particular, the crucial proof that the Weinberg–Salam theory is renormalizable has only been carried out using dimensional regularization.

The formalism of dimensional regularization will be developed in Section 10.3. In the two succeeding sections it will be used to evaluate the vacuum polarization correction and the anomalous magnetic moment of the electron. Our development will again be restricted to the lowest order of perturbation theory, i.e. to single loop integrals, but the same methods can be extended to higher orders.‡

The finite loop integrals which result after regularization can be evaluated by reducing them to certain standard forms using tricks invented for this purpose by Feynman. We shall obtain these results in Section 10.1.

10.1 MATHEMATICAL PRELIMINARIES

10.1.1 Some standard integrals

We first list the standard integrals most frequently met in the evaluation of loop integrals. Below we comment on their derivation.

$$\int \frac{\mathrm{d}^4 k}{(k^2 + s + \mathrm{i}\varepsilon)^n} = \mathrm{i}\pi^2 \frac{\Gamma(n-2)}{\Gamma(n)} \frac{1}{s^{n-2}}, \qquad n \geqslant 3, \tag{10.1}$$

$$\int \mathrm{d}^4 k \frac{k^\mu}{(k^2 + s + \mathrm{i}\varepsilon)^n} = 0, \qquad n \geqslant 3, \tag{10.2}$$

$$\int \mathrm{d}^4 k \frac{k^\mu k^\nu}{(k^2 + s + \mathrm{i}\varepsilon)^n} = \mathrm{i}\pi^2 \frac{\Gamma(n-3)}{2\Gamma(n)} \frac{g^{\mu\nu}}{s^{n-3}}, \qquad n \geqslant 4, \tag{10.3}$$

‡ For a full discussion of the technique of dimensional regularization, its applications and references, we refer the reader to G. Leibbrandt, *Rev. Mod. Phys.* **47** (1975) 849, G. 't Hooft and M. Veltman, *Nucl. Phys.* **B44** (1972) 189, and C. Nash, *Relativistic Quantum Fields*, Academic Press, London, 1978.

$$\int \frac{d^4p}{(p^2 + 2pq + t + i\varepsilon)^n} = i\pi^2 \frac{\Gamma(n-2)}{\Gamma(n)} \frac{1}{(t - q^2)^{n-2}}, \qquad n \geqslant 3, \tag{10.4}$$

$$\int d^4p \frac{p^\mu}{(p^2 + 2pq + t + i\varepsilon)^n} = -i\pi^2 \frac{\Gamma(n-2)}{\Gamma(n)} \frac{q^\mu}{(t - q^2)^{n-2}}, \qquad n \geqslant 3, \tag{10.5}$$

$$\int d^4p \frac{p^\mu p^\nu}{(p^2 + 2pq + t + i\varepsilon)^n}$$
$$= i\pi^2 \frac{\Gamma(n-3)}{2\Gamma(n)} \frac{[2(n-3)q^\mu q^\nu + (t - q^2)g^{\mu\nu}]}{(t - q^2)^{n-2}}, \qquad n \geqslant 4. \tag{10.6}$$

In the right-hand expressions of Eqs. (10.1)–(10.6) we have put $\varepsilon = 0$ which is usually permissible. If subsequently this leads to ambiguities, then we must retain $(s + i\varepsilon)$ and $(t + i\varepsilon)$ in place of s and t in these expressions. In what follows, we shall usually anticipate the limit $\varepsilon \to 0$ in this way.

The formula (10.1) for the case $n = 3$ is obtained by performing the k^0 integration as a contour integral and the subsequent integration with respect to **k** using spherical polar coordinates.‡ The general result for $n \geqslant 3$ follows by repeated differentiation with respect to s. Eq. (10.2) is obvious from symmetry. Eqs. (10.4) and (10.5) follow from Eqs. (10.1) and (10.2) respectively by changing variables from k and s to

$$p = k - q, \qquad t = q^2 + s. \tag{10.7}$$

Differentiating Eq. (10.5) with respect to q_ν leads to Eq. (10.6), and taking $q = 0$ in Eq. (10.6) gives Eq. (10.3).

Other integrals involving more complicated tensors in the numerator of the integrand are easily obtained from the above formulas by differentiation and changing variables, but the above results suffice for most purposes.

10.1.2 Feynman parameterization

The integrals (10.1)–(10.6) contain a single quadratic factor, raised to the power n, in the denominators, whereas usually one deals with integrals containing a product of several different quadratic factors in the denominator. These more general integrals are reduced to the desired form by means of an ingenious technique due to Feynman.

For a product of two quadratic factors a and b, one starts from the identity

$$\frac{1}{ab} = \frac{1}{b - a} \int_a^b \frac{dt}{t^2}. \tag{10.8}$$

‡ This derivation is given in J. J. Sakurai, *Advanced Quantum Mechanics*, Addison-Wesley, 1967, p. 315.

Defining the *Feynman parameter* z by

$$t = b + (a - b)z, \tag{10.9}$$

Eq. (10.8) can be written

$$\frac{1}{ab} = \int_0^1 \frac{dz}{[b + (a - b)z]^2}. \tag{10.10}$$

We see that by introducing the Feynman parameter z we have expressed $1/ab$ in terms of a single factor raised to the power 2. Although at this stage the integral in Eq. (10.10) may look like a complication, we shall see that Feynman parameterization allows us to evaluate all integrals straightforwardly.

The above method easily extends. For three factors, the alternative results

$$\frac{1}{abc} = 2\int_0^1 dx \int_0^x dy \frac{1}{[a + (b - a)x + (c - b)y]^3} \tag{10.11a}$$

$$= 2\int_0^1 dx \int_0^{1-x} dz \frac{1}{[a + (b - a)x + (c - a)z]^3} \tag{10.11b}$$

are proved by integrating with respect to y and z respectively and using Eq. (10.10). Eq. (10.11a) generalizes to an arbitrary number of factors, and the result

$$\frac{1}{a_0 a_1 a_2 \ldots a_n} = \Gamma(n + 1) \int_0^1 dz_1 \int_0^{z_1} dz_2 \ldots \int_0^{z_{n-1}} dz_n \times \frac{1}{[a_0 + (a_1 - a_0)z_1 + \cdots (a_n - a_{n-1})z_n]^{n+1}} \tag{10.12}$$

is established by induction.

Other useful results are obtained by differentiation with respect to one or more parameters. For example, differentiating Eq. (10.10) with respect to a gives

$$\frac{1}{a^2 b} = 2\int_0^1 dz \frac{z}{[b + (a - b)z]^3}. \tag{10.13}$$

Finally, we note that the modified photon propagator (9.21) can be written in the often useful form

$$\frac{1}{k^2 - \lambda^2 + i\varepsilon} - \frac{1}{k^2 - \Lambda^2 + i\varepsilon} = -\int_{\lambda^2}^{\Lambda^2} \frac{dt}{(k^2 - t + i\varepsilon)^2}, \tag{10.14}$$

which is the identity (10.8) again.

10.2 CUT-OFF REGULARIZATION: THE ELECTRON MASS SHIFT

As an illustration of the cut-off method of regularization, we shall calculate the electron self-energy mass shift δm in second-order perturbation theory.

From Eqs. (9.27), (9.28) and (9.4), δm is given by

$$\delta m = \mathrm{i}\bar{u}(\mathbf{p})\left\{\frac{-e^2}{(2\pi)^4}\int \mathrm{d}^4k\,\frac{\gamma^\alpha(\not p - \not k + m)\gamma_\alpha}{(p-k)^2 - m^2 + \mathrm{i}\varepsilon}\right.$$
$$\left.\times\left[\frac{1}{k^2 - \lambda^2 + \mathrm{i}\varepsilon} - \frac{1}{k^2 - \Lambda^2 + \mathrm{i}\varepsilon}\right]\right\} u(\mathbf{p}) \qquad (10.15)$$

where we have replaced the photon propagator in Eq. (9.4) by the modification (9.21) to avoid any difficulties which may arise from infra-red divergences. Eq. (10.15) is simplified by using the contraction identities (A.14b) and by setting $\not p u(\mathbf{p}) = mu(\mathbf{p})$ and $p^2 = m^2$. If we also substitute Eq. (10.14), we obtain

$$\delta m = \frac{\mathrm{i}e^2}{(2\pi)^4}\,\bar{u}(\mathbf{p})\left[\int \mathrm{d}^4\mathbf{k}\,\frac{2(\not k + m)}{k^2 - 2pk + \mathrm{i}\varepsilon}\int_{\lambda^2}^{\Lambda^2}\frac{\mathrm{d}t}{(k^2 - t + \mathrm{i}\varepsilon)^2}\right]u(\mathbf{p}), \qquad (10.16)$$

and applying Eq. (10.13) gives

$$\delta m = \frac{\mathrm{i}e^2}{(2\pi)^4}\,\bar{u}(\mathbf{p})\left[\int_{\lambda^2}^{\Lambda^2}\mathrm{d}t\int_0^1\mathrm{d}z\int \mathrm{d}^4k\right.$$
$$\left.\times\frac{4(\not k + m)z}{[k^2 - 2pk(1-z) - tz + \mathrm{i}\varepsilon]^3}\right]u(\mathbf{p}). \qquad (10.17)$$

The integral with respect to k in Eq. (10.17) is obtained from Eqs. (10.4) and (10.5), leading to

$$\delta m = \frac{me^2}{8\pi^2}\int_0^1\mathrm{d}z\int_{\lambda^2}^{\Lambda^2}\mathrm{d}t\,\frac{2z - z^2}{tz + m^2(1-z)^2}$$
$$= \frac{m\alpha}{2\pi}\int_0^1\mathrm{d}z(2-z)\ln\frac{\Lambda^2 z + m^2(1-z)^2}{\lambda^2 z + m^2(1-z)^2}. \qquad (10.18)$$

This expression remains infra-red finite in the limit $\lambda \to 0$, and we can therefore take $\lambda = 0$ in Eq. (10.18). As $\Lambda \to \infty$, the integral diverges logarithmically with the leading term given by

$$\delta m = \frac{m\alpha}{2\pi}\ln\frac{\Lambda^2}{m^2}\int_0^1\mathrm{d}z(2-z) + \mathrm{O}(1)$$
$$= \frac{3m\alpha}{2\pi}\ln\frac{\Lambda}{m} + 0(1). \qquad (10.19)$$

This is the result quoted in Section 9.3, Eqs. (9.37) and (9.28).

10.3 DIMENSIONAL REGULARIZATION

10.3.1 Introduction

The divergent loop integrals of field theory are four-dimensional integrals in energy–momentum space. Dimensional regularization consists in modifying the dimensionality of these integrals so that they become finite. In the first place, we generalize from a four-dimensional to a D-dimensional space, where D is a positive integer. The metric tensor $g^{\alpha\beta} = g_{\alpha\beta}$ of this space is defined by

$$\left.\begin{aligned} g^{00} = -g^{ii} = 1, \qquad & i = 1, 2, \ldots, D-1, \\ g^{\alpha\beta} = 0, \qquad & \alpha \neq \beta. \end{aligned}\right\} \tag{10.20}$$

Correspondingly, a four-vector k^α is replaced by a vector with D components

$$k^\alpha \equiv (k^0, k^1, \ldots, k^{D-1}), \tag{10.21}$$

and

$$k^2 = k_\alpha k^\alpha = (k^0)^2 - \sum_{i=1}^{D-1} (k^i)^2. \tag{10.22}$$

Loop integrals now become integrals in D dimensions with the volume element $\mathrm{d}^D k = \mathrm{d}k^0\, \mathrm{d}k^1 \ldots \mathrm{d}k^{D-1}$. For example, Eq. (10.1) generalizes to

$$\int \frac{\mathrm{d}^D k}{(k^2 + s + \mathrm{i}\varepsilon)^n} = \mathrm{i}\pi^{D/2} \frac{\Gamma(n - \frac{1}{2}D)}{\Gamma(n)} \frac{1}{s^{n-D/2}} \tag{10.23}$$

for integer values of $n > D/2$.‡ For $n = D/2$ (e.g. for $n = 2$ when $D = 4$), the left-hand side of Eq. (10.23) is logarithmically divergent, and the right-hand side is also singular due to the pole of $\Gamma(z)$ at $z = 0$. However, for non-integer values of D, the right-hand side of Eq. (10.23) is perfectly well-defined and finite. We can therefore use it to *define* a generalization of the integral on the left-hand side of Eq. (10.23) to D dimensions for non-integer values of D. In particular, we shall take $D = 4 - \eta$ where η is a small positive parameter. Restoring ordinary four-dimensional space (and, for example, QED) corresponds to the limit $\eta \to 0$.

Before going on to QED, we shall illustrate these ideas by a simple but unrealistic example. Suppose we were dealing with the divergent loop integral

$$\Pi(s) = \int \frac{\mathrm{d}^4 k}{(k^2 + s + \mathrm{i}\varepsilon)^2}. \tag{10.24}$$

In the cut-off method, we would multiply the integrand by, for example, $(-\Lambda^2)/(k^2 - \Lambda^2)$ and evaluate the resulting integral

$$\Pi_\Lambda(s) = \int \frac{\mathrm{d}^4 k}{(k^2 + s + \mathrm{i}\varepsilon)^2} \frac{-\Lambda^2}{k^2 - \Lambda^2} \tag{10.25}$$

‡ For the method of evaluation of such integrals, see the paper by 't Hooft and Veltman, quoted in the preamble to this chapter.

using the methods of Section 10.1. In the limit as $\Lambda \to \infty$, $\Pi_\Lambda(s)$ is of course logarithmically divergent. However, after renormalization, one would deal with a difference

$$\Pi_\Lambda(s) - \Pi_\Lambda(s_0)$$

and this has the well-defined finite limit

$$\lim_{\Lambda \to \infty} \{\Pi_\Lambda(s) - \Pi_\Lambda(s_0)\} = -\mathrm{i}\pi^2 \ln (s/s_0). \tag{10.26}$$

In the dimensional method, regularization of $\Pi(s)$ is achieved by using Eq. (10.23) to define

$$\Pi_\eta(s) = \int \frac{\mathrm{d}^{4-\eta}k}{(k^2 + s + \mathrm{i}\varepsilon)^2} = \mathrm{i}\pi^{2-\eta/2} \frac{\Gamma(\frac{1}{2}\eta)}{\Gamma(2)} s^{-\eta/2}. \tag{10.27}$$

For $\eta \to 0$, one has

$$s^{-\eta/2} = 1 - \tfrac{1}{2}\eta \ln s + \ldots \tag{10.28}$$

and

$$\Gamma\left(\frac{\eta}{2}\right) = \frac{2}{\eta} - \gamma + \ldots \tag{10.29}$$

where $\gamma = 0.5772\ldots$ is Euler's constant.[‡] Hence, one obtains

$$\Pi_\eta(s) = \mathrm{i}\frac{2\pi^2}{\eta} - \mathrm{i}\pi^2(\gamma + \ln s), \tag{10.30}$$

whence

$$\lim_{\eta \to 0} \{\Pi_\eta(s) - \Pi_\eta(s_0)\} = -\mathrm{i}\pi^2 \ln (s/s_0), \tag{10.31}$$

in agreement with the result (10.26) derived by the cut-off method.

10.3.2 General results

In order to apply dimensional regularization to QED, we must extend the above ideas in two ways. Firstly, we require other D-dimensional integrals, in addition to Eq. (10.23). Secondly, we must generalize expressions involving γ-matrices.

The relevant integrals are derived from Eq. (10.23) in much the same way in which the standard integrals (10.2)–(10.6) follow from Eq. (10.1). The only

‡ See, for example, M. Abramowitz and I. A. Stegun, *Handbook of Mathematical Functions*, Dover, New York, 1972, p. 255.

integrals we shall require, in addition to Eq. (10.23), are

$$\int d^D k \frac{k^\mu}{(k^2 + s + i\varepsilon)^n} = 0, \tag{10.32}$$

$$\int d^D k \frac{k^\mu k^\nu}{(k^2 + s + i\varepsilon)^n} = i\pi^{D/2} \frac{\Gamma(n - \frac{1}{2}D - 1)}{2\Gamma(n)} \frac{g^{\mu\nu}}{s^{n-D/2-1}}, \tag{10.33}$$

$$\int d^D k \frac{k^2}{(k^2 + s + i\varepsilon)^n} = i\pi^{D/2} \frac{\Gamma(n - \frac{1}{2}D - 1)}{2\Gamma(n)} \frac{D}{s^{n-D/2-1}}, \tag{10.34}$$

where Eq. (10.34) follows from Eq. (10.33), since

$$g_{\mu\nu} g^{\mu\nu} = D. \tag{10.35}$$

As regards the meaning of Eqs. (10.32)–(10.34), they are, in the first place, derived for integer values of D. For non-integer values, the integrals are defined by the expressions on the right-hand sides of these equations. We shall again write $D = 4 - \eta$ and shall require the limit $\eta \to 0$, i.e. $D \to 4$. The reader may feel some unease as to the meaning of $g^{\mu\nu}$ in the right-hand-side expression of Eq. (10.33) when D is not an integer. However, the singularity of this expression, in the limit $\eta \to 0$, arises from the factor $\Gamma(n - D/2 - 1)$, whereas $g^{\mu\nu}$ is non-singular in this limit. Hence only the value of $g^{\mu\nu}$ for $D = 4$ enters the final results, and only this value will be required.

We must next see how to handle expressions involving γ-matrices. In the first place, we again consider general integer values of D and introduce a set of γ-matrices $\gamma^0, \gamma^1, \ldots, \gamma^{D-1}$, which satisfy the usual anticommutation relations

$$\gamma^\mu \gamma^\nu + \gamma^\nu \gamma^\mu = 2g^{\mu\nu}. \tag{10.36}$$

From these one derives contraction and trace relations, analogous to Eqs. (A.14)–(A.18). If the γ-matrices are $f(D) \times f(D)$ matrices, and I is the $f(D) \times f(D)$ unit matrix [i.e. $f(D = 4) = 4$], one obtains the contraction identities

$$\left.\begin{aligned} \gamma_\lambda \gamma^\lambda &= DI \\ \gamma_\lambda \gamma^\alpha \gamma^\lambda &= -(D-2)\gamma^\alpha \\ \gamma_\lambda \gamma^\alpha \gamma^\beta \gamma^\lambda &= (D-4)\gamma^\alpha \gamma^\beta + 4g^{\alpha\beta} \end{aligned}\right\}, \tag{10.37}$$

etc., and the trace relations

$$\left.\begin{aligned} \operatorname{Tr}(\gamma^\alpha \gamma^\beta) &= f(D) g^{\alpha\beta} \\ \operatorname{Tr}(\gamma^\alpha \gamma^\beta \gamma^\gamma \gamma^\delta) &= f(D)[g^{\alpha\beta} g^{\gamma\delta} - g^{\alpha\gamma} g^{\beta\delta} + g^{\alpha\delta} g^{\beta\gamma}] \\ \operatorname{Tr}(\gamma^\alpha \gamma^\beta \ldots \gamma^\mu \gamma^\nu) &= 0 \end{aligned}\right\}, \tag{10.38}$$

where in the last relation $(\gamma^\alpha \gamma^\beta \ldots \gamma^\mu \gamma^\nu)$ contains an odd number of γ-matrices.

We shall now take over these γ-matrix relations uncritically to the case of

$D = 4 - \eta$ dimensions, where η is a small positive number, even though the meaning and existence of γ-matrices in a non-integer number of dimensions is far from clear. In the limit $D \to 4$, the usual relations are restored, and since these relations (unlike the integrals) are non-singular in this limit, only their behaviour at $D = 4$ enters the final results of any calculation, as we shall see explicitly in the next two sections.

10.4 VACUUM POLARIZATION

We shall now use dimensional regularization to derive the vacuum polarization expression

$$e^2\Pi_c(k^2) = -\frac{2\alpha}{\pi}\int_0^1 dz z(1-z)\ln\left[1 - \frac{k^2z(1-z)}{m^2}\right] \tag{9.68}$$

which we only quoted in Chapter 9.

We take as starting point Eq. (9.8) for the photon self-energy loop which, after dimensional regularization, reads

$$ie^2\Pi^{\mu\nu}(k) = \frac{-e^2}{(2\pi)^4}\int d^Dp \frac{N^{\mu\nu}(p,k)}{[(p+k)^2 - m^2 + i\varepsilon][p^2 - m^2 + i\varepsilon]} \tag{10.39}$$

where

$$N^{\mu\nu}(p,k) \equiv \text{Tr}\,[\gamma^\mu(\not{p} + \not{k} + m)\gamma^\nu(\not{p} + m)]. \tag{10.40}$$

Evaluating the trace by means of Eqs. (10.38) gives

$$N^{\mu\nu}(p,k) = f(D)\{(p^\mu + k^\mu)p^\nu + (p^\nu + k^\nu)p^\mu + [m^2 - p(p+k)]g^{\mu\nu}\}. \tag{10.41}$$

After Feynman parameterization and use of Eq. (10.10), we can write Eq. (10.39)

$$ie^2\Pi^{\mu\nu}(k) = \frac{-e^2}{(2\pi)^4}\int_0^1 dz \int d^Dp \frac{N^{\mu\nu}(p,k)}{[p^2 - m^2 + (k^2 + 2pk)z + i\varepsilon]^2}. \tag{10.42}$$

If we introduce the new variable

$$q = p + kz, \tag{10.43}$$

Eq. (10.42) becomes

$$ie^2\Pi^{\mu\nu}(k) = \frac{-e^2}{(2\pi)^4}\int_0^1 dz \int d^Dq \frac{N^{\mu\nu}(q - kz, k)}{[q^2 + k^2z(1-z) - m^2 + i\varepsilon]^2} \tag{10.44}$$

with

$$N^{\mu\nu}(q - kz, k) = f(D)\{[2q^\mu q^\nu - q^2 g^{\mu\nu}] + [m^2 - k^2z(1-z)]g^{\mu\nu} + [-2z(1-z)(k^\mu k^\nu - k^2 g^{\mu\nu})] + \ldots\} \tag{10.45}$$

where the dots indicate that terms linear in q have been omitted, since these terms vanish on integration, by Eq. (10.32). Combining the last two equations, we obtain

$$\mathrm{i}e^2\Pi^{\mu\nu}(k) = \frac{-e^2}{(2\pi)^4} f(D) \int_0^1 \mathrm{d}z \sum_{i=1}^{3} I_i^{\mu\nu}(k, z), \tag{10.46}$$

where, using Eqs. (10.23) and (10.32)–(10.34), we have

$$I_1^{\mu\nu}(k, z) \equiv \int \mathrm{d}^D q \frac{[2q^\mu q^\nu - q^2 g^{\mu\nu}]}{[q^2 + k^2 z(1-z) - m^2 + \mathrm{i}\varepsilon]^2}$$

$$= \frac{\mathrm{i}g^{\mu\nu}\pi^{D/2}\Gamma(1-\frac{1}{2}D)}{[k^2 z(1-z) - m^2]^{1-D/2}}(1 - \tfrac{1}{2}D), \tag{10.47a}$$

$$I_2^{\mu\nu}(k, z) \equiv [m^2 - k^2 z(1-z)]g^{\mu\nu} \int \mathrm{d}^D q \frac{1}{[q^2 + k^2 z(1-z) - m^2 + \mathrm{i}\varepsilon]^2}$$

$$= [m^2 - k^2 z(1-z)]g^{\mu\nu} \frac{\mathrm{i}\pi^{D/2}\Gamma(2-\frac{1}{2}D)}{[k^2 z(1-z) - m^2]^{2-D/2}}$$

$$= -I_1^{\mu\nu}(k, z), \tag{10.47b}$$

and

$$I_3^{\mu\nu}(k, z) \equiv [-2z(1-z)(k^\mu k^\nu - k^2 g^{\mu\nu})] \int \mathrm{d}^D q \frac{1}{[q^2 + k^2 z(1-z) - m^2 + \mathrm{i}\varepsilon]^2}$$

$$= -2z(1-z)(k^\mu k^\nu - k^2 g^{\mu\nu}) \frac{\mathrm{i}\pi^{D/2}\Gamma(2-\frac{1}{2}D)}{[k^2 z(1-z) - m^2]^{2-D/2}}. \tag{10.47c}$$

Substituting Eqs. (10.47a)–(10.47c) into Eq. (10.46) gives

$$\Pi^{\mu\nu}(k) = (k^\mu k^\nu - k^2 g^{\mu\nu})\Pi(k^2) \tag{10.48}$$

where

$$\Pi(k^2) = \frac{f(D)\Gamma(2-\frac{1}{2}D)}{2^3\pi^{4-D/2}} \int_0^1 \mathrm{d}z \frac{z(1-z)}{[k^2 z(1-z) - m^2]^{2-D/2}}. \tag{10.49}$$

It follows from Eq. (10.48) that the gauge condition

$$k_\mu \Pi^{\mu\nu}(k) = 0 \tag{10.50}$$

holds for any four-vector k.‡ Using dimensional regularization, gauge

‡ That Eq. (10.50) is a consequence of the gauge invariance condition (8.32) can be seen as follows. Inserting a vacuum polarization loop in an external field line [for example, going from Fig. 9.1 to Fig. 9.2(c)] leads to the following modification of the corresponding Feynman amplitude

$$\mathcal{M}_\alpha(\mathbf{k})A_e^\alpha(\mathbf{k}) \to \mathcal{M}'_\alpha(\mathbf{k})A_e^\alpha(\mathbf{k}) = [\mathcal{M}_\alpha(\mathbf{k}) + \mathcal{M}_\lambda(\mathbf{k})\mathrm{i}D_F^{\lambda\mu}(\mathbf{k})\mathrm{i}e^2\Pi_{\mu\alpha}(\mathbf{k})]A_e^\alpha(\mathbf{k}). \tag{10.51}$$

If $\mathcal{M}$ and $\mathcal{M}'$ satisfy the gauge condition (8.32), then Eq. (10.50) must hold.

invariance is automatically satisfied. This is a general result and follows from the fact that dimensional regularization preserves local symmetries of the Lagrangian, such as gauge invariance.

Finally we set $D = 4 - \eta$ and take the limit $\eta \to 0$. Since $f(4) = 4$, and using Eqs. (10.28) and (10.29), we obtain from Eq. (10.49) in the limit as $\eta \to 0$

$$\Pi(k^2) = \frac{1}{12\pi^2}\left(\frac{2}{\eta} - \gamma\right) - \frac{1}{2\pi^2}\int_0^1 \mathrm{d}z z(1-z)\ln[k^2 z(1-z) - m^2]. \tag{10.52}$$

Comparing Eq. (10.48) with Eqs. (9.11) and (9.15) gives

$$\Pi(k^2) = A'(0) + \Pi_c(k^2), \tag{10.53}$$

and since $\Pi_c(0) = 0$ we obtain from the last two equations

$$\begin{aligned} e^2\Pi_c(k^2) &= e^2[\Pi(k^2) - \Pi(0)] \\ &= -\frac{2\alpha}{\pi}\int_0^1 \mathrm{d}z z(1-z)\ln\left[1 - \frac{k^2 z(1-z)}{m^2}\right] \end{aligned} \tag{10.54}$$

which is the required result (9.68).

10.5 THE ANOMALOUS MAGNETIC MOMENT

As a second application of dimensional regularization we shall derive the anomalous magnetic moment of the electron to order α, Eq. (9.72).

We saw in Section 9.6.1 that this correction to the magnetic moment stems from the vertex correction, Fig. 9.15(c), given by Eq. (9.48). By dimensional regularization, retaining the infra-red cut-off λ, we can rewrite this equation

$$\begin{aligned} &e^2\Lambda^\mu(p', p) \\ &= \frac{-\mathrm{i}e^2}{(2\pi)^4}\int \mathrm{d}^D k \frac{N^\mu(p', p, k)}{(k^2 - \lambda^2 + \mathrm{i}\varepsilon)[(p'-k)^2 - m^2 + \mathrm{i}\varepsilon][(p-k)^2 - m^2 + \mathrm{i}\varepsilon]} \end{aligned} \tag{10.55}$$

where

$$N^\mu(p', p, k) = \gamma^\alpha(\not{p}' - \not{k} + m)\gamma^\mu(\not{p} - \not{k} + m)\gamma_\alpha. \tag{10.56}$$

By means of Feynman parameterization [Eq. (10.11b)], we can write Eq. (10.55) as

$$e^2\Lambda^\mu(p', p) = \frac{-\mathrm{i}e^2}{(2\pi)^4}\int_0^1 \mathrm{d}y \int_0^{1-y} \mathrm{d}z \int \mathrm{d}^D k \frac{2N^\mu(p', p, k)}{[k^2 - 2k(p'y + pz) - r + \mathrm{i}\varepsilon]^3} \tag{10.57}$$

where

$$r \equiv \lambda^2(1-y-z) - y(p'^2 - m^2) - z(p^2 - m^2). \tag{10.58}$$

Expressed in terms of the new variable

$$t^\mu = k^\mu - a^\mu \equiv k^\mu - (p'y + pz)^\mu, \tag{10.59}$$

Eq. (10.57) becomes

$$e^2\Lambda^\mu(p', p) = \frac{-\mathrm{i}e^2}{(2\pi)^4}\int_0^1 \mathrm{d}y \int_0^{1-y} \mathrm{d}z \int \mathrm{d}^D t \frac{2N^\mu(p', p, t+a)}{[t^2 - r - a^2 + \mathrm{i}\varepsilon]^3}. \tag{10.60}$$

In order to carry out the t-integration, we rearrange $N^\mu(p', p, t+a)$ as a sum of terms proportional to different powers of t

$$N^\mu(p', p, t+a) = \sum_{i=0}^{2} N_i^\mu(p', p), \tag{10.61}$$

where

$$N_0^\mu(p', p) \equiv \gamma^\alpha(\not{p}' - \not{a} + m)\gamma^\mu(\not{p} - \not{a} + m)\gamma_\alpha \tag{10.62a}$$

$$N_1^\mu(p', p) \equiv -\gamma^\alpha[\not{t}\gamma^\mu(\not{p} - \not{a} + m) + (\not{p}' - \not{a} + m)\gamma^\mu\not{t}]\gamma_\alpha \tag{10.62b}$$

$$N_2^\mu(p', p) \equiv \gamma^\alpha\not{t}\gamma^\mu\not{t}\gamma_\alpha. \tag{10.62c}$$

(For simplicity we have written $N_i^\mu(p', p)$, although in addition to being functions of p' and p, N_0^μ depends on a, N_2^μ on t, and N_1^μ on both a and t.) Eq. (10.60) can then be written

$$e^2\Lambda^\mu(p', p) = \sum_{i=0}^{2} e^2\Lambda_i^\mu(p', p) \tag{10.63a}$$

where

$$e^2\Lambda_i^\mu(p', p) = \frac{-\mathrm{i}e^2}{(2\pi)^4}\int_0^1 \mathrm{d}y \int_0^{1-y} \mathrm{d}z \int \mathrm{d}^D t \frac{2N_i^\mu(p', p)}{[t^2 - r - a^2 + \mathrm{i}\varepsilon]^3}. \tag{10.63b}$$

Of the three terms in Eq. (10.63a), Λ_1^μ vanishes since the integrand is odd in t [see Eq. (10.32)], Λ_0^μ is non-vanishing and finite in the limit $D \to 4$, and Λ_2^μ diverges in this limit.

We first consider the divergent term Λ_2^μ. Substituting Eq. (10.62c) in Eq. (10.63b), we evaluate the t-integral by means of Eq. (10.33) and obtain

$$e^2\Lambda_2^\mu(p', p) = \frac{e^2}{(2\pi)^4}\frac{\pi^{D/2}\Gamma(2-\frac{1}{2}D)}{\Gamma(3)}\int_0^1 \mathrm{d}y \int_0^{1-y} \mathrm{d}z \frac{\gamma^\alpha\gamma_\sigma\gamma^\mu\gamma^\sigma\gamma_\alpha}{[-(r+a^2)]^{2-D/2}}. \tag{10.64}$$

Setting $D = 4 - \eta$ and taking the limit $\eta \to 0$, one obtains from this equation, on account of Eqs. (10.37) and (10.28), (10.29),

$$e^2\Lambda_2^\mu(p', p) = \gamma^\mu \frac{\alpha}{2\pi}\int_0^1 \mathrm{d}y \int_0^{1-y} \mathrm{d}z \left\{\left(\frac{2}{\eta} - \gamma\right) - \ln\left[-(r + a^2)\right]\right\}. \tag{10.65}$$

In Chapter 9, the observable part $\Lambda_c^\mu(p', p)$ of the vertex correction was defined by the equation

$$\Lambda^\mu(p', p) = L\gamma^\mu + \Lambda_c^\mu(p', p). \tag{9.53}$$

We have now shown that the divergent part of $\Lambda^\mu(p', p)$ is given by the $1/\eta$ term in $\Lambda_2^\mu(p', p)$, Eq. (10.65). In identifying Eq. (9.53) with Eq. (10.63a), this $1/\eta$ term is incorporated in the term $L\gamma^\mu$ of Eq. (9.53). It follows that $\Lambda_c^\mu(p', p)$ is finite, as shown by a different approach in Section 9.5.

In order to derive the anomalous magnetic moment of the electron, we next consider the observable corrections which the vertex modification $e^2\Lambda^\mu(p', p)$ contributes to the scattering of electrons by an external static electromagnetic field. The Feynman amplitude for this is given by

$$\mathcal{M} = \mathrm{i}e\bar{u}(\mathbf{p}')e^2\Lambda^\mu(p', p)u(\mathbf{p})A_{e\mu}(\mathbf{q} = \mathbf{p}' - \mathbf{p}). \tag{10.66}$$

The most general form for this amplitude, allowing for Lorentz invariance, the Lorentz gauge condition

$$q_\mu A_e^\mu(\mathbf{q}) = 0 \tag{10.67}$$

and the Gordon identity (see Problem A.2) is

$$\mathcal{M} = \mathrm{i}e\bar{u}(\mathbf{p}')\left[\gamma^\mu F_1(q^2) + \frac{\mathrm{i}}{2m}\sigma^{\mu\nu}q_\nu F_2(q^2)\right]u(\mathbf{p})A_{e\mu}(\mathbf{q}) \tag{10.68}$$

where F_1 and F_2 are arbitrary functions of q^2. In order to calculate the magnetic moment, we need only consider the second term in Eq. (10.68). Comparing Eqs. (10.68) and (10.65), we see that Λ_2^μ makes no contribution to $F_2(q^2)$, so that the correction to the magnetic moment arises entirely from the term $\Lambda_0^\mu(p', p)$, defined by Eqs. (10.62a) and (10.63b). The t-integration in the latter equation is easily carried out, using Eq. (10.23) with $D = 4$, and leads to

$$\bar{u}(\mathbf{p}')e^2\Lambda_0^\mu(p', p)u(\mathbf{p}) = \frac{-\alpha}{4\pi}\int_0^1 \mathrm{d}y \int_0^{1-y} \mathrm{d}z \frac{\bar{u}(\mathbf{p}')N_0^\mu(p', p)u(\mathbf{p})}{(r + a^2)}. \tag{10.69}$$

Eq. (10.69) is a well-defined finite expression. The remaining analysis required to obtain the anomalous magnetic moment is straightforward but lengthy and we shall omit it. It involves evaluation of the spinor matrix element $\bar{u}(\mathbf{p}')N_0^\mu(p', p)u(\mathbf{p})$, utilizing the commutation and contraction relations of the γ-matrices, the Dirac equation, the Gordon identity and the

Lorentz condition (10.67). Retaining only terms which contribute to the magnetic moment [i.e. are of the form of the second term in Eq. (10.68)], one obtains

$$F_2(q^2) = \frac{m^2\alpha}{\pi}\int_0^1 \mathrm{d}y \int_0^{1-y} \mathrm{d}z \frac{(y+z)(1-y-z)}{\lambda^2(1-y-z)+(p'y+pz)^2}.$$

This integral is well-defined in the limit $\lambda \to 0$. Hence setting $\lambda = 0$ and $p' = p$, with $p^2 = m^2$, gives finally

$$F_2(0) = \frac{\alpha}{\pi}\int_0^1 \mathrm{d}y \int_0^{1-y} \mathrm{d}z \frac{1-y-z}{y+z} = \frac{\alpha}{2\pi} \tag{10.70}$$

which is the second-order correction to the magnetic moment quoted earlier, Eq. (9.72).

PROBLEMS

10.1 Derive Eq. (10.26) from Eq. (10.25), i.e. by using the cut-off method of regularization.

10.2 In the dimensional regularization scheme, the electron self-energy (9.4) is given by

$$\mathrm{i}e_0^2 \Sigma(p) = \frac{-e_0^2}{(2\pi)^4}\int \mathrm{d}^D k \frac{\gamma^\alpha(\not{p}-\not{k}+m)\gamma^\beta}{(p-k)^2-m^2+\mathrm{i}\varepsilon}\frac{g_{\alpha\beta}}{k^2-\lambda^2+\mathrm{i}\varepsilon}$$

in the limit as $D \to 4$, where we have also introduced a small cut-off parameter λ to guard against infra-red divergences.

Evaluate $\mathrm{i}e_0^2 \Sigma(p)$, retaining only those terms which diverge in the limit $\eta = 4 - D \to 0$, and compare your result with the expansion

$$e_0^2 \Sigma(p) = -\delta m + (\not{p}-m)(1-Z_2) + (\not{p}-m)e_0^2\Sigma_c(p).$$

Hence show that the corresponding contributions to δm and Z_2 are given by

$$\delta m = \frac{1}{\eta}\left(\frac{3\alpha m}{2\pi}\right), \qquad Z_2 = -\frac{1}{\eta}\left(\frac{\alpha}{2\pi}\right),$$

while $\Sigma_c(p)$ remains finite.

Evaluate the corresponding contribution to the vertex renormalization constant $Z_1 = (1 - e_0^2 L + \ldots)$ using Eqs. (9.53), (10.63) and (10.65). Hence, use Ward's identity to check the above value for Z_2.

CHAPTER 11

Weak interactions

11.1 INTRODUCTION

So far we have exclusively studied electromagnetic interactions and have seen that perturbation theory is spectacularly successful in handling these. Another area where perturbation theory should be valid is the weak interactions, of which historically the first and perhaps best known example is the nuclear β-decay process

$$n \to p + e^- + \bar{\nu}_e.^{\ddagger} \tag{11.1}$$

The rest of this book is devoted to the modern theory of weak interactions. This theory represents a remarkable breakthrough: the unification of the electromagnetic and weak interactions into the electro–weak interaction. In this *standard electro–weak theory*, as it is called, the weak interactions are transmitted by heavy vector bosons, analogously to the transmission of electromagnetic forces by photons. The electro–weak theory accounts successfully for the experimental data on weak interaction phenomena. Since the vector bosons predicted by the theory are very massive, they can be detected in experiments at extremely high energies only. Very recently, such experiments have been successfully carried out, i.e. they have established the existence and the predicted properties of these vector bosons.

The weak interaction processes subdivide into three classes according to the types of particles involved. Particles participating in strong interactions,

‡ For general background reading on weak interactions the reader is referred to D. H. Perkins, *Introduction to High Energy Physics*, 2nd edn, Addison-Wesley, Reading, Mass., 1982.

e.g. n, p, π, Λ, are called hadrons. Leptons are all those fermions which participate in weak and electromagnetic interactions only.[‡] In Section 7.4 we met the charged leptons $e^{\pm}$, $\mu^{\pm}$, ... which participate in both electromagnetic and weak interactions. In addition, we now have the corresponding neutrinos and antineutrinos ν_e, ν_μ, ..., $\bar{\nu}_e$, $\bar{\nu}_\mu$, ..., which participate in weak interactions only. Accordingly, we distinguish: (i) purely leptonic processes like

$$\tau^- \to e^- + \bar{\nu}_e + \nu_\tau, \tag{11.2}$$

$$\mu^- \to e^- + \bar{\nu}_e + \nu_\mu, \tag{11.3}$$

(ii) semi-leptonic processes involving hadrons and leptons like the β-decay process (11.1) and (iii) purely hadronic processes like the Λ-decay

$$\Lambda \to p + \pi^- . \tag{11.4}$$

We have only a very limited understanding of strong interactions and they cannot be treated in perturbation theory. In contrast, we believe that perturbation theory is valid for weak and electromagnetic interactions, and that we understand the latter. Consequently, purely leptonic processes afford an unambiguous and far simpler field for studying weak interactions, and we shall restrict ourselves to purely leptonic processes. This is analogous to our treatment of QED where we also did not consider hadrons.[§]

In this chapter we shall develop and apply the intermediate vector boson (IVB) theory, which is the forerunner and basis of the modern theory. We shall see that the IVB theory describes many processes successfully but that it also leads to serious difficulties and consequently does not constitute a satisfactory fundamental theory.

11.2 LEPTONIC WEAK INTERACTIONS

The electron–positron field enters the QED interaction in the bilinear combination of the electromagnetic current, i.e.

$$\mathscr{H}_{\text{QED}}(x) = -e\bar{\psi}(x)\gamma^\alpha\psi(x)A_\alpha(x). \tag{11.5}$$

The weak interaction Hamiltonian density responsible for leptonic processes is similarly constructed from bilinear forms of the lepton field operators.[¶] The experimental data on a wide range of leptonic and semi-leptonic processes are consistent with the assumption that the lepton fields enter the interaction only in the combinations

$$J_\alpha(x) = \sum_l \bar{\psi}_l(x)\gamma_\alpha(1-\gamma_5)\psi_{\nu_l}(x) \tag{11.6a}$$

[‡] We are, of course, throughout ignoring the very much weaker gravitational forces.

[§] For an excellent up-to-date comprehensive account of all aspects of weak interactions, see D. Bailin, *Weak Interactions*, 2nd edn, Adam Hilger, Bristol, 1982.

[¶] By leptonic processes we shall in future always mean purely leptonic processes.

$$J_\alpha^\dagger(x) = \sum_l \bar{\psi}_{\nu_l}(x)\gamma_\alpha(1 - \gamma_5)\psi_l(x). \tag{11.6b}$$

In Eqs. (11.6), l labels the various charged lepton fields, $l = e, \mu, \ldots$, and ν_l the corresponding neutrino fields. ψ_l and ψ_{ν_l} are the corresponding quantized fields. ψ_l is linear in the absorption operators of the l^- leptons and in the creation operators of the l^+ leptons, etc. In analogy to the electromagnetic current in Eq. (11.5), one calls $J_\alpha(x)$ and $J_\alpha^\dagger(x)$ leptonic currents since they too transform like vectors under continuous Lorentz transformations and imply certain lepton number conservation laws, as we shall see.

There is of course no unique way of constructing the leptonic interaction from the currents (11.6). In analogy with our description of the electromagnetic interaction as being transmitted by photons, we would like to describe the weak interactions as due to the transmission of quanta. These are called W particles. The QED interaction (11.5) then suggests the leptonic interaction of the IVB theory

$$\mathscr{H}_I(x) = g_W J^{\alpha\dagger}(x)W_\alpha(x) + g_W J^\alpha(x)W_\alpha^\dagger(x), \tag{11.7}$$

where g_W is a dimensionless coupling constant and the field $W_\alpha(x)$ describes the W particles. With this interaction, processes such as the muon decay process (11.3) or the neutrino scattering process

$$\nu_e + e^- \to \nu_e + e^-$$

are described, in lowest-order perturbation theory, by the Feynman graphs in Figs. 11.1 and 11.2. In each case the interaction between the two leptonic currents is brought about through the exchange of one W particle, analogously to the one-photon exchange of electron–electron scattering, Fig. 7.14.

We shall now study the interaction (11.7) in more detail. We note, first of all, that this interaction couples the field $W_\alpha(x)$ to the leptonic vector current. Hence it must be a vector field, and the W particles are vector bosons with spin 1.‡ Since each term in the leptonic currents (11.6) (i.e. each vertex in a Feynman graph) involves a charged and a neutral lepton, the W particles are electrically charged and the $W(x)$ field is non-Hermitian.

We can also infer that the W boson must be very massive. General arguments relate the range of a force to the mass of the quanta transmitting it. The long range of the electromagnetic forces results from the zero mass of the photon. In contrast, the weak interactions are of very short range and the mass of the W boson, m_W, must be very large. As we shall see in Section 14.1,

‡ We shall see in Section 11.3 that $W_\alpha(x)$ satisfies the Lorentz condition $\partial^\alpha W_\alpha(x) = 0$. Consequently, the field $W_\alpha(x)$ possesses only three independent states of polarization and describes spin 1 particles.

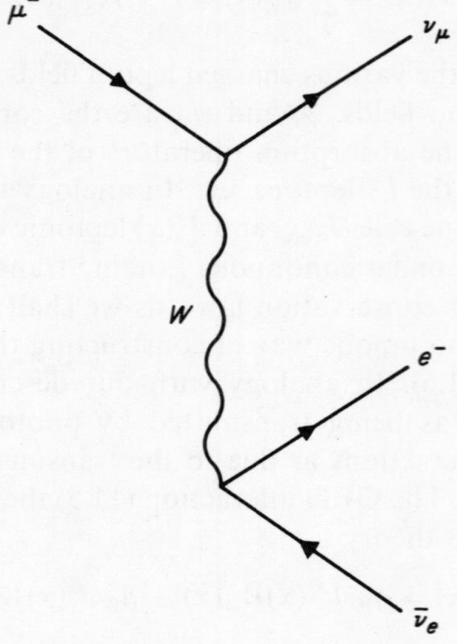

Fig. 11.1. Muon decay:
$\mu^- \to e^- + \bar{\nu}_e + \nu_\mu$.

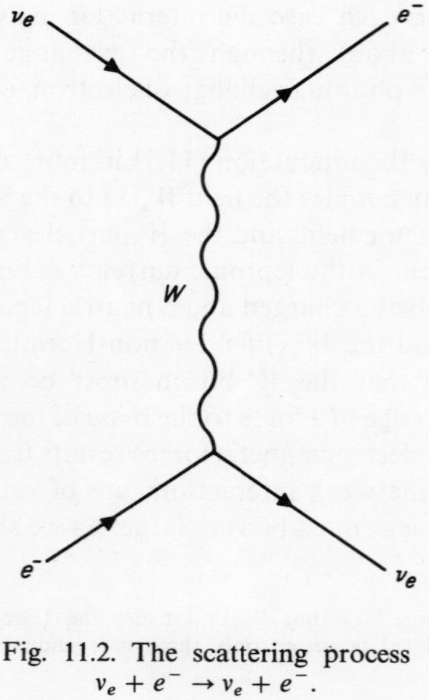

Fig. 11.2. The scattering process
$\nu_e + e^- \to \nu_e + e^-$.

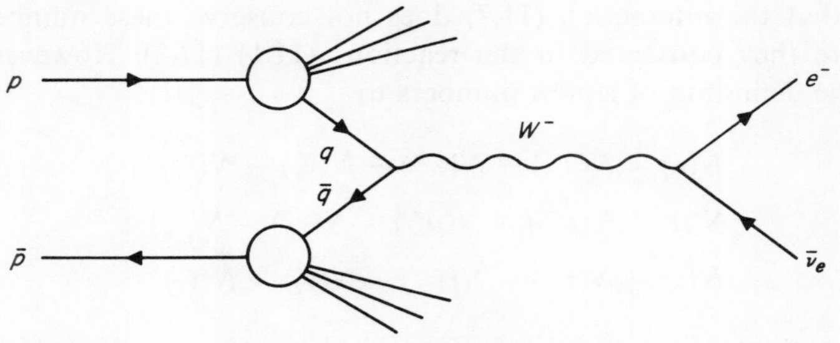

Fig. 11.3. The production of the W^- boson in a $p\bar{p}$ collision and its leptonic decay $W^- \to e^- + \bar{\nu}_e$.

the standard electro–weak theory makes the quite specific prediction

$$m_W = \left(83.0 \begin{array}{c} +2.9 \\ -2.7 \end{array}\right) \text{GeV} \tag{11.8a}$$

where the error stems from experimental uncertainties in the basic constants of the theory.

Recently, these predictions have been spectacularly confirmed in very high-energy experiments on the $p\bar{p}$ collider at CERN. In this machine, protons and antiprotons collide with a total centre-of-mass energy of 540 GeV. Such a collision can lead to a quark (q) and an antiquark ($\bar{q}$) combining to form a W boson which may decay via the weak interaction (11.7); for example, into electron plus electron-antineutrino. The Feynman graph for this process is shown in Fig. 11.3. Two experimental groups, known as the UA1 and UA2 collaborations, using different detection systems, have up-to-date detected 52 and 35 such events respectively. From the analysis of their results, they obtain the following values ‡ for the mass of the W boson

$$\left.\begin{array}{ll} m_W = (80.9 \pm 1.5 \pm 2.4)\ \text{GeV} & \text{(UA1)} \\ m_W = (81.0 \pm 2.5 \pm 1.3)\ \text{GeV} & \text{(UA2)} \end{array}\right\} \tag{11.8b}$$

where, in each result, the first error is statistical and the second systematic. (We shall also use this convention later.) The number of such events which has been observed is consistent with that expected from theoretical estimates.

In Section 7.4 we introduced lepton numbers

$$N(e) = N(e^-) - N(e^+), \tag{7.53a}$$

etc., and saw that these are conserved in electromagnetic processes. It is

‡ These data are taken from the papers by M. Spiro (UA1) and by A. Clark (UA2) in the *Proceedings of the 1983 International Symposium on Lepton and Photon Interactions at High Energies*, Cornell University, 1983.

obvious that the interaction (11.7) does not conserve these numbers, nor indeed are they conserved in the reactions (11.1)–(11.3). However, if we modify the definition of lepton numbers to

$$\left.\begin{aligned} N(e) &= N(e^-) - N(e^+) + N(\nu_e) - N(\bar{\nu}_e) \\ N(\mu) &= N(\mu^-) - N(\mu^+) + N(\nu_\mu) - N(\bar{\nu}_\mu) \\ N(\tau) &= N(\tau^-) - N(\tau^+) + N(\nu_\tau) - N(\bar{\nu}_\tau) \end{aligned}\right\} \tag{11.9}$$

then the currents $J_\alpha(x)$ and $J_\alpha^\dagger(x)$, Eqs. (11.6), do conserve lepton numbers. For example, the term

$$\bar{\psi}_e(x)\gamma_\alpha(1 - \gamma_5)\psi_{\nu_e}(x)$$

in Eq. (11.6a) is linear in electron creation and positron absorption operators, and in ν_e absorption and $\bar{\nu}_e$ creation operators. It follows that any interaction built up from these leptonic currents, as the interaction (11.7) is, conserves lepton numbers. This is in agreement with experiment, where lepton number conservation is found to hold for all processes, whatever particles or interactions are involved. For example, in bombarding nuclei with muon neutrinos, the process

$$\nu_\mu + (Z, A) \rightarrow (Z + 1, A) + \mu^- \tag{11.10}$$

is allowed by lepton number conservation, while

$$\nu_\mu + (Z, A) \rightarrow (Z - 1, A) + \mu^+ \tag{11.11}$$

is forbidden. No forbidden processes, such as (11.11), have been observed, within very small upper bounds, in experiments especially designed to detect them. The evidence for tauon number conservation is weaker—due to lack of data—but we shall assume it to hold.

In describing the neutrinos in the interaction (11.6) and (11.7) by spin $\frac{1}{2}$ Dirac fields, we are not assuming, as is sometimes done, that the neutrinos have zero mass. The experimental limits on the masses are very small:

$$m(\nu_e) \lesssim 50 \text{ eV}, \qquad m(\nu_\mu) \lesssim 0.5 \text{ MeV}, \qquad m(\nu_\tau) \lesssim 0.25 \text{ GeV}.$$

Since the corrections to the theoretical transition rates due to non-zero neutrino masses are of the order $[m(\nu_l)/m_l]^2$, one can put the neutrino masses equal to zero in comparing theory and experiment. However, we shall retain non-zero neutrino masses in our basic equations.‡

The interaction (11.7) is known as a 'V–A' interaction since the current

‡ We assume throughout that massive neutrinos are described by four-component Dirac fields. An alternative description in terms of two-component spinor fields, known as Majorana fields, is possible. For zero-mass neutrinos, these descriptions are equivalent. See T. D. Lee, *Particle Physics and Introduction to Field Theory*, Harcourt, New York, 1981, pp. 53–54.

$J^{\alpha}(x)$, from which it is built up, can be written as the difference

$$J^{\alpha}(x) = J^{\alpha}_{V}(x) - J^{\alpha}_{A}(x) \tag{11.12a}$$

of the vector current

$$J^{\alpha}_{V}(x) = \sum_{l} \bar{\psi}_{l}(x)\gamma^{\alpha}\psi_{\nu_l}(x) \tag{11.12b}$$

and the axial vector current

$$J^{\alpha}_{A}(x) = \sum_{l} \bar{\psi}_{l}(x)\gamma^{\alpha}\gamma_{5}\psi_{\nu_l}(x). \tag{11.12c}$$

Under the parity transformation $(\mathbf{x}, t) \to (-\mathbf{x}, t)$, $J^{\alpha}_{V}(x)$ changes sign, while $J^{\alpha}_{A}(x)$ does not.‡ Hence the interaction (11.7) is clearly not invariant under spatial inversion, and parity is not conserved. (Indeed, this is not peculiar to purely leptonic processes but is a characteristic of all weak interactions.) This has its most striking consequences for the neutrinos. Assume, first of all, that the neutrinos have zero mass. In this case we know from Appendix A, Eq. (A.43), that $(1 - \gamma_5)/2$ is a helicity projection operator. Since $\psi_{\nu_l}(x)$ is linear in neutrino absorption operators and in antineutrino creation operators, it follows from Eqs. (A.40) that the operator

$$\psi^{L}_{\nu_l}(x) \equiv \tfrac{1}{2}(1 - \gamma_5)\psi_{\nu_l}(x) \tag{11.13}$$

which occurs in the interaction (11.6)–(11.7) can annihilate only negative helicity neutrinos and create only positive helicity antineutrinos. Hence in weak interactions, only these states play a role, and positive helicity neutrinos and negative helicity antineutrinos do not partake in weak interaction processes.

The operators

$$\left.\begin{matrix} P_L \\ P_R \end{matrix}\right\} = \tfrac{1}{2}(1 \mp \gamma_5) \tag{11.14}$$

are of course always projection operators (since $P_L^2 = P_L$ and $P_R = 1 - P_L$), independently of the particle mass m of the fermion field. However, for particles of non-zero mass m the states projected out by P_L and P_R are helicity eigenstates only in the high-energy limit in which the particle energy is very large compared to the particle mass m. For neutrinos, this will always be a very good approximation in what follows, even if their masses are not precisely zero. However, it may also be a good approximation for high-energy charged leptons. In analogy to Eq. (11.13), we define the 'left-handed' charged lepton fields

$$\psi^{L}_{l}(x) \equiv P_L\psi_l(x) = \tfrac{1}{2}(1 - \gamma_5)\psi_l(x). \tag{11.15}$$

‡ See Appendix A, Eqs. (A.53) and Problem A.1.

One easily shows that the leptonic current (11.6a) can be written

$$J_\alpha(x) = 2 \sum_l \bar{\psi}_l^{\mathrm{L}}(x)\gamma_\alpha \psi_{\nu_l}^{\mathrm{L}}(x) \tag{11.16}$$

so that, as for the neutrinos, only the left-handed fields are involved for the charged leptons. Thus, if the electrons (positrons) emitted in a weak-interaction process are highly relativistic, they will have negative (positive) helicity. For example, this will be the case for most of the electrons and positrons in the muon decay processes

$$\mu^- \to e^- + \bar{\nu}_e + \nu_\mu, \qquad \mu^+ \to e^+ + \nu_e + \bar{\nu}_\mu. \tag{11.17}$$

11.3 THE FREE VECTOR BOSON FIELD

The simplest equation for a vector field $W^\alpha(x)$, describing particles of mass m_W and spin 1, is the Proca equation

$$\Box W^\alpha(x) - \partial^\alpha(\partial_\beta W^\beta(x)) + m_W^2 W^\alpha(x) = 0. \tag{11.18}$$

On taking the divergence of this equation, one automatically obtains the Lorentz condition

$$\partial_\alpha W^\alpha(x) = 0 \tag{11.19}$$

for $m_W \neq 0$. This is in contrast to the photon case, where the Lorentz condition must be imposed as a subsidiary condition. On account of Eq. (11.19), the Proca equation (11.18) reduces to

$$\Box W^\alpha(x) + m_W^2 W^\alpha(x) = 0. \tag{11.20}$$

A free-field Lagrangian density which leads to Eq. (11.18) is

$$\mathscr{L}(x) = -\tfrac{1}{2}F^\dagger_{W\alpha\beta}(x)F^{\alpha\beta}_W(x) + m_W^2 W^\dagger_\alpha(x)W^\alpha(x) \tag{11.21a}$$

where

$$F^{\alpha\beta}_W(x) \equiv \partial^\beta W^\alpha(x) - \partial^\alpha W^\beta(x). \tag{11.21b}$$

We are taking $W^\alpha(x)$ and $F^{\alpha\beta}_W(x)$ as non-Hermitian fields since the W bosons are electrically charged particles.

To establish the connection with the particle description, we expand the W field in the usual way in a complete set of plane waves:

$$W^\alpha(x) = W^{\alpha+}(x) + W^{\alpha-}(x) \tag{11.22a}$$

where

$$W^{\alpha+}(x) = \sum_{\mathbf{k}}\sum_r \left(\frac{1}{2V\omega_{\mathbf{k}}}\right)^{1/2} \varepsilon^\alpha_r(\mathbf{k})a_r(\mathbf{k})\,\mathrm{e}^{-ikx} \tag{11.22b}$$

$$W^{\alpha -}(x) = \sum_{\mathbf{k}} \sum_{r} \left(\frac{1}{2V\omega_{\mathbf{k}}} \right)^{1/2} \varepsilon_r^{\alpha}(\mathbf{k}) b_r^{\dagger}(\mathbf{k})\, \mathrm{e}^{\mathrm{i}kx} \tag{11.22c}$$

and

$$\omega_{\mathbf{k}} \equiv (m_W^2 + \mathbf{k}^2)^{1/2}. \tag{11.23}$$

[Eqs. (11.22) are analogous to Eqs. (3.27a) and (5.16) for charged scalar mesons and photons.] The vectors $\varepsilon_r^{\alpha}(\mathbf{k})$, $r = 1, 2, 3$, are a complete set of orthonormal polarization vectors, i.e.

$$\varepsilon_r(\mathbf{k})\varepsilon_s(\mathbf{k}) = -\delta_{rs}, \tag{11.24}$$

and the Lorentz condition (11.19) implies the conditions

$$k_{\alpha}\varepsilon_r^{\alpha}(\mathbf{k}) = 0 \tag{11.25}$$

for the polarization vectors. In the frame in which $k = (\omega_{\mathbf{k}}, 0, 0, |\mathbf{k}|)$, a suitable choice of polarization vectors is

$$\left.\begin{aligned} \varepsilon_1(\mathbf{k}) &= (0, 1, 0, 0) \\ \varepsilon_2(\mathbf{k}) &= (0, 0, 1, 0) \\ \varepsilon_3(\mathbf{k}) &= (|\mathbf{k}|, 0, 0, \omega_{\mathbf{k}})/m_W \end{aligned}\right\} \tag{11.26}$$

The completeness relation

$$\sum_{r=1}^{3} \varepsilon_r^{\alpha}(\mathbf{k})\varepsilon_r^{\beta}(\mathbf{k}) = -g^{\alpha\beta} + k^{\alpha}k^{\beta}/m_W^2 \tag{11.27}$$

follows directly from Eqs. (11.26).

It should be self-evident to the reader that quantization of the Lagrangian (11.21) by means of the canonical formalism allows us to interpret $a_r(\mathbf{k})$ and $b_r(\mathbf{k})$ as annihilation operators of vector mesons of mass m_W, momentum $\mathbf{k}$ and polarization vector $\varepsilon_r(\mathbf{k})$, with $a_r^{\dagger}(\mathbf{k})$ and $b_r^{\dagger}(\mathbf{k})$ the corresponding creation operators. In order that the interaction (11.7) implies conservation of electric charge, we shall want to associate $a_r(\mathbf{k})$ and $a_r^{\dagger}(\mathbf{k})$ with positively charged bosons (W^+), and $b_r(\mathbf{k})$ and $b_r^{\dagger}(\mathbf{k})$ with W^- bosons.

Lastly, we require the W boson propagator

$$\langle 0|\mathrm{T}\{W^{\alpha}(x)W^{\beta\dagger}(y)\}|0\rangle \equiv \mathrm{i}D_{\mathrm{F}}^{\alpha\beta}(x-y, m_W) \tag{11.28}$$

where we have explicitly shown the dependence on the mass m_W. This propagator is derived by the same method by which the scalar meson propagator was obtained in Section 3.4. We shall omit the derivation‡ and

‡ An explicit derivation is given on pp. 96–97 of D. Bailin's *Weak Interactions*, quoted in Section 11.1. The relation of the W propagator to the photon propagator is discussed in C. Itzykson and J. B. Zuber, *Quantum Field Theory*, McGraw-Hill, New York, 1980, Section 3-2-3.

only quote the result. With

$$iD_F^{\alpha\beta}(x, m_W) = \frac{1}{(2\pi)^4} \int d^4k \, e^{-ikx} iD_F^{\alpha\beta}(k, m_W), \tag{11.29}$$

the W propagator in momentum space is given by

$$iD_F^{\alpha\beta}(k, m_W) = \frac{i(-g^{\alpha\beta} + k^\alpha k^\beta / m_W^2)}{k^2 - m_W^2 + i\varepsilon}. \tag{11.30}$$

11.4 THE FEYNMAN RULES FOR THE IVB THEORY

We now easily write down the Feynman rules for treating leptonic processes in perturbation theory.

The basic vertex part which arises from the interaction (11.7) is shown in Fig. 11.4. It consists of two lepton lines (representing a lepton-number-conserving current) and a W boson line. The only restrictions at the vertex are that lepton numbers and electric charge are conserved. Hence Fig. 11.4 stands for many different processes, for example those in Fig. 11.5, where in each diagram there may occur absorption or emission of a $W^\pm$ boson of the appropriate sign to conserve charge. Substituting the leptonic interaction

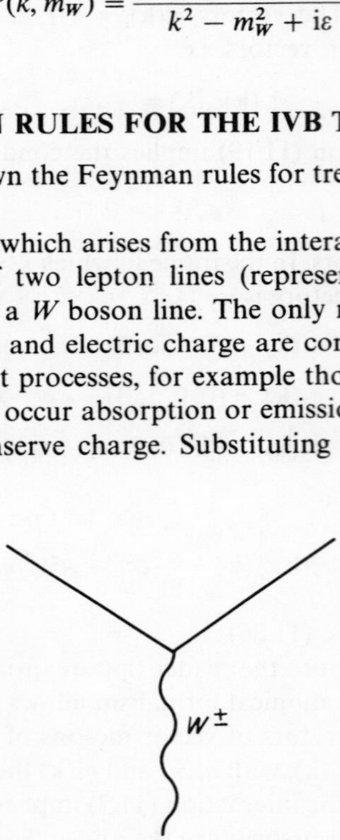

Fig. 11.4. The basic vertex part of the IVB interaction.

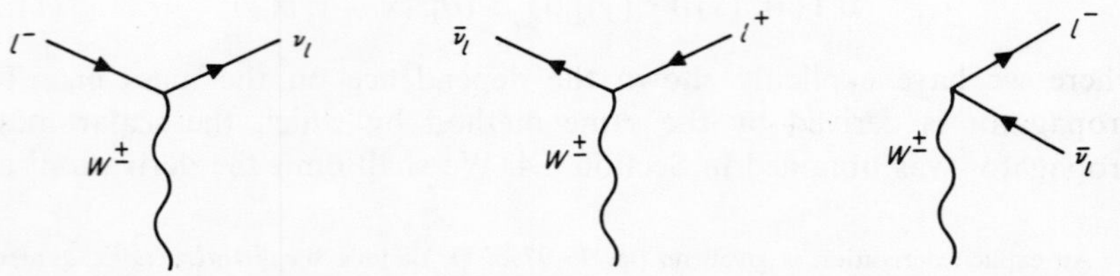

Fig. 11.5. Three particular cases of the basic vertex diagram in Fig. 11.4.

(11.7) in the S-matrix expansion (6.23), one sees that corresponding to Feynman rule 1 of QED (see Section 7.3) we now obtain the Feynman rule

10. For each vertex, write a factor

$$-\mathrm{i}g_W\gamma^\alpha(1-\gamma_5). \tag{11.31}$$

The Feynman rules for electron propagators and external lines, given in Section 7.3, only require trivial relabelling to apply to all leptons. In the Feynman rule 3 for the lepton propagator (7.48), m now stands for the mass m_l or m_{ν_l} ($l = e, \mu, \ldots$) of the lepton involved. In the Feynman rule 4 for external line factors, Eqs. (7.49a) and (7.49b) become the factors appropriate to negative leptons l^- and neutrinos ν_l, and Eqs. (7.49c) and (7.49d) the factors appropriate to the corresponding antiparticles l^+ and $\bar{\nu}_l$. Hence we only need the rules for internal and external W lines, corresponding to rules 2 and 4 for photons. These are

11. For each internal W boson line, labelled by the momentum k, write a factor

$$\mathrm{i}D_{\mathrm{F}\alpha\beta}(k, m_W) = \frac{\mathrm{i}(-g_{\alpha\beta} + k_\alpha k_\beta/m_W^2)}{k^2 - m_W^2 + \mathrm{i}\varepsilon}. \tag{11.30}$$

12. For each external initial or final W boson line, write a factor

$$\varepsilon_{r\alpha}(\mathbf{k}) \quad \begin{cases} \text{(initial)} \\ \text{(final)} \end{cases}. \tag{11.32}$$

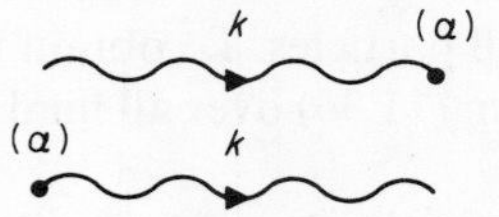

As for photons, if complex polarization vectors are used, we must write $\varepsilon^*_{r\alpha}(\mathbf{k})$ for a final-state W boson, instead of $\varepsilon_{r\alpha}(\mathbf{k})$.

11.5 DECAY RATES

Before calculating the rate for muon decay, we first derive the general expression for the decay rate of any process in terms of the corresponding Feynman amplitude. We consider a particle P decaying into N particles $P'_1, P'_2, \ldots, P'_N$:

$$P \to P'_1 + P'_2 + \cdots + P'_N. \tag{11.33}$$

The relation (8.1) between the S-matrix element and the Feynman amplitude then reduces to

$$S_{fi} = \delta_{fi} + (2\pi)^4\delta^{(4)}\left(\sum p'_f - p\right)\frac{1}{(2VE)^{1/2}}\prod_f\left(\frac{1}{2VE'_f}\right)^{1/2}\prod_l (2m_l)^{1/2}\mathscr{M} \tag{11.34}$$

where $p = (E, \mathbf{p})$ and $p'_f = (E'_f, \mathbf{p}'_f), f = 1, \ldots, N$, are the four-momenta of the initial and final particles, and the index l runs over all external fermions in the process. From Eq. (11.34) one obtains for the decay rate w (i.e. the transition probability per unit time) from a given initial to a specific final quantum state

$$w = (2\pi)^4 \delta^{(4)}(\sum p'_f - p) \frac{1}{2E} \left(\prod_f \frac{1}{2VE'_f} \right) \left(\prod_l (2m_l) \right) |\mathcal{M}|^2, \quad (11.35)$$

analogously to Eq. (8.6). The differential decay rate $\mathrm{d}\Gamma$ for the process (11.33) to final states in which the particle P'_1 has momentum in the range $\mathrm{d}^3\mathbf{p}'_1$ at $\mathbf{p}'_1$, etc., is obtained by multiplying w by the number of these states, given by Eq. (8.7), yielding

$$\mathrm{d}\Gamma = w \prod_f \frac{V \, \mathrm{d}^3\mathbf{p}'_f}{(2\pi)^3}$$

$$= (2\pi)^4 \delta^{(4)}(\sum p'_f - p) \frac{1}{2E} \left(\prod_l (2m_l) \right) \left(\prod_f \frac{\mathrm{d}^3\mathbf{p}'_f}{(2\pi)^3 2E'_f} \right) |\mathcal{M}|^2. \quad (11.36)$$

This equation is our general result. To take the analysis further we must consider a specific process. The energy–momentum conserving δ-function in Eq. (11.36) is eliminated in the usual way by integrating over the appropriate final-state variables. Eq. (11.36) gives the transition rate between definite initial and final spin states of all particles. To obtain the total decay rate Γ for the process (11.33) we must sum (11.36) over all final spin states and integrate over all final-state momenta.

So far we have considered only the decay mode (11.33) of the particle P. In general, there may be several decay modes. The branching ratio B for the decay mode (11.33) with decay rate Γ is defined by

$$B = \Gamma / \sum \Gamma \quad (11.37)$$

where $\sum \Gamma$ is the sum of the decay rates over all decay modes, i.e. $\sum \Gamma$ is the total decay rate. (It is also called the total decay width.) The life time of the particle P is then given by

$$\tau = \frac{1}{\sum \Gamma} = \frac{B}{\Gamma}. \quad (11.38)$$

11.6 APPLICATIONS OF THE IVB THEORY

In this section we shall apply the IVB theory to three leptonic processes: (i) muon decay, (iii) neutrino scattering, and (iii) the leptonic decay of the W boson. We shall give the analysis of the muon decay in some detail, which should enable the reader to perform similar calculations, and we shall therefore treat the two other cases much more concisely.

11.6.1 Muon decay

As our first application we shall consider one of the best studied leptonic processes, the muon decay

$$\mu^-(p, r) \to e^-(p', r') + \bar{\nu}_e(q_1, r_1) + \nu_\mu(q_2, r_2), \tag{11.39}$$

where p is the four-momentum of the muon and r labels its spin, etc. In lowest order of perturbation theory, this process is represented by the Feynman graph in Fig. 11.6. The corresponding Feynman amplitude follows from the Feynman rules and is given by

$$\begin{aligned}\mathscr{M} = &-g_W^2[\bar{u}(\mathbf{p}')\gamma^\alpha(1-\gamma_5)v(\mathbf{q}_1)] \\ &\times \frac{\mathrm{i}(-g_{\alpha\beta} + k_\alpha k_\beta/m_W^2)}{k^2 - m_W^2 + \mathrm{i}\varepsilon}[\bar{u}(\mathbf{q}_2)\gamma^\beta(1-\gamma_5)u(\mathbf{p})]\end{aligned} \tag{11.40}$$

where we suppressed the spin indices and

$$k = p - q_2 = p' + q_1. \tag{11.41}$$

In the limit $m_W \to \infty$, the Feynman amplitude (11.40) reduces to

$$\mathscr{M} = -\frac{\mathrm{i}G}{\sqrt{2}}[\bar{u}(\mathbf{p}')\gamma^\alpha(1-\gamma_5)v(\mathbf{q}_1)][\bar{u}(\mathbf{q}_2)\gamma_\alpha(1-\gamma_5)u(\mathbf{p})] \tag{11.42}$$

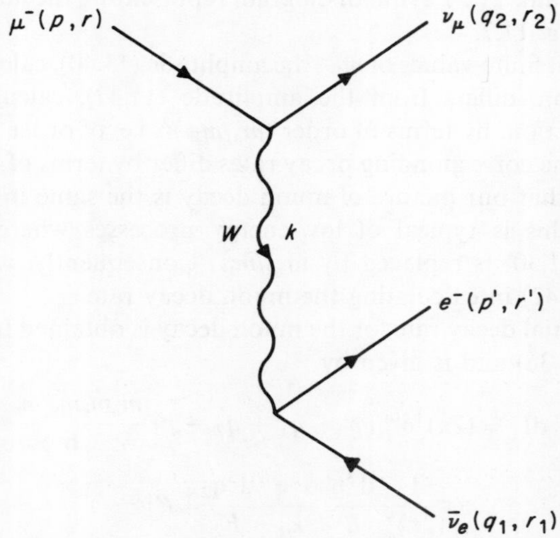

Fig. 11.6. Muon decay.

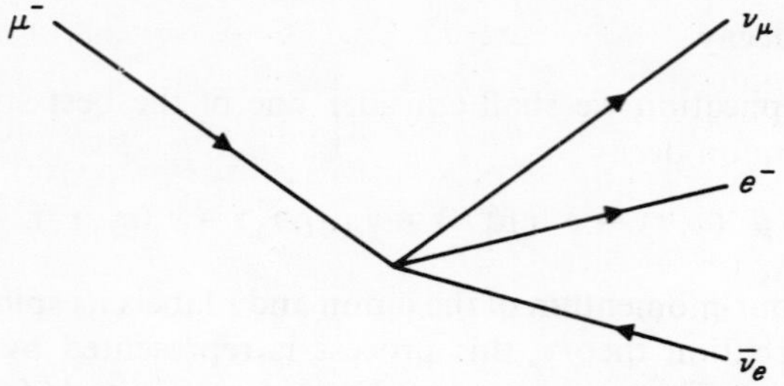

Fig. 11.7. The Feynman diagram describing muon decay in terms of the contact interaction (11.44).

where G is defined by

$$\frac{G}{\sqrt{2}} = \left(\frac{g_W}{m_W}\right)^2. \tag{11.43}$$

The same expression would have been obtained for the Feynman amplitude of this process if we had calculated it in first-order perturbation theory from the interaction

$$\mathscr{H}_1^{(\mathrm{F})}(x) = \frac{G}{\sqrt{2}} J^\alpha(x) J_\alpha^\dagger(x). \tag{11.44}$$

Eq. (11.44) represents a contact interaction of four lepton fields. Such a contact interaction was first proposed by Fermi in 1934 to describe the nuclear β-decay process (11.1), and G is known as the Fermi weak interaction coupling constant. The Feynman diagram representing the amplitude (11.42) is shown in Fig. 11.7.

For large but finite values of m_W, the amplitude (11.40), calculated from the IVB interaction, differs from the amplitude (11.42), calculated from the contact interaction, by terms of order $(m_\mu/m_W)^2$, i.e. of order 10^{-6} for the W mass (11.8). The corresponding decay rates differ by terms of the same order. We conclude that our picture of muon decay is the same in both modes of description. This is typical of low-energy processes where, in effect, the propagator (11.30) is replaced by $\mathrm{i}g_{\alpha\beta}/m_W^2$. Consequently, we shall use the amplitude (11.42) in calculating the muon decay rate.

The differential decay rate for the muon decay is obtained from the general expression (11.36) and is given by

$$\mathrm{d}\Gamma = (2\pi)^4 \delta^{(4)}(p' + q_1 + q_2 - p) \frac{m_\mu m_e m_{\nu_e} m_{\nu_\mu}}{E}$$

$$\times \frac{1}{(2\pi)^9} \frac{\mathrm{d}^3\mathbf{p}'}{E'} \frac{\mathrm{d}^3\mathbf{q}_1}{E_1} \frac{\mathrm{d}^3\mathbf{q}_2}{E_2} |\mathscr{M}|^2 \tag{11.45}$$

where $p \equiv (E, \mathbf{p})$, $p' \equiv (E', \mathbf{p}')$ and $q_i \equiv (E_i, \mathbf{q}_i)$, $i = 1, 2$.

To obtain the total decay rate from Eq. (11.45), we must sum over all final spin states and integrate over all final momenta.

We first deal with the spins. The life time of the muon is, of course, independent of its spin state. Hence we shall also average over the spin states of the initial muon, in order to express the result as a trace.‡ Using the standard techniques, developed in Section 8.2, to sum over final spin states and average over initial spin states, one obtains

$$m_\mu m_e m_{\nu_e} m_{\nu_\mu} \tfrac{1}{2} \sum_{\text{spins}} |\mathcal{M}|^2$$

$$= \frac{G^2}{64} \operatorname{Tr}\,[(\not{p}' + m_e)\gamma^\alpha(1-\gamma_5)(\not{q}_1 - m_{\nu_e})\gamma^\beta(1-\gamma_5)]$$

$$\times \operatorname{Tr}\,[(\not{q}_2 + m_{\nu_\mu})\gamma_\alpha(1-\gamma_5)(\not{p} + m_\mu)\gamma_\beta(1-\gamma_5)]$$

$$= \frac{G^2}{64} \operatorname{Tr}\,[\not{p}'\gamma^\alpha(1-\gamma_5)\not{q}_1\gamma^\beta(1-\gamma_5)] \operatorname{Tr}\,[\not{q}_2\gamma_\alpha(1-\gamma_5)\not{p}\gamma_\beta(1-\gamma_5)] \tag{11.46}$$

where in the last line we have taken the limits $m_{\nu_e} \to 0$, $m_{\nu_\mu} \to 0$.

We evaluate the first of the traces in Eq. (11.46), i.e.

$$E^{\alpha\beta} \equiv \operatorname{Tr}\,[\not{p}'\gamma^\alpha(1-\gamma_5)\not{q}_1\gamma^\beta(1-\gamma_5)]. \tag{11.47}$$

Using

$$[\gamma_5, \gamma^\alpha]_+ = 0, \qquad \alpha = 0, \ldots, 3, \qquad (1-\gamma_5)^2 = 2(1-\gamma_5),$$

we obtain

$$E^{\alpha\beta} = 2p'_\mu q_{1\nu} \operatorname{Tr}\,[\gamma^\mu\gamma^\alpha\gamma^\nu\gamma^\beta(1-\gamma_5)],$$

and using Eqs. (A.17) and (A.21) this reduces to

$$E^{\alpha\beta} = 8p'_\mu q_{1\nu} x^{\mu\alpha\nu\beta} \tag{11.48a}$$

where

$$x^{\mu\alpha\nu\beta} \equiv g^{\mu\alpha}g^{\nu\beta} - g^{\mu\nu}g^{\alpha\beta} + g^{\mu\beta}g^{\alpha\nu} + i\varepsilon^{\mu\alpha\nu\beta}. \tag{11.49}$$

It follows at once that the second of the traces in Eq. (11.46) is given by

$$M_{\alpha\beta} \equiv \operatorname{Tr}\,[\not{q}_2\gamma_\alpha(1-\gamma_5)\not{p}\gamma_\beta(1-\gamma_5)] = 8q_2^\sigma p^\tau x_{\sigma\alpha\tau\beta}. \tag{11.48b}$$

Substituting Eqs. (11.48a) and (11.48b) into Eq. (11.46) gives

$$m_\mu m_e m_{\nu_e} m_{\nu_\mu} \tfrac{1}{2} \sum_{\text{spins}} |\mathcal{M}|^2 = G^2 p'_\mu q_{1\nu} x^{\mu\alpha\nu\beta} q_2^\sigma p^\tau x_{\sigma\alpha\tau\beta}. \tag{11.50}$$

‡ For zero-mass neutrinos, the emitted $\bar{\nu}_e$ and ν_μ have definite helicities. By summing over the helicities of these neutrinos, leaving it to the helicity projection operators in the interaction to select the appropriate helicity states, one again ensures that the result is expressed as a trace.

From the definition (11.49) and Eqs. (A.14c) it follows that

$$x^{\mu\alpha\nu\beta}x_{\sigma\alpha\tau\beta} = 4g^{\mu}_{\sigma}g^{\nu}_{\tau}. \tag{11.51}$$

By means of this relation, Eq. (11.50) reduces to our final result for the spin sum

$$m_\mu m_e m_{\nu_e} m_{\nu_\mu} \tfrac{1}{2} \sum_{\text{spins}} |\mathscr{M}|^2 = 4G^2(pq_1)(p'q_2), \tag{11.52}$$

in the limit $m_{\nu_e} \to 0$ and $m_{\nu_\mu} \to 0$.

Combining Eqs. (11.45) and (11.52), we obtain the unpolarized differential decay rate

$$\mathrm{d}\Gamma = \frac{4G^2}{(2\pi)^5 E}(pq_1)(p'q_2)\delta^{(4)}(p' + q_1 + q_2 - p)\frac{\mathrm{d}^3\mathbf{p}'}{E'}\frac{\mathrm{d}^3\mathbf{q}_1}{E_1}\frac{\mathrm{d}^3\mathbf{q}_2}{E_2}. \tag{11.53}$$

We next carry out the phase space integrations, starting with integrals over the neutrino momenta, given by

$$I^{\mu\nu}(q) \equiv \int \mathrm{d}^3\mathbf{q}_1\, \mathrm{d}^3\mathbf{q}_2 \frac{q_1^\mu q_2^\nu}{E_1 E_2}\delta^{(4)}(q_1 + q_2 - q) \tag{11.54}$$

where

$$q \equiv p - p'. \tag{11.55}$$

It follows from the Lorentz covariance of the integral (11.54) that its most general form is

$$I^{\mu\nu}(q) = g^{\mu\nu}A(q^2) + q^\mu q^\nu B(q^2). \tag{11.56}$$

From this equation it follows that

$$g_{\mu\nu}I^{\mu\nu}(q) = 4A(q^2) + q^2 B(q^2) \tag{11.57a}$$

$$q_\mu q_\nu I^{\mu\nu}(q) = q^2 A(q^2) + (q^2)^2 B(q^2). \tag{11.57b}$$

From now on we shall take the neutrino masses as zero so that $q_1^2 = q_2^2 = 0$ and, on account of the δ-function in (11.54),

$$q^2 = 2(q_1 q_2). \tag{11.58}$$

In order to find $A(q^2)$ and $B(q^2)$, we calculate the expressions on the left-hand sides of Eqs. (11.57). From Eqs. (11.54) and (11.58) we obtain

$$g_{\mu\nu}I^{\mu\nu}(q) = \frac{q^2}{2}\int \frac{\mathrm{d}^3\mathbf{q}_1}{E_1}\frac{\mathrm{d}^3\mathbf{q}_2}{E_2}\delta^{(4)}(q_1 + q_2 - q) \equiv \tfrac{1}{2}q^2 I(q^2). \tag{11.59}$$

We see from its definition that the integral $I(q^2)$ is an invariant, so that it can be evaluated in any coordinate system. We shall choose the centre-of-momentum system of the two neutrinos. In this system $\mathbf{q}_1 = -\mathbf{q}_2$, so that $\mathbf{q} = 0$, and the energy ω of either neutrino is given by

$$\omega \equiv E_1 = |\mathbf{q}_1| = E_2 = |\mathbf{q}_2|. \tag{11.60}$$

Hence,

$$I(q^2) = \int \mathrm{d}^3\mathbf{q}_1 \frac{\delta(2\omega - q_0)}{\omega^2} = 2\pi \tag{11.61}$$

and from Eq. (11.59)

$$g_{\mu\nu}I^{\mu\nu}(q) = \pi q^2. \tag{11.62a}$$

Similarly, one finds from Eqs. (11.54), (11.58) and (11.61) that

$$q_\mu q_\nu I^{\mu\nu}(q^2) = (\tfrac{1}{2}q^2)^2 I = \tfrac{1}{2}\pi(q^2)^2. \tag{11.62b}$$

From Eqs. (11.57) and (11.62) we can find $A(q^2)$ and $B(q^2)$, and substituting these into Eq. (11.56) leads to

$$I^{\mu\nu}(q) = \tfrac{1}{6}\pi(g^{\mu\nu}q^2 + 2q^\mu q^\nu). \tag{11.63}$$

From Eqs. (11.63) and (11.53) we obtain the muon decay rate for emission of an electron with momentum in the range $\mathrm{d}^3\mathbf{p}'$ at $\mathbf{p}'$:

$$\mathrm{d}\Gamma = \frac{2\pi}{3}\frac{G^2}{(2\pi)^5 E}\frac{\mathrm{d}^3\mathbf{p}'}{E'}[(pp')q^2 + 2(pq)(p'q)]. \tag{11.64}$$

Finally, we must integrate Eq. (11.64) over all momenta $\mathbf{p}'$ of the emitted electron. For a muon at rest, i.e. in the rest frame of the muon, we have

$$p = (m_\mu, \mathbf{0}), \qquad q_0 = m_\mu - E', \qquad \mathbf{q} = -\mathbf{p}', \tag{11.65}$$

and in this frame Eq. (11.64) becomes

$$\mathrm{d}\Gamma = \frac{2\pi}{3}\frac{G^2}{(2\pi)^5 m_\mu}|\mathbf{p}'|\,\mathrm{d}E'\,\mathrm{d}\Omega'[(m_\mu^2 + m_e^2 - 2m_\mu E')m_\mu E' \\ + 2m_\mu(m_\mu - E')(m_\mu E' - m_e^2)] \tag{11.66}$$

where we put $\mathrm{d}^3\mathbf{p}' = |\mathbf{p}'|E'\,\mathrm{d}E'\,\mathrm{d}\Omega'$. If we neglect terms of order m_e^2/m_μ^2, Eq. (11.66) reduces to

$$\mathrm{d}\Gamma = \frac{2\pi}{3}\frac{G^2}{(2\pi)^5} m_\mu E'^2\,\mathrm{d}E'\,\mathrm{d}\Omega'(3m_\mu - 4E'). \tag{11.67}$$

Integrating Eq. (11.67) over all directions Ω' of the emitted electron and over its complete range of energies $0 \leqslant E' \leqslant \frac{1}{2}m_\mu$, we obtain the total decay rate

$$\Gamma = \frac{G^2 m_\mu^5}{192\pi^3}. \tag{11.68}$$

Taking $\mu^- \to e^- \bar{\nu}_e \nu_\mu$ as the only muon decay mode (the experimental branching ratio is 98.6 per cent‡), we obtain the muon life time

$$\tau_\mu = \frac{1}{\Gamma} = \frac{192\pi^3}{G^2 m_\mu^5}. \tag{11.69}$$

Substituting the experimental values $\tau_\mu = 2.2 \times 10^{-6}\,\text{s}$ and $m_\mu = 105.7\,\text{MeV}$ in Eq. (11.69) gives for the value of the Fermi coupling constant

$$G = 1.16 \times 10^{-5}\,\text{GeV}^{-2} = 1.02 \times 10^{-5}/m_p^2. \tag{11.70a}$$

Substituting the value (11.70a) for G into Eq. (11.43) leads to

$$\frac{g_W^2}{4\pi} = \frac{G}{4\pi\sqrt{2}}\left(\frac{m_W}{m_p}\right)^2 \approx 4 \times 10^{-3} \tag{11.71}$$

for $m_W \approx 80\,\text{GeV}$. This value is comparable to that of the fine structure constant, $e^2/4\pi = 7.3 \times 10^{-3}$, so that lowest order of perturbation theory should give a good description of weak interactions.

A much more precise value of G than Eq. (11.70a) can be obtained from the most accurate current experimental value for the muon life time

$$\tau_\mu = (2.19714 \pm 0.00007) \times 10^{-6}\,\text{s}. \tag{11.72}$$

For this purpose one must retain the terms involving the electron mass m_e in the expression for the decay rate Γ [i.e. one must derive Γ from Eq. (11.66) instead of from Eq. (11.67)], one must include the radiative corrections of QED, and one must, of course, also allow for the 98.6 per cent branching ratio of the $\mu^- \to e^- \bar{\nu}_e \nu_\mu$ mode. In this way one finds the value

$$G = (1.16632 \pm 0.00002) \times 10^{-5}\,\text{GeV}^{-2} = 1.027 \times 10^{-5}/m_p^2. \tag{11.70b}$$

The muon life time only determines the coupling constant G. A thorough test of the theory comes from comparing its predictions with detailed experiments on: (i) the energy spectrum of the emitted electrons, (ii) the energy spectrum and the angular distribution of the electrons emitted in the decay of polarized muons, and (iii) the helicity of the electrons emitted from unpolarized muons. Some of these experiments are sufficiently accurate that one must again include radiative corrections in the theory. In all cases there is good agreement between theory and experiment. For example, the mean helicity of the electrons emitted in the decay of unpolarized muons is -1.008 ± 0.057, in agreement with the theoretical predictions and with the qualitative arguments we gave following Eq. (11.16).§

‡ This and other experimental data quoted in this section are taken from Particle Data Group, *Review of Particle Properties*, *Phys. Lett.* **111B** (1982).

§ For details of these calculations involving polarized electrons or polarized muons see D. Bailin's *Weak Interactions*, quoted in Section 11.1.

Finally, we note that evidence for the universality of the leptonic interaction (11.6)–(11.7) comes from the tauon decay

$$\tau^- \to e^- + \bar{\nu}_e + \nu_\tau. \tag{11.73}$$

The tauon life time τ_τ is at once obtained from expression (11.69) for τ_μ by replacing m_μ by m_τ and allowing for the fact that the branching ratio for the process (11.73) is $B = 0.176 \pm 0.006 \pm 0.010$. Using Eq. (11.38), it follows that

$$\tau_\tau = B\frac{192\pi^3}{G^2 m_\tau^5} = B\tau_\mu\left(\frac{m_\mu}{m_\tau}\right)^5. \tag{11.74}$$

Substituting for τ_μ, B and m_μ/m_τ in Eq. (11.74) leads to the prediction $\tau_\tau = (2.8 \pm 0.2) \times 10^{-13}$ s, in agreement with the experimental life time of $(3.20 \pm 0.41 \pm 0.35) \times 10^{-13}$ s.‡

11.6.2 Neutrino scattering

Unlike the muon decay process, one would expect neutrino scattering processes at sufficiently high energies to exhibit the effects of the intermediate vector boson. We shall illustrate this for the process

$$\nu_\mu + e^- \to \nu_e + \mu^- \tag{11.75}$$

which is often called inverse muon decay. In lowest order of perturbation theory, this process is represented by the Feynman graph in Fig. 11.8, which also specifies the four-momenta of the particles. It is left as an exercise for the reader to show that the corresponding Feynman amplitude is given by

$$\begin{aligned}\mathscr{M} = &\left\{\frac{m_W^2}{m_W^2 - k^2}\right\}(-\mathrm{i})\left(\frac{g_W}{m_W}\right)^2 [\bar{u}(\mathbf{p}')\gamma_\alpha(1-\gamma_5)u(\mathbf{q})]\\ &\times [\bar{u}(\mathbf{q}')\gamma^\alpha(1-\gamma_5)u(\mathbf{p})],\end{aligned} \tag{11.76}$$

when terms of order $(m_\mu m_e/m_W^2)$ are dropped. The vector k in this equation is the momentum of the intermediate W boson and is given by

$$k = p - q' = p' - q. \tag{11.77}$$

The factor in curly brackets in the Feynman amplitude (11.76) leads to the factor

$$\left\{\frac{m_W^2}{m_W^2 - k^2}\right\}^2 \tag{11.78}$$

in the differential scattering cross-section. For $m_W \to \infty$, this factor becomes unity, and the differential cross-section becomes identical with that of the

‡ These values for B and τ_τ are taken from a recent experiment by the Mark II group at Stanford. See the paper by N. Reay in the *Proceedings of the 1983 International Symposium*, cited for Eq. (11.8b).

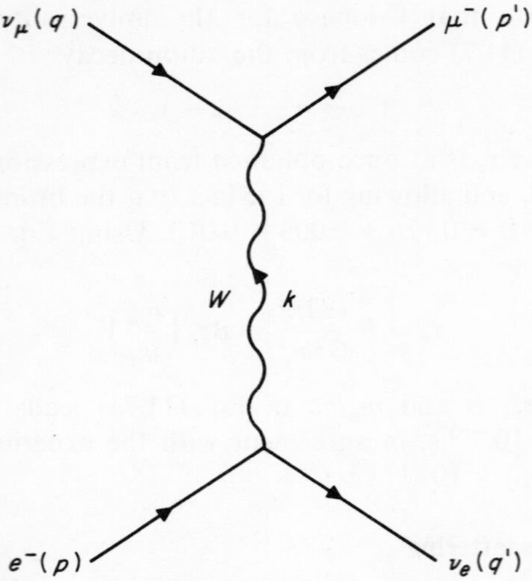

Fig. 11.8. The process $\nu_\mu + e^- \to \nu_e + \mu^-$ (inverse muon decay).

contact interaction theory if we make the identification (11.43). For finite m_W, the deviations from the latter theory are of order k^2/m_W^2. With $m_W \approx 80$ GeV, it is not surprising that experiments at the highest available neutrino energies are unable to detect such corrections, and we shall not consider the inverse muon decay process further. We only remark that the IVB theory is in good agreement with all experiments on this and similar processes, which in lowest-order perturbation theory are described by the exchange of one W boson.

11.6.3 The leptonic decay of the W boson

Lastly, we consider the leptonic decay process

$$W^+ \to l^+ + \nu_l \tag{11.79}$$

represented by the Feynman graph in Fig. 11.9. The derivation of the decay rate for this process is left as an exercise for the reader and gives

$$\Gamma_l = \frac{g_W^2 m_W}{6\pi}\left(1 - \frac{m_l^2}{m_W^2}\right)^2\left(1 + \frac{m_l^2}{2m_W^2}\right). \tag{11.80}$$

For $m_l^2 \ll m_W^2$ (which is certainly the case for the e, μ and τ leptons) this reduces to

$$\Gamma_l = \frac{g_W^2 m_W}{6\pi}. \tag{11.81}$$

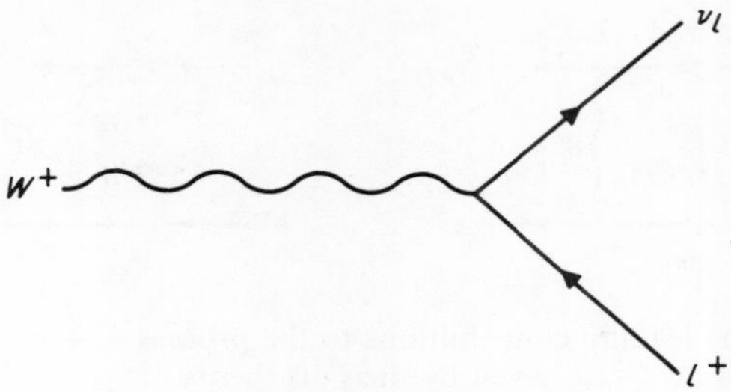

Fig. 11.9. The leptonic decay modes of the W^+ boson.

The life time τ of the W boson is related to the total decay rate Γ_{tot} for all decay channels. Apart from the three leptonic modes (11.79) for $l = e, \mu, \tau$, there may be further leptons and the W will also possess hadronic decay modes. Hence, $\Gamma_{\text{tot}} > 3\Gamma_l$ and

$$\tau = \frac{1}{\Gamma_{\text{tot}}} < \frac{2\pi}{g_W^2 m_W}. \tag{11.82}$$

From Eqs. (11.43) and (11.70), and taking $m_W \approx 80$ GeV, we obtain the upper bound $\tau \lesssim 10^{-24}$ s. This is such an extremely short life time that it is not possible to observe a free W boson directly. Instead, it has to be detected by its decay products, as in the $p\bar{p}$ collider experiment referred to earlier.

11.7 DIFFICULTIES WITH THE IVB THEORY

In spite of its successes, the IVB theory presents serious difficulties. It cannot describe such processes as

$$\nu_\mu + e^- \to \nu_\mu + e^- \tag{11.83}$$

and

$$\bar{\nu}_\mu + e^- \to \bar{\nu}_\mu + e^-, \tag{11.84}$$

which are not forbidden by any conservation laws. With the interaction (11.7), the leading contributions to the process (11.83) come from the Feynman graphs in Fig. 11.10. This is in contrast to processes such as

$$\nu_e + e^- \to \nu_e + e^- \tag{11.85}$$

and the inverse muon decay process

$$\nu_\mu + e^- \to \nu_e + \mu^- \tag{11.86}$$

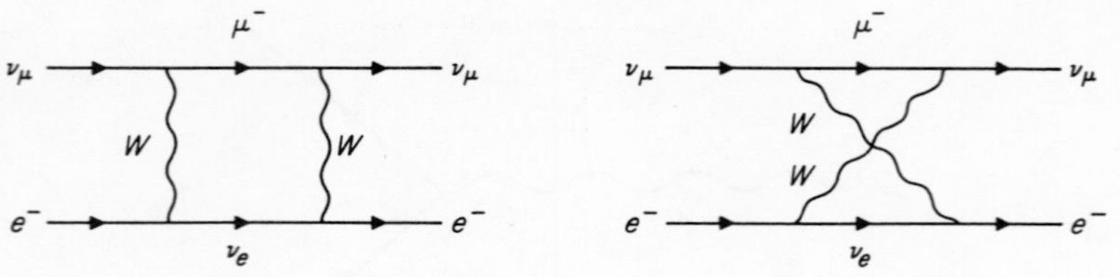

Fig. 11.10. The leading contributions to the process $\nu_\mu + e^- \rightarrow \nu_\mu + e^-$, as given by the IVB theory.

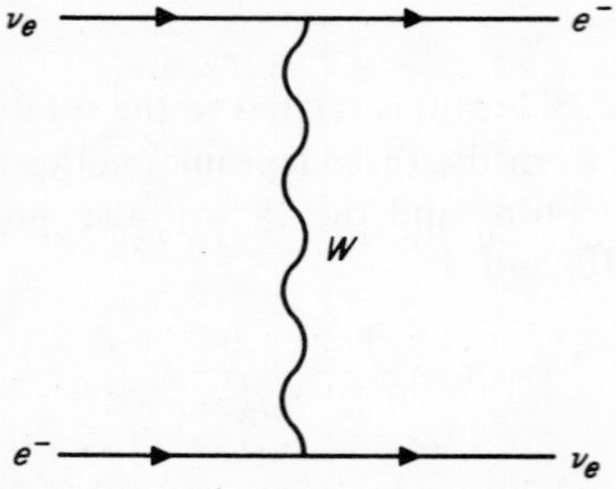

Fig. 11.11. The leading contribution to the process $\nu_e + e^- \rightarrow \nu_e + e^-$, as given by the IVB theory.

which in the lowest order involve the exchange of one W boson, as in Figs. 11.11 and 11.8. This difference is due to the fact that each term of the leptonic currents, J_α and $J_\alpha^\dagger$, Eqs. (11.6), always couples a neutral and a charged lepton. For example, the term in the interaction (11.7) which annihilates a muon-neutrino necessarily creates a μ^- or destroys a μ^+. Hence the leading contribution to the process (11.83) must involve the exchange of two W bosons, as shown in Fig. 11.10.

We found earlier that the coupling constant g_W is small. Hence one would expect the cross-sections for two-boson-exchange processes, like Eq. (11.83), to be small compared with those for one-boson-exchange processes, like Eq. (11.85). Unfortunately, the loop integrals to which the Feynman graphs 11.10 give rise are divergent, and the IVB theory is not renormalizable, as we shall discuss further in a moment. This means we simply do not know how to calculate with it sensibly if loop integrals are involved. Hence IVB theory represents a phenomenological theory only; it allows one to calculate those

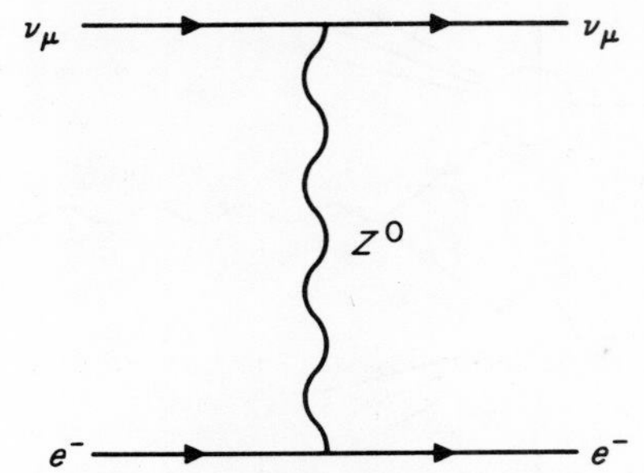

Fig. 11.12. A one-boson-exchange contribution to the process $\nu_\mu + e^- \to \nu_\mu + e^-$, not contained within the IVB theory.

processes which do not involve loop integrals in the lowest order of perturbation theory.

On the other hand, the measured cross-sections for processes like $\nu_\mu - e$ scattering, Eq. (11.83), are comparable to the cross-sections for processes like $\nu_e - e$ scattering, Eq. (11.85). This suggests that the IVB interaction (11.7) is not complete and that the leptonic interaction contains additional terms which allow the $\nu_\mu - e$ scattering to occur as a one-boson-exchange process. If we wish to retain lepton number conservation, since this is well established, these extra terms will lead to Feynman diagrams like that in Fig. 11.12, involving the exchange of a neutral vector boson Z^0 between an electron–electron current and a neutrino–neutrino current. (Currents involving the emission or absorption of a neutral vector boson are called *neutral currents*, in contrast to the *charged currents*, J_α and $J_\alpha^\dagger$, which involve the emission or absorption of a charged vector boson.) The existence of the Z^0 boson and the presence of the neutral-current terms in the interaction are required by the standard electro–weak theory, for this theory to be renormalizable. The prediction of processes like $\nu_\mu - e$ and $\bar{\nu}_\mu - e$ scattering, before they had been observed, was one of the great early successes of this theory.

Recently, the Z^0 boson has been detected in the same $p\bar{p}$ collider experiments in which the W boson was observed (see Section 11.2). The Feynman graph for Z^0 boson production is shown in Fig. 11.13. In this process, a quark (q) and an anti-quark ($\bar{q}$), produced in a $p\bar{p}$ collision, combine to form a Z^0 boson which then decays into a charged lepton pair: e^+e^-, $\mu^+\mu^-$,.... (The corresponding graph for the W boson production process was shown in Fig. 11.3.) Up to the present, a total of 14 such events have been identified in two independent experiments by the UA1 and UA2 collaborations. Analysis of just four of their events by each group leads to the

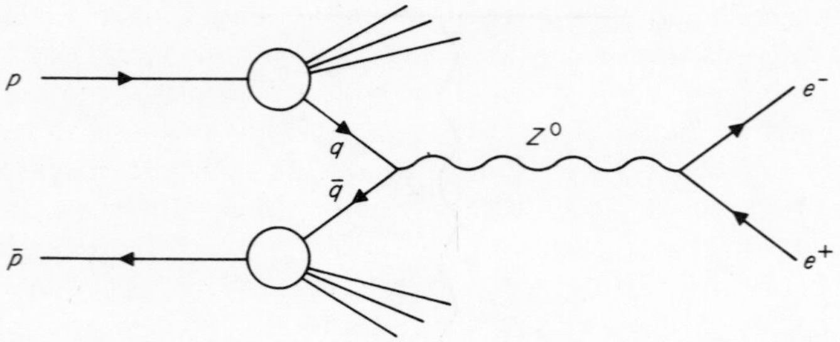

Fig. 11.13. The production of the Z^0 boson in a $p\bar{p}$ collision and its leptonic decay $Z^0 \to e^+ + e^-$.

following values ‡ for the mass of the Z^0 boson

$$\left.\begin{array}{ll} m_Z = (95.6 \pm 1.4 \pm 2.9)\ \text{GeV} & (\text{UA1}) \\ m_Z = (91.9 \pm 1.3 \pm 1.4)\ \text{GeV} & (\text{UA2}) \end{array}\right\}. \tag{11.87a}$$

(More accurate values should be obtained from the analysis of all 14 events in due course.) These values should be compared with the prediction

$$m_Z = \left(93.8 \begin{array}{l} +2.4 \\ -2.2 \end{array}\right)\ \text{GeV}, \tag{11.87b}$$

of the standard electro–weak theory (see Section 14.1).

Finally, we return to the question of the renormalizability of the IVB theory. We see from Eq. (11.30), i.e.

$$\mathrm{i}D_{\mathrm{F}\alpha\beta}(k, m_W) = \frac{\mathrm{i}(-g_{\alpha\beta} + k_\alpha k_\beta / m_W^2)}{k^2 - m_W^2 + \mathrm{i}\varepsilon}, \tag{11.30}$$

that the W propagator behaves like a constant for large momenta k. Hence the loop integrals in the amplitudes of the Feynman graphs 11.10 are quadratically divergent,

$$\int^\Lambda \frac{\mathrm{d}^4 k}{k^2} \sim \Lambda^2, \tag{11.88}$$

where Λ is a high-energy cut-off parameter. More generally, in more complicated higher-order graphs, each additional W propagator which forms part of a loop will lead to a factor of the order Λ^2/m_W^2, where the factor Λ^2 follows purely from dimensional arguments. Hence, as we consider a given process in higher orders of perturbation theory, new, progressively more

‡ See the papers by B. Sadoulet (UA1) and by A. Clark (UA2) in the Proceedings of the 1983 International Symposium, cited for Eq. (11.8b).

severe divergences arise. To cancel these, we would have to introduce additional renormalization constants at each stage, ending up with infinitely many such constants. Clearly, there is no way of determining these constants or making any predictions from such a theory, i.e. it is non-renormalizable.

As we saw in Chapter 9, the situation is quite different in QED, which contains only three renormalization constants which can be absorbed into the mass and the charge of the electron, and which is renormalizable. This difference between QED and the IVB theory is reflected in the fact that in QED the degree of divergence of a primitively divergent graph depends on the external lines only and is independent of the number of vertices n [see Eq. (9.112)]. For the IVB theory, on the other hand, the degree of divergence increases with n (see Problem 11.4). The divergence difficulties of the IVB theory stem from the term $k^\alpha k^\beta/m_W^2$ in the W propagator. This term is absent from the photon propagator $-\mathrm{i}g^{\alpha\beta}/k^2$. Hence the latter acts as a convergence factor for large k in a loop integral with respect to k. In the case of the photon propagator we were able to avoid a term in $k^\alpha k^\beta$ by exploiting gauge invariance. The renormalizability of the theory of weak interactions is similarly achieved by formulating it as a gauge theory.

PROBLEMS

11.1 Prove that for the muon decay process $\mu^- \to e^- + \bar{\nu}_e + \nu_\mu$ the maximum possible energy of the electron in the muon rest frame is

$$(m_\mu^2 + m_e^2)/2m_\mu.$$

11.2 For the leptonic process $W^+ \to l^+\nu_l$, derive the decay rate (11.80).

11.3 Show that in the IVB theory the dimensionality K of the Feynman amplitude of a Feynman graph is given by

$$K = n + 4 - \tfrac{3}{2}f_e - 2b_e,$$

where $f_e(b_e)$ is the number of external fermion (boson) lines of the graph, and n is the number of its vertices. (The corresponding result (9.112) for QED is independent of n.)

11.4 Show that the differential decay rate for a polarized muon is given by

$$\mathrm{d}\Gamma = \frac{1}{3}\frac{G^2}{(2\pi)^4}\,\mathrm{d}^3\mathbf{p}\,m_\mu[3m_\mu - 4E + \cos\theta(m_\mu - 4E)]$$

where $(E, \mathbf{p})$ is the energy–momentum four-vector of the emitted electron in the rest-frame of the muon. θ is the angle which $\mathbf{p}$ makes with the direction of the muon spin, and the electron mass has been neglected, i.e. the above formula applies only to high-energy electrons produced in the decay. The $\cos\theta$ term in the formula demonstrates the parity violation of the process.

11.5 Show that, in the centre-of-mass frame, the differential cross-section for the inverse muon decay process

$$\nu_\mu + e^- \to \nu_e + \mu^-, \qquad \text{(A)}$$

which results from the Fermi contact interaction (11.44), is given by

$$\frac{d\sigma}{d\Omega} = \frac{G^2E^2}{\pi^2}$$

where all lepton masses have been neglected compared with the centre-of-mass energy E of the incoming neutrino.

Show that for the related process

$$\bar{\nu}_e + e^- \rightarrow \mu^- + \bar{\nu}_\mu$$

the corresponding differential cross-section is given by

$$\frac{d\sigma}{d\Omega} = \frac{G^2E^2}{4\pi^2}(1 - \cos\theta)^2$$

where θ is the angle between the incoming neutrino and the outgoing muon.

For the second reaction, the cross-section vanishes in the forward direction ($\theta = 0$). Deduce this directly by considering the helicities of the leptons in the process and the contact nature of the interaction. Show that the same line of argument does not rule out forward scattering for the inverse muon decay process (A).

CHAPTER 12

A gauge theory of weak interactions

In this chapter we shall attempt to formulate a gauge theory of weak interactions. We shall first of all illustrate the characteristic features of a gauge theory for the simplest such theory, namely QED. For weak interactions, the role of the electromagnetic current is taken by the weak leptonic currents J_α and $J_\alpha^\dagger$, Eqs. (11.6), and the role of the photon is taken by the $W^\pm$ bosons. We shall find that the requirement of gauge invariance leads to the neutral leptonic current, mentioned at the end of the last chapter, and to a third, electrically neutral, vector boson. At the same time, a unification of electromagnetic and weak interactions is achieved in a natural way. The gauge invariance of the theory developed in this chapter necessitates all leptons and vector bosons to be massless. This difficulty is discussed at the end of this chapter, but it will only be resolved in the next chapter.

In what follows, we confine ourselves to the extension of gauge theories to weak interactions. The original extension was first made by Yang and Mills in 1954 in the context of strong interactions, where it eventually led to the current theory of strong interactions, quantum chromodynamics (QCD).‡

‡ For a wider discussion of gauge theories, the reader is referred to I. J. R. Aitchison, *An Informal Introduction to Gauge Field Theories*, Cambridge University Press, Cambridge, 1982; I. J. R. Aitchison and A. J. G. Hey, *Gauge Theories in Particle Physics*, A. Hilger, Bristol, 1982; D. Bailin, *Weak Interactions*, 2nd edn, A. Hilger, Bristol, 1982; C. Itzykson and J. B. Zuber, *Quantum Field Theory*, McGraw-Hill, New York, 1980; E. Leader and E. Predazzi, *An Introduction to Gauge Theories and the New Physics*, Cambridge University Press, Cambridge, 1982; T. D. Lee, *Particle Physics and Introduction to Field Theory*, Harcourt, New York, 1981; P. Ramond, *Field Theory: A Modern Primer*, Benjamin/Cummins, New York, 1981; and J. C. Taylor, *Gauge Theories of Weak Interactions*, Cambridge University Press, Cambridge, 1976.

12.1 THE SIMPLEST GAUGE THEORY: QED

In Section 4.5 we introduced the electromagnetic interaction into the free-fermion Lagrangian density

$$\mathscr{L}_0 = \bar{\psi}(x)(\mathrm{i}\gamma^\mu\partial_\mu - m)\psi(x) \tag{12.1}$$

through the minimal substitution

$$\partial_\mu \to D_\mu = [\partial_\mu + \mathrm{i}qA_\mu(x)] \tag{4.64b}$$

where q is the charge of the particle annihilated by the field $\psi(x)$. We required invariance of the resulting theory, i.e. of the Lagrangian density

$$\mathscr{L} = \bar{\psi}(x)(\mathrm{i}\gamma^\mu D_\mu - m)\psi(x) = \mathscr{L}_0 - q\bar{\psi}(x)\gamma^\mu\psi(x)A_\mu(x),$$

under gauge transformations of the electromagnetic field

$$A_\mu(x) \to A'_\mu(x) = A_\mu(x) + \partial_\mu f(x) \tag{12.2a}$$

where $f(x)$ is a real differentiable function. This invariance was ensured if, coupled with the transformation (12.2a), the Dirac fields $\psi(x)$ and $\bar{\psi}(x)$ underwent the transformations

$$\left.\begin{aligned}\psi(x) &\to \psi'(x) = \psi(x)\,\mathrm{e}^{-\mathrm{i}qf(x)}\\ \bar{\psi}(x) &\to \bar{\psi}'(x) = \bar{\psi}(x)\,\mathrm{e}^{\mathrm{i}qf(x)}\end{aligned}\right\}. \tag{12.2b}$$

Eqs. (12.2b) have the form of a *local* phase transformation. We shall refer to the *coupled* transformation (12.2a) and (12.2b) as a gauge transformation and to any theory which is invariant under such coupled transformations as a gauge theory. QED is the simplest example of a gauge theory.

In the above discussion we started from the form of the electromagnetic interaction. The invariance of the theory under the gauge transformations (12.2a) of the electromagnetic potentials then required the Dirac fields ψ and $\bar{\psi}$ simultaneously to undergo the local phase transformation (12.2b). We can try and reverse this argument and start from the invariance of the free Lagrangian density $\mathscr{L}_0$ under the *global* phase transformation

$$\left.\begin{aligned}\psi(x) &\to \psi'(x) = \psi(x)\,\mathrm{e}^{-\mathrm{i}\alpha}\\ \bar{\psi}(x) &\to \bar{\psi}'(x) = \bar{\psi}(x)\,\mathrm{e}^{\mathrm{i}\alpha}\end{aligned}\right\} \tag{12.3}$$

when α is a real number. We know from the discussion following Eq. (2.40) that this invariance ensures 'current conservation', i.e. the current

$$s^\mu(x) = q\bar{\psi}(x)\gamma^\mu\psi(x) \tag{12.4}$$

satisfies $\partial_\mu s^\mu(x) = 0$, so that the charge

$$Q = q\int \mathrm{d}^3\mathbf{x}\,\psi^\dagger(x)\psi(x) \tag{12.5}$$

is conserved.

Invariance under the global phase transformation (12.3) allows us to change the phase of the field by the same amount at each space–time point. This appears unnecessarily restrictive in a local field theory. We shall therefore demand invariance with respect to the more general local phase transformations (12.2b). Under these transformations, the free-field Lagrangian density $\mathscr{L}_0$ becomes

$$\mathscr{L}_0 \to \mathscr{L}'_0 = \mathscr{L}_0 + q\bar{\psi}(x)\gamma^\mu\psi(x)\partial_\mu f(x), \tag{12.6}$$

i.e. $\mathscr{L}_0$ is not invariant. Invariance of the theory is then restored if we can augment $\mathscr{L}_0$ by a term $\mathscr{L}_\text{I}$ such that the new Lagrangian density $\mathscr{L} = \mathscr{L}_0 + \mathscr{L}_\text{I}$ is invariant. This can be achieved by associating with the 'matter field' $\psi(x)$ a 'gauge field' $A_\mu(x)$ which transforms according to the gauge transformation (12.2a). The interaction between matter and gauge fields is then specified by making the minimal substitution (4.64b) in the free-field Lagrangian density $\mathscr{L}_0$, i.e. by replacing the ordinary derivative $\partial_\mu\psi(x)$ by the 'covariant derivative'

$$D_\mu\psi(x) = [\partial_\mu + \text{i}qA_\mu(x)]\psi(x). \tag{12.7}$$

$\mathscr{L}_0$ thus goes over into

$$\begin{aligned}\mathscr{L} &= \bar{\psi}(x)(\text{i}\gamma^\mu D_\mu - m)\psi(x)\\ &= \mathscr{L}_0 - q\bar{\psi}(x)\gamma^\mu\psi(x)A_\mu(x) \equiv \mathscr{L}_0 + \mathscr{L}_\text{I}.\end{aligned} \tag{12.8}$$

Under the coupled gauge transformation (12.2a) and (12.2b), the covariant derivative $D_\mu\psi(x)$ undergoes the transformation

$$D_\mu\psi(x) \to \text{e}^{-\text{i}qf(x)}D_\mu\psi(x), \tag{12.9}$$

i.e. it transforms in the same way as the field $\psi(x)$ itself, Eq. (12.2b). Hence $\mathscr{L}$ is invariant under gauge transformations, as required.

To summarize this approach: if one takes as the basic requirement invariance with respect to local phase transformations of the matter field, one is led to introduce a gauge field coupled to the matter field through the replacement of the ordinary derivative $\partial_\mu\psi$ by the covariant derivative $D_\mu\psi$.

None of these arguments can claim to derive the electromagnetic interaction but they suggest a gauge-invariant form for it which may or may not be confirmed by experiment. For electrodynamics where the interaction is known from the classical limit, these arguments may appear superfluous. However, we shall see that they have been extremely successful in suggesting the form of the interaction in other situations where there is no classical limit to guide us.

12.2 GLOBAL PHASE TRANSFORMATIONS AND CONSERVED WEAK CURRENTS

We shall now apply the programme outlined at the end of the last section to formulate the theory of weak interactions as a gauge theory. As a first step we must find a set of global phase transformations which leaves the free-lepton Lagrangian density invariant, leading to conservation of the weak currents $J_\alpha(x)$, $J^\dagger_\alpha(x)$, Eqs. (11.6).

To begin with, we shall assume that all leptons are massless. At the end of the chapter we shall return to the problems to which non-vanishing lepton masses give rise. The free-lepton Lagrangian density is then given by

$$\mathscr{L}_0 = \mathrm{i}[\bar{\psi}_l(x)\partial\!\!\!/\psi_l(x) + \bar{\psi}_{\nu_l}(x)\partial\!\!\!/\psi_{\nu_l}(x)] \tag{12.10}$$

where, as in the following, summation over all different kinds of leptons is understood: $l = e, \mu, \ldots$.

In Chapter 11, we found that the leptonic currents, and consequently the leptonic interaction, involve only the left-handed lepton fields, Eqs. (11.13)–(11.16). We shall therefore write Eq. (12.10) in terms of left- and right-handed fields. For any Dirac spinor $\psi(x)$, these fields are defined by

$$\left.\begin{aligned}\psi^{\mathrm{L}}(x) &= P_{\mathrm{L}}\psi(x)\\ \psi^{\mathrm{R}}(x) &= P_{\mathrm{R}}\psi(x)\end{aligned}\right\} \equiv \tfrac{1}{2}(1 \mp \gamma_5)\psi(x), \tag{12.11}$$

and Eq. (12.10) becomes

$$\begin{aligned}\mathscr{L}_0 = \mathrm{i}[&\bar{\psi}^{\mathrm{L}}_l(x)\partial\!\!\!/\psi^{\mathrm{L}}_l(x) + \bar{\psi}^{\mathrm{L}}_{\nu_l}(x)\partial\!\!\!/\psi^{\mathrm{L}}_{\nu_l}(x)\\ &+ \bar{\psi}^{\mathrm{R}}_l(x)\partial\!\!\!/\psi^{\mathrm{R}}_l(x) + \bar{\psi}^{\mathrm{R}}_{\nu_l}(x)\partial\!\!\!/\psi^{\mathrm{R}}_{\nu_l}(x)].\end{aligned} \tag{12.12}$$

We now combine the fields ψ^{L}_l and $\psi^{\mathrm{L}}_{\nu_l}$ into a two-component field

$$\Psi^{\mathrm{L}}_l(x) = \begin{pmatrix}\psi^{\mathrm{L}}_{\nu_l}(x)\\ \psi^{\mathrm{L}}_l(x)\end{pmatrix} \tag{12.13a}$$

and, correspondingly,

$$\bar{\Psi}^{\mathrm{L}}_l(x) = (\bar{\psi}^{\mathrm{L}}_{\nu_l}(x), \bar{\psi}^{\mathrm{L}}_l(x)). \tag{12.13b}$$

In terms of these fields, Eq. (12.12) becomes

$$\mathscr{L}_0 = \mathrm{i}[\bar{\Psi}^{\mathrm{L}}_l(x)\partial\!\!\!/\Psi^{\mathrm{L}}_l(x) + \bar{\psi}^{\mathrm{R}}_l(x)\partial\!\!\!/\psi^{\mathrm{R}}_l(x) + \bar{\psi}^{\mathrm{R}}_{\nu_l}(x)\partial\!\!\!/\psi^{\mathrm{R}}_{\nu_l}(x)]. \tag{12.14}$$

Although Eq. (12.12) is symmetric between left- and right-handed fields, we have written Eq. (12.14) in a very unsymmetric way, i.e. we have not introduced two-component right-handed fields. We shall see that the left–right asymmetry of weak interactions can be described in terms of different transformation properties of the left- and right-handed fields. For the two-component left-handed fields, the possibility arises of two-dimensional transformations which leave bilinear forms $\bar{\Psi}^{\mathrm{L}}_l(x)(\ldots)\Psi^{\mathrm{L}}_l(x)$ invariant. For this purpose, we introduce the three 2×2 Hermitian matrices

$$\tau_1 = \begin{pmatrix}0 & 1\\ 1 & 0\end{pmatrix}, \qquad \tau_2 = \begin{pmatrix}0 & -\mathrm{i}\\ \mathrm{i} & 0\end{pmatrix}, \qquad \tau_3 = \begin{pmatrix}1 & 0\\ 0 & -1\end{pmatrix}, \tag{12.15}$$

which satisfy the commutation relations

$$[\tau_i, \tau_j] = 2i\varepsilon_{ijk}\tau_k \tag{12.16}$$

where ε_{ijk} is the usual completely antisymmetric tensor and, as throughout, summation over the repeated index $k(=1, 2, 3)$ is implied. (The τ-matrices are just the usual Pauli spin matrices.) The operator

$$U(\boldsymbol{\alpha}) \equiv \exp(i\alpha_j\tau_j/2) \tag{12.17}$$

is unitary for any three real numbers $\boldsymbol{\alpha} \equiv (\alpha_1, \alpha_2, \alpha_3)$, and the set of transformations

$$\left.\begin{aligned} \Psi_l^{\mathrm{L}}(x) &\to \Psi_l^{\mathrm{L}\prime}(x) = U(\boldsymbol{\alpha})\Psi_l^{\mathrm{L}}(x) \equiv \exp(i\alpha_j\tau_j/2)\Psi_l^{\mathrm{L}}(x) \\ \bar{\Psi}_l^{\mathrm{L}}(x) &\to \bar{\Psi}_l^{\mathrm{L}\prime}(x) = \bar{\Psi}_l^{\mathrm{L}}(x)U^{\dagger}(\boldsymbol{\alpha}) \equiv \bar{\Psi}_l^{\mathrm{L}}(x)\exp(-i\alpha_j\tau_j/2) \end{aligned}\right\} \tag{12.18a}$$

leaves the term $i\bar{\Psi}_l^{\mathrm{L}}(x)\not{\partial}\Psi_l^{\mathrm{L}}(x)$ in $\mathscr{L}_0$, Eq. (12.14), invariant.

The operators $U(\boldsymbol{\alpha})$ are 2×2 unitary matrices with the special property that $\det U(\boldsymbol{\alpha}) = +1$. They are therefore called SU(2) transformations.‡ The transformations (12.18a) can be regarded as two-dimensional global phase transformations, i.e. as a generalization of the one-dimensional global phase transformations (12.3).

The SU(2) transformation properties of the two-component left-handed lepton fields $\Psi_l^{\mathrm{L}}(x)$ are identical with those of the two-component spinors which describe spin $\frac{1}{2}$ particles in the non-relativistic Pauli theory of spin, and with those of the two-component isospinors which describe the neutron and the proton as different charge states of the nucleon. The two-component field $\Psi_l^{\mathrm{L}}(x)$ is therefore called a weak isospinor. In addition to two-component spinors, the Pauli spin theory also gives rise to one-component scalars (e.g. singlet spin states), three-component vectors (e.g. triplet states), etc., and these entities are characterized by the transformations induced in them by the basic spinor transformations. For example, any quantity invariant under these transformations is a scalar. All these concepts carry over to weak isospin, leading to quantities being classified according to their transformation properties under SU(2) transformations as weak isoscalars, weak isospinors, and so on.

So far, we have considered the left-handed lepton fields. We shall now define each right-handed lepton field to be a weak isoscalar, i.e. to be invariant under any SU(2) transformation:

$$\left.\begin{aligned} \psi_l^{\mathrm{R}}(x) &\to \psi_l^{\mathrm{R}\prime}(x) = \psi_l^{\mathrm{R}}(x), \qquad \psi_{\nu_l}^{\mathrm{R}}(x) \to \psi_{\nu_l}^{\mathrm{R}\prime}(x) = \psi_{\nu_l}^{\mathrm{R}}(x) \\ \bar{\psi}_l^{\mathrm{R}}(x) &\to \bar{\psi}_l^{\mathrm{R}\prime}(x) = \bar{\psi}_l^{\mathrm{R}}(x), \qquad \bar{\psi}_{\nu_l}^{\mathrm{R}}(x) \to \bar{\psi}_{\nu_l}^{\mathrm{R}\prime}(x) = \bar{\psi}_{\nu_l}^{\mathrm{R}}(x) \end{aligned}\right\}. \tag{12.18b}$$

‡ The set of all SU(2) transformations forms the SU(2) group. A group is called Abelian (non-Abelian) if its elements do (do not) commute. The SU(2) group is non-Abelian, since the τ-matrices and hence the operators (12.17) are non-commuting. The terms 'SU(2)', 'non-Abelian' and some other nomenclature which we shall introduce derive from group theory, but no knowledge of group theory is required in what follows.

It follows at once that the SU(2) transformations (12.18a) and (12.18b) of the lepton fields leave the free-lepton Lagrangian density $\mathscr{L}_0$, Eq. (12.14), invariant.

From this invariance of $\mathscr{L}_0$, the conservation of the leptonic currents $J_\alpha(x)$ and $J_\alpha^\dagger(x)$, Eqs. (11.6), follows. For infinitesimal α_j, the transformations (12.18a) reduce to

$$\left.\begin{aligned} \Psi_l^{\mathrm{L}}(x) &\to \Psi_l^{\mathrm{L}\prime}(x) = (1 + \mathrm{i}\alpha_j\tau_j/2)\Psi_l^{\mathrm{L}}(x) \\ \bar{\Psi}_l^{\mathrm{L}}(x) &\to \bar{\Psi}_l^{\mathrm{L}\prime}(x) = \bar{\Psi}_l^{\mathrm{L}}(x)(1 - \mathrm{i}\alpha_j\tau_j/2) \end{aligned}\right\} \tag{12.19}$$

whereas Eqs. (12.18b) for the right-handed fields remain unchanged. An argument analogous to that following Eq. (2.39) leads to the three conserved currents

$$J_i^\alpha(x) = \tfrac{1}{2}\bar{\Psi}_l^{\mathrm{L}}(x)\gamma^\alpha\tau_i\Psi_l^{\mathrm{L}}(x), \qquad i = 1, 2, 3, \tag{12.20}$$

which are called weak isospin currents. The corresponding conserved quantities

$$I_i^W = \int \mathrm{d}^3\mathbf{x} J_i^0(x) = \tfrac{1}{2}\int \mathrm{d}^3\mathbf{x}\Psi_l^{\mathrm{L}\dagger}(x)\tau_i\Psi_l^{\mathrm{L}}(x), \qquad i = 1, 2, 3, \tag{12.21}$$

are called weak isospin charges.

The leptonic currents $J^\alpha(x)$ and $J^{\alpha\dagger}(x)$, Eqs. (11.6), in terms of which the IVB theory was formulated, can be written as linear combinations of the conserved weak isospin currents $J_1^\alpha(x)$ and $J_2^\alpha(x)$. Using Eqs. (12.11) and (12.15), one obtains

$$\left.\begin{aligned} J^\alpha(x) &= 2[J_1^\alpha(x) - \mathrm{i}J_2^\alpha(x)] = \bar{\psi}_l(x)\gamma^\alpha(1 - \gamma_5)\psi_{\nu_l}(x) \\ J^{\alpha\dagger}(x) &= 2[J_1^\alpha(x) + \mathrm{i}J_2^\alpha(x)] = \bar{\psi}_{\nu_l}(x)\gamma^\alpha(1 - \gamma_5)\psi_l(x) \end{aligned}\right\}. \tag{12.22}$$

Most remarkably, the above development necessarily led to the conservation of a *third* current, namely the weak isospin current

$$\begin{aligned} J_3^\alpha(x) &= \tfrac{1}{2}\bar{\Psi}_l^{\mathrm{L}}(x)\gamma^\alpha\tau_3\Psi_l^{\mathrm{L}}(x) \\ &= \tfrac{1}{2}[\bar{\psi}_{\nu_l}^{\mathrm{L}}(x)\gamma^\alpha\psi_{\nu_l}^{\mathrm{L}}(x) - \bar{\psi}_l^{\mathrm{L}}(x)\gamma^\alpha\psi_l^{\mathrm{L}}(x)]. \end{aligned} \tag{12.23}$$

The current $J_3^\alpha(x)$ is called a *neutral* current since it couples either electrically neutral leptons or, like the electromagnetic current

$$s^\alpha(x) = -e\bar{\psi}_l(x)\gamma^\alpha\psi_l(x), \tag{12.24}$$

electrically charged leptons. This is in contrast to the *charged* currents $J^\alpha(x)$ and $J^{\alpha\dagger}(x)$ which couple electrically neutral with electrically charged leptons. Apart from a constant factor, the last term on the right-hand side of Eq. (12.23) is a part of the electromagnetic current (12.24). We have here a first indication that, in the theory we are developing, electromagnetic and weak processes will be interconnected.

The weak hypercharge current $J^{\alpha}_{Y}(x)$ is defined by

$$\begin{aligned} J^{\alpha}_{Y}(x) &= s^{\alpha}(x)/e - J^{\alpha}_{3}(x) \\ &= -\tfrac{1}{2}\bar{\Psi}^{L}_{l}(x)\gamma^{\alpha}\Psi^{L}_{l}(x) - \bar{\psi}^{R}_{l}(x)\gamma^{\alpha}\psi^{R}_{l}(x). \end{aligned} \tag{12.25}$$

The corresponding charge

$$Y = \int d^3\mathbf{x} J^{0}_{Y}(x) \tag{12.26}$$

is called the weak hypercharge.[‡] We see from Eq. (12.25) that Y is related to the electric charge Q and the weak isocharge I^{W}_{3} by

$$Y = Q/e - I^{W}_{3}. \tag{12.27}$$

The conservation of the electric charge Q and of the weak isocharge I^{W}_{3} implies conservation of the weak hypercharge Y.

We next determine the weak isospin charge I^{W}_{3} and the weak hypercharge Y of the different leptons. From Eqs. (12.13a) and (12.15) we have

$$\tau_3 \Psi^{L}_{l}(x) = \begin{pmatrix} 1 & 0 \\ 0 & -1 \end{pmatrix} \begin{pmatrix} \psi^{L}_{\nu_l}(x) \\ \psi^{L}_{l}(x) \end{pmatrix} = \begin{pmatrix} \psi^{L}_{\nu_l}(x) \\ -\psi^{L}_{l}(x) \end{pmatrix}.$$

It follows that the weak isospin charge I^{W}_{3} has the value $+\frac{1}{2}$ for a left-handed ν_l neutrino, and the value $-\frac{1}{2}$ for a left-handed l^- lepton. A more proper derivation of these results is obtained by substituting the expansions (4.38) of the Dirac fields in Eq. (12.21). If $|l^-, \mathrm{L}\rangle$ and $|\nu_l, \mathrm{L}\rangle$ are one-particle states which respectively contain one left-handed l^- lepton and one left-handed ν_l neutrino, one finds that

$$I^{W}_{3}|l^-, \mathrm{L}\rangle = -\tfrac{1}{2}|l^-, \mathrm{L}\rangle, \qquad I^{W}_{3}|\nu_l, \mathrm{L}\rangle = +\tfrac{1}{2}|\nu_l, \mathrm{L}\rangle. \tag{12.28a}$$

Since the right-handed lepton fields are isoscalars, it similarly follows that I^{W}_{3} has the value 0 for right-handed l^- leptons and for right-handed ν_l neutrinos, i.e.

$$I^{W}_{3}|l^-, \mathrm{R}\rangle = 0, \qquad I^{W}_{3}|\nu_l, \mathrm{R}\rangle = 0. \tag{12.28b}$$

The values of the weak hypercharge Y for the different leptons follows from Eqs. (12.27) and (12.28):

$$\left.\begin{aligned} Y|l^-, \mathrm{L}\rangle = -\tfrac{1}{2}|l^-, \mathrm{L}\rangle, \qquad & Y|\nu_l, \mathrm{L}\rangle = -\tfrac{1}{2}|\nu_l, \mathrm{L}\rangle \\ Y|l^-, \mathrm{R}\rangle &= -|l^-, \mathrm{R}\rangle \\ Y|\nu_l, \mathrm{R}\rangle &= 0 \end{aligned}\right\} \tag{12.29}$$

i.e. for the left-handed states of l^- and ν_l leptons, Y has the value $-\frac{1}{2}$; for the right-handed states of l^- and ν_l leptons, it has the values -1 and 0, respectively.

[‡] These names have their origin in the formal similarity which exists with corresponding quantities and relations for the strong interactions.

Above we deduced the conservation of weak hypercharge from the conservation of Q and of I_3^W, using Eq. (12.27). The conservation of weak hypercharge also follows directly from the invariance of the free-lepton Lagrangian density (12.14) under the simultaneous transformations

$$\left.\begin{aligned} \Psi_l^{\mathrm{L}}(x) &\to \Psi_l^{\mathrm{L}\prime}(x) = \mathrm{e}^{-\mathrm{i}\beta/2}\Psi_l^{\mathrm{L}}(x) \\ \psi_l^{\mathrm{R}}(x) &\to \psi_l^{\mathrm{R}\prime}(x) = \mathrm{e}^{-\mathrm{i}\beta}\psi_l^{\mathrm{R}}(x) \\ \psi_{\nu_l}^{\mathrm{R}}(x) &\to \psi_{\nu_l}^{\mathrm{R}\prime}(x) = \psi_{\nu_l}^{\mathrm{R}}(x) \end{aligned}\right\} \tag{12.30a}$$

and the corresponding transformations of the adjoint fields, i.e.

$$\bar{\Psi}_l^{\mathrm{L}}(x) \to \bar{\Psi}_l^{\mathrm{L}\prime}(x) = \bar{\Psi}_l^{\mathrm{L}}(x)\,\mathrm{e}^{\mathrm{i}\beta/2}, \text{ etc.,} \tag{12.30b}$$

where β is an arbitrary real number. We can write the transformation (12.30) more concisely as

$$\psi(x) \to \psi'(x) = \mathrm{e}^{\mathrm{i}\beta Y}\psi(x), \qquad \bar{\psi}(x) \to \bar{\psi}'(x) = \bar{\psi}(x)\,\mathrm{e}^{-\mathrm{i}\beta Y}, \tag{12.31}$$

where $\psi(x)$ denotes any one of the lepton fields $\psi_{\nu_l}^{\mathrm{L}}$, ψ_l^{L}, $\psi_{\nu_l}^{\mathrm{R}}$ and ψ_l^{R}, and Y is the weak hypercharge of the particle annihilated by the field $\psi(x)$. These transformations are similar in form to the global phase transformations (12.3) of QED and, like the latter, are called U(1) phase transformations since $U = \mathrm{e}^{\mathrm{i}\beta Y}$ is just a one-dimensional unitary matrix, i.e. a complex number of unit modulus. (The U(1) transformations of course also form a group.)

The line of reasoning we have used above can be inverted. We can start from the invariance of the free-lepton Lagrangian density (12.14): (i) under the global SU(2) transformations (12.18) leading to conservation of the weak isospin charges I_i^W, and (ii) under the global U(1) transformations (12.30), leading to conservation of the weak hypercharge Y. The conservation of the electric charge then follows from Eq. (12.27).

12.3 THE GAUGE-INVARIANT ELECTRO–WEAK INTERACTION

We next generalize the above SU(2) and U(1) transformations from global to local phase transformations. The development will be very similar to that for QED in Section 12.1. In order to retain invariance under local phase transformations, we shall have to introduce gauge fields, and this will automatically generate the interactions.

We shall start with the SU(2) transformations and replace the global transformations (12.18) by the local phase transformations

$$\left.\begin{aligned} &\Psi_l^{\mathrm{L}}(x) \to \Psi_l^{\mathrm{L}\prime}(x) = \exp\,[\mathrm{i}g\tau_j\omega_j(x)/2]\Psi_l^{\mathrm{L}}(x) \\ &\bar{\Psi}_l^{\mathrm{L}}(x) \to \bar{\Psi}_l^{\mathrm{L}\prime}(x) = \bar{\Psi}_l^{\mathrm{L}}(x)\exp\,[-\mathrm{i}g\tau_j\omega_j(x)/2] \\ &\psi_l^{\mathrm{R}}(x) \to \psi_l^{\mathrm{R}\prime}(x) = \psi_l^{\mathrm{R}}(x), \qquad \psi_{\nu_l}^{\mathrm{R}}(x) \to \psi_{\nu_l}^{\mathrm{R}\prime}(x) = \psi_{\nu_l}^{\mathrm{R}}(x) \\ &\bar{\psi}_l^{\mathrm{R}}(x) \to \bar{\psi}_l^{\mathrm{R}\prime}(x) = \bar{\psi}_l^{\mathrm{R}}(x), \qquad \bar{\psi}_{\nu_l}^{\mathrm{R}}(x) \to \bar{\psi}_{\nu_l}^{\mathrm{R}\prime}(x) = \bar{\psi}_{\nu_l}^{\mathrm{R}}(x). \end{aligned}\right\} \tag{12.32a}$$

Here $\omega_j(x)$, $j = 1, 2, 3$, are three arbitrary real differentiable functions of x, and g is a real constant which will later be identified with a coupling constant.

Applying the transformations (12.32a) to the free-lepton Lagrangian density (12.14), the differential operator $\not{\partial}$ in the spinor term will also act on the functions $\omega_j(x)$ in the exponent. Hence Eq. (12.14) is not invariant under this transformation but transforms according to

$$\mathscr{L}_0 \to \mathscr{L}_0' = \mathscr{L}_0 + \delta\mathscr{L}_0 \equiv \mathscr{L}_0 - \tfrac{1}{2}g\bar{\Psi}_l^{\mathrm{L}}(x)\tau_j\not{\partial}\omega_j(x)\Psi_l^{\mathrm{L}}(x). \tag{12.33}$$

We shall obtain an invariant Lagrangian density if, analogously to the replacement (12.7) in QED, we replace the ordinary derivatives $\partial^\mu\Psi_l^{\mathrm{L}}(x)$ in Eq. (12.14) by the covariant derivatives,

$$\partial^\mu\Psi_l^{\mathrm{L}}(x) \to D^\mu\Psi_l^{\mathrm{L}}(x) = [\partial^\mu + ig\tau_j W_j^\mu(x)/2]\Psi_l^{\mathrm{L}}(x), \tag{12.34}$$

so that $\mathscr{L}_0$ goes over into

$$\tilde{\mathscr{L}}_0 = \mathrm{i}[\bar{\Psi}_l^{\mathrm{L}}(x)\not{D}\Psi_l^{\mathrm{L}}(x) + \bar{\psi}_l^{\mathrm{R}}(x)\not{\partial}\psi_l^{\mathrm{R}}(x) + \bar{\psi}_{\nu_l}^{\mathrm{R}}(x)\not{\partial}\psi_{\nu_l}^{\mathrm{R}}(x)]. \tag{12.35}$$

In Eq. (12.34), we had to introduce three real gauge fields $W_j^\mu(x)$, compared to the one gauge field $A^\mu(x)$ of QED, as there are now three conserved charges I_j^W and as the gauge transformation (12.32a) now contains three arbitrary functions $\omega_j(x)$.

For the modified Lagrangian density $\tilde{\mathscr{L}}_0$ to be invariant, the transformations (12.32a) of the lepton fields must be coupled to transformations of the gauge fields,

$$W_j^\mu(x) \to W_j^{\mu\prime}(x) = W_j^\mu(x) + \delta W_j^\mu(x),$$

such that the covariant derivatives $D^\mu\Psi_l^{\mathrm{L}}(x)$ transform in the same way as the fields $\Psi_l^{\mathrm{L}}(x)$ themselves, i.e.

$$D^\mu\Psi_l^{\mathrm{L}}(x) \to \exp[ig\tau_j\omega_j(x)/2]D^\mu\Psi_l^{\mathrm{L}}(x). \tag{12.36}$$

For finite functions $\omega_j(x)$, the resulting transformation law for the gauge fields $W_j^\mu(x)$ is quite complicated. However, it suffices to consider the transformations for infinitesimal functions $\omega_j(x)$. In the appendix to this chapter (Section 12.6), we shall show that the required infinitesimal transformations are given by

$$\begin{aligned} W_i^\mu(x) \to W_i^{\mu\prime}(x) &= W_i^\mu(x) + \delta W_i^\mu(x) \\ &\equiv W_i^\mu(x) - \partial^\mu\omega_i(x) - g\varepsilon_{ijk}\omega_j(x)W_k^\mu(x) \quad [\text{small } \omega_j(x)]. \end{aligned} \tag{12.32b}$$

Before discussing the implications of the transformation laws (12.32), we shall consider the global U(1) transformations (12.31). The corresponding local phase transformations are

$$\left.\begin{aligned}\psi(x) \to \psi'(x) &= \exp\,[\mathrm{i}g' Yf(x)]\psi(x) \\ \bar{\psi}(x) \to \bar{\psi}'(x) &= \bar{\psi}(x)\exp\,[-\mathrm{i}g' Yf(x)]\end{aligned}\right\} \tag{12.37a}$$

where g' is a real number which will be determined later, $f(x)$ is an arbitrary real differentiable function, and $Y = -\frac{1}{2}, -1, 0$ is the weak hypercharge associated with the fields $\Psi_l^{\mathrm{L}}(x)$, $\psi_l^{\mathrm{R}}(x)$ and $\psi_{\nu_l}^{\mathrm{R}}(x)$ respectively. The analogy with QED is even closer in this case. One obtains a Lagrangian density invariant under the local phase transformations (12.37a) if in $\mathscr{L}_0$, Eq. (12.14), one replaces the ordinary derivatives by covariant derivatives,

$$\partial^\mu \psi(x) \to D^\mu \psi(x) = [\partial^\mu + \mathrm{i}g' YB^\mu(x)]\psi(x), \tag{12.38}$$

where ψ is any one of the four lepton fields ψ_l^{L}, $\psi_{\nu_l}^{\mathrm{L}}$, ψ_l^{R} and $\psi_{\nu_l}^{\mathrm{R}}$, and the real gauge field $B^\mu(x)$, which has been introduced, transforms like

$$B^\mu(x) \to B^{\mu\prime}(x) = B^\mu(x) - \partial^\mu f(x). \tag{12.37b}$$

We have seen that making the replacement (12.34) in Eq. (12.14) yields a Lagrangian density which is invariant under the SU(2) gauge transformations (12.32a) and (12.32b), while the replacement (12.38) yields a Lagrangian density invariant under the U(1) gauge transformations (12.37a) and (12.37b). If we make both replacements (12.34) and (12.38) simultaneously in Eq. (12.14), we obtain the leptonic Lagrangian density

$$\mathscr{L}^{\mathrm{L}} = \mathrm{i}[\bar{\Psi}_l^{\mathrm{L}}(x)\not{D}\Psi_l^{\mathrm{L}}(x) + \bar{\psi}_l^{\mathrm{R}}(x)\not{D}\psi_l^{\mathrm{R}}(x) + \bar{\psi}_{\nu_l}^{\mathrm{R}}(x)\not{D}\psi_{\nu_l}^{\mathrm{R}}(x)] \tag{12.39}$$

where

$$D^\mu \Psi_l^{\mathrm{L}}(x) = [\partial^\mu + \mathrm{i}g\tau_j W_j^\mu(x)/2 - \mathrm{i}g' B^\mu(x)/2]\Psi_l^{\mathrm{L}}(x) \tag{12.40a}$$

$$D^\mu \psi_l^{\mathrm{R}}(x) = [\partial^\mu - \mathrm{i}g' B^\mu(x)]\psi_l^{\mathrm{R}}(x) \tag{12.40b}$$

$$D^\mu \psi_{\nu_l}^{\mathrm{R}}(x) = \partial^\mu \psi_{\nu_l}^{\mathrm{R}}(x). \tag{12.40c}$$

We now define the fields $W_i^\mu(x)$ to be invariant under U(1) gauge transformations, and $B^\mu(x)$ to be invariant under SU(2) gauge transformations. It then follows that the Lagrangian density (12.39) is invariant under both SU(2) and U(1) gauge transformations. Such a Lagrangian density is said to be SU(2) × U(1) gauge-invariant.

We can write the Lagrangian density (12.39) in the form

$$\mathscr{L}^{\mathrm{L}} = \mathscr{L}_0 + \mathscr{L}_{\mathrm{I}} \tag{12.41}$$

where $\mathscr{L}_0$ is the free-lepton Lagrangian density (12.14) and

$$\mathscr{L}_{\mathrm{I}} = -gJ_i^\mu(x)W_{i\mu}(x) - g'J_Y^\mu(x)B_\mu(x) \tag{12.42}$$

represents the interaction of the weak isospin currents and the weak hypercharge current, Eqs. (12.20) and (12.25), with the gauge fields $W_{i\mu}(x)$ and $B_\mu(x)$.

In order to interpret the interaction $\mathscr{L}_\mathrm{I}$ we rewrite Eq. (12.42). Using Eqs. (12.22), we write the weak isospin currents $J_1^\mu(x)$ and $J_2^\mu(x)$ in terms of the charged leptonic currents $J^\mu(x)$ and $J^{\mu\dagger}(x)$, and we introduce the non-Hermitian gauge field

$$W_\mu(x) = \frac{1}{\sqrt{2}}[W_{1\mu}(x) - \mathrm{i}W_{2\mu}(x)] \tag{12.43}$$

and its adjoint $W_\mu^\dagger(x)$, in place of $W_{1\mu}(x)$ and $W_{2\mu}(x)$. In this way, one obtains for the first two terms of $\mathscr{L}_\mathrm{I}$

$$-g\sum_{i=1}^{2} J_i^\mu(x)W_{i\mu}(x) = \frac{-g}{2\sqrt{2}}[J^{\mu\dagger}(x)W_\mu(x) + J^\mu(x)W_\mu^\dagger(x)]. \tag{12.44}$$

In the remaining two terms of $\mathscr{L}_\mathrm{I}$, we write $W_{3\mu}(x)$ and $B_\mu(x)$ as linear combinations of two different Hermitian fields $A_\mu(x)$ and $Z_\mu(x)$, defined by

$$\left.\begin{aligned} W_{3\mu}(x) &= \cos\theta_\mathrm{W} Z_\mu(x) + \sin\theta_\mathrm{W} A_\mu(x) \\ B_\mu(x) &= -\sin\theta_\mathrm{W} Z_\mu(x) + \cos\theta_\mathrm{W} A_\mu(x) \end{aligned}\right\}. \tag{12.45}$$

The angle θ_W, which specifies the mixture of Z_μ and A_μ fields in $W_{3\mu}$ and B_μ, is known as the weak mixing angle (or the Weinberg angle), and we shall consider its determination later. From Eqs. (12.45) and

$$J_Y^\mu(x) = s^\mu(x)/e - J_3^\mu(x), \tag{12.25}$$

we obtain

$$\begin{aligned} -gJ_3^\mu(x)W_{3\mu}(x) &- g'J_Y^\mu(x)B_\mu(x) \\ &= -\frac{g'}{e}s^\mu(x)[-\sin\theta_\mathrm{W} Z_\mu(x) + \cos\theta_\mathrm{W} A_\mu(x)] \\ &\quad - J_3^\mu(x)\{g[\cos\theta_\mathrm{W} Z_\mu(x) + \sin\theta_\mathrm{W} A_\mu(x)] \\ &\quad - g'[-\sin\theta_\mathrm{W} Z_\mu(x) + \cos\theta_\mathrm{W} A_\mu(x)]\}. \end{aligned} \tag{12.46}$$

We now demand that the gauge field $A_\mu(x)$, defined by Eqs. (12.45), is the electromagnetic field and is coupled to electric charges in the usual way, i.e. through the usual term $-s^\mu(x)A_\mu(x)$ in the interaction Lagrangian density. This means that in Eq. (12.46) the coefficient of $J_3^\mu(x)A_\mu(x)$ must vanish, and that of $s^\mu(x)A_\mu(x)$ must be (-1), i.e. we require

$$g\sin\theta_\mathrm{W} = g'\cos\theta_\mathrm{W} = e. \tag{12.47}$$

Substituting Eqs. (12.44) and (12.46) in Eq. (12.42), and eliminating g' by means of Eq. (12.47), we obtain as final expression for the interaction Lagrangian density

$$\mathscr{L}_I = -s^\mu(x)A_\mu(x) - \frac{g}{2\sqrt{2}}[J^{\mu\dagger}(x)W_\mu(x) + J^\mu(x)W_\mu^\dagger(x)]$$

$$-\frac{g}{\cos\theta_W}[J_3^\mu(x) - \sin^2\theta_W s^\mu(x)/e]Z_\mu(x). \quad (12.48)$$

The SU(2) × U(1) gauge-invariant interaction (12.48), first introduced by Glashow in 1961, is eminently satisfactory as a Lagrangian density describing the electromagnetic and weak interactions of leptons. The first term, obtained by imposing the conditions (12.47) on the coupling constants, is the familiar interaction of QED. The second term is just the IVB interaction Lagrangian density corresponding to Eq. (11.7), provided we set

$$g_W = \frac{g}{2\sqrt{2}}. \quad (12.49)$$

Thus the quanta of the gauge field $W(x)$ are just the $W^\pm$ vector bosons. The third term in Eq. (12.48) represents a neutral current

$$J_3^\mu(x) - \sin^2\theta_W s^\mu(x)/e$$
$$= \tfrac{1}{4}\bar{\psi}_{\nu_l}(x)\gamma^\mu(1-\gamma_5)\psi_{\nu_l}(x) - \tfrac{1}{4}\bar{\psi}_l(x)\gamma^\mu[(1-4\sin^2\theta_W)-\gamma_5]\psi_l(x) \quad (12.50)$$

coupled to a real vector field $Z_\mu(x)$. If we interpret the quanta of this field as the electrically neutral vector bosons Z^0 of Section 11.7, then this term will lead to neutral current processes of the type shown in Fig. 11.12. Such neutral current processes have been observed subsequent to their theoretical prediction. As will be discussed in Section 14.3, good agreement between theory and experiment is obtained by taking

$$\sin^2\theta_W = 0.227 \pm 0.014, \quad (12.51)$$

i.e. the last term in Eq. (12.50) is almost a pure axial current. This agreement between theory and experiment is strong support for the unified theory of electromagnetic and weak interactions. On the other hand, taking $\theta_W = 0$ in Eqs. (12.45) leads to a SU(2) gauge-invariant theory of weak interactions alone, i.e. weak and electromagnetic interactions are decoupled. Such a theory is ruled out by experiment.

12.4 PROPERTIES OF THE GAUGE BOSONS

The Lagrangian density (12.41), which we have considered, describes the free leptons and their interactions with the gauge fields. The complete Lagrangian density must also contain terms which describe these gauge bosons when no leptons are present. These terms must also be SU(2) × U(1) gauge-invariant. As for the leptons, we shall for the moment assume the gauge bosons to have zero masses, and we shall return to the question of non-vanishing masses in the next section.

For the $B^\mu(x)$ field, it is easy to construct suitable terms. The U(1) gauge transformation law (12.37b) of this field has the same form as that of the electromagnetic field $A^\mu(x)$, Eq. (12.2a). Hence in analogy to the electromagnetic case, a U(1) gauge-invariant Lagrangian density for the $B^\mu(x)$ field is given by

$$-\tfrac{1}{4}B_{\mu\nu}(x)B^{\mu\nu}(x), \tag{12.52}$$

where

$$B^{\mu\nu}(x) \equiv \partial^\nu B^\mu(x) - \partial^\mu B^\nu(x) \tag{12.53}$$

is the analogue of the electromagnetic field tensor $F^{\mu\nu}(x)$, Eq. (5.5). The SU(2) gauge invariance of the expression (12.52) follows from that of the $B^\mu(x)$ field.

The situation is more complicated for the $W_i^\mu(x)$ fields. The expression analogous to Eq. (12.52) is

$$-\tfrac{1}{4}F_{i\mu\nu}(x)F_i^{\mu\nu}(x) \tag{12.54}$$

where

$$F_i^{\mu\nu}(x) \equiv \partial^\nu W_i^\mu(x) - \partial^\mu W_i^\nu(x). \tag{12.55}$$

However, this expression is not invariant under the transformation (12.32b), on account of the term

$$-g\varepsilon_{ijk}\omega_j(x)W_k^\mu(x)$$

in Eq. (12.32b). To restore invariance, additional interaction terms must again be introduced. As is shown in the appendix to this chapter (Section 12.6), the expression obtained by replacing $F_i^{\mu\nu}(x)$ by

$$G_i^{\mu\nu}(x) \equiv F_i^{\mu\nu}(x) + g\varepsilon_{ijk}W_j^\mu(x)W_k^\nu(x) \tag{12.56}$$

in Eq. (12.54), i.e. the expression

$$\mathscr{L}_G = -\tfrac{1}{4}G_{i\mu\nu}(x)G_i^{\mu\nu}(x), \tag{12.57}$$

is SU(2) gauge-invariant. The invariance of $\mathscr{L}_G$ under U(1) transformations follows trivially from that of the $W_i^\mu(x)$ fields.

Combining expressions (12.52) and (12.57), and substituting Eq. (12.56), we obtain the complete SU(2) × U(1) gauge-invariant Lagrangian density

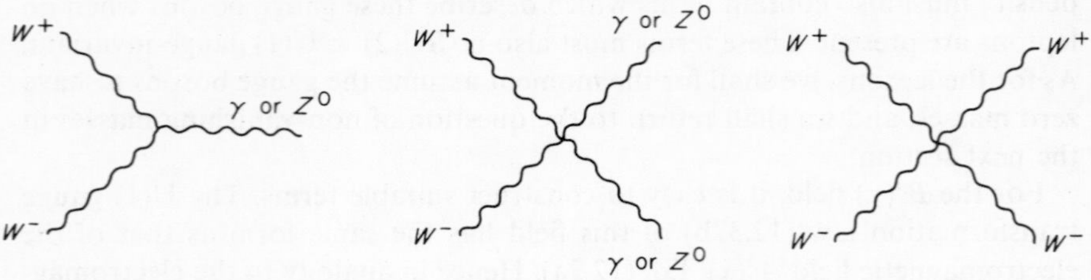

Fig. 12.1. Some examples of three-line and four-line vertices generated by the boson self-interaction terms in Eq. (12.58).

for the gauge bosons

$$\mathscr{L}^{\mathrm{B}} = -\tfrac{1}{4}B_{\mu\nu}(x)B^{\mu\nu}(x) - \tfrac{1}{4}G_{i\mu\nu}(x)G_i^{\mu\nu}(x) \tag{12.58a}$$

$$\begin{aligned} &= -\tfrac{1}{4}B_{\mu\nu}(x)B^{\mu\nu}(x) - \tfrac{1}{4}F_{i\mu\nu}(x)F_i^{\mu\nu}(x) \\ &\quad + g\varepsilon_{ijk}W_{i\mu}(x)W_{j\nu}(x)\partial^\mu W_k^\nu(x) \\ &\quad - \tfrac{1}{4}g^2\varepsilon_{ijk}\varepsilon_{ilm}W_j^\mu(x)W_k^\nu(x)W_{l\mu}(x)W_{m\nu}(x). \end{aligned} \tag{12.58b}$$

The first two terms in Eq. (12.58b) represent the Lagrangian density $\mathscr{L}_0^{\mathrm{B}}$ of the free (i.e. non-interacting) gauge fields. In terms of the fields $A^\mu(x)$, $Z^\mu(x)$ and $W^\mu(x)$, $\mathscr{L}_0^{\mathrm{B}}$ becomes

$$\mathscr{L}_0^{\mathrm{B}} = -\tfrac{1}{4}F_{\mu\nu}(x)F^{\mu\nu}(x) - \tfrac{1}{2}F^\dagger_{W\mu\nu}(x)F_W^{\mu\nu}(x) - \tfrac{1}{4}Z_{\mu\nu}(x)Z^{\mu\nu}(x), \tag{12.59}$$

where $F^{\mu\nu}(x)$ is the electromagnetic field tensor (5.5), $F_W^{\mu\nu}(x)$ is the corresponding tensor (11.21b) for the $W^\mu(x)$ field, and

$$Z^{\mu\nu}(x) \equiv \partial^\nu Z^\mu(x) - \partial^\mu Z^\nu(x) \tag{12.60}$$

is similarly associated with the $Z^\mu(x)$ field. Eq. (12.59) is thus the free-field Lagrangian density for mass zero, spin one γ, $W^\pm$ and Z^0 bosons.

In contrast, the third and fourth terms in Eq. (12.58b) represent interactions of the gauge bosons amongst themselves. In perturbation theory, these terms generate three- or four-line vertices. Some examples of such vertices are shown in Fig. 12.1. These boson self-interactions are one of the most striking features of the theory. They arise because the $W_i^\mu(x)$ fields, which transmit the interactions between the weak isospin currents, themselves are weak isospin vectors, i.e. they carry weak isospin charge. This is in contrast to QED where the electromagnetic interactions are transmitted by the photons but the latter do not carry electric charge. Consequently, there are no photon self-coupling terms in QED.

12.5 LEPTON AND GAUGE BOSON MASSES

So far we have assumed that all leptons and gauge bosons are massless. Except for the photon and possibly the neutrinos, this assumption is certainly unrealistic.

To describe massive $W^{\pm}$ and Z^0 bosons, we can add the mass terms

$$m_W^2 W_\mu^\dagger(x) W^\mu(x) + \tfrac{1}{2} m_Z^2 Z_\mu(x) Z^\mu(x) \tag{12.61}$$

to the Lagrangian density (12.59). For the $W^{\pm}$ particles, one obtains in this way the Lagrangian density (11.21a) of the IVB theory of weak interactions. Adding such mass terms results in a Lagrangian density which is not invariant under the transformations (12.32) and (12.37), i.e. it violates both SU(2) and U(1) gauge invariance. It also reintroduces all the renormalization problems associated with massive vector bosons, as discussed in Section 11.7.

We could similarly introduce non-zero lepton masses. For example, for the electron one could add the mass term,

$$-m_e \bar{\psi}_e(x) \psi_e(x) \tag{12.62a}$$

to the Lagrangian density. Unfortunately, this term is again not SU(2) × U(1) gauge invariant. This follows since the expression (12.62a) can be written

$$-m_e \bar{\psi}_e(x)[P_\mathrm{R} + P_\mathrm{L}]\psi_e(x) = -m_e[\bar{\psi}_e^\mathrm{L}(x)\psi_e^\mathrm{R}(x) + \bar{\psi}_e^\mathrm{R}(x)\psi_e^\mathrm{L}(x)], \tag{12.62b}$$

and ψ_e^R and $\bar{\psi}_e^\mathrm{R}$ are isoscalars while ψ_e^L and $\bar{\psi}_e^\mathrm{L}$ are isospinors.

We thus arrive at the conclusion that, if we wish to preserve SU(2) × U(1) gauge invariance, we must set the masses of the leptons and of the $W^{\pm}$ and Z^0 bosons equal to zero. We could at this point simply add the required lepton and boson mass terms, as outlined above, violating SU(2) × U(1) gauge invariance. The resulting model is known as the Glashow model. In lowest-order perturbation theory calculations, this model yields results which are in good agreement with present experiments for appropriately chosen values of $\sin\theta_\mathrm{W}$, m_W and m_Z. However, the Glashow model is not renormalizable. We shall therefore insist on retaining the gauge invariance of the Lagrangian density. In the next chapter, we shall learn how to introduce non-zero masses by the mechanism of spontaneous symmetry breaking. In this way we shall obtain a renormalizable theory which is in agreement with experiment and which, for given $\sin\theta_\mathrm{W}$, predicts the gauge boson masses m_W and m_Z.

12.6 APPENDIX: TWO GAUGE TRANSFORMATION RESULTS

In this appendix we shall derive two results which were only quoted earlier in this chapter: (i) the transformation law (12.32b) of the $W_i^\mu(x)$ fields, and (ii) the SU(2) gauge invariance of the Lagrangian density of the $W_i^\mu(x)$ fields, Eq. (12.57).

The derivations will gain in clarity if we introduce the 2×2 matrices

$$\omega(x) \equiv \tau_i \omega_i(x), \qquad W^\mu(x) \equiv \tau_i W_i^\mu(x). \tag{12.63}$$

For simplicity, we shall also in this appendix omit the argument x of all quantities, i.e. $\omega \equiv \omega(x)$, $\Psi_l^L \equiv \Psi_l^L(x)$, etc.

It follows from Eq. (12.16) that the matrices (12.63) satisfy the commutation relations

$$[\omega, W^\mu] = 2i\varepsilon_{ijk}\tau_i \omega_j W_k^\mu. \tag{12.64}$$

Expressed in terms of the matrices (12.63), the covariant derivative (12.34) takes the form

$$D^\mu \Psi_l^L = [\partial^\mu + igW^\mu/2]\Psi_l^L, \tag{12.65}$$

and the local phase transformation (12.32a) of Ψ_l^L becomes, for infinitesimal ω,

$$\Psi_l^L \to \Psi_l^{L\prime} = [1 + ig\omega/2]\Psi_l^L \quad \text{(small } \omega). \tag{12.66}$$

12.6.1 The transformation law (12.32b)

As we saw in Section 12.3, we require the gauge fields W^μ to undergo a transformation

$$W^\mu \to W^{\mu\prime} = W^\mu + \delta W^\mu \tag{12.67}$$

(where $\delta W^\mu \equiv \tau_i\, \delta W_i^\mu$) such that the covariant derivative $D^\mu \Psi_l^L$ transforms in the same way as the field Ψ_l^L itself, i.e.

$$\begin{aligned} D^\mu \Psi_l^L &\to [1 + ig\omega/2] D^\mu \Psi_l^L \\ &= [1 + ig\omega/2][\partial^\mu + igW^\mu/2]\Psi_l^L. \end{aligned} \tag{12.68}$$

By applying the transformations (12.66) and (12.67) in Eq. (12.65), we obtain the transformation

$$D^\mu \Psi_l^L \to [\partial^\mu + igW^\mu/2 + ig\,\delta W^\mu/2][1 + ig\omega/2]\Psi_l^L. \tag{12.69}$$

Comparing the right-hand sides of Eqs. (12.68) and (12.69), neglecting second-order terms, leads to

$$\delta W^\mu = -\partial^\mu \omega + \tfrac{1}{2} ig[\omega, W^\mu]. \tag{12.70}$$

Substituting the commutation relations (12.64) and the definitions (12.63) in Eq. (12.70), we at once obtain the transformation law

$$\delta W_i^\mu(x) = -\partial^\mu \omega_i(x) - g\varepsilon_{ijk}\omega_j(x) W_k^\mu(x), \tag{12.71}$$

quoted in Eq. (12.32b).

12.6.2 The SU(2) gauge invariance of Eq. (12.57)

We shall now show that Eq. (12.57), i.e. the Lagrangian density of the $W_i^\mu(x)$ fields

$$\mathscr{L}_G = -\tfrac{1}{4}G_{i\mu\nu}G_i^{\mu\nu}, \tag{12.72}$$

is invariant under SU(2) transformations. It will be convenient to define

$$G^{\mu\nu} \equiv \tau_i G_i^{\mu\nu} \tag{12.73}$$

and, using

$$\mathrm{Tr}\,(\tau_i\tau_j) = 2\delta_{ij}, \tag{12.74}$$

to write $\mathscr{L}_G$ as

$$\mathscr{L}_G = -\tfrac{1}{8}\,\mathrm{Tr}\,(G_{\mu\nu}G^{\mu\nu}). \tag{12.75}$$

In order to prove the invariance of $\mathscr{L}_G$, we shall first derive the transformation properties of $G^{\mu\nu}$. It follows from Eq. (12.16) that

$$2i\tau_i\varepsilon_{ijk}W_j^\mu W_k^\nu = [W^\mu, W^\nu]. \tag{12.76}$$

Substituting Eqs. (12.55) and (12.56) in Eq. (12.73) and using Eq. (12.76), we obtain

$$G^{\mu\nu} = \partial^\nu W^\mu - \partial^\mu W^\nu - \tfrac{1}{2}ig[W^\mu, W^\nu]. \tag{12.77}$$

Under a SU(2) transformation, W^μ experiences a change δW^μ which is given by Eq. (12.70). Consequently $G^{\mu\nu}$ will change by

$$\delta G^{\mu\nu} = \partial^\nu(\delta W^\mu) - \partial^\mu(\delta W^\nu) - \tfrac{1}{2}ig[\delta W^\mu, W^\nu] - \tfrac{1}{2}ig[W^\mu, \delta W^\nu]. \tag{12.78}$$

Substituting Eq. (12.70) in Eq. (12.78) we obtain, after some simplification,

$$\begin{aligned}\delta G^{\mu\nu} &= \tfrac{1}{2}ig[\omega, \partial^\nu W^\mu - \partial^\mu W^\nu] \\ &\quad + \tfrac{1}{4}g^2[[\omega, W^\mu], W^\nu] + \tfrac{1}{4}g^2[[W^\nu, \omega], W^\mu].\end{aligned} \tag{12.79}$$

By means of the Jacobi identity

$$[[A, B], C] + [[B, C], A] + [[C, A], B] \equiv 0,$$

we can rewrite Eq. (12.79) as

$$\delta G^{\mu\nu} = \tfrac{1}{2}ig[\omega, \partial^\nu W^\mu - \partial^\mu W^\nu] - \tfrac{1}{4}g^2[[W^\mu, W^\nu], \omega].$$

On account of Eq. (12.77), the last equation reduces to

$$\delta G^{\mu\nu} = \tfrac{1}{2}ig[\omega, G^{\mu\nu}]. \tag{12.80}$$

This change $\delta G^{\mu\nu}$ induces a change in $\mathscr{L}_G$ which, from Eq. (12.75), is given by

$$\begin{aligned}\delta\mathscr{L}_G &= -\tfrac{1}{8}\,\mathrm{Tr}\,(\delta G_{\mu\nu}G^{\mu\nu} + G_{\mu\nu}\,\delta G^{\mu\nu})\\ &= -\tfrac{1}{16}ig\,\mathrm{Tr}\,\{[\omega, G_{\mu\nu}]G^{\mu\nu} + G_{\mu\nu}[\omega, G^{\mu\nu}]\}\\ &= -\tfrac{1}{16}ig\,\mathrm{Tr}\,[\omega, G_{\mu\nu}G^{\mu\nu}] = 0,\end{aligned} \tag{12.81}$$

where the last step follows from the identity $\mathrm{Tr}\,[A, B] \equiv 0$. Eq. (12.81) establishes the SU(2) invariance of $\mathscr{L}_G$.

CHAPTER 13

Spontaneous symmetry breaking

In the previous chapter, a gauge-invariant and renormalizable unified theory of weak and electromagnetic interactions was obtained. However, all leptons and gauge bosons had to have zero mass. In reality, only photons and perhaps neutrinos are massless, but the charged leptons and the $W^{\pm}$ and Z^0 bosons have non-zero masses. We have seen that the *ad hoc* addition of mass terms to the Lagrangian density spoils the gauge invariance and the renormalizability of the theory. In order to obtain a renormalizable theory, it is essential to introduce the masses by a mechanism which retains the gauge invariance of the Lagrangian density. In this chapter we shall develop a quite remarkable such mechanism, that of spontaneous symmetry breaking.

As in the last chapter, we shall proceed from global to local phase invariance, i.e. to gauge invariance. In section 13.1, we shall introduce the idea of spontaneous symmetry breaking and shall consider the simplest field-theoretic example of it: the Goldstone model which is a field theory invariant under global U(1) phase transformations. The Goldstone model necessarily leads to zero-mass bosons (other than photons) which are not observed in nature. This undesirable feature is absent when applying spontaneous symmetry breaking to a theory which is invariant under *local* phase transformations, i.e. to a gauge theory. In Section 13.2, this will be demonstrated for a field theory invariant under U(1) gauge transformations (the Higgs model). It will then be easy (in Section 13.3) to generalize these results for the SU(2) × U(1) gauge-invariant electro-weak theory of the previous chapter. In this way, we shall obtain a Lagrangian density which is gauge-invariant, renormalizable, and contains mass terms for leptons and for the $W^{\pm}$ and Z^0 bosons, while the photon remains massless. The resulting theory is known as the standard electro–weak theory. It was first formulated in this way, independently, by Weinberg in 1967 and by Salam in 1968. Its interpretation and applications will form the topic of the last chapter of this book.

13.1 THE GOLDSTONE MODEL

In order to explain the idea of spontaneous symmetry breaking, we consider a system whose Lagrangian L possesses a particular symmetry, i.e. it is invariant under the corresponding symmetry transformations.‡ (For example, L might be spherically symmetric, i.e. invariant under spatial rotations.) In classifying the energy levels of this system, essentially two situations can occur. Firstly, if a given energy level is non-degenerate, the corresponding energy eigenstate is unique and invariant under the symmetry transformations of L. Secondly, the energy level may be degenerate and the corresponding eigenstates are not invariant but transform linearly amongst themselves under the symmetry transformations of L. In particular, consider the lowest energy level of the system. If it is non-degenerate, the state of lowest energy of the system (its ground state) is unique and possesses the symmetries of L. In the second case, of degeneracy, there is no unique eigenstate to represent the ground state. If we arbitrarily select one of the degenerate states as the ground state, then the ground state no longer shares the symmetries of L. This way of obtaining an asymmetric ground state is known as *spontaneous symmetry breaking*. The asymmetry is not due to adding a non-invariant asymmetric term to L but to the arbitrary choice of one of the degenerate states.

Ferromagnetism represents a familiar example of spontaneous symmetry breaking. In a ferromagnetic material, the forces which couple the electronic spins and hence the Hamiltonian of the system are rotationally invariant. However, in the ground state the spins are aligned in some definite direction resulting in a non-zero magnetization $\mathbf{M}$. The orientation of $\mathbf{M}$ is arbitrary and we are clearly dealing with a case of degeneracy. $\mathbf{M}$ could equally well point in any other direction and all properties of the system, other than the direction of $\mathbf{M}$, would remain unchanged. An important feature of this asymmetric ground state is that excited states obtained from it by small perturbations also display this asymmetry.

In field theory, the state of lowest energy is the vacuum. Spontaneous symmetry breaking is only relevant to field theory if the vacuum state is non-unique. This very bold and startling idea was first suggested by Nambu and his co-workers. It implies that some quantity in the vacuum is non-vanishing,

‡ Instead of the symmetries of L, we could talk of those of the Hamiltonian or of the equations of motion or, in the case of a field theory, of the Lagrangian density.

is not invariant under the symmetry transformations of the system, and can therefore be used to characterize a particular vacuum state as *the* ground state. In the following we shall assume this quantity to be the vacuum expectation value of a quantized field. If we require the vacuum states to be invariant under Lorentz transformations and under translations, then this field must be a scalar field, $\phi(x)$, and its vacuum expectation value must be constant:

$$\langle 0|\phi(x)|0\rangle = c \neq 0. \tag{13.1}$$

In contrast, the vacuum expectation value of any spinor field $\psi(x)$ or any vector field $V^\mu(x)$ must vanish:

$$\langle 0|\psi(x)|0\rangle = 0, \qquad \langle 0|V^\mu(x)|0\rangle = 0. \tag{13.2}$$

The simplest example of a field theory exhibiting spontaneous symmetry breaking is the Goldstone model. Its Lagrangian density is

$$\mathscr{L}(x) = [\partial^\mu\phi^*(x)][\partial_\mu\phi(x)] - \mu^2|\phi(x)|^2 - \lambda|\phi(x)|^4 \tag{13.3}$$

with

$$\phi(x) = \frac{1}{\sqrt{2}}[\phi_1(x) + i\phi_2(x)] \tag{13.4}$$

a complex scalar field, and μ^2 and λ arbitrary real parameters. To begin with, we shall consider a classical field theory, i.e. $\phi(x)$ is a classical and not a quantized field, and μ is not to be interpreted as a particle mass.

The Lagrangian density (13.3) is invariant under the global U(1) phase transformations

$$\phi(x) \to \phi'(x) = \phi(x)\,e^{i\alpha}, \qquad \phi^*(x) \to \phi^{*\prime}(x) = \phi^*(x)\,e^{-i\alpha}. \tag{13.5}$$

We shall see that this symmetry is spontaneously broken in this model.

The Hamiltonian density of this theory follows from Eq. (13.3) and the general relations (2.22) and (2.25), and is

$$\mathscr{H}(x) = [\partial^0\phi^*(x)][\partial_0\phi(x)] + [\nabla\phi^*(x)]\cdot[\nabla\phi(x)] + \mathscr{V}(\phi) \tag{13.6}$$

where

$$\mathscr{V}(\phi) = \mu^2|\phi(x)|^2 + \lambda|\phi(x)|^4 \tag{13.7}$$

is the potential energy density of the field. For the energy of the field to be bounded from below, we require $\lambda > 0$. The first two terms in Eq. (13.6) are positive definite and vanish for constant $\phi(x)$. It follows that the minimum value of $\mathscr{H}(x)$, and hence of the total energy of the field, corresponds to that constant value of $\phi(x)$ which minimizes $\mathscr{V}(\phi)$. Two different situations occur, depending on the sign of μ^2.

(i) $\mu^2 > 0$. In this case, the two terms in $\mathscr{V}(\phi)$ are also positive definite. In Fig. 13.1(a) we sketch the corresponding potential energy surface $\mathscr{V}(\phi)$ as

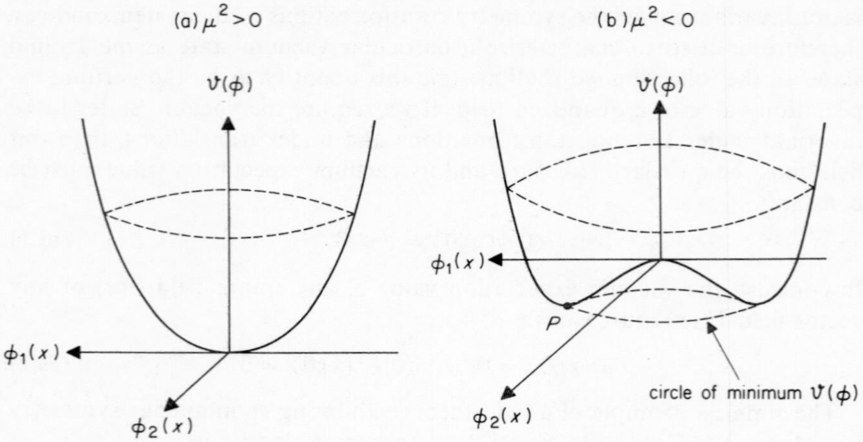

Fig. 13.1. The potential energy density $\mathscr{V}(\phi) = \mu^2|\phi(x)|^2 + \lambda|\phi(x)|^4$, Eq. (13.7), for $\lambda > 0$. In case (b), our choice $(\phi_1, \phi_2) = (v, 0)$ for the vacuum ground state corresponds to the point labelled P on the circle of minimum $\mathscr{V}(\phi)$.

a function of $\phi_1(x)$ and $\phi_2(x)$. $\mathscr{V}(\phi)$ has an absolute minimum for the unique value $\phi(x) = 0$, i.e. spontaneous symmetry breaking cannot occur. This is the situation with which we are familiar. Omitting the quartic term $\lambda|\phi(x)|^4$, the above expressions for $\mathscr{L}(x)$ and $\mathscr{H}(x)$ are those of the free complex Klein–Gordon field. Classically, they give rise to normal modes of oscillation about the stable equilibrium position $\phi(x) = 0$. On quantization, they give rise to charged spin 0 particles of mass μ (see Section 3.2). The ground state, i.e. the vacuum state, is unique and $\langle 0|\phi(x)|0\rangle = 0$. We can think of Eqs. (13.3), (13.6) and (13.7) as expansions in powers of $\phi(x)$ and $\phi^*(x)$ about the stable equilibrium configuration $\phi(x) = 0$ and, in our approach, treat $\lambda|\phi(x)|^4$ by perturbation theory. In the quantized theory, this term represents a self-interaction of the particles.

(ii) $\mu^2 < 0$. The potential energy surface for this case is shown in Fig. 13.1(b). $\mathscr{V}(\phi)$ possesses a local maximum at $\phi(x) = 0$ and a whole circle of absolute minima at

$$\phi(x) = \phi_0 = \left(\frac{-\mu^2}{2\lambda}\right)^{1/2} e^{i\theta}, \qquad 0 \leqslant \theta < 2\pi, \tag{13.8}$$

where the phase angle θ defines a direction in the complex ϕ-plane. We see that the state of lowest energy, the vacuum state, is not unique in this case. This arbitrariness in the direction θ is analogous to that in the direction of the magnetization $\mathbf{M}$ of a ferromagnet. Analogously to the latter case, spontaneous symmetry breaking will occur if we choose one particular direction θ

to represent the vacuum ground state. Because of the invariance of the Lagrangian density (13.3) under the global phase transformations (13.5), the value of θ chosen is not significant and we shall take $\theta = 0$ so that

$$\phi_0 = \left(\frac{-\mu^2}{2\lambda}\right)^{1/2} = \frac{1}{\sqrt{2}}\, v \quad (>0) \tag{13.9}$$

is purely real.

We now introduce two real fields $\sigma(x)$ and $\eta(x)$ through the equation

$$\phi(x) = \frac{1}{\sqrt{2}}\,[v + \sigma(x) + i\eta(x)]. \tag{13.10}$$

$\sigma(x)$ and $\eta(x)$ measure the deviations of the field $\phi(x)$ from the equilibrium ground state configuration $\phi(x) = \phi_0$. In terms of these fields, the Lagrangian density (13.3) becomes

$$\begin{aligned} \mathscr{L}(x) = {} & \tfrac{1}{2}[\partial^\mu \sigma(x)][\partial_\mu \sigma(x)] - \tfrac{1}{2}(2\lambda v^2)\sigma^2(x) \\ & + \tfrac{1}{2}[\partial^\mu \eta(x)][\partial_\mu \eta(x)] \\ & - \lambda v \sigma(x)[\sigma^2(x) + \eta^2(x)] - \tfrac{1}{4}\lambda[\sigma^2(x) + \eta^2(x)]^2 \end{aligned} \tag{13.11}$$

where we have omitted a constant term which is of no consequence.

Eqs. (13.3) and (13.11) are the same Lagrangian density expressed in terms of different variables. Thus they are entirely equivalent and must lead to the same physical results. This equivalence only holds for exact solutions of the theory. We shall be using perturbation theory, and for approximate solutions the picture is very different. For $\mu^2 < 0$, we cannot proceed as we did for $\mu^2 > 0$, i.e. by treating the quartic term $\lambda|\phi(x)|^4$ in Eqs. (13.3), (13.6) and (13.7) as a perturbation of the other terms which are bilinear in $\phi(x)$ and $\phi^*(x)$. For $\mu^2 < 0$, $\phi(x) = 0$ is an unstable equilibrium configuration, and one cannot carry out perturbation calculations about an unstable solution. That this procedure leads to nonsense for $\mu^2 < 0$ shows up very clearly in the quantized theory where the unperturbed system corresponds to particles of imaginary mass, and no finite order of perturbation theory can put this right.

In contrast, Eq. (13.11) suggests a different quantization procedure. We shall now treat the terms which are quadratic in $\sigma(x)$ and $\eta(x)$, i.e. the first three terms in Eq. (13.11), as the free Lagrangian density

$$\begin{aligned} \mathscr{L}_0(x) = {} & \tfrac{1}{2}[\partial^\mu \sigma(x)][\partial_\mu \sigma(x)] - \tfrac{1}{2}(2\lambda v^2)\sigma^2(x) \\ & + \tfrac{1}{2}[\partial^\mu \eta(x)][\partial_\mu \eta(x)] \end{aligned} \tag{13.12}$$

and the remaining terms, which are cubic and quartic in $\sigma(x)$ and $\eta(x)$, as interactions. The fields $\sigma(x)$ and $\eta(x)$ measure the deviations from the stable equilibrium configuration $\phi(x) = \phi_0$, and one would expect to be able to treat the interaction terms by perturbation theory about this stable solution.

The Lagrangian density (13.12) contains no terms which couple $\sigma(x)$ and $\eta(x)$, i.e. the fields $\sigma(x)$ and $\eta(x)$ are normal coordinates of Eq. (13.12). Comparing Eqs. (13.12) and (3.5), we see that $\sigma(x)$ and $\eta(x)$ are real Klein–Gordon fields. On quantization, both fields lead to neutral spin 0 particles: the σ boson with the (real positive) mass $\sqrt{(2\lambda v^2)}$, and the η boson which has zero mass since there is no term in $\eta^2(x)$ in Eq. (13.12). Since, by definition, there are no particles present in the vacuum, it follows from Eqs. (13.9) and (13.10) that

$$\langle 0|\phi(x)|0\rangle = \phi_0. \tag{13.13}$$

This is the condition for spontaneous symmetry breaking in the quantized theory, analogous to Eqs. (13.8) and (13.9) in the classical theory.

The origin of the above mass spectrum can be understood from Fig. 13.1(b) by considering small displacements $\sigma(x)$ and $\eta(x)$ from the equilibrium configuration $\phi(x) = \phi_0$. $\sigma(x)$ represents a displacement in the radial plane $\phi_2(x) = 0$ in which the potential energy density $\mathscr{V}(\phi)$ increases quadratically with $\sigma(x)$. On the other hand, $\eta(x)$ represents a displacement along the valley of minimum potential energy where $\mathscr{V}(\phi)$ is constant, so that the corresponding quantum excitations—the η bosons—are massless. Thus the zero mass of the η bosons is a consequence of the degeneracy of the vacuum. Such zero-mass bosons frequently occur in theories with spontaneous symmetry breaking and they are known as *Goldstone bosons*.

No Goldstone bosons are observed in nature. It is therefore of particular interest that gauge theories with spontaneous symmetry breaking do not generate Goldstone bosons. In the next section we shall demonstrate this for a simple model, before turning to the realistic SU(2) × U(1) gauge theory of the previous chapter.

13.2 THE HIGGS MODEL

The Goldstone model is easily generalized to be invariant under U(1) gauge transformations. As in the last chapter, we introduce a gauge field $A_\mu(x)$, replace the ordinary derivatives in the Goldstone Lagrangian density (13.3) by the covariant derivatives

$$D_\mu\phi(x) = [\partial_\mu + iqA_\mu(x)]\phi(x)$$

and add the Lagrangian density of the free gauge field

$$-\tfrac{1}{4}F_{\mu\nu}(x)F^{\mu\nu}(x) \tag{13.14}$$

where, as usual,

$$F_{\mu\nu}(x) = \partial_\nu A_\mu(x) - \partial_\mu A_\nu(x).$$

In this way we obtain the Lagrangian density

$$\mathscr{L}(x) = [D^\mu\phi(x)]^*[D_\mu\phi(x)] - \mu^2|\phi(x)|^2 - \lambda|\phi(x)|^4 \\ -\tfrac{1}{4}F_{\mu\nu}(x)F^{\mu\nu}(x) \qquad (13.15)$$

which defines the Higgs model. Eq. (13.15) is invariant under the U(1) gauge transformations

$$\left.\begin{aligned} \phi(x) &\rightarrow \phi'(x) = \phi(x)\,\mathrm{e}^{-iqf(x)} \\ \phi^*(x) &\rightarrow \phi^{*\prime}(x) = \phi^*(x)\,\mathrm{e}^{iqf(x)} \\ A_\mu(x) &\rightarrow A'_\mu(x) = A_\mu(x) + \partial_\mu f(x) \end{aligned}\right\}. \qquad (13.16)$$

The further analysis parallels that for the Goldstone model. We again start from a classical theory. Taking $\lambda > 0$, two situations arise. For $\mu^2 > 0$, the state of lowest energy corresponds to both $\phi(x)$ and $A_\mu(x)$ vanishing, so that spontaneous symmetry breaking cannot occur.

For $\mu^2 < 0$, the vacuum state is not unique, leading to spontaneous symmetry breaking. To ensure Lorentz invariance, the vector field $A_\mu(x)$ must vanish for the vacuum, but we again obtain a circle of minimum $\mathscr{H}(x)$ corresponding to $\phi(x)$ taking on the values ϕ_0, given by Eq. (13.8). As for the Goldstone model, we choose the real value (13.9) for ϕ_0 and define the real fields $\sigma(x)$ and $\eta(x)$ by Eq. (13.10). In terms of these fields, the Lagrangian density (13.15) becomes

$$\begin{aligned} \mathscr{L}(x) = {} & \tfrac{1}{2}[\partial^\mu\sigma(x)][\partial_\mu\sigma(x)] - \tfrac{1}{2}(2\lambda v^2)\sigma^2(x) \\ & - \tfrac{1}{4}F_{\mu\nu}(x)F^{\mu\nu}(x) + \tfrac{1}{2}(qv)^2 A_\mu(x)A^\mu(x) \\ & + \tfrac{1}{2}[\partial^\mu\eta(x)][\partial_\mu\eta(x)] \\ & + qvA^\mu(x)\,\partial_\mu\eta(x) + \text{‘interaction terms’} \qquad (13.17) \end{aligned}$$

where the ‘interaction terms’, which we have not given explicitly, are cubic and quartic in the fields and where an insignificant constant term has been discarded.

The direct interpretation of Eq. (13.17) leads to difficulties. The first line of this equation describes a real Klein–Gordon field which on quantization gives uncharged spin 0 bosons with mass $\sqrt{(2\lambda v^2)}$. However, the product term $A^\mu(x)\,\partial_\mu\eta(x)$ shows that $A^\mu(x)$ and $\eta(x)$ are not independent normal coordinates, and one cannot conclude that the second and third lines of Eq. (13.17) describe massive vector bosons and massless scalar bosons respectively.‡ This difficulty also shows up if we count degrees of freedom for the Lagrangian densities (13.15) and (13.17). Eq. (13.15) has four degrees of

‡ We do not include this term as a part of the interaction, to be treated in perturbation theory, since it is of the same (second) order in the fields as the first five terms in Eq. (13.17).

freedom: two from the complex scalar field $\phi(x)$, and two from the real massless vector field $A_\mu(x)$ (i.e. for massless photons there are only two independent polarization states, the third is eliminated by gauge invariance). In Eq. (13.17), the real scalar fields $\sigma(x)$ and $\eta(x)$ each represent one degree and the real massive vector field $A_\mu(x)$ contributes three degrees (corresponding to three independent polarization states), i.e. the transformed Lagrangian density (13.17) appears to have five degrees of freedom. Of course, a change of variables cannot alter the number of degrees of freedom of a system. We must conclude that the Lagrangian density (13.17) contains an unphysical field which does not represent real particles and which can be eliminated.

The scalar field $\eta(x)$ can be eliminated from Eq. (13.17). For any complex field $\phi(x)$, a gauge transformation of the form (13.16) can be found which transforms $\phi(x)$ into a real field of the form

$$\phi(x) = \frac{1}{\sqrt{2}}\,[v + \sigma(x)]. \qquad (13.18)$$

The gauge in which the transformed field has this form is called the *unitary gauge*. (We shall continue to label the transformed field $\phi(x)$ and not $\phi'(x)$, etc.) Substituting Eq. (13.18) into Eq. (13.15) gives

$$\mathscr{L}(x) = \mathscr{L}_0(x) + \mathscr{L}_{\mathrm{I}}(x) \qquad (13.19a)$$

where we have separated the quadratic terms

$$\mathscr{L}_0(x) = \tfrac{1}{2}[\partial^\mu\sigma(x)][\partial_\mu\sigma(x)] - \tfrac{1}{2}(2\lambda v^2)\sigma^2(x)$$
$$-\tfrac{1}{4}F_{\mu\nu}(x)F^{\mu\nu}(x) + \tfrac{1}{2}(qv)^2 A_\mu(x)A^\mu(x) \qquad (13.19b)$$

from the higher-order interaction terms

$$\mathscr{L}_{\mathrm{I}}(x) = -\lambda v\sigma^3(x) - \tfrac{1}{4}\lambda\sigma^4(x)$$
$$+ \tfrac{1}{2}q^2 A_\mu(x)A^\mu(x)[2v\sigma(x) + \sigma^2(x)]. \qquad (13.19c)$$

$\mathscr{L}_0(x)$ contains no terms which couple $\sigma(x)$ and $A_\mu(x)$. Hence, treating $\mathscr{L}_{\mathrm{I}}(x)$ in perturbation theory, we can interpret $\mathscr{L}_0(x)$ as the free-field Lagrangian density of a real Klein–Gordon field $\sigma(x)$ and a real massive vector field $A_\mu(x)$. On quantizing $\mathscr{L}_0(x)$, $\sigma(x)$ gives rise to neutral scalar bosons of mass $\sqrt{(2\lambda v^2)}$, and $A_\mu(x)$ to neutral vector bosons of mass $|qv|$.

This is a remarkable result! Having started from the Lagrangian density (13.15) for a *complex* scalar field and a *massless* real vector field, we have ended up with the Lagrangian density (13.19) for a *real* scalar field and a *massive* real vector field. The number of degrees of freedom is four in both cases. Of the two degrees of freedom of the complex field $\phi(x)$, one has been taken up by the vector field $A_\mu(x)$ which has become massive in the process; the other shows up as the real field $\sigma(x)$. This phenomenon by which a vector boson acquires mass without destroying the gauge invariance of the Lagrangian density is known as the *Higgs mechanism*, and the massive spin 0 boson associated with the field $\sigma(x)$ is called a Higgs boson or a Higgs scalar. The Higgs mechanism does not generate Goldstone bosons, in contrast to the spontaneous symmetry breaking of the global phase invariance of the Goldstone model. In essence, the field $\eta(x)$ in Eq. (13.17), which in the Goldstone model was associated with the massless Goldstone bosons, has been eliminated by gauge invariance, and the degree of freedom of $\eta(x)$ has been transferred to the vector field $A_\mu(x)$.

The Higgs mechanism also works for non-Abelian gauge theories. In the next section we shall apply it to the SU(2) × U(1) gauge theory of Chapter 12 and this will lead directly to the standard electro–weak theory. First, however, we comment briefly on the renormalizability of such theories.

Unlike the IVB theory or the Glashow model of weak interactions, gauge theories have the great merit of being renormalizable, allowing meaningful calculations to be made in higher orders of perturbation theory. The proof of renormalizability has been given by 't Hooft, Veltman and others. Unfortunately, it is very complicated. We shall content ourselves with briefly indicating the underlying ideas, using the Higgs model for illustration.‡

The second line of Eq. (13.19b) is identical with the Lagrangian density of a massive neutral vector boson field and leads to the propagator

$$\mathrm{i}D_{\mathrm{F}}^{\alpha\beta}(k, m) = \frac{\mathrm{i}(-g^{\alpha\beta} + k^\alpha k^\beta/m^2)}{k^2 - m^2 + \mathrm{i}\varepsilon} \tag{13.20}$$

where $m = |qv|$. We met this propagator in the IVB theory [see Eq. (11.30)], and it was suggested in Section 11.7 that the term $k^\alpha k^\beta/m^2$ in this propagator makes the IVB theory non-renormalizable. However, we saw in Section 9.8 that this kind of dimensional argument (i.e. counting powers of momenta in loop integrals) predicts the maximum possible degree of divergence. For gauge theories with spontaneously broken symmetry, the actual divergences are less severe and the theories are renormalizable. This could be due to the propagator being coupled to exactly conserved currents or to the exact cancellation of divergences arising from different Feynman graphs of the same order.

The renormalizability of the Higgs model is difficult to prove from the Lagrangian density (13.19) which employs the unitary gauge. Instead, 't Hooft proceeds in a way reminiscent of QED. There we replaced the

‡ The interested reader will find proper treatments in E. S. Abers and B. W. Lee, *Gauge Theories, Physics Reports*, 9C, No. 1 (1973), and J. C. Taylor, *Gauge Theories and Weak Interactions*, Cambridge University Press, Cambridge, 1976.

gauge-invariant Lagrangian density

$$\mathscr{L} = -\tfrac{1}{4}F_{\mu\nu}(x)F^{\mu\nu}(x) \tag{13.21}$$

by

$$\mathscr{L} = -\tfrac{1}{4}F_{\mu\nu}(x)F^{\mu\nu}(x) - \tfrac{1}{2}[\partial_\mu A^\mu(x)]^2 \tag{13.22}$$

which is equivalent to Eq. (13.21) provided we work in a Lorentz gauge, i.e.

$$\partial_\mu A^\mu(x) = 0.^{\ddagger} \tag{13.23}$$

In the present case, 't Hooft imposes the gauge condition

$$\partial_\mu A^\mu(x) - m\eta(x) = 0 \tag{13.24}$$

on the fields, where $m = |qv|$ and $\eta(x)$ is defined by Eq. (13.10). 't Hooft shows that, if condition (13.24) holds, one may add the term

$$-\tfrac{1}{2}[\partial_\mu A^\mu(x) - m\eta(x)]^2 \tag{13.25}$$

to the Lagrangian density (13.17), i.e. the modified Lagrangian density and Eq. (13.17) leads to the same predictions for observable quantities.[§] Adding the 'gauge-fixing' term (13.25) to Eq. (13.17) gives the modified Lagrangian density

$$\begin{aligned}\mathscr{L}(x) = {} & \tfrac{1}{2}[\partial^\mu\sigma(x)][\partial_\mu\sigma(x)] - \tfrac{1}{2}(2\lambda v^2)\sigma^2(x) \\ & - \tfrac{1}{4}F_{\mu\nu}(x)F^{\mu\nu}(x) + \tfrac{1}{2}m^2 A_\mu(x)A^\mu(x) - \tfrac{1}{2}[\partial_\mu A^\mu(x)]^2 \\ & + \tfrac{1}{2}[\partial^\mu\eta(x)][\partial_\mu\eta(x)] - \tfrac{1}{2}m^2\eta^2(x) \\ & + \text{'interaction terms'}\end{aligned} \tag{13.26}$$

where $m = |qv|$, as before, and we have again omitted an irrelevant four-divergence $m\,\partial_\mu[A^\mu(x)\eta(x)]$.

Eq. (13.26) no longer contains the troublesome bilinear term $A^\mu(x)\,\partial_\mu\eta(x)$ which was present in Eq. (13.17). Hence we can, in perturbation theory, treat $\sigma(x)$, $\eta(x)$ and $A_\mu(x)$ as three independent free fields which may be quantized in the usual way. We see from the first and third lines of Eq. (13.26) that $\sigma(x)$ and $\eta(x)$ are real Klein–Gordon fields which on quantization give the usual equations of motion and propagators for such fields. The second line of Eq. (13.26) differs from the Lagrangian density of a vector field in the unitary gauge by the term $-\tfrac{1}{2}[\partial_\mu A^\mu(x)]^2$. From Eq. (13.26) one obtains

$$(\Box + m^2)A^\mu(x) = 0 \tag{13.27}$$

‡ In Section 5.1 we replaced Eq. (13.21) by the Lagrangian density (5.10) and not by Eq. (13.22). However, Eqs. (5.10) and (13.22) are equivalent since they differ by a four-divergence only.

§ Eq. (13.24) represents a particular choice of gauge. 't Hooft, more generally, defined a whole class of gauges by conditions similar to Eq. (13.24). These are known as 't Hooft gauges.

as the equation of motion of $A^{\mu}(x)$. Eq. (13.27) is like the Klein–Gordon equation for a scalar field. When quantized, it leads to the propagator[‡]

$$iD_F^{\alpha\beta}(k, m) = \frac{-ig^{\alpha\beta}}{k^2 - m^2 + i\varepsilon}. \tag{13.28}$$

The troublesome term $k^{\alpha}k^{\beta}/m^2$ which occurred in the vector propagator (13.20) is absent from Eq. (13.28). For large k^2, the propagator (13.28) behaves like $1/k^2$, just like the photon propagator, and acts as a convergence factor. Dimensional arguments, like those used in Section 9.8, suggest that, like QED, the Higgs model is renormalizable. This is confirmed by the detailed analysis.

Working in a 't Hooft gauge reintroduces the $\eta(x)$ field which had been eliminated in the unitary gauge. There exist no real particles corresponding to the quantized $\eta(x)$ field although the Feynman propagator for this field can be interpreted in terms of the exchange of virtual scalar bosons. The properties of these 'ghost particles', as they are called, are analogous to those of the longitudinal and scalar photons in QED, which also do not exist as real free particles but contribute as virtual intermediate quanta to the photon propagator. The detailed properties of the $\eta(x)$ field are quite complicated and, of course, gauge-dependent. However, all observable quantities are gauge-invariant. This situation is again quite analogous to that in QED.

13.3 THE STANDARD ELECTRO–WEAK THEORY

In the last chapter we developed a unified model of electromagnetic and weak interactions of massless leptons and massless gauge bosons ($W^{\pm}, Z^0$ bosons and photons). The Lagrangian density of this model is

$$\mathscr{L} = \mathscr{L}^{L} + \mathscr{L}^{B}, \tag{13.29}$$

where $\mathscr{L}^{L}$ is the leptonic Lagrangian density (12.39) and $\mathscr{L}^{B}$ is the gauge-boson Lagrangian density (12.58). The Lagrangian density (13.29) is exactly invariant under the SU(2) × U(1) gauge transformations (12.32a), (12.32b) and (12.37a), (12.37b). We now apply the Higgs mechanism to this model to generate non-vanishing masses for the $W^{\pm}$ and Z^0 bosons, and we shall see how this also enables one to introduce lepton masses. In this way we shall finally arrive at the standard electro-weak theory of Weinberg and Salam.

The necessary formalism is an immediate extension of that of the Higgs model. To break the gauge invariance spontaneously, we must again introduce a *Higgs field*, i.e. a scalar field with non-vanishing vacuum expectation value which is not invariant under the gauge transformations.

‡ An explicit derivation of Eq. (13.28) is given in C. Itzykson and J. B. Zuber, *Quantum Field Theory*, McGraw-Hill, New York, 1980, Section 3-2-3.

Since we now want to break the SU(2) symmetry, we must introduce not a single such field but a field with several components and non-zero isospin. The simplest possibility is a weak isospin doublet

$$\Phi(x) = \begin{pmatrix} \phi_a(x) \\ \phi_b(x) \end{pmatrix} \tag{13.30}$$

where $\phi_a(x)$ and $\phi_b(x)$ are scalar fields under Lorentz transformations.

The transformation laws of $\Phi(x)$ under SU(2) × U(1) gauge transformations are, of course, the same as those of the isospin doublet $\Psi_l^L(x)$. The latter were given in Eqs. (12.32a) and (12.37a). Analogously, $\Phi(x)$ transforms under SU(2) transformations according to

$$\left.\begin{aligned} \Phi(x) &\to \Phi'(x) = \exp\,[ig\tau_j\omega_j(x)/2]\Phi(x) \\ \Phi^\dagger(x) &\to \Phi^{\dagger\prime}(x) = \Phi^\dagger(x)\exp\,[-ig\tau_j\omega_j(x)/2] \end{aligned}\right\}, \tag{13.31}$$

and under U(1) weak hypercharge transformations according to

$$\left.\begin{aligned} \Phi(x) &\to \Phi'(x) = \exp\,[ig'Yf(x)]\Phi(x) \\ \Phi^\dagger(x) &\to \Phi^{\dagger\prime}(x) = \Phi^\dagger(x)\exp\,[-ig'Yf(x)] \end{aligned}\right\} \tag{13.32}$$

where Y is the weak hypercharge of the field $\Phi(x)$. We shall determine its value shortly. [The corresponding global phase transformations are obtained from Eqs. (13.31) and (13.32) through the replacements $g\omega_j(x) \to \alpha_j$, $g'f(x) \to \beta$, where α_j and β are real constants; compare Eqs. (12.18a) and (12.31).]

We now want to generalize the Lagrangian density (13.29) to include the Higgs field $\Phi(x)$ and its interactions with the gauge-boson fields, and to continue to be SU(2) × U(1) gauge-invariant. The two terms in Eq. (13.29) already possess this invariance property. A generalization which obviously shares this property is

$$\mathscr{L} = \mathscr{L}^{\mathrm{L}} + \mathscr{L}^{\mathrm{B}} + \mathscr{L}^{\mathrm{H}} \tag{13.33}$$

where

$$\mathscr{L}^{\mathrm{H}}(x) = [D^\mu\Phi(x)]^\dagger[D_\mu\Phi(x)] - \mu^2\Phi^\dagger(x)\Phi(x) - \lambda[\Phi^\dagger(x)\Phi(x)]^2. \tag{13.34}$$

Here the covariant derivative $D^\mu\Phi(x)$ is defined by

$$D^\mu\Phi(x) = [\partial^\mu + ig\tau_j W_j^\mu(x)/2 + ig'YB^\mu(x)]\Phi(x), \tag{13.35}$$

in analogy with Eq. (12.40a) for $\Psi_l^L(x)$ which has hypercharge $-\frac{1}{2}$.

The expression $\mathscr{L}^{\mathrm{B}} + \mathscr{L}^{\mathrm{H}}$ in Eq. (13.33) is a direct generalization of the Higgs model Lagrangian density (13.15), and the further analysis closely follows that for the Higgs model. For $\lambda > 0$ and $\mu^2 < 0$, the classical energy density is a minimum for a constant Higgs field

$$\Phi(x) = \Phi_0 = \begin{pmatrix} \phi_a^0 \\ \phi_b^0 \end{pmatrix}, \tag{13.36}$$

with

$$\Phi_0^\dagger \Phi_0 = |\phi_a^0|^2 + |\phi_b^0|^2 = \frac{-\mu^2}{2\lambda}, \tag{13.37}$$

and all other fields vanishing. Choosing for the ground state a particular value Φ_0, compatible with Eq. (13.37), again leads to spontaneous symmetry breaking. Without loss of generality, we can choose

$$\Phi_0 = \begin{pmatrix} \phi_a^0 \\ \phi_b^0 \end{pmatrix} = \begin{pmatrix} 0 \\ v/\sqrt{2} \end{pmatrix} \tag{13.38}$$

where

$$v = (-\mu^2/\lambda)^{1/2} \quad (>0), \tag{13.39}$$

since any other choice of Φ_0 is related to the value (13.38) by a global phase transformation.

The Higgs field of the vacuum ground state, Eq. (13.38), is, in general, not invariant under SU(2) × U(1) gauge transformations. However, it must be invariant under U(1) electromagnetic gauge transformations, in order to ensure zero mass for the photon and conservation of the electric charge. If we assign the weak hypercharge $Y = \frac{1}{2}$ to the Higgs field, then it follows from Eq. (12.27) that the lower component $\phi_b(x)$ of the Higgs field is electrically neutral, so that spontaneous symmetry breaking occurs only in the electrically neutral component of the vacuum field (13.38), and charge conservation holds exactly. Alternatively, we see from Eqs. (13.16) and (12.27) that an electromagnetic gauge transformation of the Higgs field is given by

$$\Phi(x) \to \Phi'(x) = \exp\left[-i(Y + I_3^W)ef(x)\right]\Phi(x). \tag{13.40}$$

Applied to the vacuum field (13.38), this transformation gives

$$\Phi_0 \to \Phi_0' = \Phi_0, \tag{13.41}$$

i.e. the vacuum field is invariant under electromagnetic gauge transformations.

An arbitrary Higgs field $\Phi(x)$ can again be parameterized in terms of its deviations from the vacuum field Φ_0 in the form

$$\Phi(x) = 2^{-1/2}\begin{pmatrix} \eta_1(x) + i\eta_2(x) \\ v + \sigma(x) + i\eta_3(x) \end{pmatrix}. \tag{13.42}$$

By means of this equation, we can express the Lagrangian density $\mathscr{L}^{H}$, Eq. (13.34), in terms of the four real fields $\sigma(x)$ and $\eta_i(x)$, $i = 1, 2, 3$. The interpretation and quantization of these fields lead to the same difficulties which we met for the Higgs model. The way these difficulties are resolved and the further analysis are closely analogous to our treatment of the Higgs model. The interpretation becomes particularly simple if we employ a special gauge, the unitary gauge. We shall give the analysis in the next chapter (Section 14.1). However, the similarity to the Higgs model is so close that we can anticipate the results. We shall find that the fields $\eta_i(x)$, $i = 1, 2, 3$, are unphysical fields. In the unitary gauge they are transformed away, and the $W^{\pm}$ and Z^0 bosons are seen to acquire mass. The photon remains massless since the electromagnetic gauge symmetry has not been spontaneously broken. This is also reflected in the fact that only three unphysical fields $\eta_i(x)$ occur. In contrast to the fields $\eta_i(x)$ disappearing, the field $\sigma(x)$ survives in the unitary gauge and, on quantization, gives rise to massive, electrically neutral, spin 0 particles (Higgs scalars).

To obtain non-vanishing lepton masses, we must augment the Lagrangian density (13.33), by adding a suitable term $\mathscr{L}^{\text{LH}}$, to

$$\mathscr{L} = \mathscr{L}^{\text{L}} + \mathscr{L}^{\text{B}} + \mathscr{L}^{\text{H}} + \mathscr{L}^{\text{LH}}. \tag{13.43}$$

We shall couple the lepton and Higgs fields through Yukawa interactions,‡ described by the Lagrangian density

$$\begin{aligned}\mathscr{L}^{\text{LH}}(x) = &- g_l[\bar{\Psi}_l^{\text{L}}(x)\psi_l^{\text{R}}(x)\Phi(x) + \Phi^\dagger(x)\bar{\psi}_l^{\text{R}}(x)\Psi_l^{\text{L}}(x)] \\ &- g_{\nu_l}[\bar{\Psi}_l^{\text{L}}(x)\psi_{\nu_l}^{\text{R}}(x)\tilde{\Phi}(x) + \tilde{\Phi}^\dagger(x)\bar{\psi}_{\nu_l}^{\text{R}}(x)\Psi_l^{\text{L}}(x)].\end{aligned} \tag{13.44}$$

Here g_l and g_{ν_l} are dimensionless coupling constants, summations over $l = e, \mu, \ldots$ are implied, as usual, and $\tilde{\Phi}(x)$ is defined by

$$\tilde{\Phi}(x) = -\text{i}[\Phi^\dagger(x)\tau_2]^{\text{T}} = \begin{pmatrix} \phi_b^*(x) \\ -\phi_a^*(x) \end{pmatrix}. \tag{13.45}$$

In Eq. (13.45), τ_2 is the Pauli matrix (12.15) and T denotes the transpose.

We shall now show that the Lagrangian density $\mathscr{L}^{\text{LH}}$ is invariant under SU(2) × U(1) gauge transformations. The invariance of the first line of Eq. (13.44) follows from the transformation laws (12.32a) and (12.37a) of the lepton fields and the transformation laws (13.31) and (13.32) of the Higgs field. For the second line of Eq. (13.44), we require the transformation properties of $\tilde{\Phi}(x)$. Under U(1) transformations, it follows from Eqs. (13.32) and (13.45) that

$$\tilde{\Phi}(x) \to \tilde{\Phi}'(x) = \exp[-ig'f(x)/2]\tilde{\Phi}(x). \tag{13.46}$$

We shall show below that, under SU(2) transformations, $\tilde{\Phi}(x)$ transforms in exactly the same way as $\Phi(x)$. From these transformation properties of $\tilde{\Phi}$ and of the lepton fields, the SU(2) × U(1) gauge invariance of the second line of Eq. (13.44) follows.

‡ A term of the form $\bar{\psi}(x)\phi(x)\psi(x)$, where $\psi(x)$ and $\phi(x)$ are spinor and scalar fields respectively, is called a Yukawa interaction or a Yukawa coupling.

In order to derive the SU(2) transformation properties of $\tilde{\Phi}(x)$, it suffices to consider infinitesimal transformations. For these, we obtain from Eq. (13.31) that

$$\Phi(x) \to \Phi(x) + \delta\Phi(x) = [1 + \mathrm{i}\tfrac{1}{2}g\tau_j\omega_j(x) + \cdots]\Phi(x)$$

i.e.

$$\delta\Phi(x) = \mathrm{i}\tfrac{1}{2}g\tau_j\omega_j(x)\Phi(x). \tag{13.47}$$

Hence

$$\delta\Phi^\dagger(x) = -\mathrm{i}\tfrac{1}{2}g\omega_j(x)\Phi^\dagger(x)\tau_j$$

and from Eq. (13.45)

$$\delta\tilde{\Phi}(x) = -\mathrm{i}[\delta\Phi^\dagger(x)\tau_2]^{\mathrm{T}} = -\mathrm{i}[-\mathrm{i}\tfrac{1}{2}g\omega_j(x)\Phi^\dagger(x)\tau_j\tau_2]^{\mathrm{T}}. \tag{13.48}$$

For the Pauli matrices (12.15) $\tau_j\tau_2 = -\tau_2\tau_j^{\mathrm{T}}$. Substituting this relation in Eq. (13.48), the resulting expression simplifies to

$$\delta\tilde{\Phi}(x) = \mathrm{i}\tfrac{1}{2}g\omega_j(x)\tau_j[-\mathrm{i}\Phi^\dagger(x)\tau_2]^{\mathrm{T}} = \mathrm{i}\tfrac{1}{2}g\omega_j(x)\tau_j\tilde{\Phi}(x). \tag{13.49}$$

Comparing Eqs. (13.49) and (13.47), we see that $\tilde{\Phi}(x)$ and $\Phi(x)$ transform in the same way under SU(2) transformations.

In the next chapter, we shall transform the Lagrangian density of the electro–weak theory into the unitary gauge, and we shall see that the term (13.44) not only generates interactions between leptons and Higgs bosons but also leads to non-zero lepton masses. In their original formulation of the theory, Weinberg and Salam took $g_{\nu_l} = 0$, leading to zero neutrino masses.‡ The second line of Eq. (13.44) represents the simplest way of introducing non-vanishing neutrino masses. A generalization of considerable interest is to replace the second line in Eq. (13.44) by

$$-G_{l'l}\bar{\Psi}^{\mathrm{L}}_{l'}(x)\psi^{\mathrm{R}}_{\nu_l}(x)\tilde{\Phi}(x) - G^*_{l'l}\tilde{\Phi}^\dagger(x)\bar{\psi}^{\mathrm{R}}_{\nu_l}(x)\Psi^{\mathrm{L}}_{l'}(x) \tag{13.44a}$$

where G is a Hermitian coupling matrix. By writing Eq. (13.44a) in the unitary gauge and diagonalizing it, one obtains the eigenstate neutrinos ν_i ($i = 1, 2, \ldots$) with masses m_i (see Problem 14.5). The leptonic neutrinos ν_l ($l = e, \mu, \ldots$) are linear combinations of these eigenstate neutrinos ν_i, each of which has its own characteristic time dependence. As a result, neutrino mixing will occur. (This is just the well-known time dependence of a superposition of energy eigenstates of a system.) For simplicity, consider the

‡ In Section 11.2 the experimental upper bounds on the neutrino masses were given. For the electron neutrino, V. A. Lyubimov *et al.* have recently obtained the lower bound $m_{\nu_e} \geqslant 20$ eV; see the review paper by M. Shaevitz in the *Proceedings of the 1983 International Symposium on Lepton and Photon Interactions at High Energies*, Cornell University, 1983.

mixing of muon and electron neutrinos from two eigenstate neutrinos ν_1 and ν_2, specified in terms of a mixing angle α by

$$\begin{pmatrix} \nu_\mu \\ \nu_e \end{pmatrix} = \begin{pmatrix} \cos\alpha & \sin\alpha \\ -\sin\alpha & \cos\alpha \end{pmatrix} \begin{pmatrix} \nu_1 \\ \nu_2 \end{pmatrix}.$$

(The generalization to more neutrinos should be obvious.) If, for example, a ν_μ beam is created in an accelerator experiment, its composition will change along its path, oscillating between pure ν_μ and pure ν_e beam, with $\nu_\mu - \nu_e$ mixtures in between. The probability that a 'neutrino-flip' $\nu_\mu \to \nu_e$ has occurred, when the beam has travelled a distance L from its source, is given by[‡]

$$P(\nu_\mu \to \nu_e) = \sin^2\alpha \sin^2\left[\frac{(m_2^2 - m_1^2)L}{4E}\right]$$

where E is the beam energy (we are assuming $E \gg m_i$). Thus, for $\alpha \neq 0$, neutrino mixing will occur, and one would observe electron-neutrinos, violating lepton number conservation, at distances $L \gtrsim E/(m_1^2 - m_2^2)$ from the source.

Such effects are called neutrino oscillations. So far, there is no evidence for their existence.[§] However, many further experiments, designed to detect them, are in progress or planned. Neutrino oscillations can only occur if neutrinos have non-zero masses, and observing such oscillations would provide information on these masses. Although non-zero masses do not necessarily lead to oscillations, one might expect them from the close similarity of the theories for leptons and for quarks, since in the latter case mixing definitely occurs.

After this digression on neutrino mixing, we shall use the simpler form (13.44) for $\mathscr{L}^{\text{LH}}$ in the Lagrangian density (13.43) of the electro-weak theory.

In this book, we are restricting ourselves to the electro-weak interactions of leptons only. However, the theory is easily extended to include quarks and hence hadrons, enabling one to treat semi-leptonic processes. Indeed, this extension, as well as the gauge invariance of the Lagrangian density (13.44), is necessary in order to be able to prove the renormalizability of the theory. The proof again relies on the use of 't Hooft gauges, mentioned in Section 13.2 on the Higgs model, rather than of the unitary gauge which we shall employ in the next chapter.[¶]

We have now completed the derivation of the basic equations of the

[‡] See D. H. Perkins, *Introduction to High Energy Physics*, 2nd edn, Addison-Wesley, Reading, Mass., 1982, section 6.14.

[§] For a summary of the experimental limits on neutrino mixing, for different assumed neutrino masses, see the article by M. Shaevitz, quoted earlier in this section.

[¶] For the extension of the theory to quarks, see D. Bailin, *Weak Interactions*, 2nd edn, A. Hilger, Bristol, 1982, Section 6.5; for a full discussion of renormalizability, see the articles by Abers and Lee or the book by J. C. Taylor, cited in Section 13.2.

standard electro–weak theory for purely leptonic processes and we shall conclude the chapter by summarizing these equations.

The Lagrangian density (13.43) is from Eqs. (12.39), (12.58), (13.34) and (13.44) given by

$$\begin{aligned}\mathscr{L}(x) &= \mathscr{L}^{\mathrm{L}}(x) + \mathscr{L}^{\mathrm{B}}(x) + \mathscr{L}^{\mathrm{H}}(x) + \mathscr{L}^{\mathrm{LH}}(x) \\ &= \mathrm{i}[\bar{\Psi}_l^{\mathrm{L}}(x) \not{D} \Psi_l^{\mathrm{L}}(x) + \bar{\psi}_l^{\mathrm{R}}(x) \not{D} \psi_l^{\mathrm{R}}(x) + \bar{\psi}_{\nu_l}^{\mathrm{R}}(x) \not{D} \psi_{\nu_l}^{\mathrm{R}}(x)] \\ &\quad + [-\tfrac{1}{4} B_{\mu\nu}(x) B^{\mu\nu}(x) - \tfrac{1}{4} G_{i\mu\nu}(x) G_i^{\mu\nu}(x)] \\ &\quad + \{[D^\mu \Phi(x)]^\dagger [D_\mu \Phi(x)] - \mu^2 \Phi^\dagger(x) \Phi(x) - \lambda [\Phi^\dagger(x) \Phi(x)]^2\} \\ &\quad + \{-g_l [\bar{\Psi}_l^{\mathrm{L}}(x) \psi_l^{\mathrm{R}}(x) \Phi(x) + \Phi^\dagger(x) \bar{\psi}_l^{\mathrm{R}}(x) \Psi_l^{\mathrm{L}}(x)] \\ &\quad - g_{\nu_l} [\bar{\Psi}_l^{\mathrm{L}}(x) \psi_{\nu_l}^{\mathrm{R}}(x) \tilde{\Phi}(x) + \tilde{\Phi}^\dagger(x) \bar{\psi}_{\nu_l}^{\mathrm{R}}(x) \Psi_l^{\mathrm{L}}(x)]\}. \end{aligned} \tag{13.50}$$

Here $B^{\mu\nu}(x)$ and $G_i^{\mu\nu}(x)$ are defined in Eqs. (12.53), (12.56) and (12.55) as

$$B^{\mu\nu}(x) \equiv \partial^\nu B^\mu(x) - \partial^\mu B^\nu(x) \tag{13.51}$$

$$G_i^{\mu\nu}(x) \equiv F_i^{\mu\nu}(x) + g\varepsilon_{ijk} W_j^\mu(x) W_k^\nu(x) \tag{13.52}$$

with

$$F_i^{\mu\nu}(x) \equiv \partial^\nu W_i^\mu(x) - \partial^\mu W_i^\nu(x). \tag{13.53}$$

The covariant derivatives in Eq. (13.50) are defined in Eqs. (12.40) and (13.35) as

$$D^\mu \Psi_l^{\mathrm{L}}(x) = [\partial^\mu + \mathrm{i} g \tau_j W_j^\mu(x)/2 - \mathrm{i} g' B^\mu(x)/2] \Psi_l^{\mathrm{L}}(x) \tag{13.54}$$

$$D^\mu \psi_l^{\mathrm{R}}(x) = [\partial^\mu - \mathrm{i} g' B^\mu(x)] \psi_l^{\mathrm{R}}(x) \tag{13.55}$$

$$D^\mu \psi_{\nu_l}^{\mathrm{R}}(x) = \partial^\mu \psi_{\nu_l}^{\mathrm{R}}(x) \tag{13.56}$$

$$D^\mu \Phi(x) = [\partial^\mu + \mathrm{i} g \tau_j W_j^\mu(x)/2 + \mathrm{i} g' B^\mu(x)/2] \Phi(x). \tag{13.57}$$

The Lagrangian density (13.50) is invariant under SU(2) × U(1) gauge transformations. However, the terms $\mathscr{L}^{\mathrm{H}} + \mathscr{L}^{\mathrm{B}}$ are just the direct generalization of the Higgs model Lagrangian density (13.15), and for $\lambda > 0$ and $\mu^2 < 0$ the SU(2) × U(1) gauge symmetry is spontaneously broken. In the quantized theory, the vacuum expectation value of the Higgs field

$$\langle 0|\Phi(x)|0\rangle = \Phi_0 = \begin{pmatrix} 0 \\ v/\sqrt{2} \end{pmatrix}, \tag{13.58}$$

where

$$v = (-\mu^2/\lambda)^{1/2} \quad (>0), \tag{13.59}$$

is, in general, not invariant under SU(2) × U(1) gauge transformations. However, it is invariant under electromagnetic gauge transformations (which are contained in the full set of SU(2) × U(1) transformations), so that the photon remains massless and electric charge is conserved.

CHAPTER 14

The standard electro–weak theory

In the last chapter, we derived the basic equations of the electro–weak theory. In this chapter, we shall consider its physical interpretation and applications. In Section 14.1 we shall transform the theory into the unitary gauge which facilitates the interpretation. In particular, it will become apparent that the Lagrangian density, expressed in this gauge, describes photons, charged and neutral leptons, W and Z^0 bosons, and Higgs bosons. We shall see that, with exception of the photon, all these particles have acquired mass, and we shall obtain expressions for these masses in terms of the basic parameters of the theory. In Section 14.2 we shall derive the Feynman rules for the electro–weak theory, which are an extension of those for QED.

In the following two sections we shall consider some applications. The electro–weak theory is renormalizable and so allows calculations to be performed to all orders of perturbation theory. This opens up a rich field of phenomena for accurate calculations and asks for high-precision experiments to test the theory, particularly the radiative corrections. The latter require very elaborate calculations (for which a 't Hooft gauge would be more appropriate than the unitary gauge). We shall restrict ourselves to lowest-order calculations.‡ In Section 14.3 we shall consider neutrino–electron scattering, which leads to the determination of the weak mixing angle θ_W from experi-

‡ Further applications, including ones to semi-leptonic processes, will be found in the following books: D. Bailin, *Weak Interactions*, 2nd edn, A. Hilger, Bristol, 1982; S. M. Bilenky, *Introduction to the Physics of Weak Interactions*, Pergamon, Oxford, 1982; E. Leader and E. Predazzi, *An Introduction to Gauge Theories and the New Physics*, Cambridge University Press, Cambridge, 1982.

ment and hence to predictions of the masses of the W and Z^0 bosons. In Section 14.4 we shall discuss several electron–positron annihilation processes at increasingly higher energies. These experiments represent some of the cleanest tests of the electro–weak theory and will be of great interest in the next few years as more and better experimental results become available.

In the last section of the chapter we shall discuss the most important unresolved puzzle of the theory: the mystery of the Higgs particle which has not been observed so far; yet, its existence is essential for the renormalizability of the theory. We shall see that the nature of the interactions of the Higgs particle are such that, after all, it is not surprising that it has not yet been detected, and we shall discuss which processes are most likely to lead to its experimental discovery in the future.

14.1 THE LAGRANGIAN DENSITY IN THE UNITARY GAUGE

In the last chapter we wrote the Higgs field $\Phi(x)$ in an arbitrary gauge as

$$\Phi(x) = 2^{-1/2}\begin{pmatrix} \eta_1(x) + i\eta_2(x) \\ v + \sigma(x) + i\eta_3(x) \end{pmatrix}. \tag{13.42}$$

This isospinor can always be transformed into the form

$$\Phi(x) = 2^{-1/2}\begin{pmatrix} 0 \\ v + \sigma(x) \end{pmatrix} \tag{14.1}$$

which no longer contains the fields $\eta_i(x)$. Eqs. (13.42) and (14.1) are analogous to Eqs. (13.10) and (13.18) for the Higgs model, and the gauge in which the Higgs field has the form (14.1) is again called the unitary gauge. The gauge transformation which transforms Eq. (13.42) into Eq. (14.1) consists of an SU(2) transformation (13.31) which converts the isospinor (13.42) into a 'down'-isospinor, followed by a U(1) transformation (13.32) which makes this 'down'-isospinor a real quantity. Under this SU(2) × U(1) gauge transformation, all other fields transform according to the corresponding Eqs. (12.32a), (12.32b) and (12.37a), (12.37b), and we shall assume that all fields are already expressed in the unitary gauge.

To transform the Lagrangian density (13.50) into the unitary gauge, we substitute Eq. (14.1) into it. We shall also use Eq. (12.43), its complex conjugate and Eqs. (12.45) to replace the fields $W_j^\mu(x)$ and $B^\mu(x)$ by $W^\mu(x)$, $W^{\dagger\mu}(x)$, $Z^\mu(x)$ and $A^\mu(x)$. At this stage we are still dealing with classical fields, so that $W^{\dagger\mu}(x)$ stands for the complex conjugate field. In using this notation, we are anticipating that these fields will presently be quantized, and $W^{\dagger\mu}(x)$ will denote the Hermitian adjoint field operator, as usual. The motivation for this transformation is, of course, that the quantized fields $W^\mu(x)$, $W^{\dagger\mu}(x)$, $Z^\mu(x)$ and $A^\mu(x)$ describe $W^\pm$ bosons, Z^0 bosons and photons. The transformation of the Lagrangian density is straightforward but tedious, and we shall only quote the results.

On transforming the terms $\mathscr{L}^{\mathrm{B}} + \mathscr{L}^{\mathrm{H}}$ in Eq. (13.50) in this way one obtains‡

$$\begin{aligned}\mathscr{L}^{\mathrm{B}} + \mathscr{L}^{\mathrm{H}} = & -\tfrac{1}{4}F_{\mu\nu}F^{\mu\nu} \\ & -\tfrac{1}{2}F^{\dagger}_{W\mu\nu}F^{\mu\nu}_{W} + m_W^2 W^{\dagger}_{\mu}W^{\mu} \\ & -\tfrac{1}{4}Z_{\mu\nu}Z^{\mu\nu} + \tfrac{1}{2}m_Z^2 Z_{\mu}Z^{\mu} \\ & +\tfrac{1}{2}(\partial^{\mu}\sigma)(\partial_{\mu}\sigma) - \tfrac{1}{2}m_H^2\sigma^2 \\ & +\mathscr{L}_{\mathrm{I}}^{\mathrm{BB}} + \mathscr{L}_{\mathrm{I}}^{\mathrm{HH}} + \mathscr{L}_{\mathrm{I}}^{\mathrm{HB}}\end{aligned} \tag{14.2}$$

where we have dropped a constant term, and $\mathscr{L}_{\mathrm{I}}^{\mathrm{BB}}$, etc., are given by

$$\begin{aligned}\mathscr{L}_{\mathrm{I}}^{\mathrm{BB}} = {} & \mathrm{i}g\cos\theta_{\mathrm{W}}[(W^{\dagger}_{\alpha}W_{\beta} - W^{\dagger}_{\beta}W_{\alpha})\,\partial^{\alpha}Z^{\beta} \\ & \quad + (\partial_{\alpha}W_{\beta} - \partial_{\beta}W_{\alpha})W^{\dagger\beta}Z^{\alpha} - (\partial_{\alpha}W^{\dagger}_{\beta} - \partial_{\beta}W^{\dagger}_{\alpha})W^{\beta}Z^{\alpha}] \\ & + \mathrm{i}e[(W^{\dagger}_{\alpha}W_{\beta} - W^{\dagger}_{\beta}W_{\alpha})\,\partial^{\alpha}A^{\beta} \\ & \quad + (\partial_{\alpha}W_{\beta} - \partial_{\beta}W_{\alpha})W^{\dagger\beta}A^{\alpha} - (\partial_{\alpha}W^{\dagger}_{\beta} - \partial_{\beta}W^{\dagger}_{\alpha})W^{\beta}A^{\alpha}] \\ & + g^2\cos^2\theta_{\mathrm{W}}[W_{\alpha}W^{\dagger}_{\beta}Z^{\alpha}Z^{\beta} - W_{\beta}W^{\dagger\beta}Z_{\alpha}Z^{\alpha}] \\ & + e^2[W_{\alpha}W^{\dagger}_{\beta}A^{\alpha}A^{\beta} - W_{\beta}W^{\dagger\beta}A_{\alpha}A^{\alpha}] \\ & + eg\cos\theta_{\mathrm{W}}[W_{\alpha}W^{\dagger}_{\beta}(Z^{\alpha}A^{\beta} + A^{\alpha}Z^{\beta}) - 2W_{\beta}W^{\dagger\beta}A_{\alpha}Z^{\alpha}] \\ & + \tfrac{1}{2}g^2 W^{\dagger}_{\alpha}W_{\beta}[W^{\dagger\alpha}W^{\beta} - W^{\alpha}W^{\dagger\beta}]\end{aligned} \tag{14.3a}$$

$$\mathscr{L}_{\mathrm{I}}^{\mathrm{HH}} = -\tfrac{1}{4}\lambda\sigma^4 - \lambda v\sigma^3 \tag{14.3b}$$

$$\begin{aligned}\mathscr{L}_{\mathrm{I}}^{\mathrm{HB}} = {} & \tfrac{1}{2}vg^2 W^{\dagger}_{\alpha}W^{\alpha}\sigma + \tfrac{1}{4}g^2 W^{\dagger}_{\alpha}W^{\alpha}\sigma^2 \\ & + \frac{vg^2}{4\cos^2\theta_{\mathrm{W}}}Z_{\alpha}Z^{\alpha}\sigma + \frac{g^2}{8\cos^2\theta_{\mathrm{W}}}Z_{\alpha}Z^{\alpha}\sigma^2.\end{aligned} \tag{14.3c}$$

The parameters m_W, m_Z and m_H which have been introduced in Eq. (14.2) are defined by

$$m_W = \tfrac{1}{2}vg, \qquad m_Z = m_W/\cos\theta_{\mathrm{W}}, \qquad m_H = \sqrt{(-2\mu^2)}, \tag{14.4}$$

and v and the weak mixing angle θ_{W} are, from Eqs. (13.59) and (12.47), given by

$$v = (-\mu^2/\lambda)^{1/2} \quad (>0) \tag{14.5}$$

$$g\sin\theta_{\mathrm{W}} = g'\cos\theta_{\mathrm{W}} = e. \tag{14.6}$$

We now consider the remaining two terms $\mathscr{L}^{\mathrm{L}} + \mathscr{L}^{\mathrm{LH}}$ in Eq. (13.50). We again transform from the fields W^{μ}_{j} and B^{μ} to W^{μ}, $W^{\dagger\mu}$, Z^{μ} and A^{μ}. In

‡ In the following equations, all fields have the same argument x, and we shall therefore omit it.

addition, we use Eq. (12 11) to replace left- and right-handed lepton fields ψ^{L} and ψ^{R} by the complete fields ψ. In this way one finds that

$$\mathscr{L}^{\mathrm{L}} + \mathscr{L}^{\mathrm{LH}} = \bar{\psi}_l(\mathrm{i}\partial\!\!\!/ - m_l)\psi_l + \bar{\psi}_{\nu_l}(\mathrm{i}\partial\!\!\!/ - m_{\nu_l})\psi_{\nu_l} + \mathscr{L}_{\mathrm{I}}^{\mathrm{LB}} + \mathscr{L}_{\mathrm{I}}^{\mathrm{HL}} \tag{14.7}$$

where $\mathscr{L}_{\mathrm{I}}^{\mathrm{LB}}$ and $\mathscr{L}_{\mathrm{I}}^{\mathrm{HL}}$ are given by

$$\begin{aligned}\mathscr{L}_{\mathrm{I}}^{\mathrm{LB}} &= e\bar{\psi}_l\gamma^\alpha\psi_l A^\alpha \\ &\quad - \frac{g}{2\sqrt{2}}\left[\bar{\psi}_{\nu_l}\gamma^\alpha(1-\gamma_5)\psi_l W_\alpha + \bar{\psi}_l\gamma^\alpha(1-\gamma_5)\psi_{\nu_l}W_\alpha^\dagger\right] \\ &\quad - \frac{g}{4\cos\theta_{\mathrm{W}}}\bar{\psi}_{\nu_l}\gamma^\alpha(1-\gamma_5)\psi_{\nu_l}Z_\alpha \\ &\quad + \frac{g}{4\cos\theta_{\mathrm{W}}}\bar{\psi}_l\gamma^\alpha(1-4\sin^2\theta_{\mathrm{W}}-\gamma_5)\psi_l Z_\alpha \end{aligned} \tag{14.3d}$$

$$\mathscr{L}_{\mathrm{I}}^{\mathrm{HL}} = -\frac{1}{v}m_l\bar{\psi}_l\psi_l\sigma - \frac{1}{v}m_{\nu_l}\bar{\psi}_{\nu_l}\psi_{\nu_l}\sigma. \tag{14.3e}$$

The parameters m_l and m_{ν_l} which have been introduced in Eq. (14.7) are defined by

$$m_l = vg_l/\sqrt{2}, \qquad m_{\nu_l} = vg_{\nu_l}/\sqrt{2}. \tag{14.8}$$

Finally, combining Eqs. (14.2) and (14.7), one obtains the complete Lagrangian density of the standard electro–weak theory in the unitary gauge

$$\mathscr{L} = \mathscr{L}_0 + \mathscr{L}_{\mathrm{I}} \tag{14.9}$$

where

$$\begin{aligned}\mathscr{L}_0 &= \bar{\psi}_l(\mathrm{i}\partial\!\!\!/ - m_l)\psi_l + \bar{\psi}_{\nu_l}(\mathrm{i}\partial\!\!\!/ - m_{\nu_l})\psi_{\nu_l} \\ &\quad - \tfrac{1}{4}F_{\mu\nu}F^{\mu\nu} \\ &\quad - \tfrac{1}{2}F_{W\mu\nu}^\dagger F_W^{\mu\nu} + m_W^2 W_\mu^\dagger W^\mu \\ &\quad - \tfrac{1}{4}Z_{\mu\nu}Z^{\mu\nu} + \tfrac{1}{2}m_Z^2 Z_\mu Z^\mu \\ &\quad + \tfrac{1}{2}(\partial^\mu\sigma)(\partial_\mu\sigma) - \tfrac{1}{2}m_H^2\sigma^2 \end{aligned} \tag{14.10}$$

and

$$\mathscr{L}_{\mathrm{I}} = \mathscr{L}_{\mathrm{I}}^{\mathrm{LB}} + \mathscr{L}_{\mathrm{I}}^{\mathrm{BB}} + \mathscr{L}_{\mathrm{I}}^{\mathrm{HH}} + \mathscr{L}_{\mathrm{I}}^{\mathrm{HB}} + \mathscr{L}_{\mathrm{I}}^{\mathrm{HL}}, \tag{14.11}$$

with the individual terms $\mathscr{L}_{\mathrm{I}}^{\mathrm{LB}}, \ldots$ given by Eqs. (14.3a)–(14.3e).

Eq. (14.10) clearly admits interpretation as the Lagrangian density of free fields which can be quantized in the usual way. The two terms in the first line of this equation are just the Lagrangian densities of charged leptons with mass m_l and of neutrinos with mass m_{ν_l}. The second, third and fourth lines respectively describe photons, charged vector bosons ($W^\pm$) of mass m_W and neutral vector bosons (Z^0) of mass m_Z. The last line of Eq. (14.10) is just the Lagrangian density of a neutral Klein–Gordon field, and the quantized $\sigma(x)$ field describes neutral spin 0 bosons (Higgs scalars) of mass m_H. The mass terms of the $W^\pm$ and Z^0 bosons arise from the spontaneous symmetry breaking of the SU(2) × U(1) gauge invariance of the Lagrangian density (13.50). On the other hand, the photon remains massless [i.e. there is no term in $A_\mu A^\mu$ in Eq. (14.10)] since the electromagnetic gauge invariance is not broken spontaneously. The lepton mass terms in the first line of Eq. (14.10) have their origin in the Yukawa coupling terms $\mathscr{L}^{\rm LH}$ in Eq. (13.50).

Eq. (14.11) is the interaction Lagrangian density of the standard electroweak theory, with the individual terms (14.3a)–(14.3e) describing the interactions between pairs of particles. For example, $\mathscr{L}_{\rm I}^{\rm LB}$ is the interaction of leptons with gauge bosons, etc. These terms will be discussed in the next section where the Feynman rules for treating them in perturbation theory will be derived.

Eqs. (14.4) and (14.8) relate the boson and lepton masses to the basic parameters

$$g, g', -\mu^2, \lambda, g_l, g_{\nu_l} \tag{14.12}$$

of the theory. Rather remarkably, these relations allow the masses of the $W^\pm$ and Z^0 bosons to be determined in terms of three experimentally well known quantities: the fine structure constant

$$\alpha = e^2/4\pi = 1/137.04, \tag{14.13a}$$

the Fermi coupling constant

$$G = 1.166 \times 10^{-5}\ \text{GeV}^{-2} \tag{14.13b}$$

[see Eq. (11.70b)], and the weak mixing angle $\theta_{\rm W}$ which is determined from neutrino scattering experiments (see Section 14.3) and is given by

$$\sin^2\theta_{\rm W} = 0.227 \pm 0.014 \qquad (0 \leqslant \theta_{\rm W} \leqslant \pi/2). \tag{14.13c}$$

From Eqs. (11.43), (12.49) and (14.4), the parameter v can be expressed in terms of G as

$$v = (G\sqrt{2})^{-1/2} \tag{14.14}$$

and is therefore also known. Combining Eqs. (14.4), (14.6) and (14.14) gives

$$m_W = \left(\frac{\alpha\pi}{G\sqrt{2}}\right)^{1/2} \frac{1}{\sin\theta_{\rm W}}, \qquad m_Z = \left(\frac{\alpha\pi}{G\sqrt{2}}\right)^{1/2} \frac{2}{\sin 2\theta_{\rm W}}, \tag{14.15}$$

and substituting the above values for α, G and $\theta_{\rm W}$ leads to

$$m_W = \left(78.3 \begin{array}{l} +2.5 \\ -2.3 \end{array}\right)\text{GeV}, \qquad m_Z = \left(89.0 \begin{array}{l} +2.1 \\ -1.8 \end{array}\right)\text{GeV}. \tag{14.16a}$$

These predictions of the electro–weak theory are in good agreement with the current experimental masses, reported in Chapter 11:

$$m_W = (80.9 \pm 1.5 \pm 2.4)\ \text{GeV}, \qquad m_W = (81.0 \pm 2.5 \pm 1.3)\ \text{GeV}, \quad (11.8\text{b})$$

$$m_Z = (95.6 \pm 1.4 \pm 2.9)\ \text{GeV}, \qquad m_Z = (91.9 \pm 1.3 \pm 1.4)\ \text{GeV}. \quad (11.87\text{a})$$

This comparison is an important test of the theory. In deriving Eqs. (14.4), from which the values (14.16a) were obtained, we assumed that the Higgs field has isospin $\frac{1}{2}$. The above agreement shows that this isospin value is consistent with experiment.

Eqs. (14.15) and the values (14.16a) derived from them were obtained from the free-particle Lagrangian density and will, of course, be modified by terms of order α when radiative corrections are taken into account. The calculation of such corrections requires a discussion of renormalization which goes beyond the scope of this book, and we shall merely quote the predictions for the renormalized (i.e. physical) masses:‡

$$m_W = \left(83.0 \quad \begin{matrix} +2.9 \\ -2.7 \end{matrix}\right) \text{GeV}, \qquad m_Z = \left(93.8 \quad \begin{matrix} +2.4 \\ -2.2 \end{matrix}\right) \text{GeV}. \quad (14.16\text{b})$$

These values too are in good agreement with the experimental masses; at the same time, they differ significantly from the lowest-order values (14.16a). Clearly, these radiative corrections are important in obtaining accurate values for m_W and m_Z, and precise measurements of these masses will be of great interest since they will provide an experimental test of these radiative corrections. In the following we shall disregard these corrections and work in lowest order.

We see from Eq. (14.6) that the fine structure constant and the weak mixing angle also determine the coupling constants g and g'. Similarly, the Higgs–lepton coupling constants g_l and g_{ν_l} are known from Eqs. (14.8) and (14.14), provided the mass of the corresponding lepton is known.

This leaves only the parameter λ in the set (14.12) to be determined since, from Eq. (14.5), $(-\mu^2) = \lambda v^2$. λ occurs as coupling constant in the Higgs self-coupling terms $\mathscr{L}_{\text{I}}^{\text{HH}}$, Eq. (14.3b), but there is no chance of measuring these at present. From Eqs. (14.4) and (14.5)

$$m_{\text{H}} = \sqrt{(-2\mu^2)} = \sqrt{(2\lambda v^2)},$$

so that the mass of the Higgs boson cannot be predicted from known data. This is unfortunate, as knowing its mass would be of great help in searching for it. The existence of the Higgs boson is essential: Feynman diagrams involving the emission and subsequent reabsorption of Higgs particles contribute to higher-order corrections, and without them the

‡ See the review article by W. Marciano in the *Proceedings of the 1983 International Symposium on Lepton and Photon Interactions at High Energies*, Cornell University, 1983.

electro-weak theory would not be renormalizable. As m_H becomes very small or very large, these corrections become very large, and the success of the theory in lowest-order calculations restricts m_H to the extremely wide range‡

$$7\ \text{GeV} \lesssim m_H \lesssim 10^3\ \text{GeV}. \tag{14.17}$$

When considering applications of the electro–weak theory later in this chapter, we shall always assume that m_H lies in the range (14.17).

14.2 FEYNMAN RULES

We shall now derive the Feynman rules for treating the standard electro–weak theory in perturbation theory, restricting ourselves to the rules for obtaining the Feynman amplitudes of only the lowest-order non-vanishing graphs for a process. We shall work in the unitary gauge in which the Lagrangian density is given by Eqs. (14.9)–(14.11). As discussed in the context of QED [see Eqs. (6.9) and (6.10)], the Lagrangian densities $\mathscr{L}_0$ and $\mathscr{L}_I$ are to be interpreted as normal products in the quantized theory. In the following, normal products will always be implied. The Lagrangian density (14.9)–(14.11) is invariant under electromagnetic gauge transformations, and we shall choose to work in the Feynman gauge which has been used throughout this book. In this gauge, the Lagrangian density of the free electromagnetic field is given by Eq. (13.22), i.e. the free-field Lagrangian density (14.10) is replaced by

$$\mathscr{L}_0 - \tfrac{1}{2}[\partial_\mu A^\mu(x)]^2. \tag{14.18}$$

In the following, we shall denote this augmented Lagrangian density by $\mathscr{L}_0$.

In Section 6.2, we found that in the interaction picture the interacting fields satisfy the same equations of motion and the same commutation relations as the free fields, provided the interaction Lagrangian density $\mathscr{L}_I(x)$ does not involve derivatives of the field operators. This result enabled us to obtain the S-matrix expansion (6.23) and hence to derive the Feynman rules for the amplitudes in Chapter 7. For the electro–weak theory, the interaction Lagrangian density $\mathscr{L}_I(x)$ involves derivatives of the fields. However, irrespective of whether $\mathscr{L}_I(x)$ involves derivatives or not, a generally valid form of S-matrix expansion is

$$S = \sum_{n=0}^{\infty} \frac{i^n}{n!} \int \dots \int d^4x_1 \dots d^4x_n \mathrm{T}\{\mathscr{L}_I(x_1) \dots \mathscr{L}_I(x_n)\}, \tag{14.19}$$

provided one uses the free-field commutation relations in evaluating the matrix elements of S.§ For QED we have

$$\mathscr{L}_I(x) = -\mathscr{H}_I(x), \tag{14.20}$$

‡ These bounds are given in the review article by W. Marciano, quoted above. The lower bound is discussed in D. Bailin, *Weak Interactions*, 2nd edn, A. Hilger, Bristol, 1982, section 6.7.

§ For the derivation of this S-matrix expansion, see N. N. Bogoliubov and D. V. Shirkov, *Introduction to the Theory of Quantized Fields*, 3rd edn, Wiley, New York, 1979, Section 21.

and Eq. (14.19) reduces to our earlier expansion (6.23). From these results it follows that the same procedures which were used in Chapter 7 for QED can at once be applied to Eq. (14.19) to obtain the Feynman rules for the electro–weak theory.

With our choice of electromagnetic gauge, the Feynman rules of QED are taken over directly.[‡] The arguments used in extending these rules to the electro–weak theory are largely analogous to those for QED. We shall discuss the new features which occur. Otherwise, we shall only quote the results in Appendix B, leaving it to the reader to fill in the details.

We first consider the internal lines (propagators) and external lines (initial- and final-state particles) of the additional particles which now occur. For the neutral leptons (i.e. neutrinos and antineutrinos) the propagators and external line factors have already been included in Feynman rules 3 and 4 in Appendix B. The propagators and external line factors for $W^{\pm}$ bosons were given when we considered the IVB theory in Section 11.4 (Feynman rules 11 and 12). The extension to Z^0 bosons is trivial, and Feynman rules 11 and 12 in Appendix B state these results for both $W^{\pm}$ and Z^0 bosons. Lastly, we must deal with the Higgs particle. This is a massive neutral spin 0 particle, such as was considered in Section 3.1. The expansion of the field $\sigma(x)$ in terms of creation and annihilation operators is given by Eqs. (3.8). Since the Higgs boson has spin 0, Eqs. (3.8) contain no polarization vectors and there are no external line factors for Higgs scalars. The Feynman propagator for Higgs scalars follows from Eq. (3.59) and is given by the Feynman rule

13. For each internal Higgs line, labelled by the momentum k, write a factor

$$i\Delta_F(k, m_H) = \frac{i}{k^2 - m_H^2 + i\varepsilon} \qquad (14.21)$$

where m_H is the mass of the Higgs scalar. (We shall represent Higgs scalars by dashed lines.)

It remains to discuss the basic vertices to which the interaction (14.11) gives rise. There are now 18 types of vertices, stemming from the 18 terms in Eqs. (14.3a)–(14.3e). Let us reassure the alarmed reader that the Feynman rules for these vertices are easy to write down, and we shall shortly do so.

To begin with, we catalogue the terms of Eqs. (14.3a)–(14.3e) in Table 14.1. The first column in this table shows where in these equations each term occurs. In the last column we number the terms for easier cross-reference in Appendix B where the corresponding vertex factors are numbered (B.1) to (B.18) in the same order. In the middle column of the table we show the fields involved in each term. (For conciseness we put l for ψ_l and $\bar{l}$ for $\bar{\psi}_l$, etc.) It

[‡] In Appendix B at the end of this book we summarize the Feynman rules for QED, as well as the additional rules which will now be derived for the electro–weak theory.

Table 14.1 The 18 basic interactions of the standard electro–weak theory

Reference		Type of interaction	Term no.*
$\mathscr{L}_{\rm I}^{\rm BB}$ Eq. (14.3a)	lines 1 and 2	$W^{\dagger}WZ$	1
	lines 3 and 4	$W^{\dagger}WA$	2
	line 5	$W^{\dagger}WZ^2$	3
	line 6	$W^{\dagger}WA^2$	4
	line 7	$W^{\dagger}WZA$	5
	line 8	$(W^{\dagger}W)^2$	6
$\mathscr{L}_{\rm I}^{\rm HH}$ Eq. (14.3b)	term 1	σ^4	7
	term 2	σ^3	8
$\mathscr{L}_{\rm I}^{\rm HB}$ Eq. (14.3c)	term 1	$W^{\dagger}W\sigma$	9
	term 2	$W^{\dagger}W\sigma^2$	10
	term 3	$Z^2\sigma$	11
	term 4	$Z^2\sigma^2$	12
$\mathscr{L}_{\rm I}^{\rm LB}$ Eq. (14.3d)	line 1	$\bar{l}lA$	13
	line 2	$\bar{\nu}_l lW$ + h.c.	14
	line 3	$\bar{\nu}_l\nu_l Z$	15
	line 4	$\bar{l}lZ$	16
$\mathscr{L}_{\rm I}^{\rm HL}$ Eq. (14.3e)	term 1	$\bar{l}l\sigma$	17
	term 2	$\bar{\nu}_l\nu_l\sigma$	18

* The corresponding Feynman diagrams and vertex factors are given in Appendix B and numbered (B.1)–(B.18) in the same order.

must be remembered that each term corresponds to many possibilities, consistent with conservation of electric charge and of lepton numbers, just as the basic QED vertex stands for any one of the eight basic processes shown in Fig. 7.1. All the interactions are local interactions at one space–time point, and the type of vertices to which the 18 terms give rise are obvious from the middle column of Table 14.1 and are shown in Feynman diagrams (B.1)–(B.18). For example, terms 1 and 2 describe the interaction of three gauge bosons [Figs. (B.1) and (B.2)], terms 3–6 the interactions of four gauge bosons [Figs. (B.3)–(B.6)], and so on. The lepton–boson interaction terms 13–16 we met previously in Eq. (12.48). Term 13 is the usual QED interaction, term 14 represents the interaction of charged leptonic currents with $W^{\pm}$ bosons, and terms 15 and 16 the interaction of neutral leptonic currents with the Z^0 boson.

The remaining interactions involve Higgs particles. Terms 7 and 8 correspond to self-interactions of three or four Higgs particles, terms 9–12 to the interactions of Higgs particles with massive vector bosons. Having no charge, the Higgs particle does not couple directly to the photon. Finally, terms 17 and 18 originate from $\mathscr{L}^{\rm LH}$, Eq. (13.44), and represent the interactions of the Higgs particle with charged leptons and with neutrinos. The latter interactions only occur for non-zero neutrino masses.

The derivation of the vertex factors of these 18 interaction terms is quite similar to the corresponding calculation for QED in Section 7.2.1. We shall therefore only remind the reader of the latter calculation and discuss the new features which arise for the electro–weak interactions. The reader should then easily be able to derive the 18 vertex factors which are given in Feynman rule 14 in Appendix B.

In Section 7.2.1 we obtained the QED vertex factor $ie\gamma^{\alpha}$ by calculating the matrix element of the first-order term

$$S^{(1)} = \mathrm{i}\int \mathrm{d}^4x\mathscr{L}_{\mathrm{I}}(x) \tag{14.22}$$

in the S-matrix expansion (14.19) between free-particle states. This leads to the first-order Feynman amplitude $\mathscr{M}$ and, omitting the external line factors, to the corresponding vertex factor. For QED

$$\mathscr{L}_{\mathrm{I}}(x) = e(\bar{\psi}\gamma^{\alpha}\psi A_{\alpha})_x. \tag{14.23}$$

In Section 7.2.1 we considered the basic process $e^- \to e^- + \gamma$ and obtained the first-order Feynman amplitude

$$\mathscr{M} = \mathrm{i}e\bar{u}(\mathbf{p}')\gamma^{\alpha}u(\mathbf{p})\varepsilon_{\alpha}(\mathbf{p}' - \mathbf{p}) \tag{7.32}$$

and hence the vertex factor $ie\gamma^{\alpha}$. This vertex factor can be read off directly from Eq. (14.23) by omitting the field operators and multiplying by a factor i [which is just the factor i in Eq. (14.22)]. The same vertex factor is of course obtained from any of the other basic processes shown in Fig. 7.1.

Next, we discuss the new features which arise when deriving the vertex factors for the electro–weak theory.

First, some of the interactions in Eqs. (14.3a)–(14.3e) contain a particular field operator not linearly but to a higher power, resulting in extra combinatorial factors in the Feynman amplitude. Consider, for example, the third term in Eq. (14.3c), i.e.

$$\frac{vg^2}{4\cos^2\theta_{\mathrm{W}}} g^{\alpha\beta}Z_{\alpha}(x)Z_{\beta}(x)\sigma(x). \tag{14.24}$$

We shall calculate the vertex factor for this term by considering the first-order process

$$Z^0(k_1, r_1) + Z^0(k_2, r_2) \to H(k_3) \tag{14.25}$$

where H stands for the Higgs particle, k_i ($i = 1, 2, 3$) are the four-momenta of the three particles, and r_1 and r_2 label the polarization states of the two vector bosons. The Feynman diagram for this process is shown in Fig. 14.1. Either of

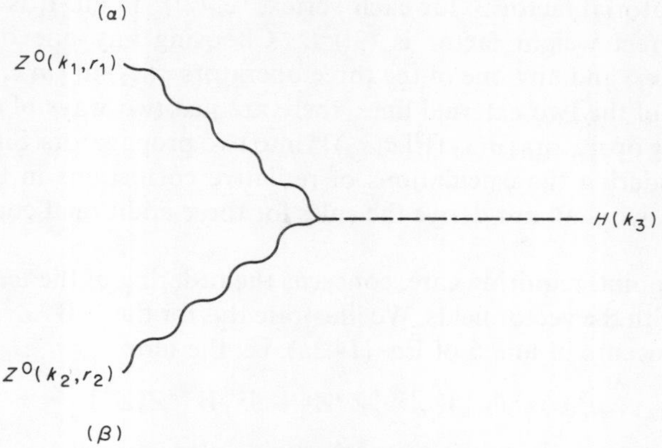

Fig. 14.1. The first-order process $Z^0(k_1, r_1) + Z^0(k_2, r_2) \to H(k_3)$.

the two operators $Z(x)$ in Eq. (14.24) can annihilate the $Z^0(k_1, r_1)$ boson, and the other operator $Z(x)$ will then annihilate the $Z^0(k_2, r_2)$ boson. This leads to a combinatorial factor 2! and to the vertex factor

$$\frac{ivg^2}{2\cos^2\theta_W} g^{\alpha\beta} \tag{14.26}$$

which is quoted in Eq. (B.11). In Figs. 14.1 and (B.11) we have attached the tensor indices of the vertex factor (14.26) to the external Z^0 boson lines to which they belong. The same vertex factor would of course have been obtained from any of the related processes, e.g. $Z^0 + Z^0 + H \to$ vacuum. The Feynman diagram of the basic vertex part, i.e. the Feynman diagram in which only the characteristic features of the vertex are retained, is shown in Fig. (B.11) in Appendix B.

The corresponding combinatorial factors for all other interaction terms are derived in the same way. For graphs containing radiative corrections, additional factors occur. Consider, for example, the modification of a Higgs boson line, shown in Fig. 14.2. In this case, the above type of argument leads

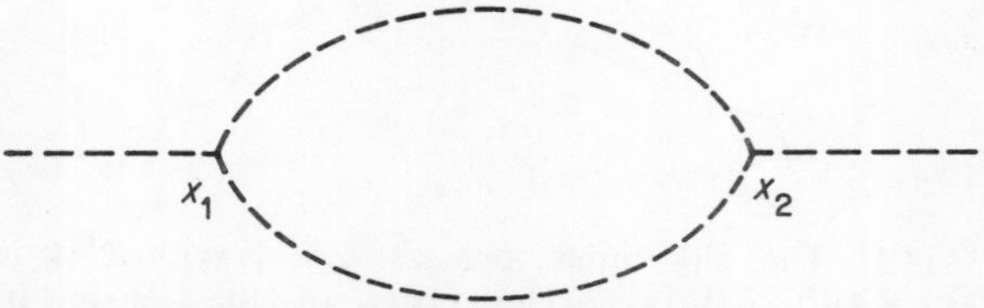

Fig. 14.2. A modification of the Higgs boson line.

to a combinatorial factor 3! for each vertex, i.e. $(3!)^2$ in all. It is easy to see that the correct weight factor is $(3!)^2/2$. Choosing any one of the three operators $\sigma(x_1)$ and any one of the three operators $\sigma(x_2)$ in $[\sigma(x_1)]^3[\sigma(x_2)]^3$ to take care of the two external lines, there are just two ways of contracting the remaining operators $[\sigma(x_1)]^2[\sigma(x_2)]^2$ into two propagators. Since we shall not be considering the calculations of radiative corrections in the electro–weak theory, we shall not derive the rules for these additional combinatorial factors.

A second point, requiring care, concerns the ordering of the tensor indices associated with the vector fields. We illustrate this for the $WW^\dagger Z^2$ interaction term which occurs in line 5 of Eq. (14.3a), i.e. the term

$$g^2 \cos^2 \theta_W [W_\alpha W^\dagger_\beta Z^\alpha Z^\beta - W_\beta W^{\dagger\beta} Z_\alpha Z^\alpha]. \tag{14.27}$$

To find the vertex part, we consider the first-order process

$$Z^0(k_1, r_1) + Z^0(k_2, r_2) \to W^+(k_1', r_1') + W^-(k_2', r_2'). \tag{14.28}$$

The corresponding Feynman diagram is shown in Fig. 14.3. We note, first of all, that the final $W^+(k_1', r_1')$ and $W^-(k_2', r_2')$ bosons must be created by the operators $W^\dagger(x)$ and $W(x)$, respectively [see Eqs. (11.22)]. On the other hand, the initial $Z^0(k_1, r_1)$ boson can be annihilated by either operator $Z(x)$ in Eq. (14.27), with $Z^0(k_2, r_2)$ being annihilated by the other operator $Z(x)$. This leads to the Feynman amplitude

$$\begin{aligned}\mathcal{M} = ig^2 \cos^2 \theta_W \{&\varepsilon_\alpha(2')\varepsilon_\beta(1')[\varepsilon^\alpha(1)\varepsilon^\beta(2) + \varepsilon^\alpha(2)\varepsilon^\beta(1)] \\ &- \varepsilon_\beta(2')\varepsilon^\beta(1')[\varepsilon_\alpha(1)\varepsilon^\alpha(2) + \varepsilon_\alpha(2)\varepsilon^\alpha(1)]\}\end{aligned} \tag{14.29}$$

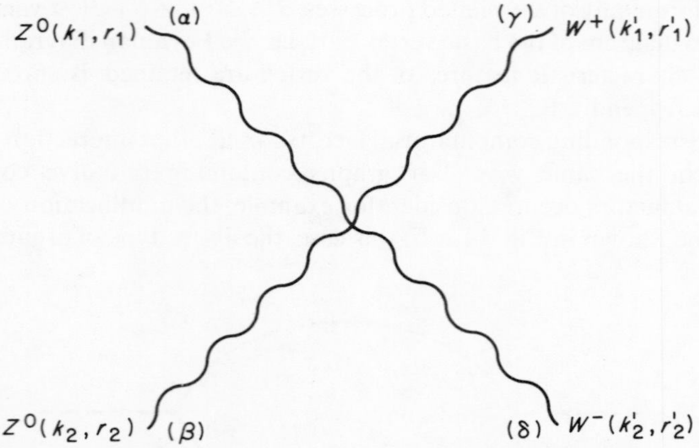

Fig. 14.3. The first-order process $Z^0(k_1, r_1) + Z^0(k_2, r_2) \to W^+(k_1', r_1') + W^-(k_2', r_2')$. The order of the tensor indices $(\alpha)\ldots(\delta)$ corresponds to the vertex factor (14.30).

where we abbreviated $\varepsilon_{r'_1\beta}(\mathbf{k}'_1)$ to $\varepsilon_\beta(1')$, etc. Relabelling tensor indices so that all terms have the common factor $\varepsilon_\alpha(1)\varepsilon_\beta(2)\varepsilon_\gamma(1')\varepsilon_\delta(2')$, Eq. (14.29) reduces to

$$\mathcal{M} = \varepsilon_\alpha(1)\varepsilon_\beta(2)\varepsilon_\gamma(1')\varepsilon_\delta(2') \times ig^2 \cos^2\theta_W[g^{\alpha\delta}g^{\beta\gamma} + g^{\alpha\gamma}g^{\beta\delta} - 2g^{\alpha\beta}g^{\gamma\delta}].$$

The corresponding vertex factor is

$$ig^2 \cos^2\theta_W[g^{\alpha\delta}g^{\beta\gamma} + g^{\alpha\gamma}g^{\beta\delta} - 2g^{\alpha\beta}g^{\gamma\delta}] \tag{14.30}$$

which is the result given in Eq. (B.3). In Figs. 14.3 and (B.3) we have attached tensor indices to the external lines in the correct order to correspond to the vertex factor (14.30).

The third new feature of the electro–weak vertices stems from the derivative couplings which occur in the $W^\dagger WZ$ and $W^\dagger WA$ terms of $\mathscr{L}_I^{BB}$, Eq. (14.3a). One sees from the plane wave expansions of the fields, e.g. Eqs. (11.22), that these derivative couplings introduce momentum factors into the vertex functions. To obtain the exact result for the term

$$ig\cos\theta_W[(W_\alpha^\dagger W_\beta - W_\beta^\dagger W_\alpha)\,\partial^\alpha Z^\beta + (\partial_\alpha W_\beta - \partial_\beta W_\alpha)W^{\dagger\beta}Z^\alpha - (\partial_\alpha W_\beta^\dagger - \partial_\beta W_\alpha^\dagger)W^\beta Z^\alpha] \tag{14.31}$$

in Eq. (14.3a), we consider the process

$$Z^0(k_1, r_1) + W^+(k_2, r_2) + W^-(k_3, r_3) \to \text{vacuum}. \tag{14.32}$$

The Feynman graph for this process is shown in Fig. 14.4(a). We have chosen the process (14.32) in which all three bosons are annihilated at the vertex on account of its greater symmetry compared with related processes, such as $Z^0 \to W^+ + W^-$. Below we shall show how to obtain the vertex factors for

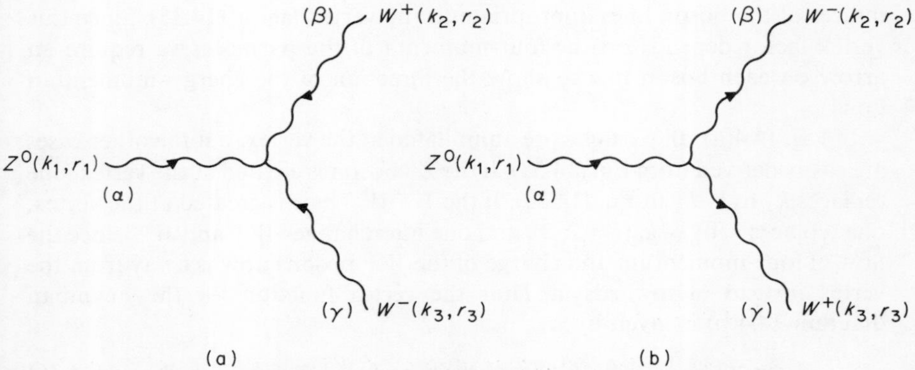

Fig. 14.4. The related first-order processes: (a) $Z^0(k_1, r_1) + W^+(k_2, r_2) + W^-(k_3, r_3) \to$ vacuum; (b) $Z^0(k_1, r_1) \to W^-(k_2, r_2) + W^+(k_3, r_3)$.

these related processes from that for the reaction (14.32). In Fig. 14.4(a), we have marked the boson lines with arrows which show the direction of the flow of energy–momentum (and, for the $W^{\pm}$ bosons, of charge) towards the vertex where the particles are annihilated.

The operators $W(x)$ and $W^{\dagger}(x)$ are linear in the absorption operators of W^{+} and W^{-} bosons respectively. It follows from Eqs. (11.22) that, in annihilating a W^{+} boson with four-momentum k and polarization index r, the operator $\partial_{\alpha} W_{\beta}(x)$ leads to a factor

$$(-ik_{\alpha})\varepsilon_{r\beta}(\mathbf{k}).$$

The same factor occurs if we replace the operator $\partial_{\alpha} W_{\beta}(x)$ by $\partial_{\alpha} W_{\beta}^{\dagger}(x)$ or $\partial_{\alpha} Z_{\beta}(x)$ and the W^{+} boson by a W^{-} or Z^{0} boson, as can be seen from the plane wave expansions of the $W^{\dagger}(x)$ and $Z(x)$ fields. It follows from these results that the Feynman amplitude for the process (14.32) is given by

$$\begin{aligned} \mathcal{M} = -g \cos \theta_{\mathrm{W}} \{ & [\varepsilon_{\alpha}(3)\varepsilon_{\beta}(2) - \varepsilon_{\beta}(3)\varepsilon_{\alpha}(2)](-ik_{1}^{\alpha})\varepsilon^{\beta}(1) \\ & + [(-ik_{2\alpha})\varepsilon_{\beta}(2) - (-ik_{2\beta})\varepsilon_{\alpha}(2)]\varepsilon^{\beta}(3)\varepsilon^{\alpha}(1) \\ & - [(-ik_{3\alpha})\varepsilon_{\beta}(3) - (-ik_{3\beta})\varepsilon_{\alpha}(3)]\varepsilon^{\beta}(2)\varepsilon^{\alpha}(1)] \} \end{aligned} \tag{14.33}$$

where $\varepsilon_{r_1}^{\alpha}(\mathbf{k}_1)$ has been abbreviated to $\varepsilon^{\alpha}(1)$, etc. Relabelling tensor indices so that all terms have the common factor $\varepsilon_{\alpha}(1)\varepsilon_{\beta}(2)\varepsilon_{\gamma}(3)$, Eq. (14.33) reduces to

$$\begin{aligned} \mathcal{M} = {} & \varepsilon_{\alpha}(1)\varepsilon_{\beta}(2)\varepsilon_{\gamma}(3) \\ & \times ig \cos \theta_{\mathrm{W}} [g^{\alpha\beta}(k_1 - k_2)^{\gamma} + g^{\beta\gamma}(k_2 - k_3)^{\alpha} + g^{\gamma\alpha}(k_3 - k_1)^{\beta}]. \end{aligned} \tag{14.34}$$

The corresponding vertex factor is

$$ig \cos \theta_{\mathrm{W}} [g^{\alpha\beta}(k_1 - k_2)^{\gamma} + g^{\beta\gamma}(k_2 - k_3)^{\alpha} + g^{\gamma\alpha}(k_3 - k_1)^{\beta}] \tag{14.35}$$

which is the result quoted in Eq. (B.1). Figs. 14.4(a) and (B.1) show the tensor indices of the boson lines appropriate to the vertex factor (14.35). Since this vertex factor depends on the four-momenta of the particles, we require an arrow on each boson line to show the direction of the energy–momentum flow.

In Fig. 14.4(a), all particles are annihilated at the vertex, but the other cases are easily derived from Eq. (14.35). If the Z^{0} boson is created at the vertex, one replaces k_1 by $-k_1$ in Eq. (14.35). If the $W^{+}W^{-}$ pair is created at the vertex, one replaces k_i by $-k_i$ $(i = 2, 3)$, and one interchanges W^{+} and W^{-} since the flow of four-momentum and charge of the $W^{\pm}$ bosons now is away from the vertex instead of towards it. Thus the vertex function for the Feynman diagram 14.4(b) is given by

$$ig \cos \theta_{\mathrm{W}} [g^{\alpha\beta}(k_1 + k_2)^{\gamma} + g^{\beta\gamma}(k_3 - k_2)^{\alpha} - g^{\gamma\alpha}(k_1 + k_3)^{\beta}]. \tag{14.36}$$

Lastly, term 6 in Table 14.1, i.e. the term in $(W^{\dagger}W)^2$ which occurs in $\mathcal{L}_{\mathrm{I}}^{\mathrm{BB}}$,

Eq. (14.3a), requires a comment. The vertex factor (B.6) for this term is obtained by the same technique by which we derived the vertex factor (14.30) for the vertex in Fig. 14.3. However, unlike all other cases involving $W^{\pm}$ bosons, the vertex factor (B.6) is altered if one exchanges the index α with β, or γ with δ. Consequently, this vertex factor depends on the direction of flow of the $W^{\pm}$ charges, towards or away from the vertex. In Fig. (B.6), these directions are indicated by the arrows on the boson lines and correspond to the vertex factor given with Fig. (B.6).

The above results enable us to write down by inspection of Eqs. (14.3a)–(14.3e) the vertex factors of all terms in Table 14.1.

This completes our discussion of the Feynman rules of the standard electro–weak theory. These rules enable one straightforwardly to calculate processes in lowest non-vanishing order of perturbation theory. The calculation of higher-order corrections involves very many graphs and, in general, becomes extremely complicated. Except for sometimes quoting results, we shall not consider such calculations further and shall limit ourselves to lowest-order calculations.

14.3 ELASTIC NEUTRINO–ELECTRON SCATTERING

As our first application of the electro–weak theory, we shall consider the four elastic neutrino–electron scattering processes

$$\nu_\mu + e^- \to \nu_\mu + e^-, \qquad \bar{\nu}_\mu + e^- \to \bar{\nu}_\mu + e^-, \tag{14.37a}$$

$$\nu_e + e^- \to \nu_e + e^-, \qquad \bar{\nu}_e + e^- \to \bar{\nu}_e + e^-. \tag{14.37b}$$

More briefly, we shall refer to the first of these reactions as $(\nu_\mu e)$ scattering, etc. If lepton numbers are conserved—as in the electro–weak theory—the leading one-boson-exchange contributions to the $(\nu_\mu e)$ and $(\nu_e e)$ processes arise from the Feynman graphs in Figs. 14.5 and 14.6 respectively, with similar graphs in which neutrinos are replaced by antineutrinos for the other two processes. The significance of these four processes is due to the fact that all of them involve neutral currents [diagrams (a) in Figs. 14.5 and 14.6] and can therefore be employed to determine the weak mixing angle θ_W. The $(\nu_e e)$ and

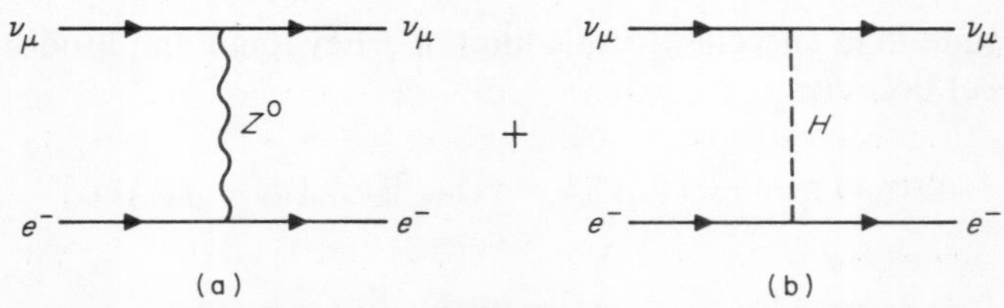

Fig. 14.5. The leading contributions to $(\nu_\mu e)$ scattering.

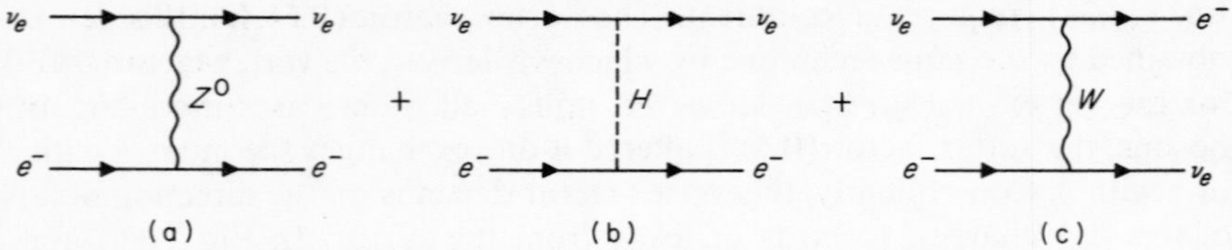

Fig. 14.6. The leading contributions to $(\nu_e e)$ scattering.

$(\bar{\nu}_e e)$ processes have additional contributions from the exchange of one W boson, Fig. 14.6(c).

We first consider the $(\nu_\mu e)$ process. From the Feynman graphs 14.5 and the Feynman rules in Appendix B, its Feynman amplitude is given by

$$\mathscr{M}(\nu_\mu e) = \mathscr{M}_Z(\nu_\mu e) + \mathscr{M}_H(\nu_\mu e) \tag{14.38a}$$

where

$$\mathscr{M}_Z(\nu_\mu e) = \frac{-g^2}{8\cos^2\theta_W}[\bar{u}'_{\nu_\mu}\gamma^\alpha(1-\gamma_5)u_{\nu_\mu}]\mathrm{i}D_{F\alpha\beta}(k, m_Z)[\bar{u}'_e\gamma^\beta(g_V - g_A\gamma_5)u_e] \tag{14.38b}$$

and

$$\mathscr{M}_H(\nu_\mu e) = \frac{-1}{v^2}m_{\nu_\mu}m_e(\bar{u}'_{\nu_\mu}u_{\nu_\mu})\mathrm{i}\Delta_F(k, m_H)(\bar{u}'_e u_e). \tag{14.38c}$$

Here u and $\bar{u}'$ are the spinors of the initial and final leptons respectively, with the kind of lepton (electron or ν_μ) labelled, but momenta and spin quantum numbers suppressed, and we have introduced the abbreviations

$$g_V \equiv 2\sin^2\theta_W - \tfrac{1}{2}, \qquad g_A \equiv -\tfrac{1}{2}. \tag{14.39}$$

The momentum k of the intermediate boson is given by

$$k = q - q' = p' - p, \tag{14.40}$$

where q and q' (p and p') are the momenta of the initial and final neutrino (electron) respectively.

With $m_Z \approx 94$ GeV and our assumption (14.17) for m_H, we have

$$k^2 \ll m_Z^2, \qquad k^2 \ll m_H^2 \tag{14.41a}$$

even for quite high energies. In this limit, the Feynman amplitudes (14.38b) and (14.38c) become

$$\mathscr{M}_Z(\nu_\mu e) = \frac{-\mathrm{i}G}{\sqrt{2}}[\bar{u}'_{\nu_\mu}\gamma^\alpha(1-\gamma_5)u_{\nu_\mu}][\bar{u}'_e\gamma_\alpha(g_V - g_A\gamma_5)u_e] \tag{14.42a}$$

$$\mathscr{M}_H(\nu_\mu e) = \mathrm{i}G\sqrt{2}\frac{1}{m_H^2}m_{\nu_\mu}m_e(\bar{u}'_{\nu_\mu}u_{\nu_\mu})(\bar{u}'_e u_e) \tag{14.42b}$$

where we used Eqs. (14.4) and (14.14). We see from Eqs. (14.42a) and (14.42b) that $\mathcal{M}_H(\nu_\mu e)$ is of order $m_{\nu_\mu} m_e/m_H^2$ relative to $\mathcal{M}_Z(\nu_\mu e)$ and so can be neglected. We shall also assume that the lepton masses are negligible; more precisely, that

$$s \equiv (p + q)^2 \gg m_e^2. \tag{14.41b}$$

Here s is the square of the centre-of-mass momentum and is given by

$$s = 4E_{\text{CoM}}^2 = 2m_e E_{\text{Lab}} + m_e^2, \tag{14.43}$$

where E_{CoM} and E_{Lab} are the neutrino energy in the centre-of-mass system and in the laboratory system in which the target electron is at rest.

If the conditions (14.41a) and (14.41b) hold, a straightforward calculation leads to the total elastic $(\nu_\mu e)$ cross-section

$$\sigma_{\text{T}}(\nu_\mu e) = \frac{G^2 s}{3\pi}(g_{\text{V}}^2 + g_{\text{V}} g_{\text{A}} + g_{\text{A}}^2). \tag{14.44a}$$

In the same way, one obtains the total elastic $(\bar{\nu}_\mu e)$ cross-section

$$\sigma_{\text{T}}(\bar{\nu}_\mu e) = \frac{G^2 s}{3\pi}(g_{\text{V}}^2 - g_{\text{V}} g_{\text{A}} + g_{\text{A}}^2). \tag{14.44b}$$

We next obtain the Feynman amplitude for the $(\nu_e e)$ process

$$\mathcal{M}(\nu_e e) = \mathcal{M}_Z(\nu_e e) + \mathcal{M}_H(\nu_e e) + \mathcal{M}_W(\nu_e e) \tag{14.45}$$

where the three terms stem from the three Feynman graphs in Fig. 14.6. The first two of these graphs lead to the same amplitudes [Eqs. (14.42a) and (14.42b)] as before, with m_{ν_μ} replaced by m_{ν_e}, so that the Higgs contribution $\mathcal{M}_H(\nu_e e)$ is again negligible. Assuming

$$k^2 \ll m_W^2 \tag{14.41c}$$

and that lepton masses may again be neglected leads to the amplitude

$$\mathcal{M}_W(\nu_e e) = \frac{\text{i}G}{\sqrt{2}}[\bar{u}'_e \gamma^\alpha (1 - \gamma_5) u_{\nu_e}][\bar{u}'_{\nu_e} \gamma_\alpha (1 - \gamma_5) u_e]. \tag{14.46}$$

[The relative signs of the amplitudes (14.46) and (14.42a) follow from the Feynman rule 8 by arranging the lepton operators, which lead to $\mathcal{M}_Z(\nu_e e)$ and $\mathcal{M}_W(\nu_e e)$, in the same order.] Using the Fierz identity‡

$$(\bar{u}_1 \gamma^\alpha (1 - \gamma_5) u_2)(\bar{u}_3 \gamma_\alpha (1 - \gamma_5) u_4) = -(\bar{u}_1 \gamma^\alpha (1 - \gamma_5) u_4)(\bar{u}_3 \gamma_\alpha (1 - \gamma_5) u_2), \tag{14.47}$$

‡ For the derivation of this and other Fierz identities, see C. Itzykson and J. B. Zuber, *Quantum Field Theory*, McGraw-Hill, 1980, pp. 160–162.

where $u_1, \ldots, u_4$ are arbitrary spinors, we can write the amplitude (14.46) as

$$\mathcal{M}_W(\nu_e e) = \frac{-iG}{\sqrt{2}} [\bar{u}'_{\nu_e} \gamma^\alpha (1 - \gamma_5) u_{\nu_e}][\bar{u}'_e \gamma_\alpha (1 - \gamma_5) u_e]. \tag{14.48}$$

Combining these results in Eq. (14.45), it follows that the Feynman amplitude $\mathcal{M}(\nu_e e)$ is obtained from Eq. (14.42a) by replacing ν_μ by ν_e, together with the replacements

$$g_V \to g_V + 1, \qquad g_A \to g_A + 1. \tag{14.49}$$

Making the replacements (14.49) in Eq. (14.44a) leads to the total cross-section for elastic $(\nu_e e)$ scattering. A similar argument shows that the same replacements made in Eq. (14.44b) give the total cross-section for elastic $(\bar{\nu}_e e)$ scattering.

Experimentally, the above processes are investigated by scattering neutrinos from atomic electrons and detecting the recoil electrons. The $(\nu_\mu e)$ and $(\bar{\nu}_\mu e)$ reactions have been studied by several groups using ν_μ and $\bar{\nu}_\mu$ beams originating from pion decay. The $(\bar{\nu}_e e)$ process has been studied using a $\bar{\nu}_e$ beam produced from neutron decays in a nuclear reactor. The best value of $\sin^2 \theta_W$ comes from an experiment in which the total $(\nu_\mu e)$ and $(\bar{\nu}_\mu e)$ cross-sections are measured using the same detector. Their ratio determines the weak mixing angle with an experimental uncertainty which is smaller than in a measurement of a single cross-section. Comparing this experimental result with the prediction of the electro–weak theory leads to the value [F. Bergsma *et al.*, *Phys. Lett.* **117B** (1982), 272]

$$\sin^2 \theta_W = 0.215 \pm 0.040 \pm 0.015 \tag{14.50}$$

where the first error is statistical and the second systematic. The latter is small because both reactions are measured in the same apparatus and many systematic effects cancel. A fit of g_V and g_A, treated as parameters to be determined from experiment, to all available data on the reactions $(\nu_\mu e)$, $(\bar{\nu}_\mu e)$ and $(\bar{\nu}_e e)$ yields the values [Krenz, 1982(N)][‡]

$$\sin^2 \theta_W = 0.235 \pm 0.04, \qquad g_A = -0.51 \pm 0.06, \tag{14.51}$$

in agreement with the value (14.50) and the theoretical prediction $g_A = -\frac{1}{2}$ which is independent of the value of $\sin^2 \theta_W$.

A more precise value of the mixing angle is obtained from neutrino scattering by nucleons. We shall not discuss this in detail and only quote the

‡ Except where stated otherwise, the experimental and theoretical data, quoted in the remainder of this chapter, and their sources are given in two review papers by W. Marciano and by B. Naroska in the *Proceedings of the 1983 International Symposium on Lepton and Photon Interactions at High Energies*, Cornell University, 1983. These two review articles will be identified by (M) and (N).

resulting value, obtained by averaging over several experiments and estimating systematic errors [see Sirlin and Marciano (1981)(M)],

$$\sin^2 \theta_W = 0.227 \pm 0.014. \qquad (14.52)$$

This value is seen to be consistent with the values (14.50) and (14.51) obtained from purely leptonic processes.

14.4 ELECTRON–POSITRON ANNIHILATION

In Section 8.4 we discussed electron–positron annihilation in the context of QED. We now reconsider this process in the wider framework of the electroweak theory, starting with the reaction

$$e^+ + e^- \to l^+ + l^- \qquad (l = \mu, \tau, \ldots, \qquad l \neq e). \qquad (14.53)$$

In lowest order, this process is described by the Feynman graphs in Fig. 14.7. The corresponding Feynman amplitude is given by

$$\mathscr{M} = \mathscr{M}_\gamma + \mathscr{M}_Z + \mathscr{M}_H \qquad (14.54)$$

where

$$\mathscr{M}_\gamma = ie^2(\bar{u}_l\gamma^\alpha v_l)\frac{1}{k^2 + i\varepsilon}(\bar{v}_e\gamma_\alpha u_e) \qquad (14.55a)$$

$$\mathscr{M}_Z = \frac{ig^2}{4\cos^2\theta_W}[\bar{u}_l\gamma^\alpha(g_V - g_A\gamma_5)v_l]\frac{1}{k^2 - m_Z^2 + i\varepsilon}[\bar{v}_e\gamma_\alpha(g_V - g_A\gamma_5)u_e] \qquad (14.55b)$$

$$\mathscr{M}_H = \frac{-i}{v^2}m_e m_l(\bar{u}_l v_l)\frac{1}{k^2 - m_H^2 + i\varepsilon}(\bar{v}_e u_e) \qquad (14.55c)$$

and $\bar{v}_e, u_e$ and $v_l, \bar{u}_l$ are the spinors of the initial e^+e^- pair and the final l^+l^- pair. In writing Eq. (14.55b), we have omitted the term $k_\alpha k_\beta/m_Z^2$ from the boson propagator $D_{F\alpha\beta}(k, m_Z)$, since this term gives contributions of order $m_e m_l/m_Z^2$.‡

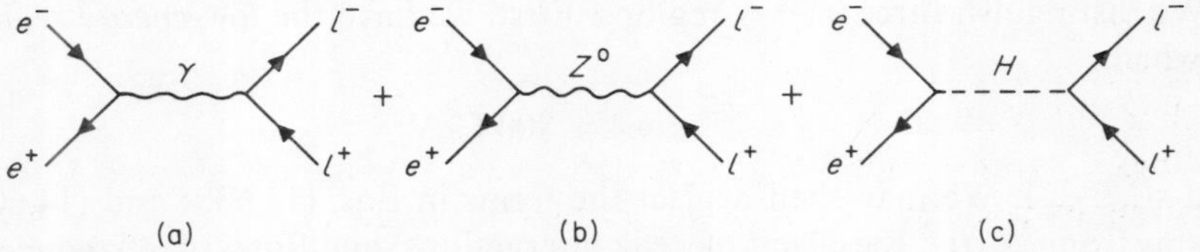

Fig. 14.7. The leading contributions to $e^+ + e^- \to l^+ + l^-$.

‡ To see this, one writes the intermediate boson momenta in $k_\alpha k_\beta/m_Z^2$ as

$$k = p_1 + p_2 = p_1' + p_2', \qquad (14.56)$$

where $p_1(p_2)$ and $p_1'(p_2')$ are the momenta of the positive (negative) initial and final leptons respectively, and uses the Dirac equation to replace $\bar{v}_e \not{p}_1$ by $(-\bar{v}_e m_e)$, etc.

We see from Eqs. (14.55b), (14.55c) and (14.4) that $\mathcal{M}_H/\mathcal{M}_Z$ is of order

$$\frac{m_e m_l}{m_Z^2}\frac{k^2 - m_Z^2}{k^2 - m_H^2}. \tag{14.57}$$

Hence $\mathcal{M}_H$ may be neglected, unless k^2 is very close to m_H^2, and the Feynman amplitude (14.54) reduces to

$$\mathcal{M} = \mathcal{M}_\gamma + \mathcal{M}_Z. \tag{14.58}$$

A long but straightforward calculation gives the cross-section for the process $e^+e^- \to l^+l^-$. Assuming that the electron's energy E in the CoM system is sufficiently high so that

$$s \equiv k^2 = 4E^2 \gg m_l^2 \tag{14.59}$$

(i.e. all lepton masses may be neglected), one obtains the CoM differential cross-section

$$\sigma(\theta) = F(s)(1 + \cos^2\theta) + G(s)\cos\theta. \tag{14.60a}$$

Here θ is the angle (in the CoM system) between the initial positron direction and the final l^+ direction (compare Fig. 8.1), and $F(s)$ and $G(s)$ are given by

$$F(s) = \frac{\alpha^2}{4s}\left[1 + \frac{g_V^2}{\pi\sqrt{2}}\frac{m_Z^2}{s - m_Z^2}\left(\frac{sG}{\alpha}\right) + \frac{(g_V^2 + g_A^2)^2}{8\pi^2}\left(\frac{m_Z^2}{s - m_Z^2}\right)^2\left(\frac{sG}{\alpha}\right)^2\right] \tag{14.60b}$$

$$G(s) = \frac{\alpha^2}{4s}\left[\frac{\sqrt{2}g_A^2}{\pi}\frac{m_Z^2}{s - m_Z^2}\left(\frac{sG}{\alpha}\right) + \frac{g_A^2 g_V^2}{\pi^2}\left(\frac{m_Z^2}{s - m_Z^2}\right)^2\left(\frac{sG}{\alpha}\right)^2\right]. \tag{14.60c}$$

In these equations the terms proportional to a^2, αG and G^2 stem respectively from $|\mathcal{M}_\gamma|^2$, the $\mathcal{M}_\gamma - \mathcal{M}_Z$ interference term and $|\mathcal{M}_Z|^2$ in the expression $|\mathcal{M}_\gamma + \mathcal{M}_Z|^2$ from which the cross-section is derived.

We distinguish three energy regions. First, we have the *low-energy region*, in which

$$s \equiv k^2 \ll m_Z^2 \approx 800\ \text{GeV}^2$$

and $sG/\alpha \ll 1$. We may then neglect the terms in Eqs. (14.60b) and (14.60c) arising from $\mathcal{M}_Z$ (i.e. the effects of weak interactions) and Eqs. (14.60) reduce to the cross-section (8.46) of pure QED.

As s increases, $|\mathcal{M}_\gamma|$ falls off like $1/s$ while $|\mathcal{M}_Z|$ increases like $|s - m_Z^2|^{-1}$, i.e. the expansion parameter

$$\frac{m_Z^2}{|s - m_Z^2|}\left(\frac{sG}{\alpha}\right) \tag{14.61}$$

in Eqs. (14.60b) and (14.60c) increases. Hence we reach an *intermediate energy region* where the $\mathcal{M}_\gamma - \mathcal{M}_Z$ interference term is no longer negligible but $|\mathcal{M}_Z|^2$ is; in Eqs. (14.60b) and (14.60c) we retain the terms in αG and drop those in G^2. In particular, $G(s)$ is different from zero so that the cross-section (14.60a) contains a term in $\cos\theta$ and displays a forward–backward asymmetry

$$A(\theta) \equiv \frac{\sigma(\theta) - \sigma(\pi - \theta)}{\sigma(\theta) + \sigma(\pi - \theta)} = \frac{G(s)}{F(s)} \frac{\cos\theta}{1 + \cos^2\theta} \tag{14.62}$$

where $0 \leqslant \theta \leqslant \pi/2$. From Eqs. (14.60b) and (14.60c) this becomes, on neglecting terms proportional to G^2,

$$A(\theta) = \frac{\sqrt{2}}{\pi} g_{\mathrm{A}}^2 \frac{m_{\mathrm{Z}}^2}{s - m_{\mathrm{Z}}^2} \left(\frac{sG}{\alpha}\right) \frac{\cos\theta}{1 + \cos^2\theta}. \tag{14.63}$$

For $s < m_{\mathrm{Z}}^2$, $A(\theta)$ is negative (i.e. there is excess backward scattering) and its magnitude increases rapidly with energy. To gain an idea of the size of the asymmetry, we calculate $A(\theta = 60°)$. With $g_{\mathrm{A}} = -\frac{1}{2}$ and $m_{\mathrm{Z}} \approx 94$ GeV, $A(\theta = 60°)$ has the approximate values -0.03, -0.07 and -0.14 for the total CoM energies $\sqrt{s} = 2E = 20$, 30 and 40 GeV respectively.

The asymmetry $A(\theta)$ has been measured in this energy range for both $\mu^+\mu^-$ and $\tau^+\tau^-$ production. In comparing theory with experiment, it is essential to take into account electromagnetic radiative corrections associated with soft photons which are typically of the order of 10 per cent in this region and which can themselves generate positive asymmetries of the order of 2 per cent. As noted in Section 8.9, the precise value of the radiative corrections depends on the experimental set-up. For this reason, it is usual to subtract the radiative corrections from the experimental data before comparing with theory, so that the results of different experiments can be compared more easily.‡

Fig. 14.8 shows a typical result: the 'radiatively corrected' angular distribution for $\mu^+\mu^-$ production at $\sqrt{s} = 34$ GeV, obtained by the TASSO Collaboration [R. Brandelik *et al.*, *Phys. Lett.* **110B** (1982), 173]. The agreement of the experimental points with the electro–weak theory (the continuous curve) is seen to be excellent, and the deviation from the symmetric dashed curve, which corresponds to the lowest-order QED distribution, is apparent. Similar results have been obtained in several other experiments and at different energies for both $\mu^+\mu^-$ and $\tau^+\tau^-$ production. For a discussion of these, see the review article by B. Naroska, quoted in the previous section.

Going to higher energies, we reach the Z^0 *resonance region*: $s \approx m_{\mathrm{Z}}^2$, i.e. the total CoM energy $\sqrt{s}$ is in the vicinity of the mass m_{Z} of the Z^0 boson. In this region, sG/α is of order unity and it follows that, for $s \approx m_{\mathrm{Z}}^2$, the cross-section for

‡ For a detailed discussion, giving typical results, see F. A. Berends and R. Gastmans, '*Radiative Corrections in e^+e^- Collisions*', in: *Electromagnetic Interactions of Hadrons*, vol. 2 (edited by A. Donnachie and G. Shaw), Plenum Press, New York, 1978.

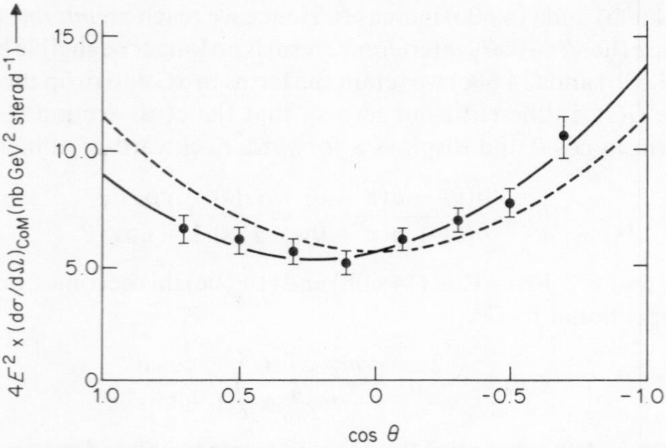

Fig. 14.8. The differential cross-section $(d\sigma/d\Omega)_{\text{CoM}}$ for the process $e^+e^- \to \mu^+\mu^-$, at the total CoM energy $2E = 34$ GeV. [After R. Brandelik *et al.*, *Phys. Lett.* **110B** (1982), 173.] ⫯: experimental data; —— electro–weak theory; ---- QED.

the process $e^+e^- \to l^+l^-$ is dominated by the terms proportional to G^2 in Eqs. (14.60b) and (14.60c), i.e. by the terms originating from $|\mathscr{M}_Z|^2$. Retaining only these terms, we obtain from Eqs. (14.60a)–(14.60c) the total cross-section

$$\sigma_T(e^+e^- \to l^+l^-) = \frac{1}{6\pi}(g_V^2 + g_A^2)^2 \frac{G^2 m_Z^6}{(s - m_Z^2)^2 + \varepsilon^2} \tag{14.64}$$

for $s \approx m_Z^2$, where we have temporarily restored the infinitesimal parameter ε which occurs in the denominator of the amplitude $\mathscr{M}_Z$, Eq. (14.55b). Eq. (14.64) can be written

$$\sigma_T(e^+e^- \to l^+l^-) = \frac{12\pi\Gamma(Z^0 \to e^+e^-)\Gamma(Z^0 \to l^+l^-)}{(s - m_Z^2)^2 + \varepsilon^2} \quad (s \approx m_Z^2) \tag{14.65}$$

where we have substituted the expression

$$\Gamma(Z^0 \to l^+l^-) = \frac{1}{\pi 6\sqrt{2}} G m_Z^3 (g_V^2 + g_A^2). \tag{14.66}$$

This is the width for the decay process

$$Z^0 \to l^+ + l^- \qquad (l = e, \mu, \ldots), \tag{14.67}$$

represented by the Feynman graph in Fig. 14.9, assuming that terms of order m_l^2/m_Z^2 are negligible. In this approximation, the width (14.66) does not depend on the type of lepton pair, $l = e, \mu, \ldots$. The derivation of the decay rate (14.66), using the method of Section 11.5, is straightforward and is left as an exercise for the reader.

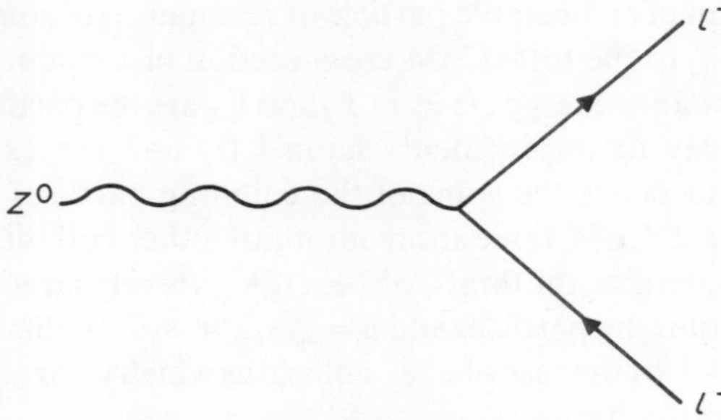

Fig. 14.9. The decay process $Z^0 \to l^+ + l^-$.

Eq. (14.65) corresponds to an infinitesimally narrow peak in the cross-section at the resonance energy $\sqrt{s} = m_Z$. Of course, the peak in the experimental cross-section is not infinitesimally narrow. From the uncertainty principle, it must have a width of order $\tau^{-1} = \Gamma_t$, where τ and Γ_t are the life time and total decay width of the Z^0 boson. More formally, near $s = m_Z^2$, higher-order corrections to the denominator in Eq. (14.65) cannot be neglected, since in lowest order this denominator vanishes as $\varepsilon \to 0$. The relevant corrections arise from the propagator modifications, shown in Fig. 14.10, which involve intermediate fermion–antifermion states, like l^+l^-, $\nu_l\bar{\nu}_l$, to which the Z^0 boson can decay. The calculation of these terms is similar to that for the vacuum polarization graph, Fig. 9.8, given in section 10.4. We shall not discuss it here, but merely quote the final result, which is that Eq. (14.65) becomes modified to

$$\sigma_T(e^+e^- \to l^+l^-) = \frac{12\pi\Gamma(Z^0 \to e^+e^-)\Gamma(Z^0 \to l^+l^-)}{(s - m_Z^2)^2 + m_Z^2\Gamma_t^2} \qquad (s \approx m_Z^2) \qquad (14.68)$$

where we have again set $\varepsilon = 0$.

Eq. (14.68) is a special case of the one-level Breit–Wigner formula‡

$$\sigma_T(i \to X) = 4\pi \frac{m^2}{p^2} \frac{2J + 1}{(2s_1 + 1)(2s_2 + 1)} \frac{\Gamma_i\Gamma_X}{(s - m^2)^2 + m^2\Gamma_t^2} \qquad (s \approx m^2)$$

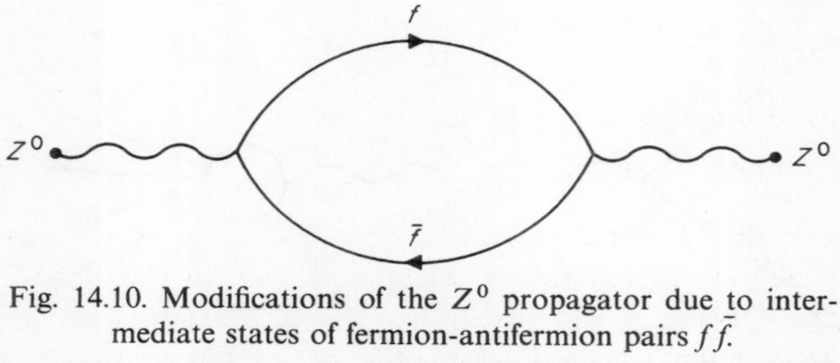

Fig. 14.10. Modifications of the Z^0 propagator due to intermediate states of fermion-antifermion pairs $f\bar{f}$.

‡ See H. M. Pilkuhn, *Relativistic Particle Physics*, Springer, New York, 1979, p. 168.

for the contribution of an unstable particle or resonance (of spin J, mass m and total decay with Γ_t) to the total CoM cross-section of a reaction $i \to X$, in the vicinity of the resonance energy $\sqrt{s} = m$. Γ_i and Γ_X are the partial widths of this resonance for decay to the incident channel (i) and the exit channel (X) respectively, s_1 and s_2 are the spins of the colliding particles in the incident channel, and p is the CoM three-momentum of either colliding particle. For highly relativistic energies, the total CoM energy $\sqrt{s}$ is very large compared with the masses of the colliding particles and $p = \sqrt{s}/2 = m/2$. In this approximation and with $s_1 = s_2 = \frac{1}{2}$ for the case of e^+e^- collisions which we are considering, the Breit–Wigner formula becomes

$$\sigma_T(i \to X) = \frac{4\pi(2J+1)\Gamma_i\Gamma_X}{(s-m^2)^2 + m^2\Gamma_t^2} \qquad (s \approx m^2). \qquad (14.69)$$

We can apply Eq. (14.69) to any reaction $e^+e^- \to X$ which can proceed via an intermediate Z^0 boson: $e^+e^- \to Z^0 \to X$. At energies in the vicinity of the resonance energy $\sqrt{s} = m_Z$, where the Feynman graph 14.11 will dominate, the total cross-section is, from Eq. (14.69), given by

$$\sigma_T(e^+e^- \to X) = \frac{12\pi\Gamma(Z^0 \to e^+e^-)\Gamma(Z^0 \to X)}{(s-m_Z^2)^2 + m_Z^2\Gamma_t^2} \qquad (s \approx m_Z^2). \quad (14.70)$$

We see from Eq. (14.68) that the total cross-section for the reaction $e^+e^- \to l^+l^-$ exhibits a peak at $s = m_Z^2$, with half-width $m_Z\Gamma_t$ and a height which is proportional to the square of the branching ratio

$$B(Z^0 \to l^+l^-) = \frac{\Gamma(Z^0 \to l^+l^-)}{\Gamma_t} \qquad (l = e, \mu, \ldots). \qquad (14.71)$$

Hence precision measurements of the cross-section (14.68) in the region of the peak will yield accurate values for the mass of the Z^0 boson, its lifetime and the branching ratio (14.71). (In the analysis of such experiments, the radiative corrections associated with the emission of soft photons must again be included!) Similar measurements for the corresponding peaks in the cross-

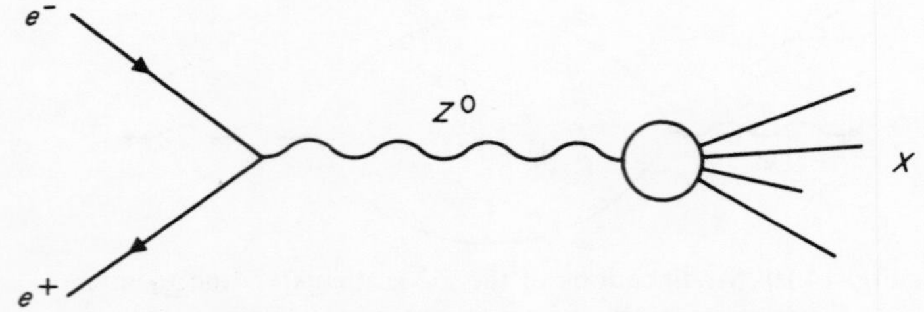

Fig. 14.11. The dominant contribution to the reaction $e^+ + e^- \to X$, in the vicinity of the Z^0 resonance peak.

section (14.70) for other processes will yield the branching ratios for other decay modes of the Z^0 boson.

Thus an electron–positron colliding beam facility, operating at a total CoM energy in the region of $\sqrt{s} = m_Z \approx 94$ GeV, is ideally suited for studying the properties of the Z^0 boson. At present, no such facilities exist; however, two are under construction: the Stanford Linear Collider (SLC) which should be working in 1986, and the LEP electron–positron storage rings at CERN which should become operational in 1988. In the latter machine in particular, the beam intensities are such that, at the Z^0 peak, several thousand Z^0 production events per hour are expected. (This should be contrasted with the present world total of 14 identified Z^0 decays, discussed in Section 11.7.) With such high event rates, detailed studies of the kind discussed above will become practicable. In addition, it may well be possible to detect other decay processes, which occur only in higher orders of perturbation theory, despite their small branching ratios.

So far, we have considered those reactions $e^+e^- \to X$ which, in lowest order of perturbation theory, depend only on the lepton–boson interactions (14.3d) and (14.3e), represented by the Feynman graphs (B.13)–(B.18) in Appendix B. The boson self-coupling terms (14.3a), Figs. (B.1)–(B.6), are a most striking and characteristic feature of a non-Abelian gauge theory, and electron–positron annihilation processes allow us to explore the three-boson self-coupling terms [lines 1–4 of Eq. (14.3a) and Figs. (B.1) and (B.2)]. One way would be to produce Z^0 bosons in the LEP electron–positron storage ring at CERN and study the decays

$$Z^0 \to W^+ + l^- + \bar{\nu}_l, \qquad Z^0 \to W^- + l^+ + \nu_l \tag{14.72}$$

at the Z^0 resonance energy. In lowest order, the first of these processes is represented by the Feynman graph in Fig. 14.12, with a similar graph for the second process. However, these Z^0 decays are second-order processes, and the predicted branching ratios are very small, of the order of 10^{-8} [W. Alles *et al.*,

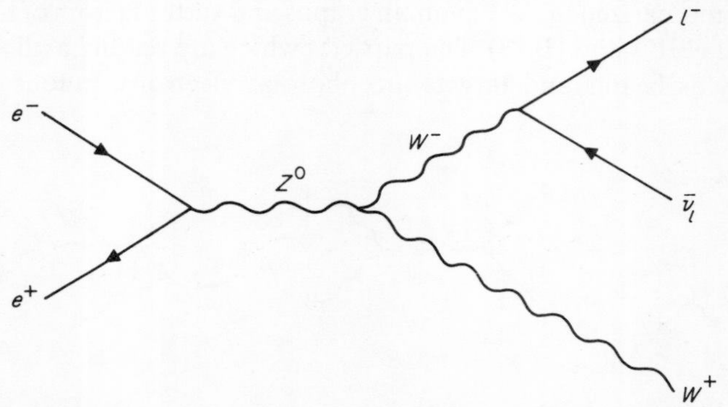

Fig. 14.12. The reaction $e^+ + e^- \to Z^0 \to W^+ + l^- + \bar{\nu}_l$.

Nucl. Phys. **B119** (1977), 125]. This will make it difficult to study these reactions, even with the beam intensities expected at LEP.

A more promising possibility is the reaction

$$e^+ + e^- \rightarrow W^+ + W^-. \tag{14.73}$$

In lowest order, the main contribution to this process comes from the Feynman diagrams in Fig. 14.13. (There is an additional diagram in which the photon in Fig. 14.13(b) is replaced by a Higgs boson. This contribution can be neglected because of the very weak Higgs–lepton coupling; see Section 14.5.) The cross-section predicted by the electro–weak theory for the reaction (14.73) has been calculated [O. P. Sushkov *et al.*, *Sov. J. Nucl. Phys.* **20** (1975), 537, and W. Alles *et al.*, cited above]. In the immediate vicinity of the threshold, the cross-section is dominated by the contribution from diagram 14.13(a). However, at somewhat higher energies, the amplitudes for the Feynman graphs 14.13(a)–(c) become of comparable importance, and the cross-section becomes extremely sensitive to the large interference terms between these amplitudes. For this reason, the reaction (14.73) will, in the future when yet higher energies become available, present us with an ideal way of testing the detailed forms of the $\gamma W^+ W^-$ and $Z^0 W^+ W^-$ interactions predicted by the standard electro–weak theory.

14.5 THE HIGGS BOSON

The recent detection of the W and Z^0 bosons, with the expected masses, represents impressive evidence in support of the electro–weak theory. However, its ultimate confirmation and triumph will have to await the experimental discovery of the Higgs particle. In studying neutrino–electron scattering and electron–positron annihilation, we found that the contributions involving Higgs particles are negligibly small. Unfortunately, these examples are typical. The Higgs boson is a most elusive creature which will be difficult to observe.

To understand this, we must look at the coupling of the Higgs boson to other particles, summarized in the Feynman graphs and vertex factors of Figs. (B.9)–(B.12), and (B.17) and (B.18). The particles which are readily available in the laboratory as beams and targets are photons, electrons, muons and some

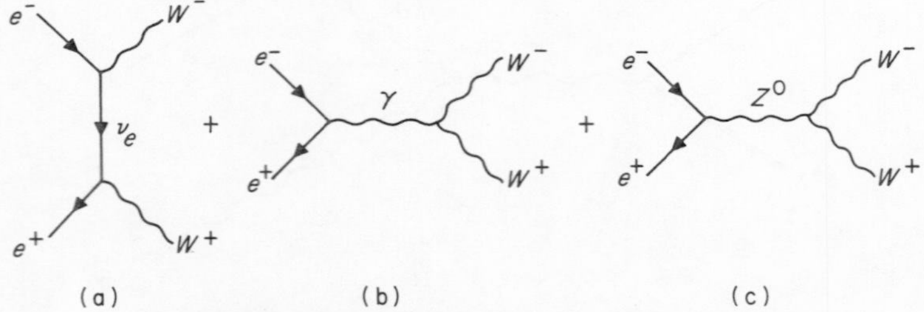

Fig. 14.13. The reaction $e^+ + e^- \rightarrow W^+ + W^-$.

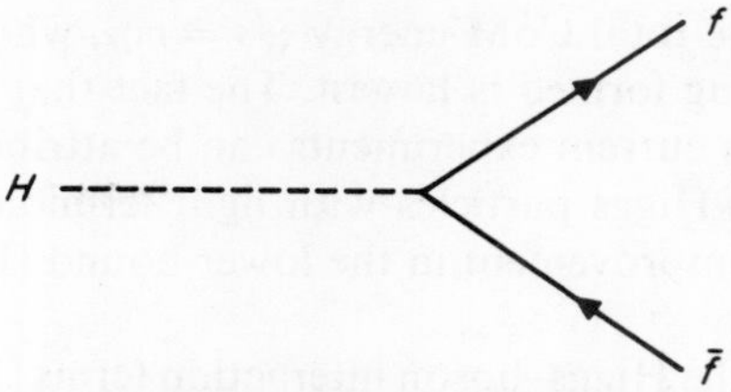

Fig. 14.14. The basic vertex part of the Higgs–fermion interaction, where $f = l, \nu_l$ or q.

hadrons (pions, kaons and nucleons). The Higgs particle does not couple directly to photons. The couplings to leptons are specified by the vertex factor $-\mathrm{i}m/v$, where m is the relevant lepton mass. The coupling of Higgs particles to quarks $q(= u, d, \ldots)$ is of the same form, so that the basic vertex part shown in Fig. 14.14 and the associated vertex factor

$$\frac{-\mathrm{i}}{v} m_f \qquad (f = l, \nu_l, q) \tag{14.74}$$

apply generally to leptons and quarks of mass m_f.‡ Using Eqs. (14.4) and (14.6), we can write this vertex factor

$$\frac{-\mathrm{i}}{v} m_f = \frac{-\mathrm{i}e}{2 \sin \theta_\mathrm{W}} \frac{m_f}{m_W}, \tag{14.75}$$

i.e. the Higgs–fermion coupling is of order m_f/m_W compared with the QED coupling strength and is weak for $m_f \ll m_W$, a condition which is satisfied for neutrinos, electrons and muons, and the light quarks (u, d, s) of which pions, kaons and nucleons are composed. Hence, although it is possible to produce Higgs bosons in reactions initiated by these particles, the production rates are extremely small. For example, the contribution to the process $e^+e^- \to f\bar{f}$ coming from the Feynman diagram 14.15 will be extremely small, even for

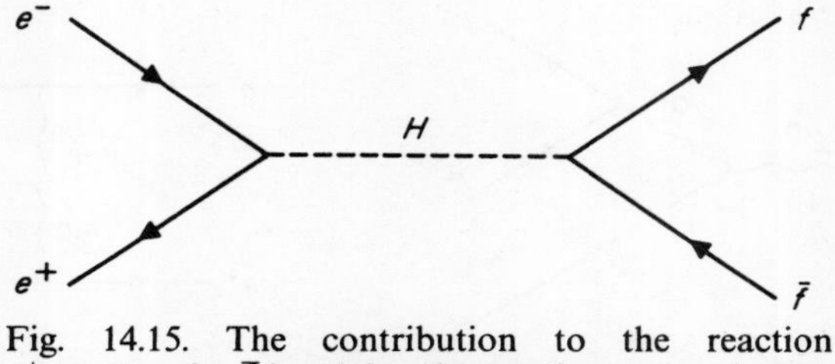

Fig. 14.15. The contribution to the reaction $e^+ + e^- \to f + \bar{f}$, involving the creation and reabsorption of one Higgs particle.

‡ See D. Bailin, *Weak Interactions*, 2nd edn, A. Hilger, Bristol, 1982, Section 6.5.

e^+e^- collisions at the total CoM energy $\sqrt{s} = m_H$, where the probability of the H resonance being formed is largest. The fact that Higgs particles have not been observed in current experiments can be attributed to the weakness of the interaction of Higgs particles with light fermions ($m_f \ll m_W$) and so does not lead to an improvement in the lower bound (14.17) for the mass of the Higgs boson.

We next consider the Higgs–boson interaction terms (14.3c). Expressing the parameters v and g in terms of the Fermi coupling constant and of the boson masses m_W and m_Z, we see that these terms are proportional to the squares of these masses. Consequently, the fact that these bosons are very massive now works in our favour, in contrast to the Higgs–fermion coupling (14.75).

In looking for the Higgs boson with an electron–positron colliding beam facility, the most promising reaction to search for and investigate is the process

$$e^+ + e^- \to H + l^+ + l^-. \tag{14.76}$$

The l^+l^- pair produced in this reaction allows particularly clear identification, and the analysis gives the mass of the Higgs particle. In lowest order, the leading contributions to this process are described by the Feynman diagrams in Fig. 14.16. For $l \neq e$ (e.g. for $e^+e^- \to H\mu^+\mu^-$), only diagram (a) occurs. However, even for $l = e$ the contribution from the diagram (a) will dominate at energies near the resonance energy $\sqrt{s} = m_Z$.

Both the SLC and LEP electron–positron colliding beam facilities, discussed in Section 14.4, are designed to operate at the Z^0 resonance energy $\sqrt{s} \approx m_Z$. Near this energy, the total cross-section for the process (14.76) is given by the Breit–Wigner formula (14.70), with $X = (Hl^+l^-)$ as final state. We can write this cross-section in the form

$$\sigma_T(e^+e^- \to Hl^+l^-) = \frac{12\pi\Gamma(Z^0 \to e^+e^-)\Gamma_t}{(s - m_Z^2)^2 + m_Z^2\Gamma_t^2} \times \frac{\Gamma(Z^0 \to l^+l^-)}{\Gamma_t} \times \frac{\Gamma(Z^0 \to Hl^+l^-)}{\Gamma(Z^0 \to l^+l^-)}. \tag{14.77}$$

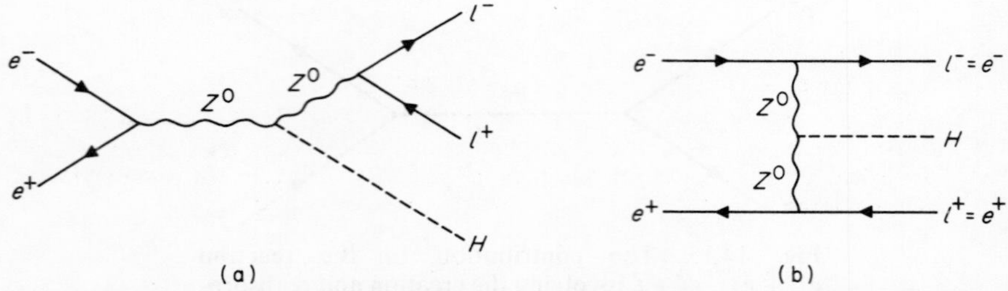

Fig. 14.16. The leading contributions, in lowest order, to the process $e^+e^- \to Hl^+l^-$. For $l \neq e$, only diagram (a) contributes. For $l = e$, diagram (a) dominates near the resonance energy $\sqrt{s} = m_Z$.

The first factor on the right-hand side of this equation is the total cross-section for the production of Z^0 bosons in electron–positron collisions. At LEP, the corresponding event rate for Z^0 production, at the resonance energy $\sqrt{s} = m_Z$, is expected to be about 2000 per hour.

The second factor in Eq. (14.77) we estimate as follows. The partial width $\Gamma(Z^0 \to l^+l^-)$ in the numerator is given by Eq. (14.66). Taking $m_Z \approx 94$ GeV, one obtains the value $\Gamma(Z^0 \to l^+l^-) \approx 0.09$ GeV. The total width Γ_t in the denominator is calculated by considering the decays $Z^0 \to f\bar{f}$ for all known leptons and quarks. In this way one obtains the estimate $\Gamma_t \approx 3$ GeV‡ and hence the branching ratio $\Gamma(Z^0 \to l^+l^-)/\Gamma_t \approx 0.03$.

The last factor in Eq. (14.77) has been calculated by J. D. Bjorken [see R. N. Cahn *et al.*, *Phys. Lett.* **82B** (1979), 113] who obtains the range of values

$$3 \times 10^{-3} \gtrsim \frac{\Gamma(Z^0 \to Hl^+l^-)}{\Gamma(Z^0 \to l^+l^-)} \gtrsim 10^{-4}$$

for the range of Higgs masses

$$10 \text{ GeV} \lesssim m_H \lesssim 50 \text{ GeV}.$$

Combining these results in Eq. (14.77), we obtain the estimate that the number of $e^+e^- \to Hl^+l^-$ events per hour expected at LEP, operating at the Z^0 resonance energy, lies in the range 0.01 per hour (for $m_H = 50$ GeV) to 0.2 per hour (for $m_H = 10$ GeV). These are quite reasonable event rates, certainly for the lower mass values. Provided the Higgs boson is not too heavy, one might hope that it will be detected in this way at LEP towards the end of this decade.

PROBLEMS

14.1 For the process $Z^0 \to l^+l^-$, derive the decay rate (14.66), assuming that terms of order m_l^2/m_Z^2 are negligible.

Show that the decay rate for the process $Z^0 \to \nu_l\bar{\nu}_l$ is given by

$$\Gamma(Z^0 \to \nu_l\bar{\nu}_l) = \frac{1}{\pi 12\sqrt{2}} G m_Z^3$$

which, for $m_Z = 94$ GeV, has the value 0.18 GeV.

14.2 The Higgs boson decays to quark–antiquark and lepton–antilepton pairs. In lowest order of perturbation theory these decays, $H \to f\bar{f}$ $(f = l, \nu_l, q)$, are described by the Feynman diagram in Fig. 14.14. Derive the corresponding decay widths

$$\Gamma(H \to f\bar{f}) = \frac{1}{\pi 4\sqrt{2}} G m_H m_f^2 \left(1 - 4\frac{m_f^2}{m_H^2}\right)^{3/2}$$

‡ See J. Ellis's article in *Gauge Theories in High Energy Physics, Les Houches Session 37*, Ed. by M. K. Gaillard and R. Stora, North-Holland, 1983.

14.3 Derive the total cross-sections (14.44a) and (14.44b) for $\nu_\mu e^-$ and $\bar{\nu}_\mu e^-$ scattering.

14.4 Starting from the Feynman amplitude, Eqs. (14.58) and (14.55), derive the differential cross-section for the process $e^+e^- \to l^+l^-$ $(l \neq e)$, making the reasonable simplifying approximation $\sin^2\theta_W = 0.25$, i.e. $g_V = 0$ and $g_A = -\frac{1}{2}$. Check your result against Eqs. (14.60a)–(14.60c).

14.5 Show that, in the unitary gauge, the generalized Higgs–neutrino coupling term (13.44a) reduces to

$$\sum_j \frac{-1}{\sqrt{2}} \lambda_j \bar{\psi}_j(x)\psi_j(x)[v + \sigma(x)]$$

where

$$\psi_j(x) = \sum_l U_{jl}\psi_{\nu_l}(x)$$

and U is the unitary matrix which diagonalized the Hermitian coupling matrix G:

$$(UGU^\dagger)_{ij} = \lambda_i \delta_{ij}.$$

Hence show that Eq. (13.44a) leads to eigenstate neutrinos ν_j associated with the fields $\psi_j(x)$, with masses

$$m_j = \lambda_j v/\sqrt{2},$$

and that the Higgs–neutrino interactions are given by vertices of the form of Fig. 14.14, where now $f = \nu_j$ are eigenstate neutrinos, and by vertex factors

$$\frac{-i}{v} m_f \qquad (f = \nu_j).$$

For $G_{ij} = \delta_{ij}$, the eigenstate neutrinos ν_j become identical with the leptonic neutrinos, $\nu_e, \nu_\mu, \ldots$, and we regain the results of Chapter 14.

14.6 Write down the Feynman amplitude for the decay process

$$Z^0(p, r) \to H(p') + e^+(q_1, r_1) + e^-(q_2, r_2),$$

and show that, taking $\sin^2\theta_W = 0.25$ as a reasonable first approximation and neglecting electron masses, it reduces to

$$\mathcal{M} = \frac{-iC}{k^2 - m_Z^2} \bar{u}(\mathbf{q}_2)\gamma^\alpha\gamma_5 v(\mathbf{q}_1)\varepsilon_{r\alpha}(\mathbf{p})$$

where $k = p - p'$ and

$$C = vg^3/(8\cos^3\theta_W).$$

Hence show that, neglecting electron masses,

$$(2m_e)^2 \tfrac{1}{3} \sum_r \sum_{r_1 r_2} |\mathcal{M}|^2 = \frac{C^2}{(k^2 - m_Z^2)^2} \tfrac{4}{3}\left[(q_1 q_2) + \frac{2}{m_Z^2}(pq_1)(pq_2)\right].$$

From this expression, the differential decay rate $d\Gamma$ to final states in which the momentum of the Higgs particle lies in the range $d^3\mathbf{p}'$ at $\mathbf{p}'$ can be obtained by integrating over the lepton momenta q_1, q_2 in a manner closely analogous to that used to integrate over the neutrino momenta in muon decay, Eqs. (11.53)–(11.63). Exploit this to show that

$$d\Gamma = \frac{C^2}{36(2\pi)^4} \frac{d^3\mathbf{p}'}{EE'} \frac{1}{(k^2 - m_Z^2)^2}\left[2k^2 + \frac{1}{m_Z^2}(pk)^2\right].$$

APPENDIX A

The Dirac equation

In this appendix we shall derive the main results relating to the Dirac equation.[‡]

A.1 THE DIRAC EQUATION

The Dirac equation can be written

$$i\hbar \frac{\partial \psi(x)}{\partial t} = [c\boldsymbol{\alpha}\cdot(-i\hbar\nabla) + \beta mc^2]\psi(x) \tag{A.1}$$

where $\boldsymbol{\alpha} = (\alpha_1, \alpha_2, \alpha_3)$ and β are 4×4 Hermitian matrices satisfying

$$[\alpha_i, \alpha_j]_+ = 2\delta_{ij}, \qquad [\alpha_i, \beta]_+ = 0, \qquad \beta^2 = 1, \qquad i, j = 1, 2, 3. \tag{A.2}$$

With

$$\gamma^0 = \beta, \qquad \gamma^i = \beta\alpha_i \tag{A.3}$$

the Dirac equation becomes

$$i\hbar\gamma^\mu \frac{\partial \psi(x)}{\partial x^\mu} - mc\psi(x) = 0 \tag{A.4}$$

[‡] The reader is assumed familiar with the elementary ideas of the Dirac equation as given in, for example, L. I. Schiff, *Quantum Mechanics*, 3rd edn, McGraw-Hill, New York, 1968, pp. 472–488. A more complete treatment, suitable for complementary background reading to this appendix, will be found in H. A. Bethe and R. W. Jackiw, *Intermediate Quantum Mechanics*, 2nd edn, Benjamin, New York, 1968, pp. 349–377.

with the 4 × 4 matrices γ^μ, $\mu = 0, \ldots, 3$, satisfying the anticommutation relations

$$[\gamma^\mu, \gamma^\nu]_+ = 2g^{\mu\nu} \tag{A.5}$$

and the Hermiticity conditions

$$\gamma^{\mu\dagger} = \gamma^0\gamma^\mu\gamma^0. \tag{A.6}$$

Except for Section A.8 at the end of the appendix, the following properties are consequences of Eqs. (A.4)–(A.6) only and do not depend on choosing a particular representation for the γ-matrices.

A fifth anticommuting γ-matrix is defined by

$$\gamma^5 \equiv i\gamma^0\gamma^1\gamma^2\gamma^3, \tag{A.7}$$

and γ^5 has the properties

$$[\gamma^\mu, \gamma^5]_+ = 0, \qquad (\gamma^5)^2 = 1, \qquad \gamma^{5\dagger} = \gamma^5. \tag{A.8}$$

Note that Greek indices will *always* stand for the values 0, ..., 3 only, and not for 5.

The 4 × 4 spin matrices

$$\sigma^{\mu\nu} = \frac{i}{2}[\gamma^\mu, \gamma^\nu] \tag{A.9}$$

satisfy

$$\sigma^{\mu\nu\dagger} = \gamma^0\sigma^{\mu\nu}\gamma^0 \tag{A.10}$$

and, with $\boldsymbol{\sigma} = (\sigma^{23}, \sigma^{31}, \sigma^{12})$ and $i, j, k = 1, 2, 3$ in cyclic order, we can write

$$\sigma^{ij} = -\gamma^0\gamma^5\gamma^k. \tag{A.11}$$

So far the γ-matrices, and matrices derived from them, have been defined with upper indices: γ^μ, $\sigma^{\mu\nu}$, etc. We now *define* corresponding matrices with lower indices by the relation

$$\gamma_\mu = g_{\mu\nu}\gamma^\nu, \tag{A.12}$$

etc.‡

We also define the matrix γ_5 through

$$\gamma_5 \equiv \frac{i}{4!}\varepsilon_{\lambda\mu\nu\pi}\gamma^\lambda\gamma^\mu\gamma^\nu\gamma^\pi = \gamma^5, \tag{A.13}$$

‡ However, these matrices do not transform as tensors. We shall see below that it is bilinear forms of spinors containing these matrices which have the transformation properties of tensors. See Eqs. (A.53).

where the completely antisymmetric alternating symbol $\varepsilon_{\lambda\mu\nu\pi}$ is equal to $+1$ for (λ, μ, ν, π) an even permutation of (0, 1, 2, 3), is equal to -1 for an odd permutation, and vanishes if two or more indices are the same.

A.2 CONTRACTION IDENTITIES

The manipulation of expressions involving γ-matrices is often greatly facilitated by the use of the following algebraic identities, which follow easily from the anticommutation relations (A.5):

$$\left.\begin{aligned} \gamma_\lambda\gamma^\lambda = 4, \qquad & \gamma_\lambda\gamma^\alpha\gamma^\lambda = -2\gamma^\alpha \\ \gamma_\lambda\gamma^\alpha\gamma^\beta\gamma^\lambda = 4g^{\alpha\beta}, \qquad & \gamma_\lambda\gamma^\alpha\gamma^\beta\gamma^\gamma\gamma^\lambda = -2\gamma^\gamma\gamma^\beta\gamma^\alpha \\ \gamma_\lambda\gamma^\alpha\gamma^\beta\gamma^\gamma\gamma^\delta\gamma^\lambda = 2(\gamma^\delta\gamma^\alpha & \gamma^\beta\gamma^\gamma + \gamma^\gamma\gamma^\beta\gamma^\alpha\gamma^\delta) \end{aligned}\right\}. \tag{A.14a}$$

If $A, B, \ldots$ denote four-vectors and 'A slash' is defined by $\not{A} \equiv \gamma^\alpha A_\alpha$, etc., we obtain the following contraction identifies from Eqs. (A.14a):

$$\left.\begin{aligned} \gamma_\lambda\not{A}\gamma^\lambda = -2\not{A} & \\ \gamma_\lambda\not{A}\not{B}\gamma^\lambda = 4AB, \qquad & \gamma_\lambda\not{A}\not{B}\not{C}\gamma^\lambda = -2\not{C}\not{B}\not{A} \\ \gamma_\lambda\not{A}\not{B}\not{C}\not{D}\gamma^\lambda = 2(\not{D}\not{A}\not{B}\not{C} & + \not{C}\not{B}\not{A}\not{D}) \end{aligned}\right\}. \tag{A.14b}$$

The completely antisymmetric alternating symbol $\varepsilon^{\alpha\beta\gamma\delta}$, introduced in Eq. (A.13), satisfies the following contraction identities:

$$\left.\begin{aligned} \varepsilon^{\alpha\beta\mu\nu}\varepsilon_{\alpha\beta\sigma\tau} &= -2(g^\mu_\sigma g^\nu_\tau - g^\mu_\tau g^\nu_\sigma) \\ \varepsilon^{\alpha\beta\gamma\nu}\varepsilon_{\alpha\beta\gamma\tau} &= -6g^\nu_\tau \\ \varepsilon^{\alpha\beta\gamma\delta}\varepsilon_{\alpha\beta\gamma\delta} &= -24 \end{aligned}\right\}. \tag{A.14c}$$

A.3 TRACES

We next list some rules and relations which are extremely useful in evaluating the trace of a product of γ-matrices. Some comments on the derivation of these results are given at the end of the list.

(i) For any two $n \times n$ matrices U and V

$$\mathrm{Tr}\,(UV) = \mathrm{Tr}\,(VU). \tag{A.15}$$

(ii) If $(\gamma^\alpha\gamma^\beta \ldots \gamma^\mu\gamma^\nu)$ contains an odd number of γ-matrices, then

$$\mathrm{Tr}\,(\gamma^\alpha\gamma^\beta \ldots \gamma^\mu\gamma^\nu) = 0. \tag{A.16}$$

(iii) For a product of an even number of γ-matrices:

$$\left.\begin{aligned} \mathrm{Tr}\,(\gamma^\alpha\gamma^\beta) = 4g^{\alpha\beta}, \qquad & \mathrm{Tr}\,\sigma^{\alpha\beta} = 0 \\ \mathrm{Tr}\,(\gamma^\alpha\gamma^\beta\gamma^\gamma\gamma^\delta) = 4(g^{\alpha\beta}g^{\gamma\delta} - g^{\alpha\gamma} & g^{\beta\delta} + g^{\alpha\delta}g^{\beta\gamma}) \end{aligned}\right\}, \tag{A.17}$$

and from Eqs. (A.17):

$$\operatorname{Tr}(\not{A}\not{B}) = 4(AB) \tag{A.18a}$$

$$\operatorname{Tr}(\not{A}\not{B}\not{C}\not{D}) = 4\{(AB)(CD) - (AC)(BD) + (AD)(BC)\}, \tag{A.18b}$$

and generally, if $A_1, A_2, \ldots, A_{2n}$ are four-vectors, then

$$\operatorname{Tr}(\not{A}_1\not{A}_2 \ldots \not{A}_{2n}) = \{(A_1A_2)\operatorname{Tr}(\not{A}_3 \ldots \not{A}_{2n}) - (A_1A_3)\operatorname{Tr}(\not{A}_2\not{A}_4 \ldots \not{A}_{2n}) + \cdots + (A_1A_{2n})\operatorname{Tr}(\not{A}_2\not{A}_3 \ldots \not{A}_{2n-1})\}. \tag{A.18c}$$

In many specific cases one can evaluate traces more simply than by direct repeated use of Eq. (A.18c). The contraction relations, Eqs. (A.14), are particularly useful in this connection, as is

$$\not{A}\not{B} = AB - \mathrm{i}\sigma^{\alpha\beta}A_\alpha B_\beta = 2AB - \not{B}\not{A}, \tag{A.19a}$$

and the particular cases of this equation:

$$\not{A}\not{A} = A^2; \qquad \not{A}\not{B} = -\not{B}\not{A}, \quad \text{if } AB = 0. \tag{A.19b}$$

(iv) For any product of γ-matrices

$$\operatorname{Tr}(\gamma^\alpha\gamma^\beta \ldots \gamma^\mu\gamma^\nu) = \operatorname{Tr}(\gamma^\nu\gamma^\mu \ldots \gamma^\beta\gamma^\alpha) \tag{A.20a}$$

whence

$$\operatorname{Tr}(\not{A}_1\not{A}_2 \ldots \not{A}_{2n}) = \operatorname{Tr}(\not{A}_{2n} \ldots \not{A}_2\not{A}_1). \tag{A.20b}$$

(v) The above results can be extended to products involving the γ^5 matrix, the most important relations being

$$\left.\begin{aligned} \operatorname{Tr}\gamma^5 = \operatorname{Tr}(\gamma^5\gamma^\alpha) = \operatorname{Tr}(\gamma^5\gamma^\alpha\gamma^\beta) = \operatorname{Tr}(\gamma^5\gamma^\alpha\gamma^\beta\gamma^\gamma) = 0 \\ \operatorname{Tr}(\gamma^5\gamma^\alpha\gamma^\beta\gamma^\gamma\gamma^\delta) = -4\mathrm{i}\varepsilon^{\alpha\beta\gamma\delta} \end{aligned}\right\}. \tag{A.21}$$

Other results involving γ^5 are easily obtained from Eqs. (A.15)–(A.21), using Eqs. (A.7) and (A.8) which define γ^5 and state its main properties.

The following comments should suffice to enable the reader to derive Eqs. (A.16)–(A.21).

We obtain Eq. (A.16) by using $(\gamma^5)^2 = 1$ and Eq. (A.15), whence

$$\operatorname{Tr}(\gamma^\alpha \ldots \gamma^\nu) = \operatorname{Tr}[(\gamma^5)^2\gamma^\alpha \ldots \gamma^\nu] = \operatorname{Tr}(\gamma^5\gamma^\alpha \ldots \gamma^\nu\gamma^5).$$

In the last trace, we use $[\gamma^\mu, \gamma^5]_+ = 0$ to commute the left-hand γ^5 matrix through to the right-hand side, with the result $-\operatorname{Tr}(\gamma^\alpha \ldots \gamma^\nu)$ for an odd number of factors in $(\gamma^\alpha \ldots \gamma^\nu)$. Hence Eq. (A.16) follows.

Eqs. (A.17) are derived by repeated use of the anticommutation relations (A.5) of the γ-matrices and of the cyclic property (A.15) of traces.

To derive Eq. (A.20a), we introduce the transposed matrices $\gamma^{\alpha\mathrm{T}}$. It follows from Eqs. (A.5) and (A.6) that the matrices $(-\gamma^{\alpha\mathrm{T}})$ also satisfy the anticommutation relations (A.5) and the Hermiticity conditions (A.6). Hence by Pauli's fundamental theorem (A.61) there exists a unitary matrix C such that

$$C\gamma^{\alpha}C^{-1} = -\gamma^{\alpha \mathrm{T}}. \tag{A.22}$$

Eq. (A.20a) then follows from

$$\begin{aligned}\mathrm{Tr}\,(\gamma^{\alpha}\gamma^{\beta}\ldots\gamma^{\nu}) &= \mathrm{Tr}\,[(-C^{-1}\gamma^{\alpha\mathrm{T}}C)(-C^{-1}\gamma^{\beta\mathrm{T}}C)\ldots(-C^{-1}\gamma^{\nu\mathrm{T}}C)] \\ &= \mathrm{Tr}\,(\gamma^{\alpha\mathrm{T}}\gamma^{\beta\mathrm{T}}\ldots\gamma^{\nu\mathrm{T}}) = \mathrm{Tr}\,[(\gamma^{\nu}\ldots\gamma^{\beta}\gamma^{\alpha})^{\mathrm{T}}].\end{aligned}$$

Lastly, Eqs. (A.21) follow from $\gamma^5 = \mathrm{i}\gamma^0\gamma^1\gamma^2\gamma^3$, $[\gamma^{\mu}, \gamma^5]_+ = 0$, the cyclic property (A.15) of traces, and Eqs. (A.16) and (A.17). E.g.

$$\mathrm{Tr}\,\gamma^5 = \mathrm{Tr}\,[(\gamma^5\gamma^0)\gamma^0] = \mathrm{Tr}\,[(-\gamma^0\gamma^5)\gamma^0] = -\mathrm{Tr}\,\gamma^5 = 0.$$

A.4 PLANE WAVE SOLUTIONS

The Dirac equation (A.4) possesses plane wave solutions

$$\psi(x) = \text{const.}\begin{Bmatrix} u_r(\mathbf{p}) \\ v_r(\mathbf{p}) \end{Bmatrix} \mathrm{e}^{\mp \mathrm{i}px/\hbar} \tag{A.23}$$

where $p = (E_{\mathbf{p}}/c,\ \mathbf{p})$ and $E_{\mathbf{p}} = +(m^2c^4 + c^2\mathbf{p}^2)^{1/2}$. In the single-particle theory, $u_r(\mathbf{p})$ corresponds to a particle of momentum $\mathbf{p}$ and positive energy $E_{\mathbf{p}}$, $v_r(\mathbf{p})$ to momentum $-\mathbf{p}$ and negative energy $-E_{\mathbf{p}}$. The index $r = 1, 2$ labels two independent solutions for each four-momentum p, which we shall choose to be orthogonal.

The constant four-spinors $u_r(\mathbf{p})$ and $v_r(\mathbf{p})$, and their adjoints

$$\bar{u}_r(\mathbf{p}) = u_r^{\dagger}(\mathbf{p})\gamma^0, \qquad \bar{v}_r(\mathbf{p}) = v_r^{\dagger}(\mathbf{p})\gamma^0, \tag{A.24}$$

satisfy the equations

$$(\not{p} - mc)u_r(\mathbf{p}) = 0 \qquad (\not{p} + mc)v_r(\mathbf{p}) = 0 \tag{A.25}$$

$$\bar{u}_r(\mathbf{p})(\not{p} - mc) = 0 \qquad \bar{v}_r(\mathbf{p})(\not{p} + mc) = 0. \tag{A.26}$$

With the normalization of these spinors defined by

$$u_r^{\dagger}(\mathbf{p})u_r(\mathbf{p}) = v_r^{\dagger}(\mathbf{p})v_r(\mathbf{p}) = \frac{E_{\mathbf{p}}}{mc^2}, \tag{A.27}$$

one derives the following *orthonormality relations* from Eqs. (A.25) and (A.26):

$$\left.\begin{aligned} u_r^{\dagger}(\mathbf{p})u_s(\mathbf{p}) = v_r^{\dagger}(\mathbf{p})v_s(\mathbf{p}) &= \frac{E_{\mathbf{p}}}{mc^2}\,\delta_{rs} \\ u_r^{\dagger}(\mathbf{p})v_s(-\mathbf{p}) &= 0 \end{aligned}\right\} \tag{A.28}$$

and

$$\left.\begin{aligned}\bar{u}_r(\mathbf{p})u_s(\mathbf{p}) &= -\bar{v}_r(\mathbf{p})v_s(\mathbf{p}) = \delta_{rs}\\ \bar{u}_r(\mathbf{p})v_s(\mathbf{p}) &= \bar{v}_r(\mathbf{p})u_s(\mathbf{p}) = 0\end{aligned}\right\}. \tag{A.29}$$

The spinors $u_r(\mathbf{p})$ and $v_r(\mathbf{p})$, $r = 1, 2$, satisfy the *completeness relation*

$$\sum_{r=1}^{2} [u_{r\alpha}(\mathbf{p})\bar{u}_{r\beta}(\mathbf{p}) - v_{r\alpha}(\mathbf{p})\bar{v}_{r\beta}(\mathbf{p})] = \delta_{\alpha\beta}. \tag{A.30}$$

This relation can be established by showing that it holds for the four basis states $u_s(\mathbf{p})$ and $v_s(\mathbf{p})$, $s = 1, 2$.

A.5 ENERGY PROJECTION OPERATORS

The energy projection operators are defined by

$$\Lambda^{\pm}(\mathbf{p}) = \frac{\pm \not{p} + mc}{2mc}. \tag{A.31}$$

They have the properties, which follow from Eqs. (A.25) and (A.26), of projecting out the positive/negative energy solutions from a linear combination of the four plane wave states $u_r(\mathbf{p})$ and $v_r(\mathbf{p})$, i.e.

$$\left.\begin{aligned}\Lambda^{+}(\mathbf{p})u_r(\mathbf{p}) &= u_r(\mathbf{p}), &\quad \Lambda^{-}(\mathbf{p})v_r(\mathbf{p}) &= v_r(\mathbf{p})\\ \bar{u}_r(\mathbf{p})\Lambda^{+}(\mathbf{p}) &= \bar{u}_r(\mathbf{p}), &\quad \bar{v}_r(\mathbf{p})\Lambda^{-}(\mathbf{p}) &= \bar{v}_r(\mathbf{p})\end{aligned}\right\}, \tag{A.32}$$

and

$$\Lambda^{+}(\mathbf{p})v_r(\mathbf{p}) = \Lambda^{-}(\mathbf{p})u_r(\mathbf{p}) = 0, \qquad \bar{v}_r(\mathbf{p})\Lambda^{+}(\mathbf{p}) = \bar{u}_r(\mathbf{p})\Lambda^{-}(\mathbf{p}) = 0. \tag{A.33}$$

From Eqs. (A.31) one verifies directly the property characteristic of projection operators

$$[\Lambda^{\pm}(\mathbf{p})]^2 = \Lambda^{\pm}(\mathbf{p}) \tag{A.34a}$$

(since $\not{p}\not{p} = p^2 = m^2c^2$), as well as

$$\Lambda^{\pm}(\mathbf{p})\Lambda^{\mp}(\mathbf{p}) = 0, \qquad \Lambda^{+}(\mathbf{p}) + \Lambda^{-}(\mathbf{p}) = 1. \tag{A.34b}$$

Using the completeness relation (A.30), one easily shows that the projection operators $\Lambda^{\pm}(\mathbf{p})$ can be written

$$\Lambda^{+}_{\alpha\beta}(\mathbf{p}) = \sum_{r=1}^{2} u_{r\alpha}(\mathbf{p})\bar{u}_{r\beta}(\mathbf{p}), \qquad \Lambda^{-}_{\alpha\beta}(\mathbf{p}) = -\sum_{r=1}^{2} v_{r\alpha}(\mathbf{p})\bar{v}_{r\beta}(\mathbf{p}). \tag{A.35}$$

A.6 HELICITY AND SPIN PROJECTION OPERATORS

In Section 4.2 we chose the plane wave solutions $u_r(\mathbf{p})$ and $v_r(\mathbf{p})$ of the Dirac equation as eigenstates of the 4×4 spin matrix

$$\sigma_{\mathbf{p}} = \frac{\boldsymbol{\sigma} \cdot \mathbf{p}}{|\mathbf{p}|}, \tag{A.36}$$

satisfying the equations

$$\sigma_{\mathbf{p}} u_r(\mathbf{p}) = (-1)^{r+1} u_r(\mathbf{p}), \qquad \sigma_{\mathbf{p}} v_r(\mathbf{p}) = (-1)^r v_r(\mathbf{p}), \qquad r = 1, 2. \tag{4.35}$$

Correspondingly, we now define the operators

$$\Pi^{\pm}(\mathbf{p}) = \tfrac{1}{2}(1 \pm \sigma_{\mathbf{p}}) \tag{A.37}$$

which are easily seen to have the properties

$$[\Pi^{\pm}(\mathbf{p})]^2 = \Pi^{\pm}(\mathbf{p}), \qquad \Pi^{\pm}(\mathbf{p})\Pi^{\mp}(\mathbf{p}) = 0, \qquad \Pi^{+}(\mathbf{p}) + \Pi^{-}(\mathbf{p}) = 1, \tag{A.38}$$

and

$$[\Lambda^{+}(\mathbf{p}), \Pi^{\pm}(\mathbf{p})] = [\Lambda^{-}(\mathbf{p}), \Pi^{\pm}(\mathbf{p})] = 0. \tag{A.39}$$

It follows from Eqs. (4.35) and (A.37) that the spinors $u_r(\mathbf{p})$ and $v_r(\mathbf{p})$ satisfy the equations

$$\left.\begin{aligned} \Pi^{+}(\mathbf{p})u_r(\mathbf{p}) &= \delta_{1r}u_r(\mathbf{p}), \qquad \Pi^{+}(\mathbf{p})v_r(\mathbf{p}) = \delta_{2r}v_r(\mathbf{p}) \\ \Pi^{-}(\mathbf{p})u_r(\mathbf{p}) &= \delta_{2r}u_r(\mathbf{p}), \qquad \Pi^{-}(\mathbf{p})v_r(\mathbf{p}) = \delta_{1r}v_r(\mathbf{p}) \end{aligned}\right\} \quad r = 1, 2. \tag{A.40}$$

We see from Eqs. (A.38) that the operators $\Pi^{\pm}(\mathbf{p})$ are projection operators of mutually orthogonal states. From the first of Eqs. (4.35), it follows that $u_1(\mathbf{p})[u_2(\mathbf{p})]$ represents a positive energy electron with spin parallel [antiparallel] to its direction of motion $\mathbf{p}$, i.e. it is a positive [negative] helicity state, as defined in Section 4.3, following Eq. (4.48). Hence the operators $\Pi^{\pm}(\mathbf{p})$ are called *helicity projection operators.*

The corresponding interpretation for the spinors $v_r(\mathbf{p})$ is also possible in the single-particle theory, either in terms of negative energy states of electrons or in terms of hole theory. Although our discussion of positrons in the quantized field theory in Chapter 4 does not depend on hole theory, a reader familiar with the latter may like to see the connection. Consider, for example, the spinor $v_1(\mathbf{p})$. In the language of negative energy states, $v_1(\mathbf{p})$ represents a negative energy electron with momentum $-\mathbf{p}$ and, according to Eqs. (4.35), spin parallel to $-\mathbf{p}$, so that $v_1(\mathbf{p})$ is a positive helicity state of the negative energy electron. Translated into hole theory language, the absence of this negative energy electron represents a positron with momentum $+\mathbf{p}$ and spin parallel to $+\mathbf{p}$, i.e. it continues to be a positive helicity state of the positron.

For a zero mass Dirac particle, we can express the helicity projection operators in terms of the γ^5 matrix. For $m = 0$, we have $p_0 = |\mathbf{p}|$, and Eqs. (A.25) become

$$\gamma^0|\mathbf{p}|w_r(\mathbf{p}) = -\gamma^k p_k w_r(\mathbf{p}) = \gamma^k p^k w_r(\mathbf{p}), \tag{A.41}$$

where $w_r(\mathbf{p})$ stands for either $u_r(\mathbf{p})$ or $v_r(\mathbf{p})$. Premultiplying the last equation by $\gamma^5\gamma^0$ and using Eq. (A.11), we obtain

$$\gamma^5 w_r(\mathbf{p}) = \sigma_{\mathbf{p}} w_r(\mathbf{p}) \tag{A.42}$$

i.e. the helicity projection operators (A.37) become, for $m = 0$,

$$\Pi^{\pm}(\mathbf{p}) = \tfrac{1}{2}(1 \pm \gamma^5). \tag{A.43}$$

For particles of non-zero mass m, this result holds to $O(m/p_0)$ in the high-energy limit.

So far we have considered helicity projection operators only. For a Dirac particle, the spin component in an arbitrary direction is a good quantum number in the rest frame of the particle only. We shall see that in the rest frame one can define spin projection operators for an arbitrary axis of quantization in a covariant way, and by carrying out a Lorentz transformation to an arbitrary coordinate frame one can then define spin projection operators in this frame.

With the axis of quantization in the rest frame of the particle specified by the unit vector $\mathbf{n}$, we define the unit vector n^μ to be given in the rest frame by

$$n^\mu = (0, \mathbf{n}). \tag{A.44}$$

From the invariance of scalar products, it follows that in any other frame also

$$n^2 = -1, \qquad np = 0, \tag{A.45}$$

where p is the four-momentum of the particle in this frame. The required spin projection operators are then given by

$$\Pi^{\pm}(n) = \tfrac{1}{2}(1 \pm \gamma^5 \not{n}). \tag{A.46}$$

One verifies easily that these operators satisfy equations like Eqs. (A.38), characteristic of projection operators, and that they commute with the energy projection operators $\Lambda^{\pm}(\mathbf{p})$ for all vectors p satisfying Eq. (A.45). It is left to the reader to verify that, in the rest frame of the particle, the operators (A.46) have the desired properties. A matrix representation suitable for this purpose is given below, Eqs. (A.63)–(A.66).

A.7 RELATIVISTIC PROPERTIES

The Dirac equations for $\psi(x)$ and its adjoint $\bar{\psi}(x)$,

$$i\hbar\gamma^\mu \frac{\partial\psi(x)}{\partial x^\mu} - mc\psi(x) = 0, \qquad i\hbar \frac{\partial\bar{\psi}(x)}{\partial x^\mu}\gamma^\mu + mc\bar{\psi}(x) = 0, \tag{A.47}$$

are Lorentz-covariant, provided the spinors $\psi(x)$ and $\bar{\psi}(x)$ transform appropriately. We shall consider homogeneous orthochronous Lorentz transformations

$$x^\mu \to x'^\mu = \Lambda^\mu{}_\nu x^\nu, \tag{A.48}$$

i.e. $\Lambda^0{}_0 > 0$ and $\det \Lambda^\mu{}_\nu = \pm 1$, so that the sense of time is not reversed, but the transformation may or may not involve spatial inversion. It can be shown ‡ that corresponding to each such transformation one can construct a non-singular 4×4 matrix $S = S(\Lambda)$ with the properties

$$\gamma^\nu = \Lambda^\nu{}_\mu S \gamma^\mu S^{-1} \tag{A.49}$$

and

$$S^{-1} = \gamma^0 S^\dagger \gamma^0. \tag{A.50}$$

If the transformation properties of the Dirac spinor $\psi(x)$ are defined by

$$\psi(x) \to \psi'(x') = S\psi(x), \tag{A.51}$$

then the covariance of the Dirac equations (A.47) is easily established.

From Eqs. (A.50) and (A.51) one derives the corresponding transformation property of the adjoint spinor $\bar{\psi}(x)$ as

$$\bar{\psi}(x) \to \bar{\psi}'(x') = \bar{\psi}(x) S^{-1}. \tag{A.52}$$

From Eqs. (A.49), (A.51) and (A.52), one obtains the five basic bilinear covariants of the Dirac theory. Under a Lorentz transformation

$$\left.\begin{array}{l} \bar{\psi}\psi \\ \bar{\psi}\gamma^\mu\psi \\ \bar{\psi}\sigma^{\mu\nu}\psi \\ \bar{\psi}\gamma^5\gamma^\mu\psi \\ \bar{\psi}\gamma^5\psi \end{array}\right\} \text{transforms as a} \left\{\begin{array}{l} \text{scalar} \\ \text{vector} \\ \text{antisymmetric second-rank tensor} \\ \text{pseudo-vector} \\ \text{pseudo-scalar} \end{array}\right\}. \tag{A.53}$$

Finally, we obtain the explicit form of the transformation (A.51) for an infinitesimal Lorentz rotation (2.46), i.e.

$$x_\mu \to x'_\mu = \Lambda_{\mu\nu} x^\nu = (g_{\mu\nu} + \varepsilon_{\mu\nu}) x^\nu \tag{A.54}$$

where $\varepsilon_{\mu\nu} = -\varepsilon_{\nu\mu}$. Eq. (2.47) now becomes

$$\psi_\alpha(x) \to \psi'_\alpha(x') = S_{\alpha\beta}\psi_\beta(x) = (\delta_{\alpha\beta} + \tfrac{1}{2}\varepsilon_{\mu\nu} S^{\mu\nu}_{\alpha\beta})\psi_\beta(x)$$

or in matrix form

$$\psi(x) \to \psi'(x') = S\psi(x) = (1 + \tfrac{1}{2}\varepsilon_{\mu\nu} S^{\mu\nu})\psi(x) \tag{A.55}$$

where $S^{\mu\nu}$ is antisymmetric.

‡ See, for example, Bethe and Jackiw, quoted at the beginning of this appendix, pp. 360–365.

Using the orthogonality relation (2.6) for $\Lambda_{\mu\nu}$, we can rewrite Eq. (A.49) as

$$S\gamma^\lambda S^{-1} = \gamma^\nu \Lambda_\nu{}^\lambda. \tag{A.56}$$

Using Eq. (A.55) we can write the left-hand side of Eq. (A.56), to first order in $\varepsilon_{\mu\nu}$, as

$$\gamma^\lambda + \tfrac{1}{2}\varepsilon_{\mu\nu}[S^{\mu\nu}, \gamma^\lambda]. \tag{A.57}$$

From (A.54) the right-hand side of Eq. (A.56) can be written

$$\begin{aligned}\gamma^\nu(g_\nu{}^\lambda + \varepsilon_\nu{}^\lambda) &= \gamma^\lambda + \gamma^\nu g^{\lambda\mu}\varepsilon_{\nu\mu} \\ &= \gamma^\lambda + \tfrac{1}{2}\varepsilon_{\mu\nu}(\gamma^\mu g^{\lambda\nu} - \gamma^\nu g^{\lambda\mu}).\end{aligned} \tag{A.58}$$

Equating the last two expressions, we obtain

$$[S^{\mu\nu}, \gamma^\lambda] = \gamma^\mu g^{\lambda\nu} - \gamma^\nu g^{\lambda\mu},$$

and one verifies directly that this equation has the solution

$$S^{\mu\nu} = \tfrac{1}{2}\gamma^\mu\gamma^\nu. \tag{A.59}$$

Hence, from Eqs. (A.55) and (A.59), and the definition (A.9) of $\sigma^{\mu\nu}$, one finds that under Lorentz transformations a Dirac spinor transforms as

$$\psi(x) \to \psi'(x') = \psi(x) - \frac{\mathrm{i}}{4}\,\varepsilon_{\mu\nu}\sigma^{\mu\nu}\psi(x). \tag{A.60}$$

A.8 PARTICULAR REPRESENTATIONS OF THE γ-MATRICES

So far we have developed the Dirac theory in a representation-free way, relying only on the anticommutation relations (A.5) and the Hermiticity conditions (A.6) of the γ-matrices. There are many ways of writing γ^μ, $\mu = 0, \ldots, 3$, as 4×4 matrices such that Eqs. (A.5) and (A.6) hold. If γ^μ, $\mu = 0, \ldots, 3$, and $\tilde{\gamma}^\mu$, $\mu = 0, \ldots, 3$, are two such sets of matrices, i.e. each set satisfies Eqs. (A.5) and (A.6), then Pauli's fundamental theorem‡ states that

$$\tilde{\gamma}^\mu = U\gamma^\mu U^\dagger, \tag{A.61}$$

where U is a unitary matrix.

We shall state two particular representations which are useful in practice.

(i) *Dirac–Pauli representation.* This representation has a simple non-relativistic limit.

In terms of the Pauli 2×2 spin matrices

$$\sigma_1 = \begin{pmatrix} 0 & 1 \\ 1 & 0 \end{pmatrix} \qquad \sigma_2 = \begin{pmatrix} 0 & -\mathrm{i} \\ \mathrm{i} & 0 \end{pmatrix} \qquad \sigma_3 = \begin{pmatrix} 1 & 0 \\ 0 & -1 \end{pmatrix} \tag{A.62}$$

‡ For its derivation, see Bethe and Jackiw, quoted at the beginning of this appendix, pp. 358–359.

the Dirac matrices can in this representation be written as

$$\left.\begin{aligned} \gamma^0 = \beta = \begin{pmatrix} 1 & 0 \\ 0 & -1 \end{pmatrix}, \quad \alpha_k = \begin{pmatrix} 0 & \sigma_k \\ \sigma_k & 0 \end{pmatrix}, \quad k = 1, 2, 3 \\ \gamma^k = \beta\alpha_k = \begin{pmatrix} 0 & \sigma_k \\ -\sigma_k & 0 \end{pmatrix}, \quad k = 1, 2, 3 \end{aligned}\right\}, \tag{A.63}$$

whence

$$\sigma^{ij} = \begin{pmatrix} \sigma_k & 0 \\ 0 & \sigma_k \end{pmatrix}, \qquad i, j, k = 1, 2, 3 \text{ in cyclic order}, \tag{A.64}$$

$$\sigma^{0k} = \mathrm{i}\alpha_k = \mathrm{i}\begin{pmatrix} 0 & \sigma_k \\ \sigma_k & 0 \end{pmatrix}, \qquad k = 1, 2, 3, \tag{A.65}$$

and

$$\gamma^5 = \begin{pmatrix} 0 & 1 \\ 1 & 0 \end{pmatrix}. \tag{A.66}$$

A complete set of plane wave states is now easily constructed. With the two-component non-relativistic spinors defined by

$$\chi_1 \equiv \chi_2' \equiv \begin{pmatrix} 1 \\ 0 \end{pmatrix}, \qquad \chi_2 \equiv \chi_1' \equiv \begin{pmatrix} 0 \\ 1 \end{pmatrix}, \tag{A.67}$$

the positive and negative energy solutions of the Dirac equation for a particle at rest can be written

$$u_r(\mathbf{0}) = \begin{pmatrix} \chi_r \\ 0 \end{pmatrix}, \qquad v_r(\mathbf{0}) = \begin{pmatrix} 0 \\ \chi_r' \end{pmatrix}, \qquad r = 1, 2. \tag{A.68}$$

Since

$$(mc \pm \not{p})(mc \mp \not{p}) = (mc)^2 - p^2 = 0,$$

it follows that

$$u_r(\mathbf{p}) = \frac{(mc + \not{p})}{\sqrt{(2mE_\mathbf{p} + 2m^2c^2)}} u_r(\mathbf{0}), \qquad r = 1, 2, \tag{A.69}$$

and

$$v_r(\mathbf{p}) = \frac{(mc - \not{p})}{\sqrt{(2mE_\mathbf{p} + 2m^2c^2)}} v_r(\mathbf{0}), \qquad r = 1, 2, \tag{A.70}$$

are solutions of the Dirac equation with energy–momentum vectors $\pm p = (\pm E_\mathbf{p}/c, \pm \mathbf{p})$ respectively. The denominators in Eqs. (A.69) and (A.70) ensure the normalization (A.27).

From the representation (A.63) we can write Eq. (A.69) as

$$u_r(\mathbf{p}) = A\begin{pmatrix} \chi_r \\ B\mathbf{p}\cdot\boldsymbol{\sigma}\chi_r \end{pmatrix}, \qquad r = 1, 2, \tag{A.71}$$

where

$$A \equiv \left(\frac{E_{\mathbf{p}} + mc^2}{2mc^2}\right)^{1/2}, \qquad B \equiv \frac{c}{E_{\mathbf{p}} + mc^2}. \tag{A.72}$$

Using Eqs. (A.62) and (A.67) we finally obtain from (A.71)

$$u_1(\mathbf{p}) = A\begin{pmatrix} 1 \\ 0 \\ Bp^3 \\ B(p^1 + ip^2) \end{pmatrix}, \qquad u_2(\mathbf{p}) = A\begin{pmatrix} 0 \\ 1 \\ B(p^1 - ip^2) \\ -Bp^3 \end{pmatrix}. \tag{A.73}$$

In the same way, the negative energy spinors (A.70) can be written as

$$v_r(\mathbf{p}) = A\begin{pmatrix} B\mathbf{p}\cdot\boldsymbol{\sigma}\chi'_r \\ \chi'_r \end{pmatrix}, \qquad r = 1, 2, \tag{A.74}$$

and

$$v_1(\mathbf{p}) = A\begin{pmatrix} B(p^1 - ip^2) \\ -Bp^3 \\ 0 \\ 1 \end{pmatrix}, \qquad v_2(\mathbf{p}) = A\begin{pmatrix} Bp^3 \\ B(p^1 + ip^2) \\ 1 \\ 0 \end{pmatrix}. \tag{A.75}$$

We see that the spinors (A.73) and (A.75) are, in general, not helicity eigenstates but eigenstates of the z-component of spin in the rest frame of the particle.

The behaviour of these solutions for non-relativistic velocities v (i.e. $v/c \approx |\mathbf{p}|/mc \ll 1$) is easily seen. For the positive energy solutions u_r, the upper two components are very large compared to the lower two components, while for the negative energy solutions it is the lower two components which dominate.

(ii) *Majorana representation.* We showed in Section 4.3 that the symmetry of the quantized Dirac field between particles and antiparticles becomes particularly obvious if one works in a Majorana representation, distinguished by the property that the four γ-matrices are pure imaginary, i.e. using a subscript M to denote a Majorana representation we require

$$\gamma_{\mathrm{M}}^{\mu *} = -\gamma_{\mathrm{M}}^{\mu}, \qquad \mu = 0, \ldots, 3. \tag{A.76}$$

A particular Majorana representation $\gamma^{\mu}_{\mathrm{M}}$ is obtained from the Dirac–Pauli representation, Eqs. (A.63)–(A.66), by the unitary transformation [see Eq. (A.61)]

$$\gamma^{\mu}_{\mathrm{M}} = U\gamma^{\mu}U^{\dagger} \tag{A.77}$$

with

$$U = U^{\dagger} = U^{-1} = \frac{1}{\sqrt{2}}\gamma^0(1 + \gamma^2). \tag{A.78}$$

Explicitly, the matrices in this Majorana representation are given by

$$\left.\begin{aligned}
&\gamma^0_{\mathrm{M}} = \gamma^0\gamma^2 = \begin{pmatrix} 0 & \sigma_2 \\ \sigma_2 & 0 \end{pmatrix}, \qquad \gamma^1_{\mathrm{M}} = \gamma^2\gamma^1 = \mathrm{i}\sigma^{12} = \begin{pmatrix} \mathrm{i}\sigma_3 & 0 \\ 0 & \mathrm{i}\sigma_3 \end{pmatrix} \\
&\gamma^2_{\mathrm{M}} = -\gamma^2 = \begin{pmatrix} 0 & -\sigma_2 \\ \sigma_2 & 0 \end{pmatrix}, \qquad \gamma^3_{\mathrm{M}} = \gamma^2\gamma^3 = -\mathrm{i}\sigma^{23} = \begin{pmatrix} -\mathrm{i}\sigma_1 & 0 \\ 0 & -\mathrm{i}\sigma_1 \end{pmatrix} \\
&\gamma^5_{\mathrm{M}} = -\mathrm{i}\gamma^0\gamma^1\gamma^3 = \gamma^0\sigma^{31} = \begin{pmatrix} \sigma_2 & 0 \\ 0 & -\sigma_2 \end{pmatrix}
\end{aligned}\right\}. \tag{A.79}$$

We see that in this representation all five γ-matrices are pure imaginary, since the Pauli matrices σ_1 and σ_3 are real and σ_2 is pure imaginary.

PROBLEMS

A.1 From Eq. (A.49) prove that

$$S^{-1}\gamma^5 S = \gamma^5 \det \Lambda$$

and hence that $\bar{\psi}(x)\gamma^5\psi(x)$ transforms as a pseudo-scalar under Lorentz transformations. Establish the transformation properties of the other four covariants in Eq. (A.53) in a similar way.

A.2 For any two positive energy solutions $u_r(\mathbf{p})$ and $u_s(\mathbf{p}')$ of the Dirac equation prove that

$$2m\bar{u}_s(\mathbf{p}')\gamma^{\mu}u_r(\mathbf{p}) = \bar{u}_s(\mathbf{p}')[(p' + p)^{\mu} + \mathrm{i}\sigma^{\mu\nu}(p' - p)_{\nu}]u_r(\mathbf{p}). \tag{A.80}$$

Eq. (A.80) is known as Gordon's identity. [*Hint*: Consider the identity

$$\bar{u}_s(\mathbf{p}')[\not{a}(\not{p} - m) + (\not{p}' - m)\not{a}]u_r(\mathbf{p}) = 0$$

for an arbitrary four-vector a_{μ}.]

APPENDIX B

Feynman rules and formulae for perturbation theory

In this appendix we collect together the principal results of covariant perturbation theory, i.e. the expressions for cross-sections, life times and Feynman amplitudes, and the Feynman rules for writing down these amplitudes directly from the Feynman diagrams. This summary is intended for readers who have assimilated these methods and wish to apply them. Explanations have been kept to a minimum but the frequent cross-references should help readers in difficulty.

(i) *Feynman amplitude.* The Feynman amplitude $\mathscr{M}$ for the transition $|i\rangle \to |f\rangle$ is defined in terms of the corresponding S-matrix element S_{fi} by

$$S_{fi} = \delta_{fi} + (2\pi)^4\delta^{(4)}\left(\sum p'_f - \sum p_i\right) \times \prod_i \left(\frac{1}{2VE_i}\right)^{1/2} \prod_f \left(\frac{1}{2VE'_f}\right)^{1/2} \prod_l (2m_l)^{1/2}\mathscr{M} \qquad (8.1)$$

where $p_i = (E_i, \mathbf{p}_i)$ and $p'_f = (E'_f, \mathbf{p}'_f)$ are the four-momenta of the initial and final particles and l runs over all charged and neutral external leptons in the process. m_l is the mass of lepton l. We are retaining non-vanishing neutrino masses in all basic equations. The limit $m_{\nu_l} \to 0$ is easily obtained for transition rates, etc.

(ii) *Cross-section.* The differential cross-section for the collision of two particles ($i = 1, 2$) moving collinearly with relative velocity v_{rel} and resulting in N final particles ($f = 1, 2, \ldots, N$) is given by

$$\mathrm{d}\sigma = (2\pi)^4\delta^{(4)}\left(\sum p'_f - \sum p_i\right)\frac{1}{4E_1E_2v_{\text{rel}}}\left(\prod_l (2m_l)\right)\left(\prod_f \frac{\mathrm{d}^3\mathbf{p}'_f}{(2\pi)^3 2E'_f}\right)|\mathscr{M}|^2. \qquad (8.8)$$

In Eq. (8.8) all initial and final particles are in definite spin/polarization states. Eq. (8.8) holds in any Lorentz frame and

$$v_{\text{rel}} = [(p_1 p_2)^2 - m_1^2 m_2^2]^{1/2}/(E_1 E_2); \tag{8.9}$$

in particular:

$$v_{\text{rel}} = |\mathbf{p}_1| \frac{E_1 + E_2}{E_1 E_2} \quad \text{(CoM system)} \tag{8.10a}$$

$$v_{\text{rel}} = |\mathbf{p}_1|/E_1 \quad \text{(laboratory system).} \tag{8.10b}$$

(iii) *Life time.* The differential decay rate $d\Gamma$ for the decay of a particle P with four-momentum $p = (E, \mathbf{p})$ into N particles with four-momenta $p'_f = (E'_f, \mathbf{p}'_f)$ is given by

$$d\Gamma = (2\pi)^4 \delta^{(4)}\left(\sum p'_f - p\right) \frac{1}{2E} \left(\prod_l (2m_l)\right) \left(\prod_f \frac{d^3\mathbf{p}'_f}{(2\pi)^3 2E'_f}\right) |\mathscr{M}|^2. \tag{11.36}$$

Eq. (11.36) refers to definite initial and final spin/polarization states for all particles.

The life time τ of the particle P is given by

$$\tau = B/\Gamma \tag{11.38}$$

where Γ is the total decay rate for the above decay mode, and B is the branching ratio for this mode, Eq. (11.37).

(iv) *The Feynman rules for QED.* The Feynman amplitude for a given graph in QED is obtained from the Feynman rules of Sections 7.3 and 8.7. Some of the Feynman rules for charged leptons $l^\pm$ apply with trivial changes to neutral leptons ν_l and $\bar{\nu}_l$. To avoid lengthy repetition later, we write these rules at once for both charged and neutral leptons.

1. For each QED vertex, write a factor

$$ie\gamma^\alpha.$$

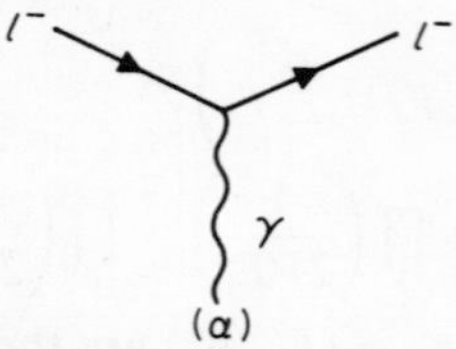

2. For each internal photon line, labelled by the momentum k, write a factor

$$iD_{F\alpha\beta}(k) = i\frac{-g_{\alpha\beta}}{k^2 + i\varepsilon}. \tag{7.47}$$

3. For each internal lepton line, labelled by the momentum p, write a factor

$$iS_F(p) = i\frac{1}{\not{p} - m + i\varepsilon}. \tag{7.48}$$

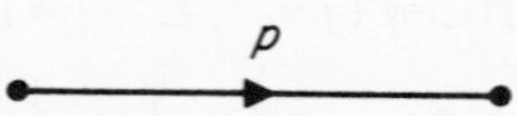

Here m stands for the mass m_l or m_{ν_l} of the particular lepton considered.

4. For each external line, write one of the following factors:

(a) for each initial lepton l^- or ν_l: $u_r(\mathbf{p})$ 

(7.49a)

(b) for each final lepton l^- or ν_l: $\bar{u}_r(\mathbf{p})$ 

(7.49b)

(c) for each initial lepton l^+ or $\bar{\nu}_l$: $\bar{v}_r(\mathbf{p})$ 

(7.49c)

(d) for each final lepton l^+ or $\bar{\nu}_l$: $v_r(\mathbf{p})$ 

(7.49d)

(e) for each initial photon: $\varepsilon_{r\alpha}(\mathbf{k})$ 

(7.49e)

(f) for each final photon: ‡ $\varepsilon_{r\alpha}(\mathbf{k})$ 

(7.49f)

In Eqs. (7.49) $\mathbf{p}$ and $\mathbf{k}$ denote the three-momenta of the external particles, and r $(=1, 2)$ their spin and polarization states.

5. The spinor factors (γ-matrices, S_F-functions, four-spinors) for each fermion line are ordered so that, reading from right to left, they occur in the same sequence as following the fermion line in the direction of its arrows.

6. For each closed fermion loop, take the trace and multiply by a factor (-1).

7. The four-momenta associated with the lines meeting at each vertex satisfy energy–momentum conservation. For each four-momentum q which is not fixed by energy–momentum conservation carry out the integration $(2\pi)^{-4}\int d^4q$. One such integration with respect to an internal momentum variable q occurs for each closed loop.

8. Multiply the expression by a phase factor δ_P which is equal to $+1$ (-1) if an even (odd) number of interchanges of neighbouring fermion operators is required to write the fermion operators in the correct normal order.

To allow for the interaction with an *external static electromagnetic field* $A_{e\alpha}(\mathbf{x})$:

(a) In Eq. (8.1), relating $\mathscr{M}$ to S_{fi}, make the replacement

$$(2\pi)^4\delta^{(4)}\left(\sum p'_f - \sum p_i\right) \rightarrow (2\pi)\,\delta\left(\sum E'_f - \sum E_i\right). \qquad (8.89)$$

(b) Add the following Feynman rule:

‡ For linear polarization states, $\varepsilon_{r\alpha}(\mathbf{k})$ is real. In general it is complex and we must then replace $\varepsilon_{r\alpha}(\mathbf{k})$ by $\varepsilon^*_{r\alpha}(\mathbf{k})$ for a final-state photon.

9. For each interaction of a charged particle with an external static field $A_e(\mathbf{x})$, write a factor

$$A_{e\alpha}(\mathbf{q}) = \int \mathrm{d}^3\mathbf{x}\, \mathrm{e}^{-\mathrm{i}\mathbf{q}\cdot\mathbf{x}} A_{e\alpha}(\mathbf{x}). \qquad (8.90)$$

Here $\mathbf{q}$ is the momentum transferred from the field source (×) to the particle.

(v) *Additional Feynman rules for the standard electro–weak theory.* The following rules allow calculations to be carried out in lowest non-vanishing order of perturbation theory only.‡

11. For each internal massive vector boson line, labelled by the momentum k, write a factor

$$\mathrm{i}D_{\mathrm{F}\alpha\beta}(k, m) = \frac{\mathrm{i}(-g_{\alpha\beta} + k_\alpha k_\beta/m^2)}{k^2 - m^2 + \mathrm{i}\varepsilon} \qquad (11.30)$$

where $m = m_W$ for a $W^\pm$ boson, and $m = m_Z$ for a Z^0 boson.

12. For each external line, representing an initial or final $W^\pm$ or Z^0 vector boson, write a factor

$$\varepsilon_{r\alpha}(\mathbf{k}) \qquad (11.32)$$

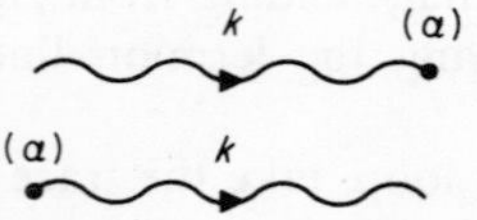

(initial)
(final)

If complex polarization vectors are used, we must instead write $\varepsilon^*_{r\alpha}(\mathbf{k})$ for a final-state vector boson.

13. For each internal Higgs line, labelled by the momentum k, write a factor

$$\mathrm{i}\Delta_{\mathrm{F}}(k, m_H) = \frac{\mathrm{i}}{k^2 - m_H^2 + \mathrm{i}\varepsilon} \qquad (14.21)$$

where m_H is the mass of the Higgs scalar.

14. The standard electro–weak theory gives rise to 18 basic vertex parts, resulting from the terms in Eqs. (14.3a)–(14.3e), which were listed in Table 14.1. Below we give the Feynman diagrams of these basic vertex parts and the corresponding vertex factors, numbered (B.1)–(B.18) in the same order as the terms in Table 14.1.§

‡ Feynman rule 10 of Section 11.4 dealt with the vertex factor of the IVB theory. This rule will now be incorporated in rule 14 for the vertex factors of the electro–weak theory.

§ Feynman rule 14 lists all the interactions of the electro–weak theory. In order to have a complete, self-contained list of Feynman rules for QED in (iv) above, we also included the QED vertex factor in Feynman rule 1.

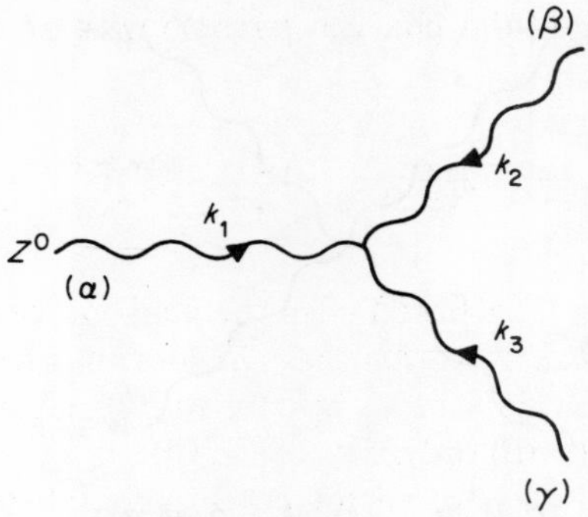

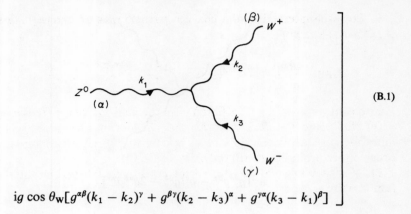

$$ig\cos\theta_{\mathrm{W}}[g^{\alpha\beta}(k_1 - k_2)^{\gamma} + g^{\beta\gamma}(k_2 - k_3)^{\alpha} + g^{\gamma\alpha}(k_3 - k_1)^{\beta}]$$

(B.1)

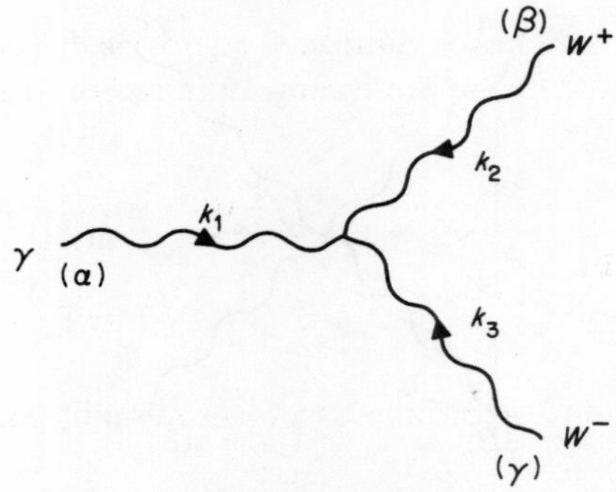

$$\mathrm{i}e[g^{\alpha\beta}(k_1 - k_2)^{\gamma} + g^{\beta\gamma}(k_2 - k_3)^{\alpha} + g^{\gamma\alpha}(k_3 - k_1)^{\beta}]$$

(B.2)

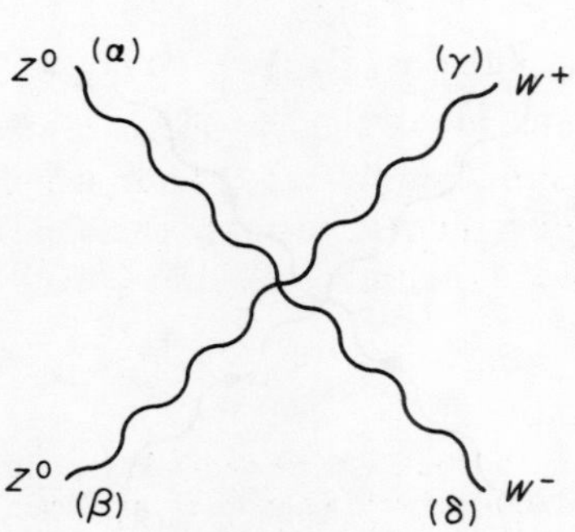

$$ig^2\cos^2\theta_{\mathrm{W}}[g^{\alpha\delta}g^{\beta\gamma} + g^{\alpha\gamma}g^{\beta\delta} - 2g^{\alpha\beta}g^{\gamma\delta}]$$

(B.3)

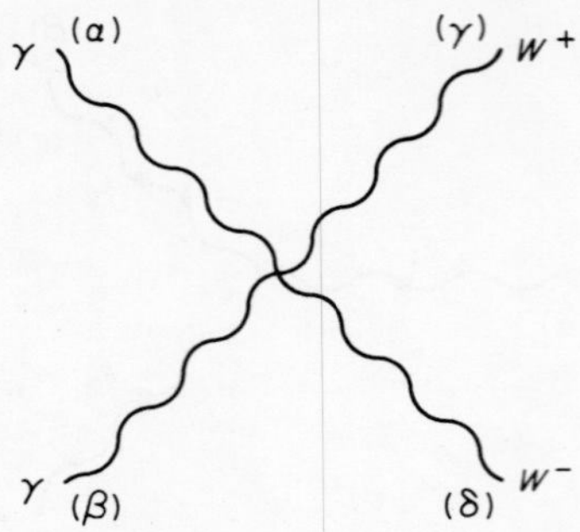

$$ie^2[g^{\alpha\delta}g^{\beta\gamma} + g^{\alpha\gamma}g^{\beta\delta} - 2g^{\alpha\beta}g^{\gamma\delta}] \qquad \text{(B.4)}$$

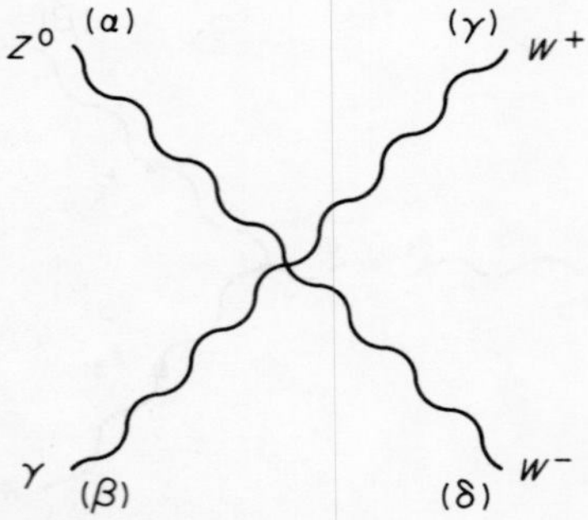

$$ieg\cos\theta_W[g^{\alpha\delta}g^{\beta\gamma} + g^{\alpha\gamma}g^{\beta\delta} - 2g^{\alpha\beta}g^{\gamma\delta}] \qquad \text{(B.5)}$$

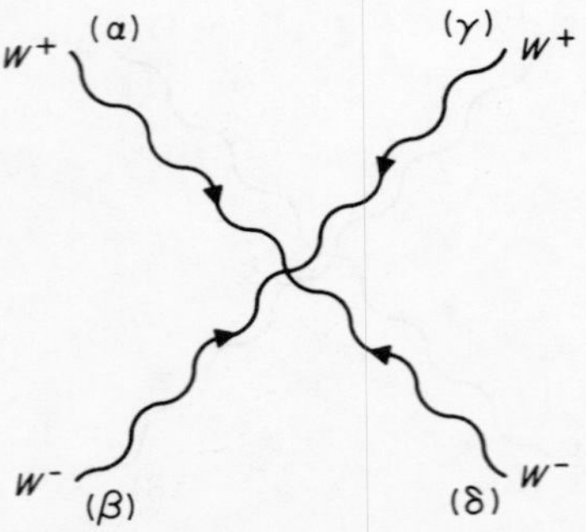

$$ig^2[2g^{\alpha\gamma}g^{\beta\delta} - g^{\alpha\beta}g^{\gamma\delta} - g^{\alpha\delta}g^{\beta\gamma}] \qquad \text{(B.6)}$$

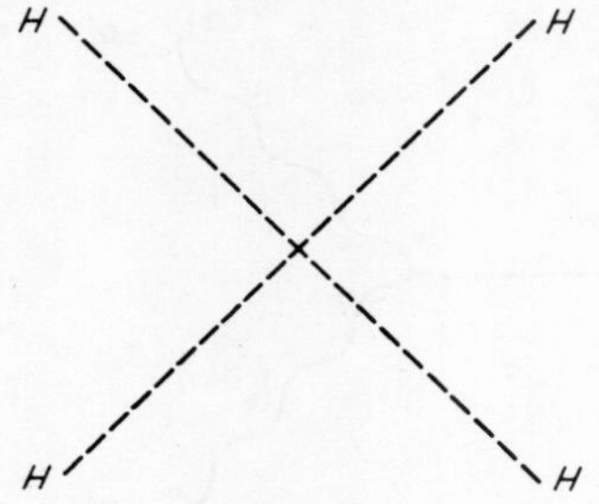

$-6i\lambda$ (B.7)

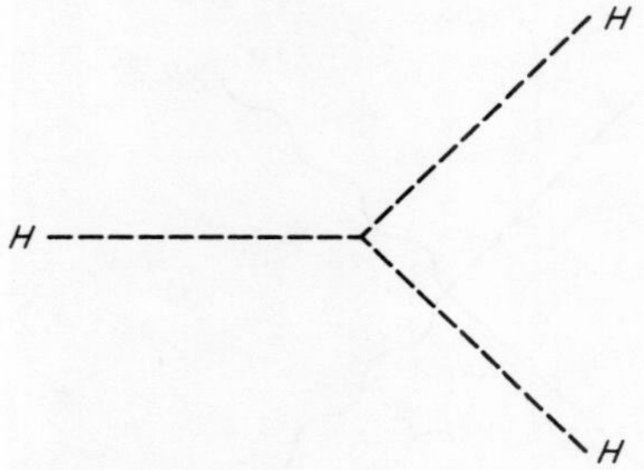

$-6i\lambda v$ (B.8)

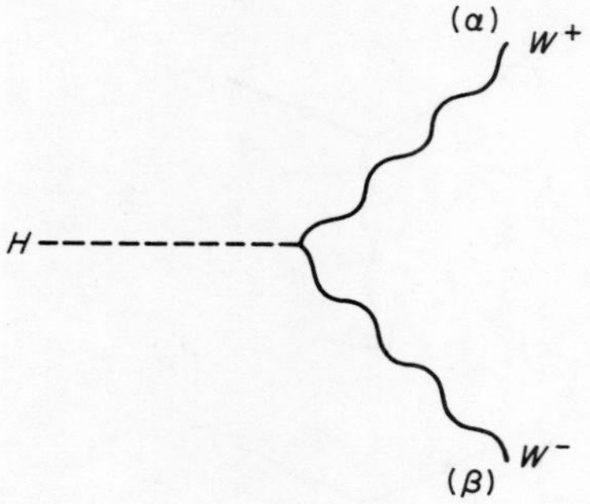

$\frac{1}{2}ivg^2g^{\alpha\beta}$ (B.9)

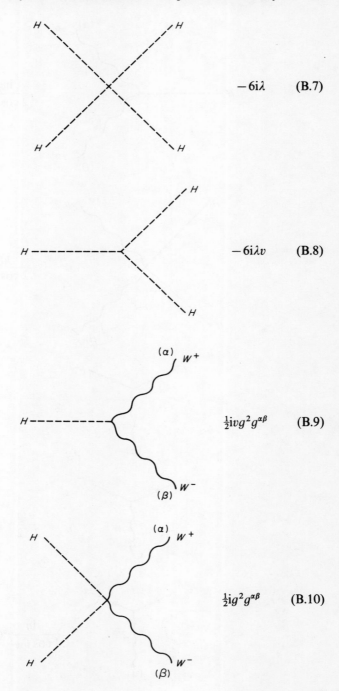

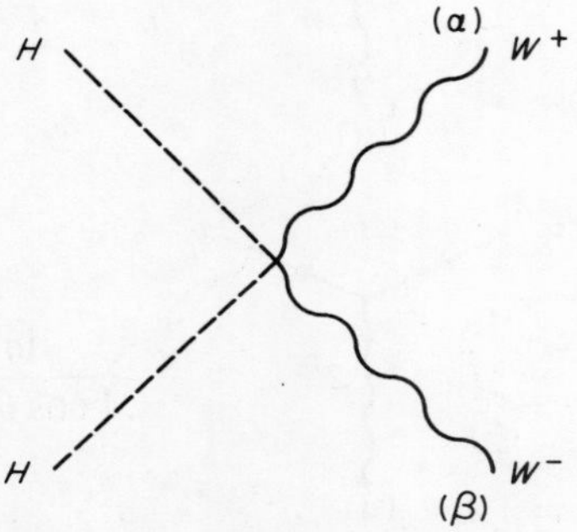

$\frac{1}{2}ig^2g^{\alpha\beta}$ (B.10)

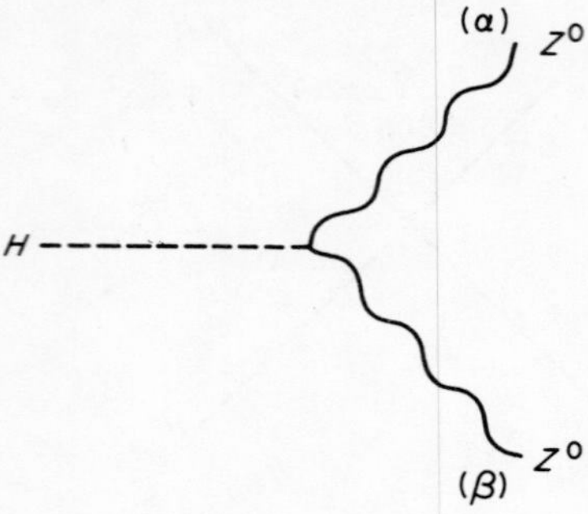

$$\frac{ivg^2}{2\cos^2\theta_W}g^{\alpha\beta} \quad \text{(B.11)}$$

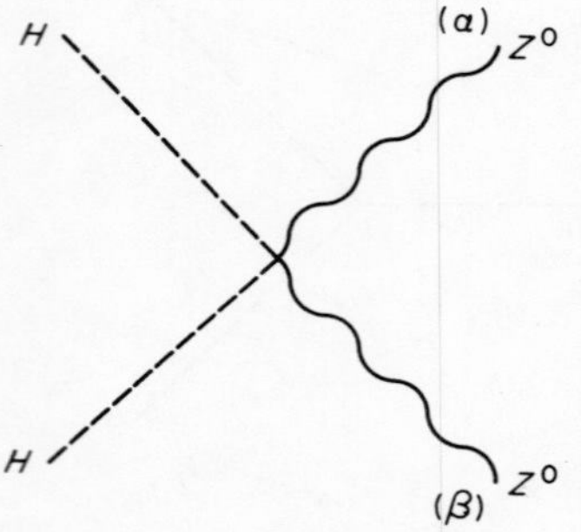

$$\frac{ig^2}{2\cos^2\theta_W}g^{\alpha\beta} \quad \text{(B.12)}$$

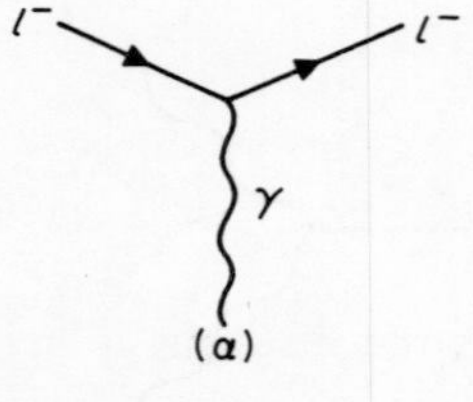

$$ie\gamma^\alpha \quad \text{(B.13)}$$

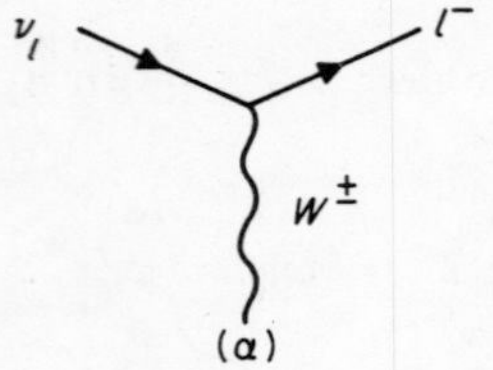

$$\frac{-ig}{2\sqrt{2}}\gamma^\alpha(1-\gamma_5) \quad \text{(B.14)}$$

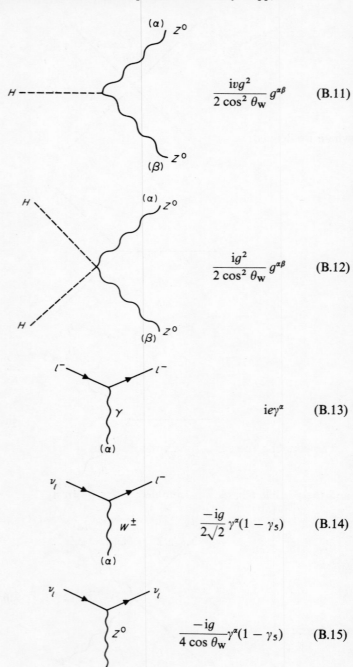

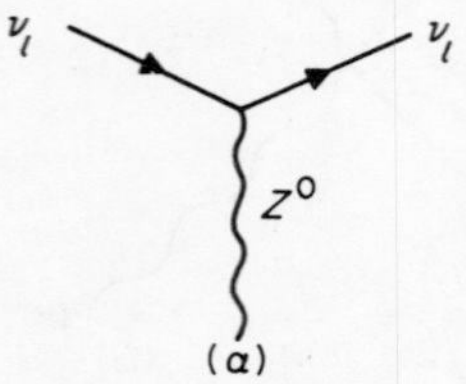

$$\frac{-ig}{4\cos\theta_W}\gamma^\alpha(1-\gamma_5) \quad \text{(B.15)}$$

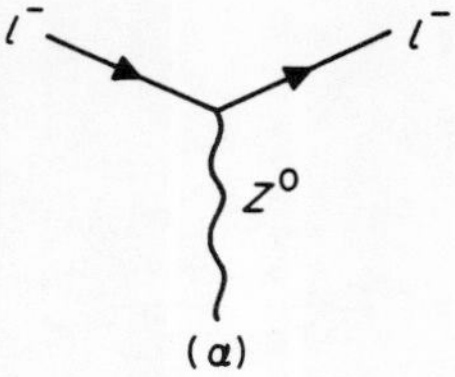

$$\frac{\mathrm{i}g\gamma^{\alpha}}{4\cos\theta_{\mathrm{W}}}(1-4\sin^2\theta_{\mathrm{W}}-\gamma_5) \quad \text{(B.16)}$$

$$= \frac{-\mathrm{i}g\gamma^{\alpha}}{2\cos\theta_{\mathrm{W}}}(g_{\mathrm{V}}-g_{\mathrm{A}}\gamma_5) \quad \text{(B.16a)}$$

where we defined

$$g_{\mathrm{V}} = 2\sin^2\theta_{\mathrm{W}} - \tfrac{1}{2}, \qquad g_{\mathrm{A}} = -\tfrac{1}{2} \quad \text{(B.16b)}$$

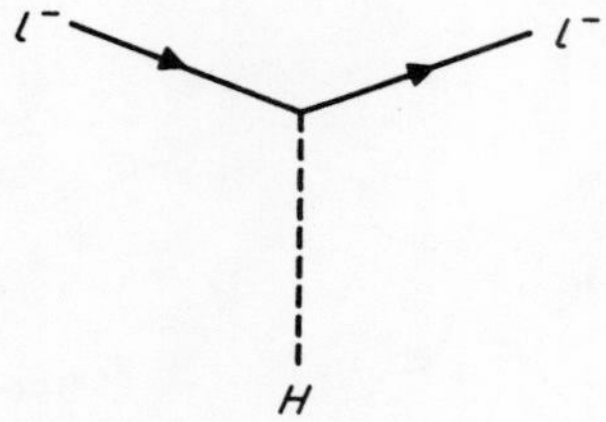

$$\frac{-\mathrm{i}}{v}m_l \quad \text{(B.17)}$$

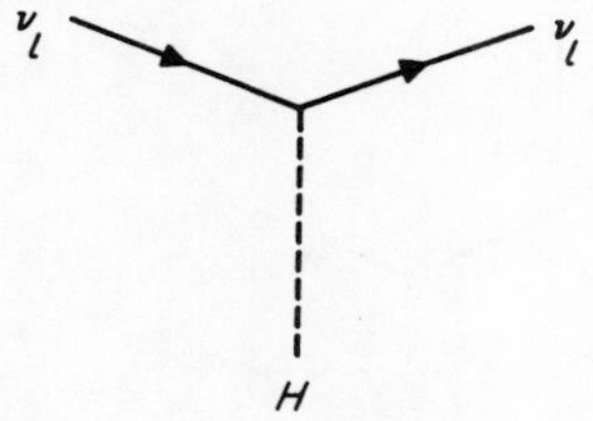

$$\frac{-\mathrm{i}}{v}m_{\nu_l} \quad \text{(B.18)}$$

The coupling constants satisfy the relations

$$g\sin\theta_{\mathrm{W}} = g'\cos\theta_{\mathrm{W}} = e \quad (14.6)$$

and the particle masses are, in lowest order, given by

$$m_W = \tfrac{1}{2}vg, \qquad m_Z = m_W/\cos\theta_{\mathrm{W}}, \qquad m_H = \sqrt{(-2\mu^2)}, \quad (14.4)$$

$$m_l = vg_l/\sqrt{2}, \qquad m_{\nu_l} = vg_{\nu_l}/\sqrt{2}, \quad (14.8)$$

where

$$v = \left(\frac{-\mu^2}{\lambda}\right)^{1/2} = (G\sqrt{2})^{-1/2} \quad (>0), \quad (14.5)\text{ and }(14.14)$$

and hence

$$\frac{G}{\sqrt{2}} = \frac{g^2}{8m_W^2}.$$

Index